AF693303

Current Mammalogy

Editorial Board

Robert J. Baker, *Texas Tech University, Lubbock, Texas*
Mary R. Dawson, *Carnegie Museum of Natural History, Pittsburgh, Pennsylvania*
John F. Eisenberg, *University of Florida, Gainesville, Florida*
Jerran T. Flinders, *Brigham Young University, Provo, Utah*
Charles W. Fowler, *National Marine Mammal Laboratory, Seattle, Washington*
William B. Jackson, *Bowling Green State University, Bowling Green, Ohio*
Michael A. Mares, *University of Oklahoma, Norman, Oklahoma*
James L. Patton, *University of California, Berkeley, California*
Carleton J. Phillips, *Hofstra University, Hempstead, New York*
Galen B. Rathbun, *U. S. Fish and Wildlife Service, San Simeon, California*
Don E. Wilson, *National Fish and Wildlife Laboratory, Washington, D. C.*
Bruce A. Wunder, *Colorado State University, Fort Collins, Colorado*

A Continuation Order Plan is available for this series. A continuation order will bring delivery of each new volume immediately upon publication. Volumes are billed only upon actual shipment. For further information please contact the publisher.

CURRENT MAMMALOGY

VOLUME 1

Edited by

HUGH H. GENOWAYS

University of Nebraska State Museum
Lincoln, Nebraska

PLENUM PRESS • NEW YORK AND LONDON

Library of Congress Cataloging in Publication Data

Current mammalogy.

Includes bibliographies and index.
1. Mammals. I. Genoways, Hugh H.
QL703.C87 1987 599 86-30384
ISBN 0-306-42430-4 (v. 1)

© 1987 Plenum Press, New York
A Division of Plenum Publishing Corporation
233 Spring Street, New York, N.Y. 10013

All rights reserved

No part of this book may be reproduced, stored in a retrieval system, or transmitted in any form or by any means, electronic, mechanical, photocopying, microfilming, recording, or otherwise, without written permission from the Publisher

Printed in the United States of America

CONTRIBUTORS

ROBERT J. BAKER, Department of Biological Sciences and The Museum, Texas Tech University, Lubbock, Texas 79409

ANTHONY D. BARNOSKY, Section of Vertebrate Fossils, Carnegie Museum of Natural History, Pittsburgh, Pennsylvania 15213

PATRICIA J. BERGER, Department of Biology, University of Utah, Salt Lake City, Utah 84112

DARYL J. BONESS, Department of Zoological Research, National Zoological Park, Smithsonian Institution, Washington, D.C. 20008

WARREN P. CLARY, U.S. Forest Service Forestry Science Laboratory, Boise, Idaho 83704

MARY R. DAWSON, Section of Vertebrate Fossils, Carnegie Museum of Natural History, Pittsburgh, Pennsylvania 15213

JERRAN T. FLINDERS, Wildlife and Range Resources Program, Brigham Young University, Provo, Utah 84602

CHARLES W. FOWLER, National Marine Mammal Laboratory, Northwest and Alaska Fisheries Center, National Marine Fisheries Services, NOAA, Seattle, Washington 98115

CRAIG S. HOOD, Department of Biological Sciences and The Museum, Texas Tech University, Lubbock, Texas 79409

A. BLAIR IRVINE, Dolphin Biology Research Associates, Sarasota, Florida 34242, and Department of School and Community Health Education, University of Oregon, Eugene, Oregon 97401

WILLIAM B. JACKSON, Department of Biological Sciences, Bowling Green State University, Bowling Green, Ohio 43403

LEONARD KRISHTALKA, Section of Vertebrate Fossils, Carnegie Museum of Natural History, Pittsburgh, Pennsylvania 15213

THOMAS E. LACHER, JR., Huxley College of Environmental Studies, Western Washington University, Bellingham, Washington 98225

LI CHUAN-KUEI, Institute of Vertebrate Paleontology and Paleoanthropology, Academia Sinica, Beijing, China

MICHAEL A. MARES, Stovall Museum and Department of Zoology, University of Oklahoma, Norman, Oklahoma 73019

NORMAN C. NEGUS, Department of Biology, University of Utah, Salt Lake City, Utah 84112

OLAV T. OFTEDAL, Department of Zoological Research, National Zoological Park, Smithsonian Institution, Washington, D.C. 20008

G. TIMOTHY O'NEAL, Wildlife and Range Resources Program, Brigham Young University, Provo, Utah 84602

CARLETON J. PHILLIPS, Department of Biology, Hofstra University, Hempstead, New York 11550

MAZIN B. QUMSIYEH, Department of Biological Sciences and The Museum, Texas Tech University, Lubbock, Texas 79409

KENT H. REDFORD, Museum of Comparative Zoology, Harvard University, Cambridge, Massachusetts 02138, and Department of Zoological Research, National Zoological Park, Smithsonian Institution, Washington, D.C. 20008. *Present address*: Center for Latin American Studies, University of Florida, Gainesville, Florida 32611

MICHAEL D. SCOTT, Dolphin Biology Research Associates, Sarasota, Florida 34242, and InterAmerican Tropical Tuna Commission, Scripps Institute of Oceanography, La Jolla, California 92038

BERNARD TANDLER, Department of Oral Biology, School of Dentistry, Case Western Reserve University, Cleveland, Ohio 44106

RAYMOND A. TEDMAN, Department of Anatomy and Histology, University of Adelaide, Adelaide, South Australia 5001, Australia

MANFRED TEMME, Department of Biological Sciences, Bowling Green State University, Bowling Green, Ohio 43403. *Present address*: Alter Horst 18, 2982 Norderney, West Germany

RANDALL S. WELLS, Dolphin Biology Research Associates, Sarasota, Florida 34242, and Long Marine Laboratory, Institute of Marine Sciences, University of California, Santa Cruz, California 95060

ROBERT W. WILSON, Museum of Natural History, University of Kansas, Lawrence, Kansas 66045

PREFACE

When I first proposed a series entitled *Current Mammalogy* to the publishers, they were reluctant to undertake such a project because they viewed the field of mammology as overly fragmented. At first I found this idea to be difficult to accept; however, upon reflection, I came near to agreeing with it. Although many of us work on mammals, we generally feel more allegiance to our specialties, such as systematics, genetics, cytogenetics, ecology, behavior, pest control, paleontology, wildlife management, primatology, and marine mammalogy, than we do to the general field of mammalogy. However, rather than becoming discouraged from pursuing this project, I became more certain than ever that a series such as *Current Mammalogy* was needed. We hope to make this series a place where specialists can present their ideas not only to other members of their specialty, but to those outside the area as well. Hopefully, this exchange of ideas will be a mutually beneficial exercise.

The Editorial Board of *Current Mammalogy* has decided to keep the range of subjects in each volume as broad as possible rather than concentrating on one or two topics, in the hope that this will keep the series as useful as possible to the broadest range of readers. Papers in *Current Mammalogy* will be broad reviews of a particular subject area, in depth, long-term studies, accounts of exciting new discoveries that may open new areas of research or syntheses of data that give new insight into a subject. In this way, we hope not to compete with the existing professional journals, but to complement them. Although each chapter will be reviewed by two outside reviewers, it is our intention to allow the authors more latitude in expressing their opinions than would normally be possible in a professional journal.

It was not an easy task to group the 13 chapters of Volume 1 into any meaningful sequence that would be apparent to the reader. In many cases, the chapters overlap in several areas of mammalogy, so that they could be placed in a number of places in the sequence. As they appear in this volume, the first four chapters deal with evolutionary mechanisms and theory. The chapter by Phillips and Tandler deals with evolution at the cellular level and tries to draw together the diverse areas of cellular biology and systematics and evolutionary theory. Baker and his coauthors summarize and evaluate the evolutionary models that have resulted from recent work in mammalian cytogenetics. The other two chapters in this group are based upon paleontological work. Li and his coauthors synthesize recent work on the origins of rodents and lagomorphs and conclude that the concept of the Cohort Glires may have validity. In Chapter 4, Barnosky presents data from Quaternary mammals relating to the controversy between punctuated equilibrium and phyletic gradualism as the mode of the evolutionary development of life.

Chapters 5 (Negus and Berger) and 6 (Oftedal and coauthors) deal with various aspects of mammalian reproduction, but both take a strong evolutionary view in interpretation of their data. Wells and his coauthors present in Chapter 7 the results of one of the first long-term studies of social structure in a free-ranging cetacean.

Chapters 7–11 are generally ecological in scope. Mares and Lacher, in Chapter 8, present an in-depth review of convergence among rock-dwelling mammals and Redford gives the same treatment to ant- and termite-feeding mammals in Chapter 9. In Chapter 10, Fowler reviews the literature on the population dynamics of large mammals to identify those factors that are density dependent. This chapter and Chapter 11, which is an intensive study of the kit fox on a managed desert rangeland in Utah by O'Neal and co-workers, should be of considerable interest to wildlife managers as well as ecologists.

The short Chapter 12 presents the exciting discovery by Temme of a somatic mutation in the Polynesian rat on the Enewetak Nuclear Test Site. This discovery may have broad implications in mammalian genetics and in the area of the use of mammals as bioindicators of environmental problems. The final chapter is a review by Jackson of those tools that are available and in use to control mammalian pests, particularly commensal rodents. Jackson also gives a view of why control measures are necessary and desirable.

Papers will appear in *Current Mammalogy* only by invitation of the Editorial Board. The Editorial Board will be actively soliciting articles for future volumes. However, if mammalogists would like to be invited to prepare a manuscript for *Current Mammalogy*, they should submit a prospectus of no more than one page for consideration by the Board. Authors should *not* submit full manuscripts for initial consideration by the Board. In this way authors will not be wasting their valuable time and the Editorial Board will be free to balance the content of each volume. If there are particular areas that readers would like to have the Editorial Board include in future volumes, they should send their suggestions to me. We hope that readers will find the results of these efforts to be interesting and stimulating.

I wish to give special acknowledgment to Marion A. Burgwin, who assisted in the preparation of this first volume of *Current Mammalogy*. She spent the long hours that are necessary to trace out and correct the numerous small errors that tend to become part of a major volume such as this. I also wish to thank Winifred Woodland, who provided clerical assistance in the preparation of this volume.

Hugh H. Genoways

Lincoln, Nebraska

CONTENTS

CHAPTER 2

Role of Chromosomal Banding Patterns in Understanding Mammalian Evolution

Robert J. Baker, Mazin B. Qumsiyeh, and Craig S. Hood

CHAPTER 3

The Origin of Rodents and Lagomorphs

Li Chuan-kuei, Robert W. Wilson, Mary R. Dawson, and Leonard Krishtalka

CHAPTER 4

Punctuated Equilibrium and Phyletic Gradualism: Some Facts from the Quaternary Mammalian Record

Anthony D. Barnosky

CHAPTER 7

THE SOCIAL STRUCTURE OF FREE-RANGING BOTTLENOSE DOLPHINS

RANDALL S. WELLS, MICHAEL D. SCOTT, AND A. BLAIR IRVINE

CHAPTER 8

Ecological, Morphological, and Behavioral Convergence in Rock-Dwelling Mammals

Michael A. Mares and Thomas E. Lacher, Jr.

CHAPTER 9

Ants and Termites As Food: Patterns of Mammalian Myrmecophagy

Kent H. Redford

CHAPTER 10

A REVIEW OF DENSITY DEPENDENCE IN POPULATIONS OF LARGE MAMMALS

CHARLES W. FOWLER

CHAPTER 11

BEHAVIORAL ECOLOGY OF THE NEVADA KIT FOX (*VULPES MACROTIS NEVADENSIS*) ON A MANAGED DESERT RANGELAND

G. TIMOTHY O'NEAL, JERRAN T. FLINDERS, AND WARREN P. CLARY

CHAPTER 12

SOMATIC MUTATION IN THE POLYNESIAN RAT (*RATTUS EXULANS*) AT ENEWETAK NUCLEAR TEST SITE

MANFRED TEMME

CHAPTER 13

CURRENT MANAGEMENT STRATEGIES FOR COMMENSAL RODENTS

WILLIAM B. JACKSON

CHAPTER 1

MAMMALIAN EVOLUTION AT THE CELLULAR LEVEL

CARLETON J. PHILLIPS
and BERNARD TANDLER

1. INTRODUCTION

The development of new techniques in molecular biology has greatly advanced our understanding of the evolutionary and systematic biology of mammals. Before 1965 our comprehension of mammalian evolution and our approach to systematics rested upon our ability—sometimes strictly intuitive—to interpret the significance of interspecific differences in gross anatomic features. Since then, we have been increasingly able to study proteins, chromosomal structure and organization, comparisons of individual genes, gene loci, and nuclear and mitochondrial DNA. Evolutionary and systematic investigations in mammalogy now routinely include biochemistry of proteins as well as gross anatomy, but often neglect everything in between.

Despite exciting technological developments, several basic scientific questions concerning evolution have remained virtually unanswered and others have been brought to our attention by data that are difficult to reconcile with more traditional ideas. Perhaps the most fundamental and complex of these is the question of how genetic changes

CARLETON J. PHILLIPS • Department of Biology, Hofstra University, Hempstead, New York 11550. BERNARD TANDLER • Department of Oral Biology, School of Dentistry, Case Western Reserve University, Cleveland, Ohio 44106.

actually have been translated into evolutionary adaptations in mammalian organ systems. This area is left untouched even by Eisenberg's (1981) already classic survey of mammalian radiations; indeed, the disparity between our ability to detect base pair substitutions in nuclear DNA and our ability to comprehend the evolution of highly integrated, functionally complicated organ systems—not to mention the organism as a whole—seems enormous. In purely practical terms, it is now necessary to begin to elucidate the relationships between genomes and phenotypic traits before rapid advances in molecular technology cause us either to lose sight of the broader picture, that is, the evolutionary biology of mammals as organisms, or miss opportunities to use new technology to the best advantage in addressing evolutionary questions, or both.

The knowledge of gross anatomic features of mammals is highly developed and many of the organ systems in mammals have been investigated. It is on this basis that contemporary concepts of mammalian phylogeny are founded. However, the gross anatomy of the skeleton, dentition, brain, and digestive tract is actually far removed from the initial site of action of the genome, so such anatomic data in the end only serve to highlight the overall structural diversity in organ systems. One solution to this problem has been to take a "systems analysis" approach in which organs are investigated as an integrated functional unit (Ulinski, 1980) or considered in terms of their ontogeny at both the tissue and cellular levels (Phillips and Oxberry, 1972; Gans and Northcutt, 1983; Maderson, 1983; Oster and Alberch, 1982; Liem and Wake, 1985). We hold the view that microscopic (especially transmission electron microscopic) comparisons used in concert with molecular probes of certain organs and certain "types" of cells constitute another logical approach to some of these issues, because it is at the level of the individual cell that the genome is initially expressed. In particular we think that such an examination of various types of secretory cells, which use a small segment of their nuclear DNA as a blueprint for the synthesis, packaging, and exportation of a product, might help us fill in the existing gaps in our knowledge. This is especially true when the cell product by itself can be studied as a complete structure, as is the case with extracellular enamel prisms in the teeth of mammals (Greenberg et al., 1984), or investigated when it causes a physiological stimulus, such as in the case of regulatory peptides or neuropeptides (Mennone et al., 1986) or as a possible pheromone, as in the case of some salivary gland secretions (Nagato et al., 1984).

A major obstacle to our thesis is the difficulty in developing an interdisciplinary approach to the study of evolutionary biology. There is an unfortunate tendency for organismal biologists to view microscopic and molecular investigations as excessively reductionist and unlikely to produce data that can be correlated with the biology of an intact, functioning mammal. This is because many find it difficult to imagine that an understanding of cell structure and function could further the understanding of mammals in an ecological or behavioral context. At the other extreme, our colleagues in anatomy and especially some of those in cell biology occasionally forget that the Class Mammalia encompasses more than *Homo sapiens* and an assortment of domestic and laboratory species. The refusal of our colleagues to employ traditional taxonomy or modern systematics as a means of identifying candidate species of special interest and as a foundation for interpreting comparative data is a barrier to such research. In illustrating the possibilities afforded by comparative microscopic investigations, we hope to convince both organismal biologists and anatomists that the various disciplines can be brought together to address basic questions of universal interest—questions that must be answered if we are ever to bridge the gap between genes and organisms.

2. MICROSCOPIC STUDIES

2.1. Light Microscopy

The investigation of organ systems by means of a light microscope represents a significant improvement over the use of gross anatomic comparisons in the attempt to ascertain the relationships between genes, evolution of structure–function complexes, and organismal biology. Although the basic histology of the generalized mammal is well known, comparative histology, which is a necessary component of systematic and evolutionary studies, has been relatively neglected. Even so, a full review of the comparative histology of mammals is far beyond the scope of this paper. However, selected systematic investigations of the integumentary glands and alimentary tract of rodents and lagomorphs (Quay, 1954; Dearden, 1966, 1969; Naumova, 1981; Forman and Phillips, 1987a,b) and the teeth, salivary glands, stomachs, tongues, urinary system, and reproductive tracts of bats (Phillips, 1971; Greenbaum and Phillips, 1974; Phillips and Steinberg, 1976; Phillips et al., 1977; Kamiya and Pirlot, 1975; Okon, 1977; Forman et al., 1979; Smith and Madkour, 1980; Hood and Smith, 1983; Studier et al., 1983; Yamada et al., 1984; Ishikawa et al., 1985) are good examples of this approach in mammalogy and introduce the reader to the questions that might be answered with comparative histology, as well as to the thought processes behind the technique.

One of the potential uses of comparative histological data is illustrated by studies of teeth in rodents (Phillips and Oxberry, 1972; Oxberry, 1975; Golenishchev, 1976). In two microtine rodents (*Clethrionomys gapperi* and *Microtus pennsylvanicus*), comparative histology was used in an attempt to delineate the aspects of dental ontogeny that might have been altered in the evolution of evergrowing teeth from the more typical "rooted" molar teeth (Phillips and Oxberry, 1972; Phillips, 1985b). The underlying idea was that comparative data about dental tissues and their cellular organization might serve to highlight differences between evergrowing and nonevergrowing teeth. Through histology, evergrowing teeth thus might be analyzed in terms of their components, particularly those components that differed from homologous tissues in nonevergrowing teeth. Phillips and Oxberry (1972) compared these histological data to general information about dental ontogeny and then hypothesized that the evolution of evergrowing molars could have involved seemingly minor changes in the timing of the typical dental ontogenetic sequence that resulted in continuous production of ameloblasts and odontoblasts and the elaboration of enamel and dentin by these cells. Moreover, they listed a set of three specific features of mammalian teeth that might have been significant preadaptive factors in the evolution of evergrowing molars. Since the publication of this work, there has been increasing interest in the role of developmental pathways in morphological evolution and ideas and data about developmental constraints, developmental pathways, and heterochrony (developmental timing) now are becoming commonplace in the literature, although most of the data have not come from mammalian studies (Alberch, 1980; Oster and Alberch, 1982; Straney, 1984; Liem and Wake, 1985; Maynard Smith et al., 1985). The use of comparative histology to address evolutionary questions and to discover possible evolutionary mechanisms also shows how the use of the light microscope enables us to link otherwise disparate data bases—developmental biology, systematic mammalogy, gross anatomy, and paleontology—and ultimately to fill some of the gap between genetics and gross morphology. Descriptive histological data such as these can be interpreted in the light of experimental evidence from a variety of disciplines while at the same time drawing attention to particular features that could be used to establish testable hypotheses for experimentation.

Systematic mammalogy and ecological data can be used to identify species that might be of special interest for comparative histology. Certain species are most likely to produce data that could illuminate some evolutionary trends and some aspects of the evolutionary process. An excellent example of this is the descriptive histological investigation of the stomach of the grasshopper mouse, *Onychomys torridus*, by Horner et al. (1964). This rodent has a very unusual diet that includes a variety of insects and even scorpions. Horner et al. (1964) compared their data for *Onychomys* to a systematically logical assortment of rodent species and used their histological data to discuss the evolution of the extensive nonglandular stomach and the restricted gastric glands common in rodents. Our recent investigation (Phillips and Tandler, 1985; Phillips et al., 1987a,b) of the accessory submandibular salivary glands in two unrelated species of frog-eating bats is another example of this approach. In this instance, we used histology and histochemistry to compare the salivary glands in *Trachops cirrhosus*, a Neotropical phyllostomid bat, to those in two species of *Megaderma*, which occur only in the Old World tropics. Although these genera now share an unusual diet, they were derived independently from very distantly related insectivorous ancestors. In both *Trachops* and *Megaderma* we found that the accessory submandibular salivary glands consisted of large, folliclelike structures rather than of the typical secretory endpieces seen in these glands in other bats and in all other mammals studied as well (Phillips et al., 1987a,b). The submandibular glands in *Trachops* and *Megaderma* also shared glycoprotein histochemistry, with the exception of the presence of scattered mucous cells in *Trachops* and not in *Megaderma*. Such histological comparisons are significant in that they provide us with fundamental data about convergence or about the extent of divergence at the tissue level, and identify excellent examples of structure–function complexes that might next be investigated experimentally and then related to the biology of mammals at the organismal level.

Histological comparisons are especially interesting when undertaken in concert with physiological or behavioral investigations. An outstanding example is the work of Studier et al. (1983), who combined data on urine-concentrating ability and diet with histological analysis of the kidneys in several species of bats and found microscopically detectable correlations between their structure and functional capacity. A study by Braniŝ (1981) also is an excellent model for combining histological, functional, and ecological analyses. Braniŝ (1981) examined the eyes in three genera of soricid shrews (*Sorex*, *Crocidura*, and *Neomys*) and was able to compare rod–cone ratios among the genera and to relate those ratios to the densities of photoreceptor cells, retinal area, relative size of the eyeball, and equatorial diameter of the lens. These data not only provided an index of divergence among these shrews, but also became a foundation for interpreting the observed behaviors of live animals under different conditions of light.

In some instances histological comparisons might shed light on systematic problems as well as ecology. For example, Sands et al. (1977) compared the foot pads of wolves, domestic dogs, and two geographically separated populations of coyotes. They found striking interspecific histological differences in both the density of eccrine sweat glands and the presence or absence of apocrine sweat glands, both of which also appeared to be correlated with the different thermal sweating responses seen in these various canids and thus might be of significance in the physiological ecology of wolves and coyotes (Sands et al., 1977).

Although clearly valuable, it should be noted that the comparative histological approach has several weaknesses that ultimately limit the usefulness of the technique in the study of evolutionary biology. The most obvious problem is that while cell populations and their organization into tissues often can be determined by histology, study of the cells themselves remains beyond the reach of the light microscope except in a general sense or under special circumstances that are discussed next. This is significant because it is within the cells that the genetic material is actually located and it is therefore the cell to which our attention must be directed if we are ultimately to establish a foundation for understanding organ systems and their evolution in mammals. Homologous cells that appear to be identical when general interspecific histological comparisons are made might in fact be found to be very different when they are compared to one another by some other means, such as histochemistry, immunocytochemistry, or ultrastructure.

Histochemistry and immunocytochemistry are two related ways to improve upon general histological comparisons. Basic histochemistry has been practiced for many years and essentially is a collection of procedures that enable us to identify groups or classes of chemicals within tissues or cells. Although usually somewhat imprecise, this technique nevertheless does allow us to categorize in a general way some of the gene products manufactured within a tissue. Histochemistry has considerable potential for comparative studies and it is thus surprising to learn that comparative histochemical research in systematic relationships is not commonly undertaken in mammalogy. Several such studies have been carried out, including the comparative work of Junqueira and Fava de Moraes (1965), Junqueira et al. (1967, 1973), and Pinkstaff (1981), who have used histochemistry to compare the nature of salivary gland secretions in various mammals, and of Forman (1971,

1972), who has used histochemistry extensively to compare glycoproteins produced by a variety of mucous cells found in the glandular gastric mucosa of microchiropteran bats that exhibit strikingly divergent dietary habits.

Immunohistochemistry involves the use of a molecular probe; it differs from classical histochemistry in its greater precision. Until very recently immunohistochemistry was far from routine and the technique has hardly been applied to problems in evolutionary or systematic mammalogy. However, a variety of polyclonal and monoclonal antisera are now commercially available and if care is taken to ascertain their specificity, these can be used in several different ways to probe for and possibly identify specific gene products in cells or tissues. It is the precision of this technique, especially when monoclonal antibodies are used, that in some ways is also its weakness, at least for our purposes. Fairly minor evolutionary changes in the molecular structure of a protein that affect its hydrophilicity or mobility and might involve as few as five amino acids could render a peptide unrecognizable by the antiserum (Berzofsky, 1985). Immunohistochemical techniques thus tend to reveal evolutionarily conserved gene products rather than the interspecific differences that might be more readily recognizable through histochemistry or some other technique. Ultimately, the value of immunohistochemistry in systematic and evolutionary research might be confined to its use as a probe to locate specific products and as a means of demonstrating different distributions of the same gene product in different species of mammals. The distribution of particular gene products, as well as their presence or absence, also can be investigated ontogenetically within a species as well as among species. The literature contains a variety of indications of major interspecific differences in the distribution of peptide hormones and neuropeptides in mammals. Perhaps most importantly, some newly acquired immunohistochemical data provide the first examples of interspecific differences within families of mammals.

Recently it was shown in certain species of phyllostomid bats that significant interspecific differences in the abundance (density) of gastrin-producing G cells were correlated with other histological features in the pylorus of the stomach (Mennone et al., 1986). Gastrin is a peptide hormone that stimulates acid production by parietal cells in the gastric glands; it also might act as a trophic hormone. Mennone et al. (1986) thought that the occurrence of interspecific differences in the numbers of cells that produce gastrin suggested that evolutionary selection of quantity and distribution of enteroendocrine cells, rather than selection

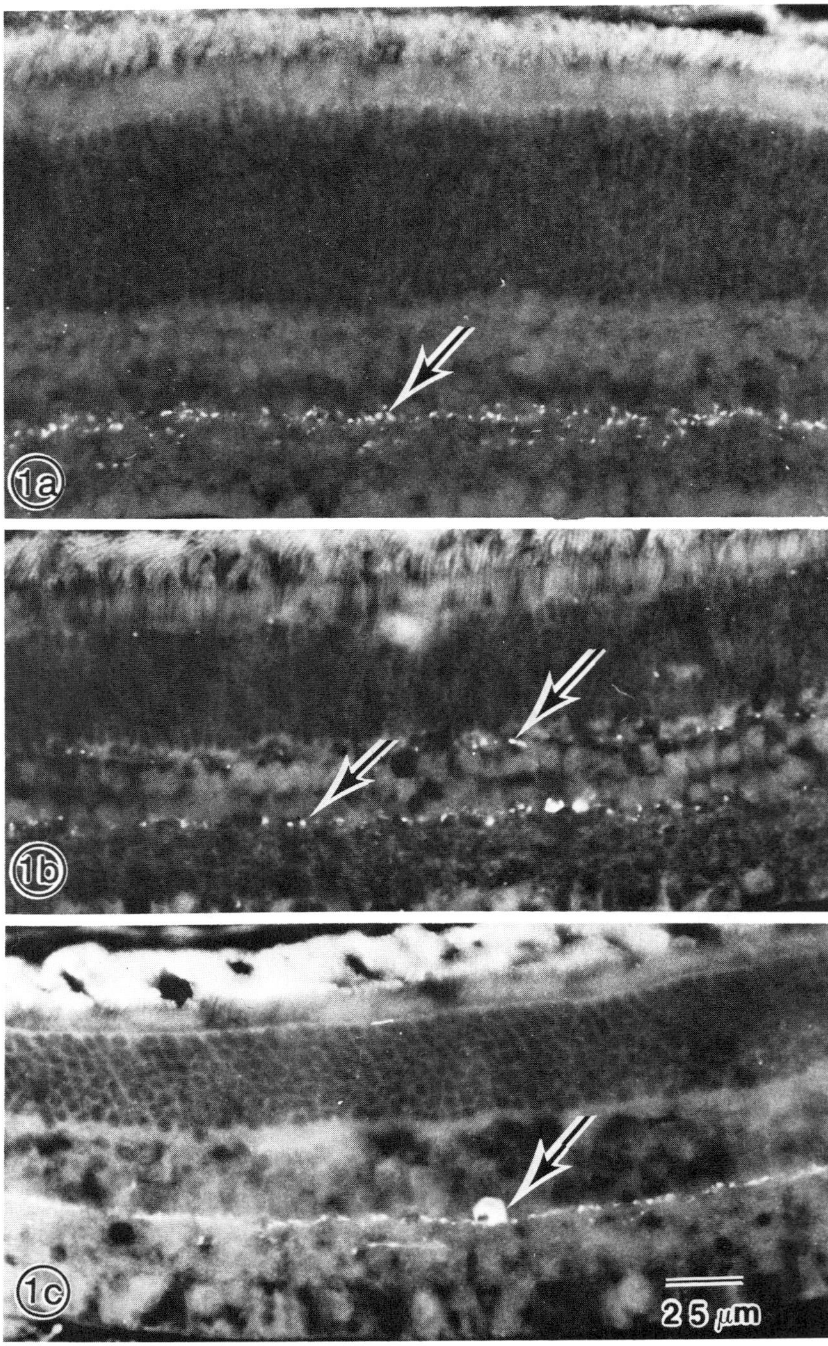
1a
1b
1c
25 μm

of new peptide hormones, might have been an important feature of the evolution of physiological differences in bats with divergent dietary habits.

Interspecific differences also have been reported in the distribution of several neurotransmitters and markers for possible neurotransmitters in the retinas of ecologically distinct phyllostomid bats (Studholme et al., 1987). This study provides an example of significant diversity within a family of mammals. Immunoreactivity to tyrosine hydroxylase (TOH) and endogenous 5-hydroxytryptamine (serotonin) varied interspecifically. This variation correlated with the taxonomy of the phyllostomid genera in the sense that *Macrotus* was distinctive from the others, *Glossophaga* and *Monophyllus* were exactly the same, and *Artibeus* resembled the two glossophagines (Fig. 1).

C. J. Phillips and G. L. Forman (unpublished data) found that the distribution of VIP (vasoactive intestinal polypeptide)-like immunoreactivity (Fig. 2) in the enteric nervous system of the colon differed interspecifically in cricetid and heteromyid rodents of the genera *Oryzomys, Neotoma, Peromyscus, Chaetodipus,* and *Dipodomys*. VIP is a neuropeptide that is widely distributed in both the CNS and peripheral nerves in mammals. It has been implicated in such physiological actions as control of transepithelial water transport, ion exchange, increased blood flow, and relaxation of smooth muscle. The distribution of this neuropeptide in the colon of rodents appeared to be correlated with ecological and physiological factors such as drinking water requirements, diet, and habitat.

These three studies illustrate one of the new means that we have for bridging the gap between molecular and organismal mammalogy. In all of these projects, immunohistochemistry was used as a molecular probe to reveal interspecific differences in the distribution of highly conserved gene products. Interspecific functional differences in organs might arise in part from differences in the numbers or the presence or

FIGURE 1. Examples of interspecific differences in the distribution of a conserved gene product: tyrosine hydroxylase (TOH)-like immunoreactivity in retinas from three species of phyllostomid bats (TOH is the biosynthetic enzyme for dopamine). (a) Immunoreactivity (arrow) in the inner plexiform layer in *Glossophaga soricina*; (b) immunoreactivity (arrows) in both the inner and outer plexiform layers in *Artibeus jamaicensis*; (c) an immunoreactive cell body (arrow) in *Macrotus waterhousii*. Scale bar applies to all three micrographs. These specimens were collected in Jamaica, West Indies, and field-fixed in 4% paraformaldehyde with 0.1 M cacodylate buffer at pH 7.2. Sheep anti-tyrosine hydroxylase (1 : 1000) and anti-rabbit IgG conjugated to fluorescein isothiocyanate were used for the preparations. (Micrographs courtesy of K. M. Studholme and S. Yazulla.)

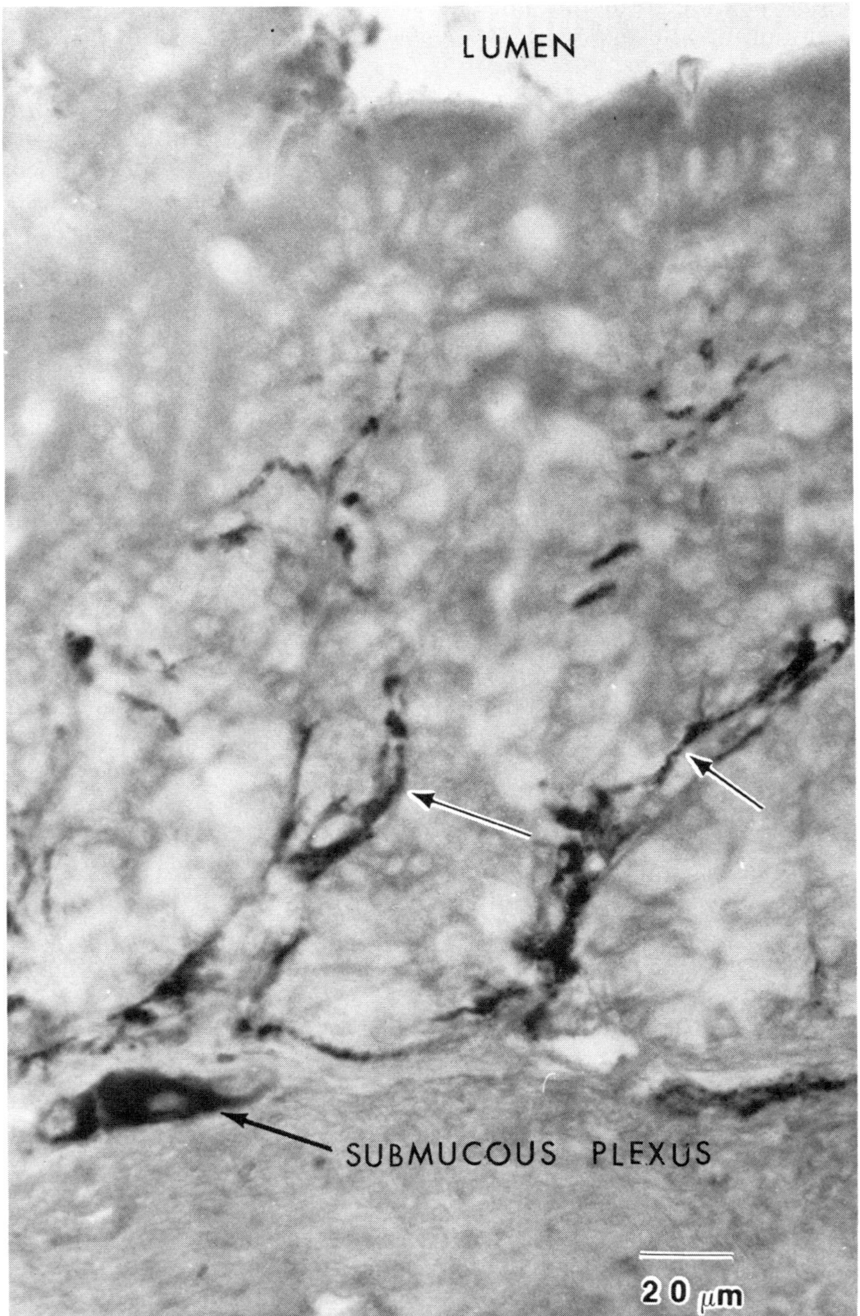
LUMEN
SUBMUCOUS PLEXUS
20 μm

absence of cells that produce particular peptide hormones, or from regional differences in the presence or absence of particular neuropeptides. Such examples indicate how immunohistochemical localization of specific gene products eventually may increase our understanding of the evolution of mammals, especially when the interspecific comparisons are based on a systematic or ecological rationale.

2.2. Transmission Electron Microscopy

Another level of investigation, the cellular level, has become the focus of our joint work, so it is the main subject of the remainder of this paper. The use of comparative ultrastructural analysis to address questions of evolutionary and systematic significance is a fairly new idea, but was discussed in some detail at a symposium sponsored by the American Society of Zoologists (Tyler, 1979). It is noteworthy, however, that none of the papers presented at that symposium dealt specifically with mammals and, in fact, most of the presentations (cf. Storch, 1979) focused on invertebrates. One exception was the work of Eakin (1979), who for many years has used ultrastructure to investigate the evolution of various types of photoreceptor cells. Comparative ultrastructural investigations that have been specifically focused on either mammalian evolution or systematics, or both, are relatively rare. The work of Padykula and Taylor (1977) on the reproductive system represents a pioneering effort. These authors specifically used transmission electron microscopy to examine the chorioallantoic placenta in bandicoots (Peramelidae) in order to elucidate the evolutionary history of marsupials and eutherians. More recently, Feldman and Phillips (1984) used transmission electron microscopy to investigate the retina of pocket gophers (*Geomys*) in order to determine the degree of their divergence from other rodents. Phillips and Studholme (1982), Phillips et al. (1984), and Studholme et al. (1986) used TEM to delineate cellular level patterns of divergence in the gastric mucosa of insectivorous, animalivorous, and frugivorous species of microchiropteran bats. Ultrastructural differences in gastric gland cells to some extent appear to correlate with both diet and systematic relationships. Tandler et al. (1986) used com-

FIGURE 2. Vasoactive intestinal polypeptide (VIP)-like immunoreactivity in nerve bundles (arrows) and in nerve cell bodies of the submucous plexus in the colon of a pocket mouse, *Chaetodiphs baileyi*. This heteromyid rodent lives in desert habitat, feeds on seeds, and can survive without drinking water. VIP is a neuropeptide that has been associated with transepithelial transport of water. Peroxidase–antiperoxidase (PAP) technique; tissue field-fixed in 10% nonbuffered formalin.

parative ultrastructure of seromucous submandibular salivary gland cells to study systematic relationships among Neotropical fruit bats of the genus *Artibeus*. They found interspecific differences that grouped the species according to their genetic relationships. Lastly, a number of excellent examples of the use of comparative ultrastructure in systematics involve studies of mammalian spermatozoa. Rouse and Robson (1986) recently reviewed the data available for primates and concluded that "spermatozoal ultrastructure appears to represent a valid phylogenetic tool . . . . it is species specific and presently allows the separation of taxa to at least the sub-ordinal and in some cases to infraordinal level." Comparative data also have been obtained for a number of marsupials and used to make phylogenetic inferences (Harding et al., 1981, 1982). Finally, spermatozoan ultrastructure has been used to address the question of the evolutionary origin of bats. Forman (1968), Forman et al. (1968), and Forman and Genoways (1979) had pointed out previously that chiropteran spermatozoa varied morphologically and that light microscopic comparisons of these cells might be valuable to resolution of systematic problems. Rouse and Robson (1986) then used transmission electron microscopy to compare spermatozoa from bats and other mammals and concluded that differences within the Chiroptera were "not inconsistent" with the suggestion of Smith (1976) and Smith and Madkour (1980) that bats might have a diphyletic origin.

The remainder of this paper will outline a theoretical foundation for the selection of cell types for additional comparative study, and will discuss the use of comparative ultrastructure to ascertain how the intracellular expression of the genome is ultimately linked to organ function and the biology of mammals. We also have considered three distinct questions. The first, simply put, is, "what is ultrastructure and why use it in systematic and evolutionary studies of mammals?" The second is, "which cell types might be the most valuable to the study of the evolutionary process, and why?" The final one is, "what types of data are available?"

2.2.1. The Technology

The technology for investigating cell architecture has been available for at least three decades, but has improved greatly over the past 10 years because of the increasingly sophisticated techniques developed for study of intracellular events at the molecular level. Conventional and high-voltage transmission electron microscopes provide a view of cell structure that essentially is a representation of intracellular molecular events that have been captured in situ. Our understanding of cells has progressed greatly in recent years and as Gershon et al. (1985) have noted, "the concept of the cell as a tiny bag of enzymes containing a few membrane-limited structures is being laid to rest." Instead, it is now appreciated that cells are highly organized and that their individual structural characteristics are often directly related to their function. Electron microscopy thus does not simply give better resolution than does light microscopy, but rather makes a quantum leap to the point where molecular organization and events can be visualized to a degree. At the same time, the variable magnification available in the transmission electron microscope provides a viewing range that allows these molecular events to be seen in the context of cells and tissues rather than as isolated, purely biochemical reactions occurring in vitro.

Fixation and preparation of tissues are of critical importance in electron microscopy and are rather difficult to accomplish well. Indeed, problems of fixation account for the delay in the application of transmission electron microscopy to systematic mammalogy, because specimens of the most interesting species are not necessarily available on demand and examples stored in conventional museum collections are not suitable for ultrastructural study.

The literature on the ultrastructure of cells in mammals is based almost entirely on domestic and laboratory species. Early efforts to collect wild species and then fix and store tissues under field conditions met with only sporadic success and were discouraging to those who attempted it. Some species of wild mammals that could be brought alive to laboratories were investigated, but comparative studies directed at systematic problems were not undertaken because the necessary samples were difficult to obtain. One simple approach, tried in the mid-1970s, involved placing pieces of tissue in buffered 3% glutaraldehyde and then storing the samples in buffer at ambient temperature. The tissues thus prepared were typically of poor quality for transmission electron microscopy because some of the organelles, especially the mitochondria, would swell greatly. Apparently this was due to the fact that cells are still osmotically reactive after simple glutaraldehyde fixation, and buffer strength is therefore critical (Lee et al., 1981; Phillips, 1985a). By 1979, however, a procedure developed by Kalt and Tandler (1971) was modified slightly and then used successfully under some of the most extreme field conditions (Phillips, 1985a; Forman and Phillips, 1987a). This technique involved the use of a trialdehyde (glutaraldehyde, paraformaldehyde, and acrolein) mixture in a dilute cacodylate buffer with dimethylsulfoxide (DMSO). This mixture allowed for extremely rapid penetration of tissue, obviating the necessity of time-consuming vascular perfusion. The use of low concentrations of cacodylate buffer also meant that tissue specimens could be stored in buffer for several weeks, even at tropical ambient temperature, without incurring problems with osmolarity (Phillips, 1985a). Later, upon return to the laboratory, the tissues could be transferred to fresh 3% glutaraldehyde and then stored at 4°C for at least several years (Phillips, 1985a).

This breakthrough in technique has opened the door to the study of mammalian systematics and evolution at the cellular level because samples for research now can be obtained nearly as easily as more traditional museum research specimens, such as skeletons, skins, and formalin-fixed whole organs. Additionally, the availability of field procedures for TEM fixation has provided us with a means of establishing baselines for the broader subject of variation in cell structure among mammals and for actually establishing museum collections suitable for TEM studies (Phillips, 1985a).

The importance of fixation to the preparation of materials for comparative study is underscored when one considers our previous comment that transmission electron microscopy allows direct visualization of the supramolecular organization and events within cells. It is the fixation process, then, that determines how the cells will appear. In a sense, ultrastructure as viewed with the transmission electron microscope is an artifact of how the chemistry of the cell interacts with the fixative at a particular moment in time. Simson et al. (1978) have clearly documented how pH as well as the basic chemistry and osmolarity of the fixative used for transmission electron microscopy can affect the appearance of secretory granule substructure, while other authors (Minassian and Huang, 1979; Candipan and Sjöstrand, 1984) have demonstrated how various fixation procedures influence some organelles, especially mitochondria. Unfortunately, inattention to the quality and consistency of fixation characterized some earlier comparative work and resulted in a number of incorrect descriptions of cellular features and cell products.

2.2.2. Interpretation of Ultrastructure

What does ultrastructure tell us? How does one actually make interpretations from transmission electron micrographs? Neither of these questions can be answered simply, and hard and fast rules are not available in the literature, primarily because micrographs are the main source of data and, although they provide permanent documentation, they are, at best, qualitative, even when cytochemical or immunocytochemical techniques are used. On the other hand, the techniques and equipment for comparative quantitative analysis of cells now are more generally available than they were before 1980. Image analyzers hold considerable promise for interspecific investigations, even though most of the examples involve light microscopy rather than ultrastructure. The mathematics, sampling strategy, and analytical processes for stereology largely have been derived from fields other than biology, but have been applied successfully to studies of cells and cell structure (Cruze-Orive and Weibel, 1981; Kristensen and Papadimitriou, 1981; Aherne and Dunnill, 1982). The technology is available for making interspecific comparisons of homologous cells and for determining such measures as volume fractions V_v of particular organelles, surface area per unit volume, and the spatial organization of organelles (Hemon et al., 1981; Aherne and Dunnill, 1982). One example of the use of quantitative comparisons of ultrastructural features can be found in Feldman and Phillips (1984), who compared cell height and the depth of infolding of the basal plasma membrane in retinal pigment epithelial cells in two species of rodents that have strikingly different behavioral patterns and ecology. Significant interspecific quantitative differences between homologous, differentiated cells can be used as a foundation for experiments designed to determine the functional importance of ultrastructural differences.

Regardless of technology for quantification, the actual interpretation of transmission electron micrographs still is more of an art than a science because the investigator can only make intuitive judgments about the appropriateness of fixation, the quality of the material, the "best" portion of the tissue block to examine, and the presence and elimination of undesirable artifacts. The problem of selecting appropriate materials upon which to base interpretation is particularly difficult, but we think that the investigator should use published examples as a guide to recognizing high-quality fixation and should test various fixation protocols and embedding procedures on some previously well-characterized tissue from a laboratory mammal before proceeding to other species.

Aside from the variety of factors that can produce structural artifacts, interpretation is still difficult because no criteria or guidelines have been proposed, and the significance of variations in ultrastructural features in homologous cells therefore remains largely unknown. Two of our solutions to this problem are presented in the next sections. Section 3 provides a rationale for selecting cell types that might be profitably compared and then reviews some of the available information that could serve as a foundation for comparative interpretation of ultrastructural features in endocrine cells, neurons, and exocrine cells. Section 4 offers a series of examples in which systematic mammalogy is used as a logical basis upon which a set of ultrastructural data can be interpreted.

Finally, it should be mentioned that the point of obtaining ultrastructural images of cells is not that the descriptive data in themselves are the final desiderata. Instead, the descriptive ultrastructural data are to be used as a foundation for logical experimentation by providing guidelines to the selection of species, organs, and cell types whose study may illuminate some aspects of the evolutionary biology of mammals. Random, uninformed choice of species, organs, and cell types clearly is unwarranted and will in most instances yield ultrastructural data that cannot be interpreted.

3. CELL TYPES: SPECIAL CONSIDERATIONS

Not all organs and certainly not all cell types are equally good candidates for ultrastructural comparison. The arbitrary selection of particular organs or particular types of cells within organs for comparative study might seem to introduce bias into the comparison, but comparisons of randomly chosen cells would be impossible to evaluate from an evolutionary, systematic, or functional perspective. For our joint work, we have concentrated on salivary glands and especially on their secretory cells. Before discussing these cells and the types of data that might be obtained from them, it is worthwhile to establish a general rationale for making interspecific comparisons at the cellular level.

Our use of ultrastructure to visualize molecular events and ultimately to understand the transition from nuclear DNA to structure–function complexes provides us with criteria for selecting cell types for investigation. The ideal cell type thus might be one that actively synthesizes a product for export, because this process in itself involves cellular elements of interest. However, although there is a wide variety of such cell types, not all of them are suitable for comparative study. For example, fibroblasts, chondrocytes, macrophages, and neutrophils are constitutive, nonpolarized secretory cells. Constitutive cells are cells that synthesize proteins, but do not store them in an intracellular compartment (Tartakoff and Vassalli, 1978). Instead, the newly synthesized products are transported relatively quickly (minutes to hours) by vesicles to the plasma membrane, where they are released by exocytosis. The intracellular transport mechanism may involve a form of bulk flow that does not require some type of intracellular protein sorting

(Kelly, 1985). In such cells it is the rate of protein synthesis that determines secretion rate, that is, the greater the rate of synthesis, the greater the rate of secretion. Furthermore, the process of exocytosis does not seem to require an increase in some cytoplasmic second messenger triggered by an external stimulus (Tartakoff and Vassalli, 1978).

Constitutive, nonpolarized secretory cells are not particularly good candidates for comparative ultrastructural studies because the absence of stored secretory protein precludes visualization of the secretory product except by chance. The lack of polarization within the cytoplasm of these cells is another drawback. The cytoplasm of the nonpolarized secretory cells is not randomly organized, but nevertheless lacks the clearly defined intracellular architecture that is characteristic of polarized secretory cells. In the latter, the cell products are manufactured in an orderly manner and then follow a set, visible pathway toward the cell surface where they are released. The highly organized interior of the polarized secretory cell provides a distinctive pattern that can be visualized with TEM; this characteristic design allows easy recognition of cytoplasmic landmarks, providing a frame of reference that is essential in making interspecific comparisons. It also is worth noting that the constitutive, nonpolarized cells appear in general to be highly conservative in both ultrastructure and secretory proteins. This characteristic in itself raises interesting evolutionary questions, but such cells do not lend themselves very well to interspecific ultrastructural comparisons among species of mammals. Lastly, in an evolutionary sense constitutive secretion apparently is the closest to the original type of protein secretion in eukaryotic cells (Kelly, 1985). This means that we might view constitutive secretion as a conservative system that has not been adapted for the particular needs of specific mammals.

Regulated, polarized secretory cells are an alternative to the constitutive, nonpolarized cells. In regulated secretory cells, protein is synthesized at the rough endoplasmic reticulum (RER), packaged (often with modification) at the Golgi complex, and then stored as membrane-bound granules within an area of the cytoplasm (Fig. 3). Ordinarily, the mature secretory granules are stored for some period of time (hours to days) before being released. The release of these products occurs as the result of an externally stimulated alteration of the level of a cytoplasmic messenger such as calcium. Because these cells are polarized, the mature product is released from the cell only at specific sites, such as the apical plasma membrane or the membrane that lines intercellular canaliculi. The area of plasma membrane where release takes place might have some special properties, or, alternatively, the secretory granules may be directed to a particular zone of plasma membrane by the

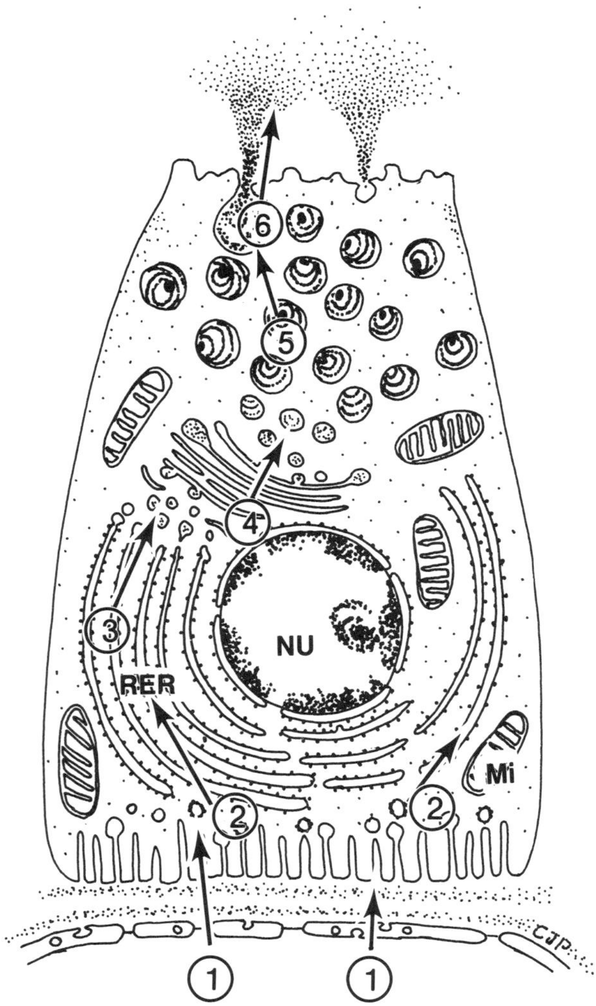

FIGURE 3. Diagram of an exocrine cell, showing the generalized secretory pathway. Raw materials are delivered by capillaries (1) and incorporated by diffusion or endocytosis; protein synthesis takes place at the RER (2); the newly synthesized proteins are transferred to the Golgi complex via the GERL (3); newly packaged product appears in condensing vesicles at the Golgi face (4); as the product matures it is briefly stored and transferred to the apical cell membrane (5), from which it is released by exocytosis (6).

intracellular matrix. Occasionally, the mature secretory granules are not released and under such circumstances they are destroyed by an autophagic mechanism. Those regulated secretory cells that are polarized typically have distinguishing characteristics that can be visualized easily with the transmission electron microscope. For example, the plasma membrane of the apical surface might have microvilli, the membrane of the lateral surface might have various specializations related

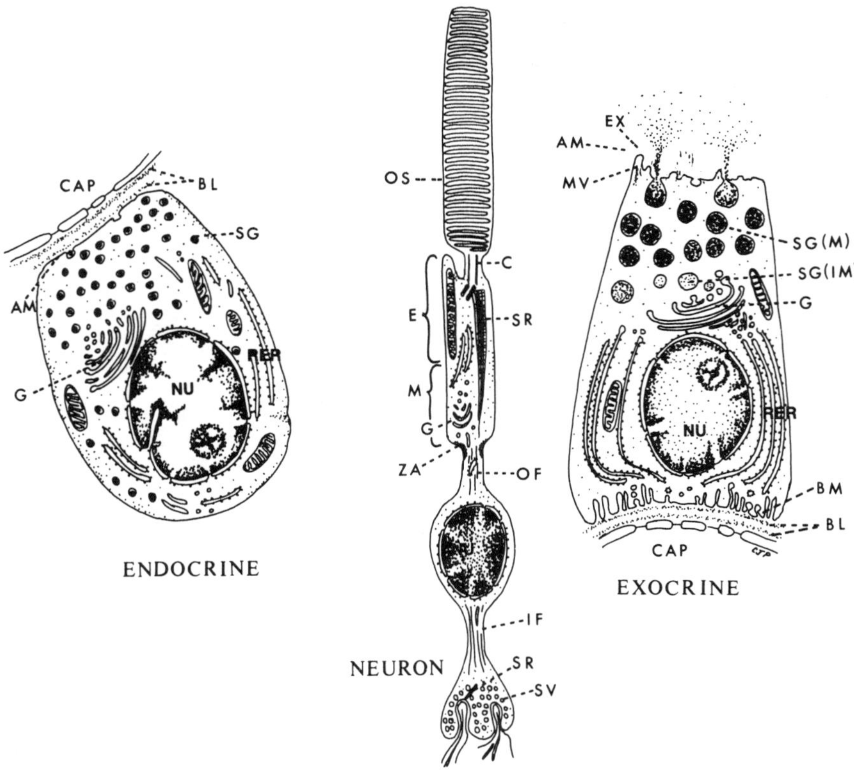

FIGURE 4. Examples of three classes of regulated, polarized secretory cells. Endocrine cell: note the relationship of this example to a fenestrated capillary (CAP). Abbreviations are: AM, apical membrane; BL, basal lamina; SG, mature secretory granule; G, Golgi; NU, nucleus; RER, rough endoplasmic reticulum. Neuron: this example is a mammalian rod photoreceptor. Abbreviations not previously explained are: OS, outer segment; C, cillium; SR, striated rootlet; E, ellipsoid; M, myoid; ZA, zonula adherens; OF, outer fiber; IF, inner fiber; SR, synaptic ribbon; SV, synaptic vesicles. Exocrine cell: compare this generalized exocrine cell with the endocrine cell. New abbreviations are: EX, exocytosis; MV, microvilli; SG(M), mature secretory granule; SG(IM), immature (condensing) secretory granule; GERL, Golgi endoplasmic reticulum lysosomes; BM, basal membrane.

to lateral adhesion of contiguous cells, and the basal surface might have elaborate infoldings where raw materials enter the cell by diffusion or by endocytosis.

Interspecific comparisons of regulated, polarized secretory cells may have special relevance to the study of mammalian systematics and evolution. A small portion of their nuclear genome not only is read and translated into synthesis of a protein or proteins, but, more important for our purposes, these proteins in their final form are packaged in membrane-bound secretory granules and then stored long enough to be visualized by transmission electron microscopy. Indeed, many aspects of the process of protein secretion by regulated, polarized cells can be visualized with this technique and the product thus can be viewed from its earliest stages of formation to its release from the cell. The general ultrastructural features of these cells actually are the morphological reflection of certain molecular activities of the cell and may be regarded as at least partly indicative of the specific requirements for the production, storage, and eventual release of the product. Comparisons of homologous cells can be made following the criteria established for this purpose by Rieger and Tyler (1979), and interspecific comparisons, if properly made, thus can provide a wealth of data about cellular evolution.

A wide variety of regulated, polarized secretory cells is found in mammals. Some, for example pancreatic acinar cells, have been the objects of intense scrutiny, and an assortment of information about their differentiation, life cycle, physiology, and cytochemistry is available in literature. The cells that might be most valuable in infraclass, interspecific comparisons are the endocrine cells, neurons, and exocrine cells (Fig. 4). We are of the opinion that certain of the exocrine cells—especially those associated with the oral cavity—are of considerable value in studying mammalian evolution at the cellular level. Before discussing these in more detail and providing some examples, we first consider the other two cell categories because each offers significant data.

3.1. Endocrine Cells

Endocrine cells are the primary constituents of a variety of discrete glands, such as the thyroid, parathyroid, pituitary, pineal, and adrenal, found in mammals. Endocrine cells also are found within organs, such as the pancreas, that have exocrine as well as endocrine features. The most poorly understood of the endocrine cells, at least insofar as nonspecialists are concerned, are those that comprise the so-called "diffuse" endocrine system. These cells once were thought to be derived embryologically from the neural crest ectoderm, but most workers now think otherwise. Because these endocrine cells are scattered throughout many organs, they have been difficult to categorize, but many of them are classified as amine precursor uptake and decarboxylation (APUD) cells (Solcia et al., 1981).

The ultrastructure of many endocrine cells (Figs. 4 and 5), especially those found in discrete glands, has been described and illustrated for several species of mammals and some interspecific comparisons have been made. For example, Karesek et al. (1982) compared the pineal gland of the eastern chipmunk, *Tamias striatus*, to that of other rodents and reported that several ultrastructural features differed among the species. In particular they noted that pinealocytes in *Tamias* contained large numbers of "synaptic" ribbons, exceeding by 5–30 times the numbers found in homologous cells in other mammals.

The diffuse endocrine system in some ways is more complex than the discrete glands because it includes a great variety of cell types, and, moreover, one cannot predict what kind of cell might be found in a given location, such as within particular areas of the gastrointestinal tract, in a given species. The cells of the diffuse endocrine system were classified according to their microscopic appearance and the size of their secretory granules (Grube and Forssmann, 1979; Solcia et al., 1981) and named before the chemical composition of their products was known (Fig. 5). However, over the past decade the use of polyclonal and even monoclonal antibodies has clarified the relationship between ultrastructure of the secretory granules and the presence of particular peptides or groups of peptides for which the molecular structure is known.

Because the peptide hormones produced by diffuse endocrine cells play an important role in the regulatory biology of animals, they have attracted the attention of biochemists and endocrinologists. As a result, the molecular structure has been determined for many of the peptides from a variety of vertebrate species and data on function have accumulated very rapidly over the past decade (Miller, 1984). These data, in combination with comparative ultrastructural analysis, form a substantial foundation for evolutionary studies of mammals at the cellular level. Indeed, Solcia et al. (1981) have already used molecular homologies among the peptide products and ultrastructural similarities of diffuse endocrine cells to argue for a common origin of the cells in the glucagon–secretin family of peptides. The potential value of cells in the diffuse endocrine system for our purposes can be further illustrated

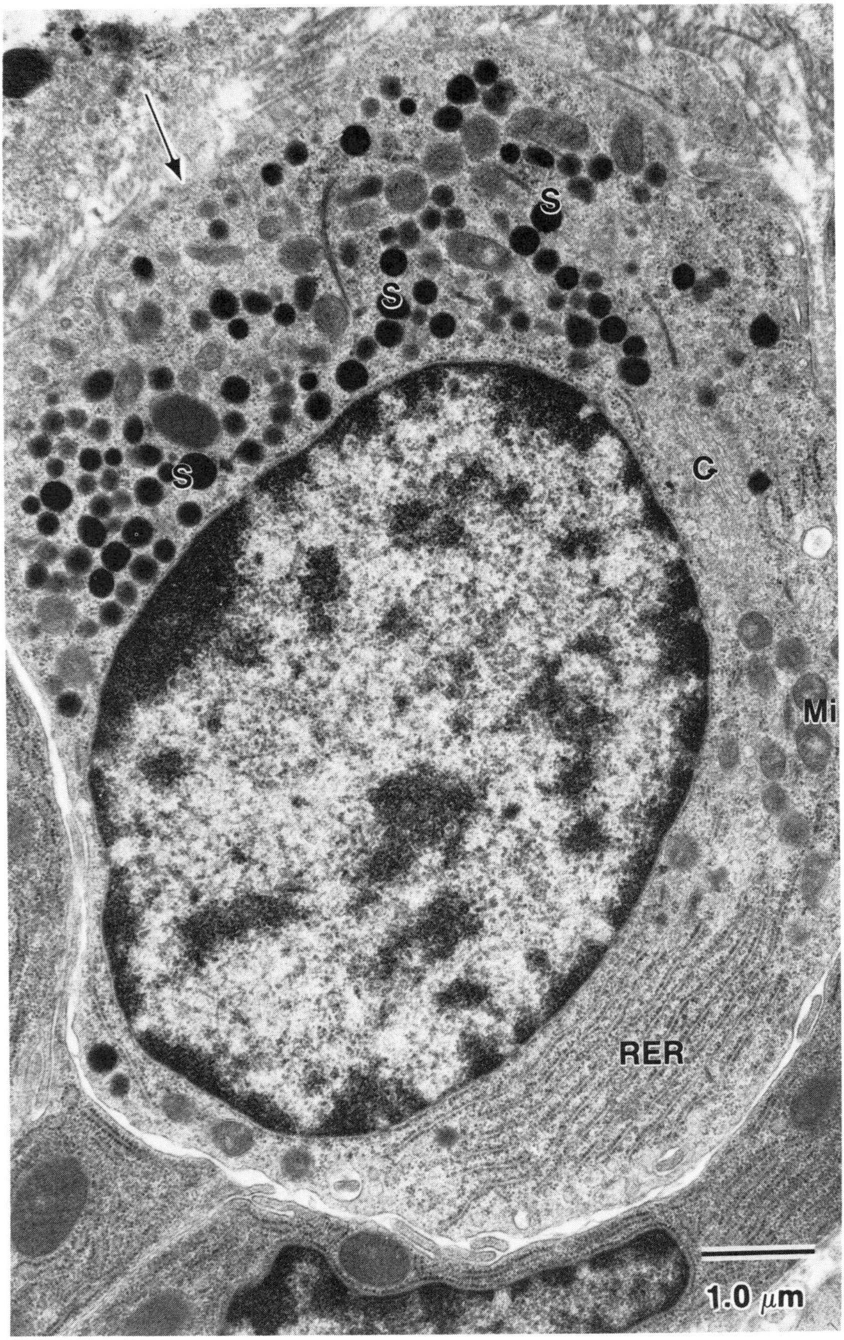
S
S
S
G
Mi
RER
1.0 μm

by considering in some detail the endocrine cells that produce such peptides as somatostatin, gastrin, cholecystokinin (CCK), glucagon, and pancreatic polypeptide (PP).

Somatostatin is a peptide hormone that exhibits biological effects ranging from the stimulation of behavioral changes in some species to inhibition of other gastrointestinal hormones such as gastrin (Patel et al., 1981). Its actions are clearly of interest to mammalogists because these different effects can each be studied in the intact organism in a particular ecological setting. Immunoreactivity of somatostatinlike molecules has been noted in diffuse endocrine cells that originally were classified as "D cells" on the basis of the ultrastructure of the secretory granules. Morphologically the D cells are highly variable and sometimes have long cytoplasmic processes with a bulblike terminus resembling an axon (Fig. 6). At the molecular level, somatostatin has been found to be heterogenous in molecular structure even within individuals, although one tetradecapeptide sequence, termed SRIF, has been isolated from teleost fish, birds, and a variety of mammals (Patel et al., 1981) and is highly conserved in at least the higher vertebrates. Two different high-molecular weight, biologically active SRIF precursors (12,000–15,000 and 3000–6000 daltons) have been found in mammals; they occur in different proportions in different tissues within individual mammals; the larger molecular weight form might differ from the smaller one mainly in its being glycosylated. These intertissue molecular differences in somatostatin seem to correspond to ultrastructural differences in stored secretory granules in D cells in various tissues (Solcia et al., 1981), supporting the notion that subtle differences visualized in D-cell secretory granules may reflect real differences in the molecular structure of the product. This type of data is important to our understanding of the possible significance of interspecific differences found when ultrastructural comparisons are made.

Another peptide hormone, gastrin, is produced by G cells and is best known for its stimulation of the acid-producing parietal cells in the stomach of mammals. As with somatostatin, the gastrin peptide hormone seems to occur in several molecular forms; these have been labeled according to their respective numbers of amino acid residues,

←

FIGURE 5. Enteroendocrine cell in the gastric mucosa of a specimen of an insectivorous phyllostomid bat, *Tonatia bidens*. Based on ultrastructural criteria, this probably is a glucagon-secreting A cell; note the electron-dense secretory product (S), which probably is released at the surface marked with an arrow. Field-fixed in trialdehyde. Abbreviations are: G, Golgi complex; Mi, mitochondria; RER, rough endoplasmic reticulum.

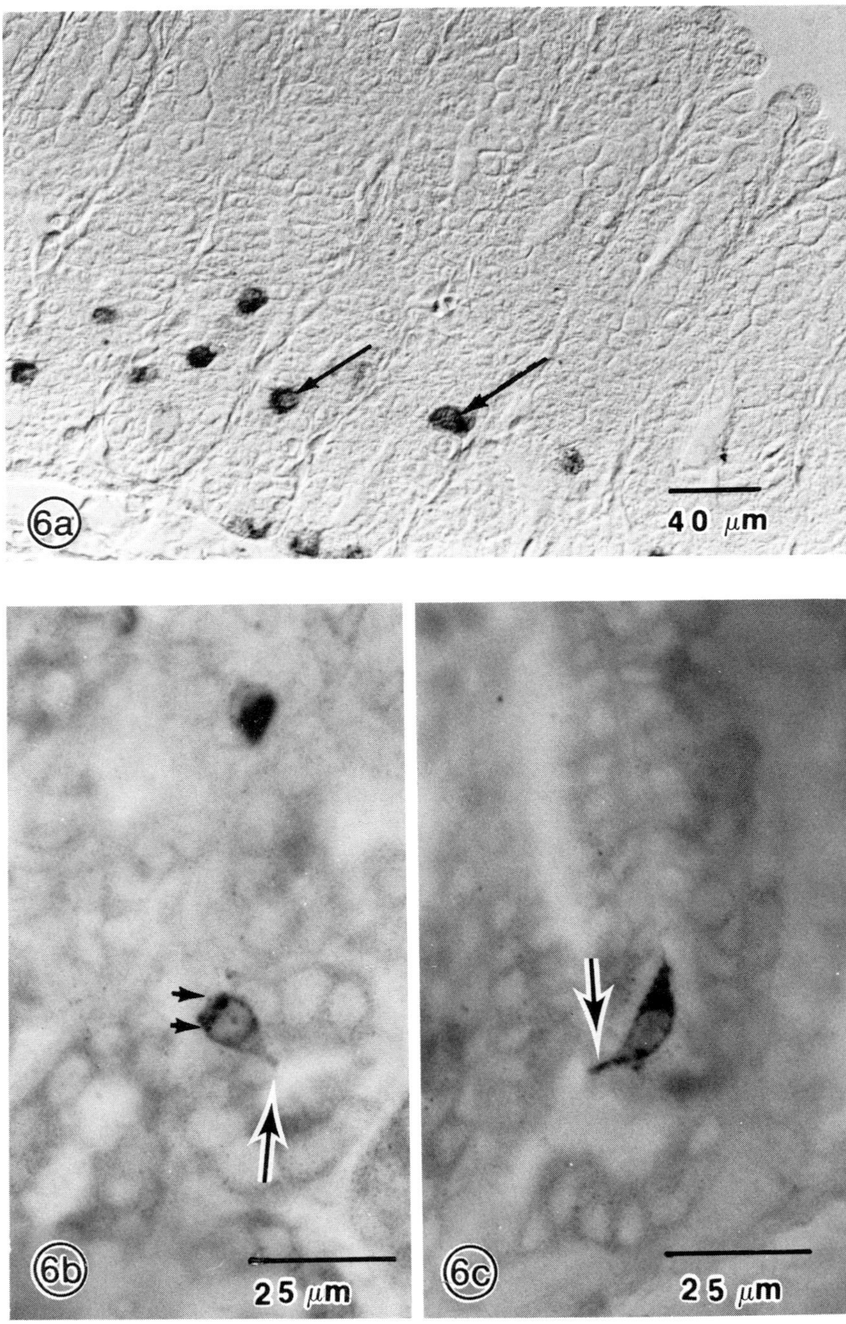
6a
40 μm
6b
25 μm
6c
25 μm

for example, G17 and G34, and differ in being sulfated or nonsulfated (Walsh, 1981). Different molecular forms of gastrin apparently occur within the same cell and in different proportions in different tissues. The molecular structure of gastrin is closely related to that of another peptide, cholecystokinin. These two molecules have the same COOH-terminal structure (Dockray, 1981) and to some extent have similar biological actions.

An evolutionary perspective for gastrin and CCK is offered by the work of Larsson and Rehfeld (1977). These authors made elegant use of immunohistochemical techniques to compare molecular forms of both peptides in mammals, birds, reptiles, amphibians, and teleost fish. They found that in the amniotes various types of gastrin and CCK-like molecules occurred in separate cell systems, whereas both types of molecules were synthesized in a single cell type in both amphibians and bony fish (Larsson and Rehfeld, 1977). They concluded that gastrin and CCK probably evolved from a common ancestral molecule similar to caerulein, which possibly occurred in early vertebrates. This study provides us with an idea of the origin of particular types of peptide hormones, the details of the molecular structure of the hormones, an intimation of the molecular variation within each type of hormone, and a means of localizing the hormones to a particular type of endocrine cell that in turn can be visualized in detail by electron microscopy.

Gastrinlike immunoreactivity has been localized in cells ultrastructurally defined as G cells. The secretory granules tend to be highly variable, even within an individual cell. Typically, gastrin-type molecules are packaged in granules that measure approximately 300 nm in diameter; the content of these membrane-bound vesicles varies from pale and finely particulate to compact with a lucent halo. This variation is indicative of the great range of variability in gastrin-type molecules, particularly with respect to size. Cholecystokinin (CCK) is in a separate group of molecules but has a terminus that is identical to that in gastrin. Immunohistochemically labeled cells containing CCK differ ultrastructurally from the G cells. The CCK-type molecules are found in cells

FIGURE 6. Somatostatinlike immunoreactivity in D cells in the pylorus of the stomach in the flower bat, *Erophylla sezekorni*. (a) A group of immunoreactive cells (arrows) positioned in the lower portion of the pyloric gland unit. (b) In this D cell most of the somatostatinlike immunoreactivity is localized in the base of the cell (black arrows), whereas the apical surface, which reaches the gland lumen, is not as immunoreactive (white arrow). (c) Immunoreactive process on a D cell (arrow). Specimens were field-fixed in 4% paraformaldehyde with 0.1 M cacodylate buffer at pH 7.2; peroxidase–antiperoxidase (PAP) technique with DAKO antisomatostatin.

that originally were termed "I cells," which are easily distinguished from G cells by electron microscopy. In the I cells, the secretory granules are about 260 nm in diameter and uniformly electron-dense. In contrast to gastrin, the molecular variations between the several forms of CCK (CCK 8, CCK 33, CCK 39) do not appear to be mirrored by differences in the appearance of the secretory granules containing them in the I cells.

These first two examples illustrate instances in which differences in the ultrastructural appearance of stored secretory product seem to be correlated with molecular differences in the peptides. These data may contribute to our understanding of how particular types of cells might have come to produce slightly different peptides in different tissues, as in the case of somatostatin, and also may enable us to understand how cells such as those secreting gastrin and CCK became specialized to produce specific peptide hormones. The G and I cells are of particular interest because they may have evolved from some common type of precursor cell in the anamniote vertebrates and study of these cells may eventually reveal how new cell types originate and how ultrastructural differences correspond to evolutionary divergence.

Additional data valuable to a foundation for interpretation of ultrastructural differences can be found in investigations of glucagon and pancreatic polypeptide (PP). Immunocytochemical and immunohistochemical techniques have been used to demonstrate that glucagon and pancreatic polypeptide usually are synthesized and released from separate cell types (Larsson et al., 1974, 1976). Both cells are found in the pancreas, but their secretory granules are so different that one can readily distinguish between them with electron microscopy. Glucagon is synthesized in A cells; when granules are viewed with electron microscopy, it appears as a small, electron-dense core within each granule. This core is surrounded by a paler outer material that contains glicentinlike molecules (Ravazzola and Orci, 1980). Glucagon often is stored in a form that is identical to the crystalline form it assumes in vitro (Thomas, 1970). This is important because on these occasions one can identify glucagon by its crystalline structure when it is visualized by electron microscopy. The ultrastructure of the secretory granules is a distinguishing feature of A cells, which have been described from a wide variety of mammalian species (Solcia et al., 1981; Phillips et al., 1984). Immunocytochemistry has been used to show that pancreatic polypeptide usually is found in endocrine cells in which secretory product is packaged in somewhat variably sized (130–360 nm) granules that are decidedly ultrastructurally different from the granules in the

A cells (Kaung, 1985). The PP secretory granules are not dense-cored; instead, they have a somewhat paler, grainy appearance. However, Kaung (1985) found a type of pancreatic endocrine cell in the rat that synthesizes both glucagon and pancreatic polypeptide; this cell is ultrastructurally distinct from typical A cells and PP-producing cells; not surprisingly, the secretory granules are distinctive.

Comparative data such as these, from cells that synthesize a glucagon–glicentin product or pancreatic polypeptide or some combination of the two, help us recognize the potential significance of ultrastructural differences. Furthermore, the molecular structure of glucagon and pancreatic polypeptide indicates that these two products have broad molecular homology, that they might share a common precursor molecule during their intracellular formation, and that glucagon and PP may have evolved from some common type of ancestral molecule in earlier vertebrates. It is possible, because glucagon is synthesized in some mammalian cells and pancreatic polypeptide in others, and on occasion both molecules are formed in the same cell, that these various cells might serve as additional excellent models for studying the evolution of cell "types" in mammals. Additionally, it should be remembered that information is available on both glucagon and pancreatic polypeptide function and that the glucagon "gene" has been investigated in considerable detail (Lopez et al., 1984).

In summary, the regulated, polarized cells of the diffuse endocrine system are an extremely valuable group for interspecific studies because (1) some of the secretory products have been characterized in detail, (2) antisera are available for localization of types of peptides in particular cells, (3) extensive information on the biological function of these peptides is becoming available, and (4) variations in ultrastructural appearance sometimes can be correlated with variations in the molecular structure of peptides as they are packaged and stored in the cell cytoplasm. The only apparent obstacle is that these endocrine cells and their products appear to be fairly well conserved within mammals. For example, the ultrastructure of a variety of endocrine cells seen in the stomach of bats closely matches those seen in primates, rodents, and ungulates (Phillips et al., 1984; Studholme et al., 1986; Mennone et al., 1986; Solcia et al., 1981; Domeneghini et al., 1985). Infraclass comparisons may therefore prove less interesting than comparisons among all of the vertebrate classes. For now, however, the regulated, polarized secretory cells in the diffuse endocrine system clearly hold great promise as a key to our understanding of the evolutionary process at the cellular level in mammals.

3.2. Neurons

Neurons are an exceedingly complex group of regulated, polarized secretory cells. Neurons are found in many morphological forms, but typically consist of a nerve cell body (perikaryon) with one or more processes (axons and dendrites) that extend from the cell body. The nomenclature of neurons is based on the morphology of these processes. A wide variety of molecules is synthesized and packaged in the perikaryon, which contains RER, smooth endoplasmic reticulum (SER), and a Golgi complex. These molecules then are translocated to the terminus of the axon by means of an axonal transport system involving microtubules and microfilaments. The secretory or synaptic vesicles of neurons can be readily visualized by electron microscopy and, as in other regulated secretory cells, these vesicles contain cell products in a stored form. It was originally assumed that the synaptic vesicles of neurons contained either acetylcholine (ACh) or norepinephrine (NE), which were then thought to be the only kinds of neurotransmitter; it is now known that these vesicles can contain one or more of a great variety of molecules. In addition to ACh and NE, this list includes 5-hydroxytryptamine (serotonin), dopamine, gamma aminobutyric acid, substance P, vasoactive intestinal polypeptide (VIP), somatostatin, neurotensin, a gastrin-CCK-like molecule, enkephalin, and bombesin. Most of these molecules apparently are neurotransmitters, but some doubt remains as to whether all of them meet the appropriate criteria (see Gershon, 1981).

The great variety of molecules found in secretory granules in neurons offers many opportunities for comparative study within the Mammalia. Indeed, the ultrastructural variation in these granules as well as the fact that some granules resemble in a general way the granules seen in endocrine cells has led some observers to suggest that some nerves are "peptidergic" (Baumgarten et al., 1970). However, unlike the secretory granules of the endocrine cells just discussed, the composition of secretory granules produced by neurons is not nearly so readily linked to particular ultrastructural images, and direct ultrastructural comparison of neurons in particular tissues in different species is therefore greatly limited. That is, the synaptic vesicles differ with respect to appearances (clear to granular to dense core) and size, but these particular appearances cannot be consistently correlated with the presence of the various amines or neuropeptides that can be localized to an axon terminal by means of immunohistochemistry and immunocytochemistry. The various neurotransmitters thus cannot as yet be identified by means of TEM alone.

One particularly interesting set of comparisons can be made between diffuse endocrine cells and neurons. Somatostatin, as mentioned in the previous section, is a peptide hormone known to be synthesized and stored in D cells; this same peptide, or at least a very closely related molecule, can also be immunochemically demonstrated in certain neurons. However, the neuronal synaptic vesicles that seem to contain somatostatin do not resemble the stored secretory granules seen in D cells. This lack of resemblance has held true for all of the peptides that so far have been localized in both neurons and endocrine cells in vertebrates (Solcia et al., 1981), and this makes their recognition by TEM alone currently impossible.

The difficulty in relating TEM images to the presence of particular neuropeptides is illustrated in a recent paper by Loesch and Burnstock (1985). These authors used the peroxidase–antiperoxidase (PAP) technique to localize vasoactive intestinal polypeptide (VIP)-like immunoreactivity in nerve fibers in the myenteric plexus of the laboratory rat. VIP, the first neuropeptide to be localized by immunocytochemistry and transmission electron microscopy to synaptic vesicles (Larsson, 1977), is of great physiological interest because it is involved with relaxation of smooth muscle, dilation of blood vessels, and transepithelial transport of water. In their study, Loesch and Burnstock (1985) found that in the nerve fibers that exhibited VIP-like immunoreactivity, 93% of the immunoreactive synaptic vesicles were spherical, empty-appearing, and measured between 40 and 55 nm in diameter. The immunoreactivity (which was traced by formation of a precipitate) seemed to be localized over the membrane of the vesicles rather than within their pale interior. The remaining 7% of the immunoreactive synaptic vesicles were larger (100–144 nm) and contained an electron-dense core. In this instance, it was the dense core of the vesicle that was immunoreactive (Loesch and Burnstock, 1985) rather than the surrounding membrane. Berezin et al. (1985) used the electron-immunohistochemical protein A–gold technique to localize VIP-like molecules in nerve varicosities in dog ileum. They found that immunoreactivity was located in granular vesicles measuring 50–90 nm. Probert et al. (1981, 1983) reported that in neurons in guinea pig myenteric plexus, VIP-like immunoreactivity is only found in large (55–129 nm) dense-core vesicles. These discrepancies might be due to differences in technique or may reflect a difference among animal species (Loesch and Burnstock, 1985). The variation in ultrastructural appearance of secretory vesicles in individual neurons might be largely due to the packaging in each cell of different combinations of neurotransmitters. In other words, whereas in some cells the synaptic vesicles might include combinations of ACh and VIP, in other neurons VIP and ACh might be combined with 5-HT or substance P and so forth. In a light-microscopic immunohistochemical investigation of intramural neurons of the small intestine in laboratory mice, rats, and domestic pigs, Ekblad et al. (1984) were able to colocalize VIP- and neuropeptide Y (NPY)-like immunoreactivity in nonadrenergic neurons in these species. How these combinations of molecules influence ultrastructural images is unknown and the extent to which various neuropeptides occur in combination also is uncertain. For whatever reasons, synaptic vesicles and neurons, in marked contrast to the various diffuse endocrine cells discussed in the previous section, do not lend themselves well to TEM comparisons, despite the fact that some of the same or nearly identical molecules are synthesized and exported by both neurons and endocrine cells. Indeed, in the absence of immunocytochemistry, relatively little can be said about these synaptic vesicles, except that they vary greatly in structure and function from one neuron to another.

Aside from that concerning photoreceptors, which are discussed below, relatively little comparative ultrastructural data for neurons are available for mammals, even though a huge body of literature on neuron structure, including general ultrastructure, and function has been generated in recent years. One of the few examples of comparative ultrastructure of neurons undertaken from an evolutionary perspective may be found in the work of D'yachkova (1984), who investigated the olfactory cerebral cortex in primates, rodents, carnivores, and lagomorphs. The general scarcity of interspecific comparisons at the ultrastructural level is due mainly to the great difficulty in identifying homologous nerve cells and in obtaining the necessary chemical and neurophysiological data. Light-microscopic comparisons using immunohistochemistry (for example, Schultzberg et al., 1980; Keast et al., 1985) clearly suggest, however, that TEM comparisons among rodents and other groups eventually could be of great interest if approached from a systematic perspective. The enteric nervous system in particular should be interesting because of its degree of developmental independence, its accessibility, the availability of much ancillary data (Gershon, 1981), and the unvarying location of clusters of specific neuron types, affording great opportunity for recognizing homologous neurons in different species.

It also is worth noting that broader phylogenetic comparisons of neurons—among vertebrates and protochordates, for example—have been made, albeit on a limited basis, either by chemically localizing 5-HT or by inferring its presence based on nerve cell autofluorescence (Feher et al., 1980; Goodrich et al., 1980). That serotinergic neurons are found in a variety of mammalian species should be of interest to mammalogists because mammals can thus be compared to other vertebrates in this regard.

The photoreceptor cell is one type of mammalian neuron that has been investigated ultrastructurally from a comparative, systematic perspective. Photoreceptor cells, which are highly specialized bipolar neurons in the retina (Fig. 4), in many ways are ideal for interspecific studies because (1) they are readily recognizable *in vivo*, (2) they are located in a fairly accessible position for correlative ultrastructural and neurophysiological analysis, (3) they are a major component of the visual system, which is highly variable in mammals, and (4) they clearly play a role in determining behavior and potential responses to the environment. The origin and evolution of mammals seems to have involved major modification of the reptilian visual system, especially of the photoreceptors, and associated components of the brain (Walls, 1942; Jerison, 1973). These fundamental changes in the eye seem to have set the stage for great evolutionary diversity in the mammalian retina. Photoreceptors are in many ways the most completely studied of all neurons and are the cell type probably most frequently compared both within mammals and among all other vertebrates (Walls, 1942; Eakin, 1963, 1968, 1979; Levine, 1985; Snyder et al., 1986).

Photoreceptor cells not only are polarized, but also are perhaps the most dramatically regionalized of all of the various regulated secretory cells (Figs. 4 and 7). The photoreceptor cell has an outer segment, the end of which contacts the single layer of retinal pigment epithelial (RPE) cells, and a complex inner segment; the two segments are connected by a nonmotile cilium. The inner segment can be further subdivided into an ellipsoid zone with abundant mitochondria and a myoid zone with RER, SER, and a Golgi complex. An outer fiber leads to the nucleus and an inner fiber (= short axon) that terminates in a synaptic connection. It is this complicated, but characteristic, organization of the photoreceptor cell that greatly facilitates comparison among species. Specific portions of the cell can be compared and contrasted under controlled conditions, using systematic relationships, behavior, ecology, or any combination of the three to select species of interest and to interpret any observed patterns.

Mammalian photoreceptor cells typically are placed into one of two categories—rods and cones. On the basis of ultrastructural criteria alone, the existence of these two types of cells can be established by comparing relative length, shape, breadth, and membrane of the outer segment, size and shape of the ellipsoid, location of the nucleus, relative amount of heterochromatin in the nucleus, and the type of synaptic

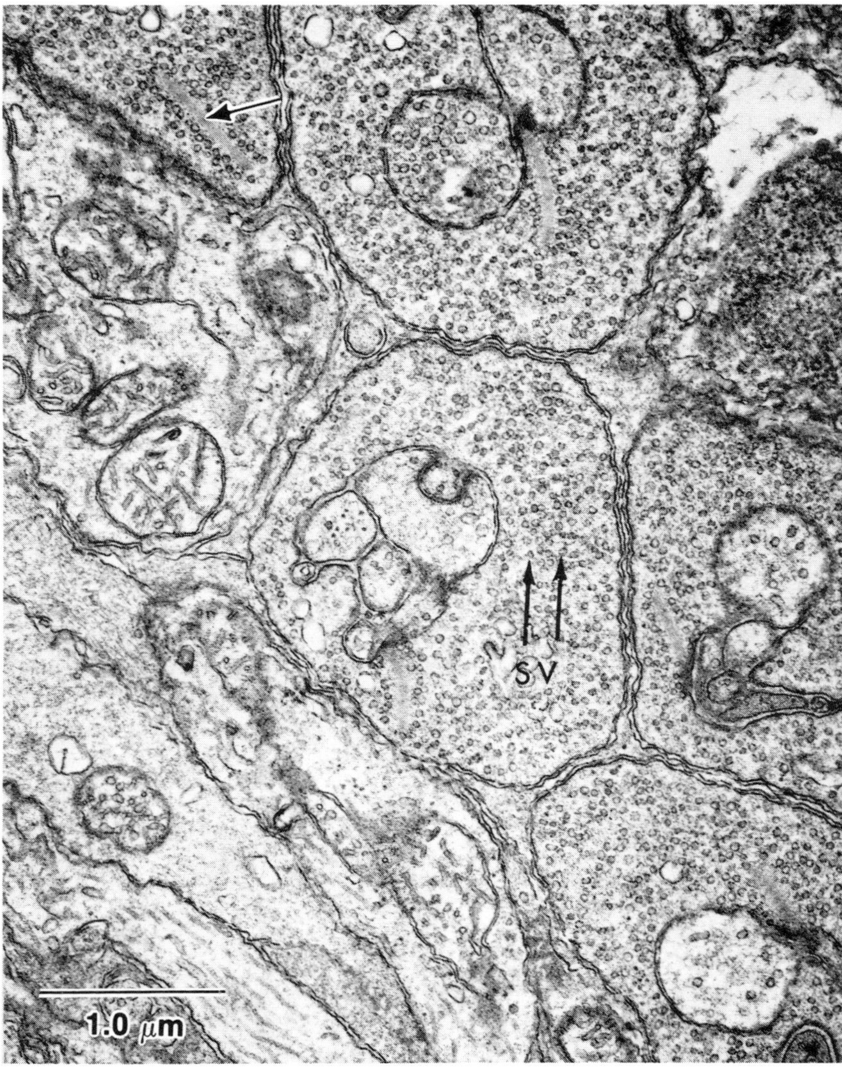

FIGURE 7. Synaptic vesicles (SV) and a synaptic ribbon (white arrow) in rod photoreceptor cells in the retina of a megachiropteran fruit bat, *Megaerops ecaudatus*. Specimen field-fixed in trialdehyde. (Micrograph courtesy of K. M. Studholme.)

connection (spherule versus pedicle). Although these criteria are commonly used to distinguish between rods and cones, in the absence of electrophysiological or other functional evidence the cells probably should be regarded as conelike or rodlike (West and Dowling, 1975). Enough ultrastructural comparisons of photoreceptor cells have been made for mammals, rodents in particular, to allow us to begin to recognize some evolutionary patterns. This is especially true for the squirrels (Sciuridae) because data are available for ground-dwelling diurnal and arboreal diurnal and nocturnal species (Cohen, 1964; West and Dowling, 1975; Fisher et al., 1976; Jacobs et al., 1976; Long and Fisher, 1983). Feldman and Phillips (1984) compared the ultrastructure of photoreceptors in a nocturnal cricetid rodent (*Neotoma floridana*) and a fossorial pocket gopher (*Geomys bursarius*) and found that the pocket gopher resembled several of the diurnal squirrels in that respect. This correspondence in ultrastructure was in many ways surprising because pocket gophers spend considerable time in dark burrows and are said to have reduced, that is, relatively small, eyes. These authors thought it unlikely that the neural retina of pocket gophers had evolved to resemble that of diurnal squirrels, preferring instead to hypothesize that gophers had retained a retinal structure that was relatively unchanged from that of their ancestors. On this basis it was suggested that pocket gophers might have evolved from some type of diurnal, ground-dwelling ancestral rodent. In this way, Feldman and Phillips (1984) used ultrastructure to deduce something about the probable behavior and ecology of an otherwise unknown ancestral group.

The cells of the retinal pigment epithelium (RPE) are not neurons, but constitutive, polarized secretory cells. However, they are considered here because, in contrast to most constitutive cells, their polarized structure and their intimate association with the photoreceptors combine to make them ideal for comparative ultrastructural analysis (Fig. 8). The RPE is directly involved in the rhodopsin cycle. Rhodopsin is the visual pigment of the rod photoreceptors; it is a conjugate of opsin (a protein synthesized in the photoreceptor myoid region) and 11-*cis*-retinal, which is its chromophore. After 11-*cis*-retinal is hydrolyzed from opsin, it is reduced to retinol, transferred by some unknown means to the RPE cells, stored, and then eventually returned to the rod outer segment through the tips of the elongate apical microvilli that characterize the RPE cells (Williams and Penn, 1985). The RPE also degrades the distal membrane discs of the outer segment of rod photoreceptors, and the RPE cytoplasm is therefore typically filled with phagosomes, lysosomes, and melanolysosomes (Leuenberger and Novikoff, 1975). This dynamic interaction between the RPE and the photoreceptors has

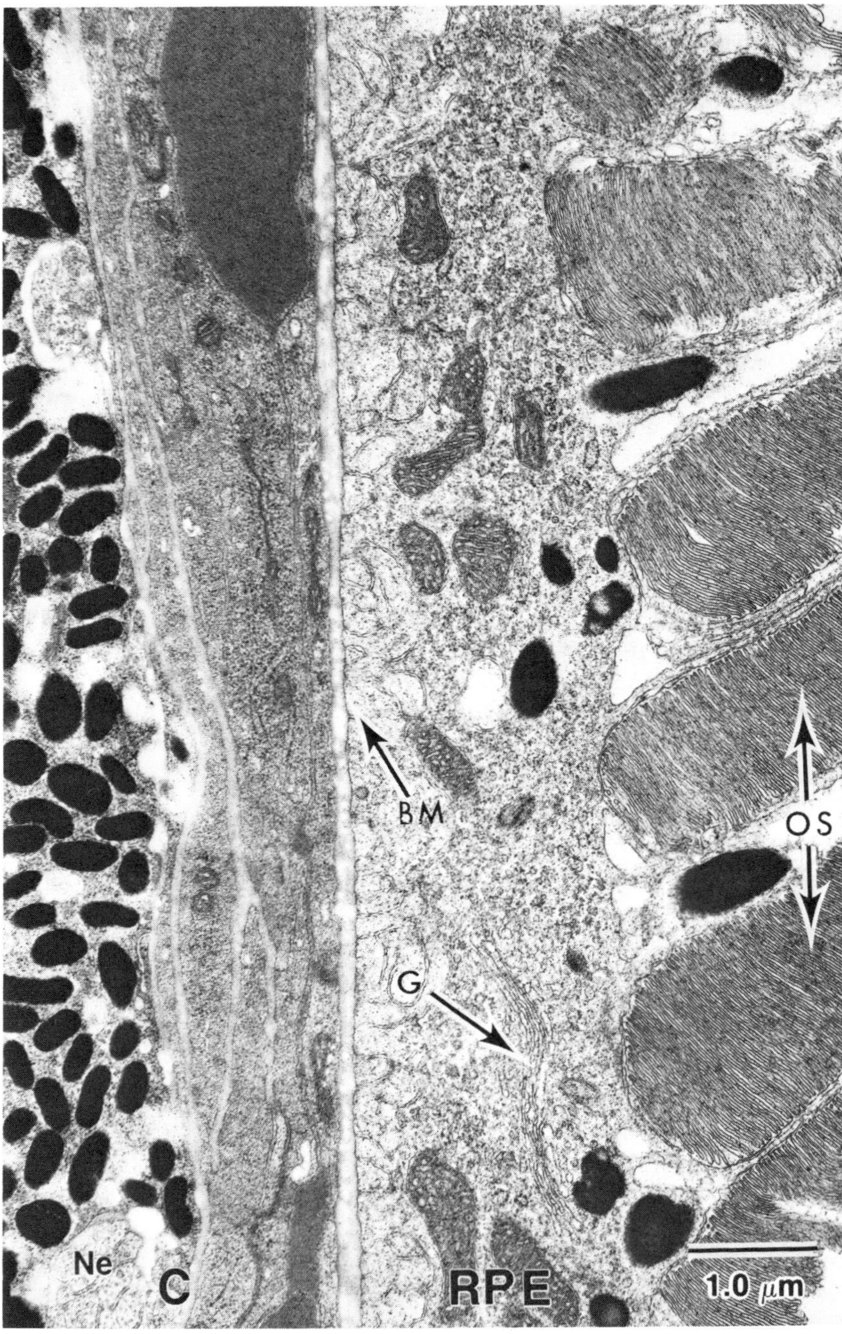
BM
OS
G
Ne
C
RPE
1.0 μm

attracted considerable attention, and comparative data are now available on the ultrastructure of RPE cells in a variety of mammals (Kuwabara, 1979; Feldman and Phillips, 1984; Phillips, 1985b). Among rodents, significant interspecific ultrastructural differences have been found in such features as the relationship between the depth of basal membrane infolding to cell height (Feldman and Phillips, 1984), relative length of apical microvilli, presence or absence of lipidlike droplets, and the relative abundance of smooth endoplasmic reticulum (Kuwabara, 1979; Phillips, 1985b). At this time, it seems likely that the interspecific differences recorded for the RPE of rodents relate in some way to differences in behavior, for example, diurnal versus nocturnal, and ultimately in ecology. An examination of the comparative ultrastructure of the eye in rodents thus might help elucidate the factors involved in the exploitation of microhabitats by various species. Finally, the RPE cells that are the most unusual in ultrastructure have been described in megachiropteran fruit bats of the genus *Pteropus* (Pedler and Tilley, 1969). The significance of these unusual cells is unclear, but eventually their ultrastructure might be integrated with the interesting functional data recently obtained by Murphy et al. (1983), Calford et al. (1985), and Pettigrew (1986).

In summary, neurons are a group of regulated, polarized secretory cells that are extraordinarily complex in both structure and function. Neurons have great potential in the study of the evolutionary history of mammals at the cellular level, but they are more difficult to investigate than diffuse endocrine cells. One of the current difficulties is our inability to correlate the ultrastructural appearance of synaptic vesicles with their contents. This problem, however, can be solved when a qualitative technique such as immunocytochemistry is employed. Another problem is the difficulty in selecting homologous neurons. Regardless of these difficulties, some comparative data, particularly from immunohistochemical studies, do suggest the existence of major interspecific differences in the distribution of some neurotransmitters among mammalian species (Keast et al., 1985) and investigations of homologous superior cervical ganglion cells have revealed notable interspecific differences in neuronal geometry, at least among laboratory rodents and

FIGURE 8. An electron micrograph illustrating the choroid (C), retinal pigment epithelium (RPE), and outer segments (OS) of photoreceptor cells in the eye of a white-lined bat, *Chiroderma villosum* (Phyllostomidae). This specimen was collected in Suriname, South America and field-fixed in trialdehyde. Other abbreviations are: Ne, nerve; BM, basal membrane of the RPE cells; G, Golgi complex.

lagomorphs (Purves and Lichtman, 1985). For the time being, however, photoreceptors seem to hold the most promise for study of the comparative ultrastructure of neurons.

3.3. Exocrine Cells

Exocrine cells are found in a large variety of mammalian glands. This third general group of regulated, polarized secretory cells can be used in infraclass ultrastructural comparisons because these cells are characterized by a recognizable set of architectural features (Fig. 4), they synthesize and export a large variety of complex products, and their products can be correlated with various aspects of behavior, diet, reproduction, and thermoregulation. The study of exocrine cells should yield considerable data on evolution at the cellular level because, as discussed in this section, numerous ultrastructural differences between species have been observed. In some instances even congeners are known to exhibit ultrastructural differences that correlate with systematic relationships. From the perspective of systematic mammalogy, ultrastructural comparisons of exocrine cells might reveal evolutionary patterns at the cellular level.

The details of the entire secretory process are probably better known for exocrine gland cells than for any other type of secretory cell. Autoradiography, cell fractionation, and transmission electron microscopy have been used to elucidate the secretory process (Jamieson and Palade, 1971; Castle et al., 1975; Palade, 1975). Amino acids, which enter the cell via the basal plasma membrane, migrate to the rough endoplasmic reticulum, where they are incorporated into peptides. The newly synthesized proteins proceed to the Golgi complex, where they are glycosylated and packaged into immature condensing vacuoles. The incipient secretory granules mature as the contents become condensed, and make their way to the cell apex, where they are stored and eventually released (Fig. 3). Since this basic pathway was determined, further details have been added. For example, the transfer of the new proteins from the RER to the Golgi apparatus involves an intermediate organelle complex termed the "GERL" (Novikoff, 1976; Hand, 1980a; Hand and Oliver, 1984). The problems of protein sorting and the transfer of proteins through the stack of Golgi membranes and from the Golgi to secretory vesicles have been attracting considerable attention and is perhaps a subject that will be elucidated fairly soon (Kelly, 1985; Rothman, 1986). Details of some of the means by which secretory products exit these cells also have been obtained. For example, Tandler and Poulsen (1976) have demonstrated fusion of the membrane surrounding mucous granules with the luminal plasma membrane of acinar cells in the submandibular salivary gland of the domestic cat and the dehiscence of the granule.

Both the current and older literature on integumentary glands of mammals contain numerous hints that the ultrastructure of exocrine cells might be of considerable evolutionary and systematic interest. Quay (1954) reported that interspecific differences in tarsal (Meibomian) glands could be detected in histological studies of the eyelids from species of microtine rodents and Saki and van Lennep (1984) described histological and histochemical differences in these glands among 12 species of Australian marsupials. Quay (1968) also described interspecific differences in the histology, distribution, and presence and absence of skin glands in many species of microtines. These glands have been shown to be influenced by androgens (Jannett, 1975) and to be involved in communication through scent marking (Müller-Schwarze, 1983; Ralls, 1971; Wolff and Johnson, 1979). Indeed, skin glands now are commonly regarded as important sources of products that have chemobehavioral roles in nearly all of the major groups of mammals (see Müller-Schwarze, 1983, for a review). Evidence of infraclass variation in exocrine cells and their secretory products can be found in histochemical and histological comparisons of salivary glands (Junqueira and Fava de Moraes, 1975; Junqueira et al., 1967, 1973; Phillips et al., 1977; Pinkstaff, 1981) and interspecific histochemical comparisons of the glandular mucosa of the stomach (Forman, 1972).

Although the ultrastructure of a great variety of exocrine cells has been described and illustrated in the literature, homologous cells in related species rarely have been deliberately compared. The few studies available demonstrate the kinds of information that may be obtained. In her analysis of the cells comprising the nasal glands in marsupials, Kratzing (1984) examined the ultrastructure of exocrine cells in three species, the honey possum (*Tarsipes rostratus*), northern brown bandicoot (*Isoodon macrourus*), and agile wallaby (*Macropus agilis*), that had been selected for comparison on the basis of variation in body size, habitat, and diet. A fourth species, the koala (*Phascolarctos cinereus*) also was examined, but was found to lack nasal glands, except during a transitory developmental phase. The basic fine structure of the exocrine cells in the nasal glands was compared to that of homologous cells in laboratory rats, hamsters, and dogs (Kratzing, 1984). Wong et al. (1985) recently described the ultrastructure of exocrine cells in the prostate in the hopping mouse (*Notomys alexis*) and made comparisons to the laboratory mouse (*Mus musculus*). Secretory cells in the prostate of *Notomys* exhibited numerous differences from those in *Mus*, including the appearance and intracellular distribution of secretory granules, possible modes of formation of secretory products (some might be packaged at the Golgi, others might form directly in the RER), and the ultrastructure of intracellular organelles (Wong et al., 1985).

A final example of comparative ultrastructure of exocrine cells may be found in a series of papers that describe the divergence in ultrastructure of the gastric mucosa in related species of microchiropteran bats that have different diets (Phillips and Studholme, 1982; Phillips et al., 1984; Studholme et al., 1986). Pepsinogen-producing chief cells in frugivorous microchiropteran bats were compared to those of insectivorous or insectivorous/animalivorous bats and found to be dramatically different. In the latter group the chief cells produce electron-dense granules. In the fruit bats the chief cell product consist of pale, somewhat speckled granules that are irregular in shape and that sometimes coalesce (Phillips and Studholme, 1982). Because specimens from all of these species of bats had been collected under the same circumstances and handled, fixed, and processed in the same manner, these ultrastructural differences in the chief cells were thought to be nonartifactual and of functional significance, and to represent an evolutionary divergence of the chief cells in frugivores from those of their probable insectivorous ancestors.

4. SALIVARY GLANDS: A MODEL SYSTEM

4.1. Introduction

Salivary glands are a group of exocrine (and, to a degree, endocrine) glands associated with the mucosa of the oral cavity. They are among the most variable organs in mammals; major cellular differences have been frequently documented among congeneric species. Although salivary glands often are described as digestive organs, evidence is accumulating that this is far too simplistic a view. The data on mammalian salivary glands are far from complete, but it can be said with some certainty that these organs are involved in (1) digestion, (2) excretion, (3) water and electrolyte balance, (4) protection of the oral mucosa and teeth, (5) production of several growth factors, (6) production of specialized enzymes, such as that used as an "anticoagulant" by vampire bats, (7) production of a variety of vasoactive substances, and (8) production of chemicals that play a role in behavior and, ultimately, in reproduction.

Salivary glands are an excellent model system for investigating the evolution of exocrine cell ultrastructure because of the availability of an abundant general literature on these organs, of basic textbook descriptions and summaries of structure and physiology (Tandler, 1972, 1978; Tandler and Riva, 1986; Dawes, 1978; Hand, 1980b), and of compendia on mammals (Young and van Lennep, 1978; Pinkstaff, 1980). Considerable information concerning the histology, histochemistry, and ultrastructure of salivary glands of bats has been accumulated (Wimsatt, 1956; DiSanto, 1960; Phillips, 1976; Phillips et al., 1977, 1987a,b; Pinkstaff et al., 1982; Tandler and Cohan, 1984). We believe that bats are ideal subjects on which to test the hypothesis that ultrastructural variation in exocrine cells might reveal something of the mechanisms by which the evolution of cells occurs, as well as the extent of variation in the secretory process, because of the diversity in their feeding habits and the extent to which their morphological features correlate with their ecology. Findley and Wilson, (1982), among others, have used the term "ecomorphology" to denote this correspondence.

4.2. Comparative Ultrastructure of Acinar Secretory Products

All of the major epithelial components of salivary glands essentially consist of regulated, polarized secretory cells. Although this is most obviously true of exocrine cells that form secretory endpieces at the beginnings of the duct system, it also holds for the intercalated duct cells, the granular ducts of rodents and insectivores, and the striated ducts as well, depending upon gland and species. In this paper, we will confine ourselves to a consideration of the acinar cells, which are located at the beginning of the duct system (Fig. 9).

Terminal acinar cells can be either serous, mucous, or seromucous in nature. These cell types often can be recognized by means of conventional histology, or by use of TEM, or by application of certain carbohydrate histochemical tests. This classification scheme is not entirely adequate, because some cells that appear by TEM to be serous produce secretory products that can be shown by histochemistry to contain abundant glycoproteins (Pinkstaff et al., 1982) and thus to be seromucous.

It is noteworthy that although whole saliva has been obtained from the main excretory ducts of some salivary glands in some species and then characterized biochemically, not much is known about any of the secretory products, proteinaceous or otherwise, produced by specific salivary gland acinar cells. Parotid acinar secretory granules have been isolated from only a few mammals, and biochemical analysis has been restricted to identification and quantitation of relatively few proteins,

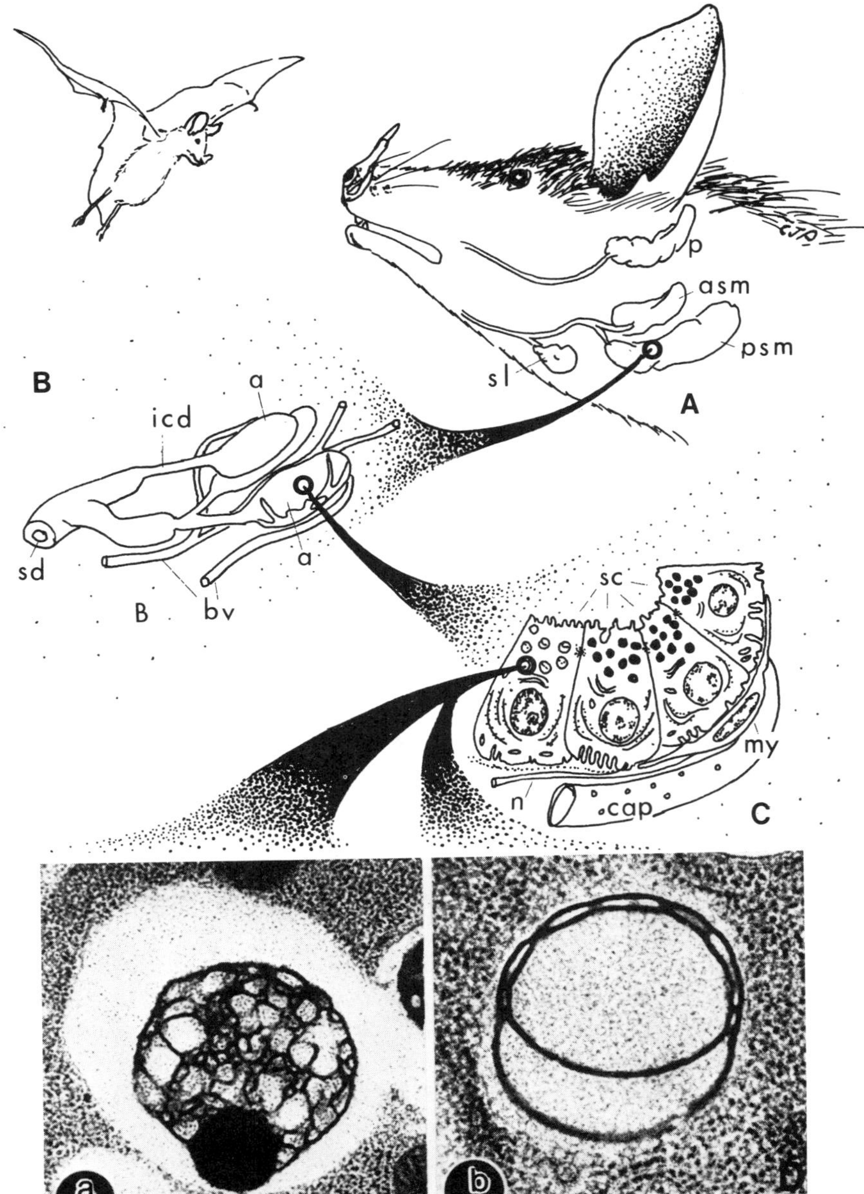

FIGURE 9. (A) Typical gross anatomy of the parotid (p), principal and accessory submandibular (psm, asm), and sublingual (sl) salivary glands in the Neotropical phyllostomid bat. (B) Typical salivary gland, showing the striated and intercalated ducts (sd, icd), acinus (a), and blood vessels (bv). (C) Magnified view of two types of secretory cells (sc), a myoepithelial cell (my), nerve (n), and fenestrated capillary (cap). (D) Electron micrographs of two examples (a,b) of mature secretory granules illustrating the dramatic geometry sometimes seen in phyllostomid bats (examples from the parotid salivary gland of an omnivorous phyllostomid bat, *Carollia perspicillata*).

most from rats (Robinovitch and Sreebny, 1969; Ball, 1974; Wallach et al., 1975). Acinar secretory granules from the submandibular gland have not been successfully isolated from any mammal. While there is little information concerning the composition of the granules in salivary gland exocrine cells, there is, as mentioned earlier, abundant data on the granules of APUD cells; for many of these latter cells the products are known, some of the peptides are recognizable by their ultrastructural images alone, and even the variation in amino acid residues in a number of products has been described. There are several reasons for this. First, unlike salivary gland acinar cells, each endocrine cell produces a very specific product. Second, the products, such as the peptide hormone gastrin, of endocrine cells are simple in comparison to the polypeptides, complex glycoproteins, and mucosubstances produced by salivary glands. Third, in contrast to the products of salivary glands, those of endocrine cells appear to be highly conserved.

It is the complexity of the exocrine cell product in salivary glands that on the one hand makes biochemical and molecular data extremely difficult to obtain while on the other hand makes these cells so interesting in infraclass and even intrageneric comparisons. For salivary gland acinar cells to be of interest in the study of the evolution of mammals, they must synthesize and package products that are too variable and complex for ready assessment at the molecular level. The secretory granules of salivary gland acinar cells are chemically complex and poorly characterized and the nature of the granules cannot be determined by their ultrastructure alone. Given this, why should we make ultrastructural comparisons and what, if anything, do such comparisons tell us?

A survey of the literature in which salivary gland acinar cells are described ultrastructurally reveals a great range of secretory granule morphology (Young and van Lennep, 1978; Pinkstaff, 1980). This variation can be attributed to one or more of the following factors: differences in technique, especially fixation; the degree of maturity of the granules in different studies; and interspecific differences in chemical composition and packaging of granule components. Some of the variations seen in published micrographs clearly are a result of suboptimal technique; some of the earlier investigations must be repeated using better techniques and consistent procedures. As stated in the previous section, fixation, embedding, and selection of regions within a tissue block all can significantly affect the final ultrastructural image. Immature secretory granules can be distinguished from mature ones mainly by their location within the cell. Immature granules are far more likely to be found adjacent to the Golgi complex than elsewhere in the cy-

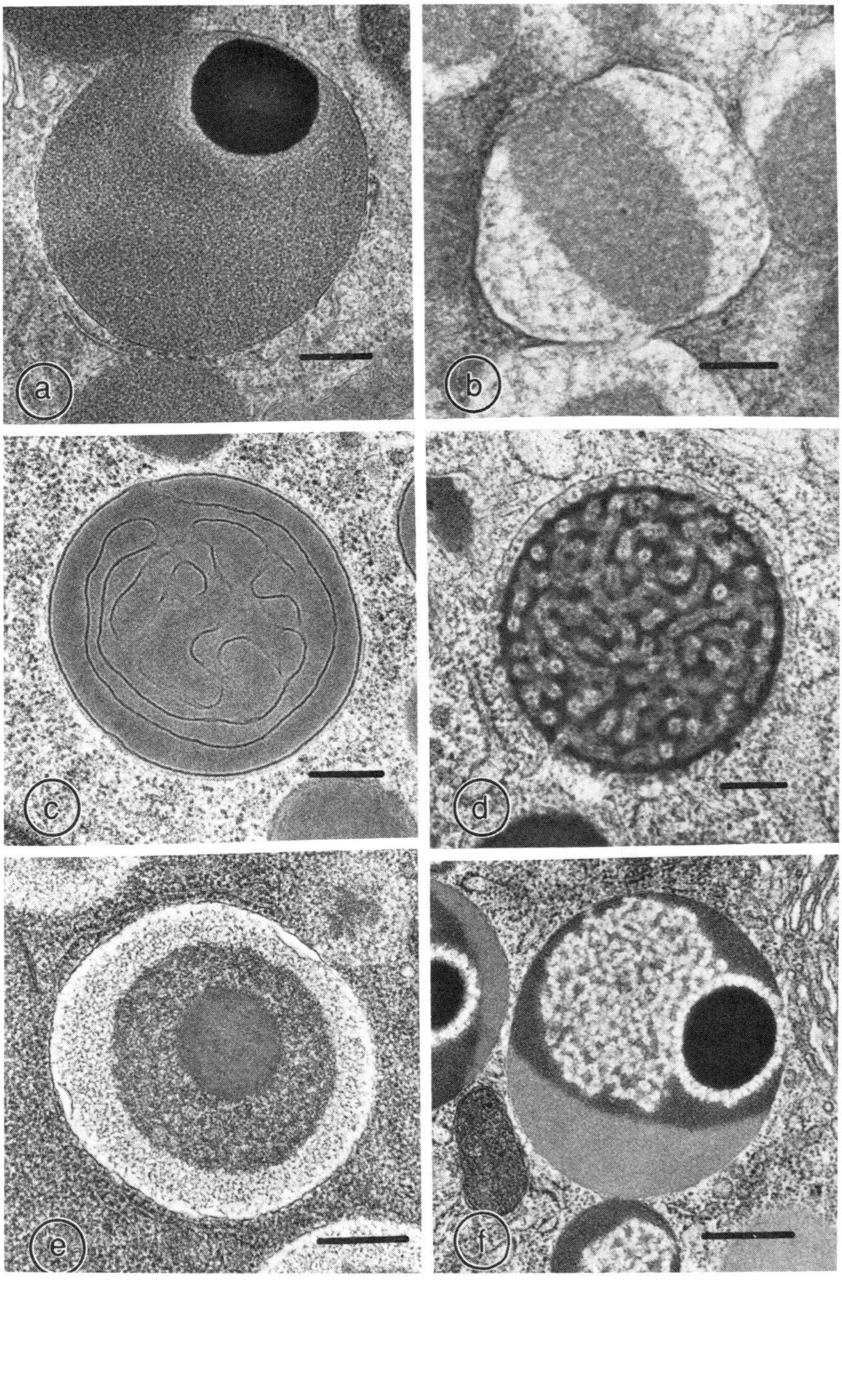
a
b
c
d
e
f

toplasm; those ready for export are usually positioned in the cytoplasm immediately below the apical plasma membrane. The structure of mature granules can be definitively determined by examining granules captured in the act of exocytosis.

The anatomic literature contains numerous examples of interspecific variation in salivary secretory granules, which range from homogeneously dense to strikingly patterned, some with concentric shells or crystalline arrays (Figs. 9 and 10). These different geometric patterns represent differences in molecular organization within the secretory granules. Apparently the macromolecules within the granules are sorted out during the condensing process that follows granule formation at the exit face of the Golgi complex; these macromolecules are sorted with respect to several factors, including their chemical affinities, molecular structure, and charge. The three-dimensional arrangement of polypeptide chains determines the conformation of a protein and this not only is its biologically functional state, but is also the structure that interacts with fixative and produces the image seen by electron microscopy. Folding of the protein into its final conformation apparently results from a spontaneous kinetic process that has been described as a "second translation" of the genetic message (Goldberg, 1985). It is reasonable to conclude that variation in electron microscopic image reflects variation in protein conformation and that this variation is based on actual genetic variation. Actual interspecific variations related to the synthesis and packaging of a specific product in a particular cell are obviously of interest and they can be employed in establishing guidelines for making use of comparative ultrastructural data from an evolutionary perspective.

Sequestration of macromolecules within the mature secretory granules is another cause of variation in the appearance of the granule. Experimental evidence of segregation of individual types of macromolecules within granules is sparse, but the several examples available

←

FIGURE 10. Interspecific variation in the ultrastructure of mature secretory granules in salivary gland acinar cells: (a) Submandibular salivary gland serous granule in a megachiropteran fruit bat, *Rousettus leschenaulti*; scale bar 0.5 μm; (b) lingual salivary gland seromucous granule in the common vampire bat, *Desmodus rotundus*; scale bar 0.25 μm; (c) parotid salivary gland serous granule in an insectivorous phyllostomid bat, *Phyllostomus elongatus*; scale bar 0.5 μm; (d) parotid salivary gland serous granule in the frog-eating bat, *Trachops cirrhosus*; scale bar 0.25 μm; (e) parotid salivary gland serous granule in a South American rodent, *Zygodontomys* sp.; scale bar 0.5 μm; (f) submandibular salivary gland serous granule in a rhesus macaque monkey, *Macaca mulatta*; scale bar 0.5 μm. All tissues fixed in trialdehyde.

clearly substantiate this idea. Through the use of immunocytochemistry, proline-rich proteins have been localized within the dense spherules within salivary secretory granules (Kousvelari et al., 1982), lysozyme has been localized in the dense cores of serous cell granules in nasal glands (Machino et al., 1984), and glucagon and glicentin have been found to occupy separate regions of the granules in A cells (Ravazzola and Orci, 1980). In summary, we think that the substructure seen in the secretory granules of salivary gland acinar cells is a consequence of molecular interaction and segregation of the granule contents.

The anatomic literature contains many examples of interspecific differences in salivary gland acinar cells from most orders of eutherians and at least some data are available from monotremes and marsupials (Young and van Lennep, 1978). Many of these observations are difficult to compare because although most groups are represented, the species investigated form an almost random array with respect to systematic relationships. Because for some species only one or another of the three major glands has been characterized, it is difficult to draw comparisons. To determine if existing variations in gland structure are related to systematics, it will be necessary to use a methodical taxonomic approach. We also want to ascertain whether the observed types of structures can be correlated with variations in some other biological parameters, such as diet, ecology, or behavior. To test our hypothesis that variation in secretory granule substructure is correlated with some aspect of the biology or genetic relationships of the organisms, we have used bat species because they are ecologically diverse and their systematic relationships have already been investigated intensively.

Homologous acinar cells in the parotid salivary gland of 16 Neotropical species (13 genera from four families: Mormoopidae, Phyllostomidae, Vespertillionidae, and Molossidae) were fixed and processed uniformly and then compared by TEM (Phillips et al., 1987a,b; Tandler et al., 1985, 1986). A great variety of sometimes bizarre secretory granules was observed in the parotid acinar cells and, surprisingly, no two species were found to possess identical granules. The range of variation found in these secretory granules was so extensive as to seem artifactual, were it not for the fact that no microscopic variation was detected within multiple samples of a given species from a particular geographic location (Phillips et al., 1987a,b). Intergeneric differences always were distinct; when two species within each of the genera *Phyllostomus*, *Tonatia*, and *Eptesicus* were compared, subtle differences were consistently seen (Fig. 11).

Certain trends or patterns can be detected in the ultrastructure of secretory granules. In the aforementioned studies, it was found that

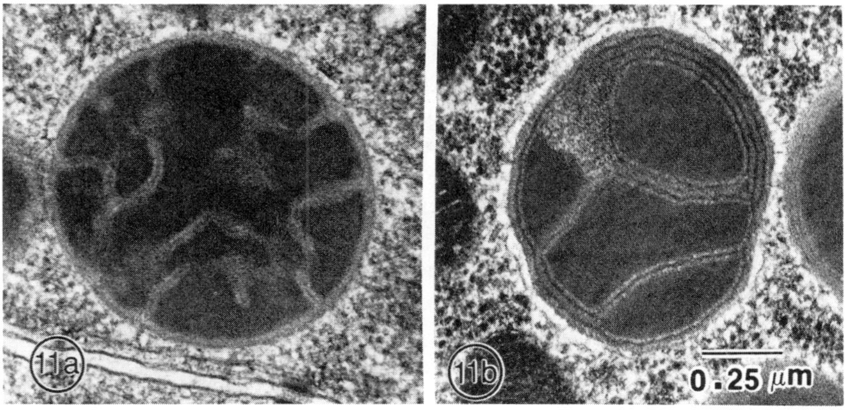

FIGURE 11. Interspecific differences in the mature serous granules in the parotid acinar cells in two congeneric insectivorous vespertilionid bats, (a) *Eptesicus lynni* and (b) *E. brasiliensis*. Field-fixed in trialdehyde; scale bar applies to both.

secretory granules in the parotid acinar cells in some of the frugivorous species, for example, *Artibeus jamaicensis* and *Ariteus flavescens*, contained far less electron-dense material than did the granules produced by homologous cells in the insectivorous/animalivorous species, for example, *Tonatia* or *Trachops* (Fig. 12), both those related at the family

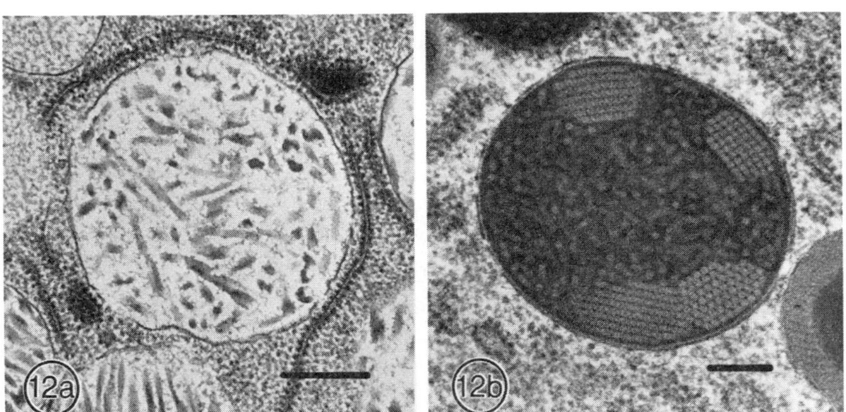

FIGURE 12. Comparison of mature secretory granules in parotid acinar cells in (a) a phyllostomid frugivrous species, *Ariteus flavescens* (scale bar, 0.5 μm), and (b) an insectivorous species, *Tonatia bidens* (scale bar, 0.25 μm). Pale granules such as those produced in *Ariteus* possibly have a lower enzyme content than do electron-dense granules. Note the unusual crystalloid material in the *Tonatia* secretory product. Specimens field-fixed in trialdehyde.

level and in essentially unrelated insectivorous species such as *Eptesicus* or *Tadarida* (Phillips et al., 1987a,b). Junqueira et al. (1973) earlier had reported that enzyme production by the parotid salivary gland in *Artibeus* was very low relative to that of other tested mammals, so it appears that there is a correlation between electron density and enzyme content of secretory granules.

The extensive interspecific variation in parotid acinar secretory granules—to the extent that species and certainly genera can be distinguished from one another by the ultrastructure of the product exported by this one type of cell—provided us with a meaningful set of data with regard to the usefulness and significance of ultrastructural comparisons. We compared acinar cell product in seromucous cells in the principal submandibular salivary gland in five related species of frugivorous bats of the genus *Artibeus* (Phillips et al., 1977; Tandler et al., 1985, 1986). This genus is widely distributed in the Neotropics, is diverse specifically, and has been investigated intensively over the past decade. We compared *Artibeus cinereus, A. phaeotis, A. jamaicensis, A. lituratus,* and *A. concolor* and found diversity of substructure in secretory granules produced in the submandibular gland seromucous cells (Tandler et al., 1986). *Artibeus cinereus* and *A. phaeotis* were virtually indistinguishable from each other, as were *A. jamaicensis* and *A. lituratus,* whereas *A. concolor* fell into a third, completely separate category (Fig. 13). The groupings that we observed among these particular species of bats correspond perfectly to those arrived at independently by Koop and Baker (1983), who used data on nonsalivary isozymes to determine genic relationships. Using data from systematic mammalogy, we were able to determine that in these particular bats at least, the ultrastructural image of secretory granules produced by a particular exocrine cell varied in correspondence with the genetic relationships of the animals themselves.

At first it might seem surprising that the electron microscopic images of exocrine cell products would vary in accord with genetic relationships among species, but actually this finding is reasonable when one considers that conformation of the secretory proteins is determined by a kinetic process that "translates" amino acid sequences into a three-dimensional molecular structure that affects the electron microscopic image of the secretory granule. The amino acid sequences are encoded genetically. That congeneric species would differ might also be surprising, but saliva is known to vary considerably even among individuals of the same species. Genetic markers are well known in human saliva (for example, Azen, 1972, 1973; Azen and Oppenheim, 1973) and in laboratory mice amylase and other salivary proteins have been

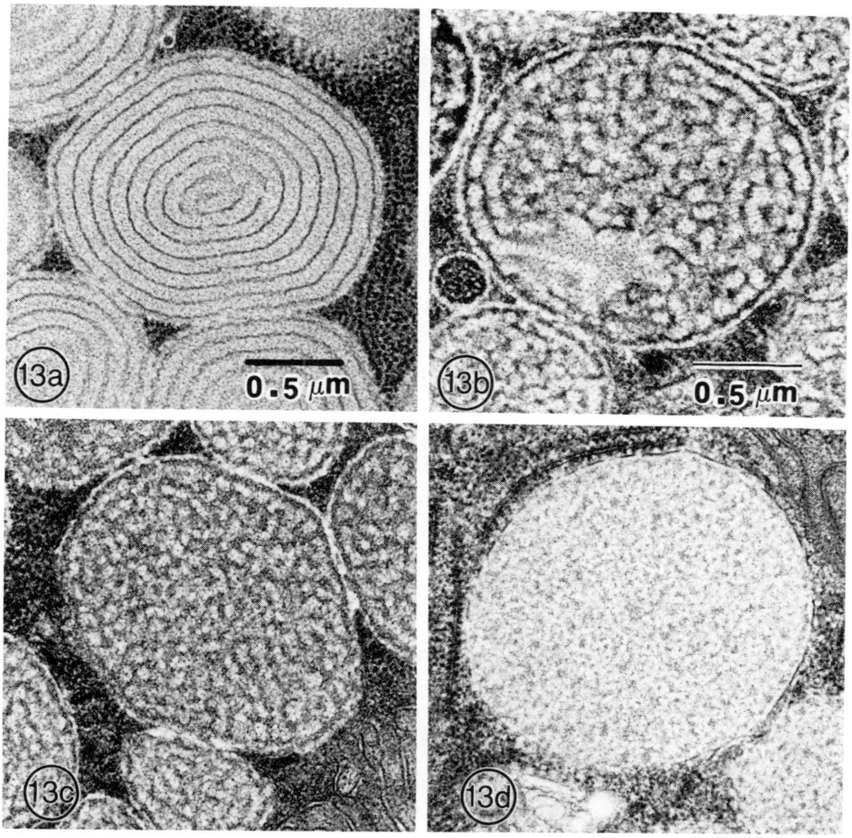

FIGURE 13. Pattern of interspecific variation in mature seromucous secretory granules in the submandibular salivary gland in five species in phyllostomid fruit bats of the genus *Artibeus*. (a) *Artibeus cinereus (A. phaeotis* is identical); (b) *A. lituratus;* (c) *A. jamaicensis*; (d) *A. concolor*. The morphology of these granules groups the species in exactly the same way as does independently derived genic data (Tandler et al., 1986). Scale bar in (a) applies also to (d); scale bar in (b) applies also to (c). Specimens field-fixed in trialdehyde.

shown to vary among genetic strains and even between sexes (Nielsen and Sick, 1964; Ikemoto and Matsushima, 1984).

The ultrastructural data and the various other data on genetic variability of saliva show that salivary gland exocrine cells synthesize, package, and secrete highly variable products. Why do these cells exhibit not only so much interspecific variation, but also such variability in comparison to other types of regulated, polarized secretory cells, such as endocrine cells? What are the differences in macromolecules that account for ultrastructural differences among congeneric species? These are just a few of the evolutionary questions raised by comparative ultrastructural studies of exocrine cells. These ultrastructural comparisons show how microscopic data in concert with systematic mammalogy have provided a rational basis for future biochemical and physiological comparisons of particular types of cells.

Secretory products formed by salivary gland exocrine cells are collinear with a small portion of the operational genome in these cells in particular species. The proteins obtained from isolated secretory granules eventually can be used to learn about nuclear DNA in exocrine cells just as has been done for many of the APUD endocrine cells. Because of their interspecific variability, the salivary gland exocrine cells might also serve systematic mammalogy as a means of making intrageneric comparisons and as a model for understanding diversity at the cellular level within the Class Mammalia.

4.3. Variation in the Secretory Process and Acinar Cells

The basic secretory process for exocrine cells already has been summarized; it may be regarded as a conserved process, in that the same sequence of intracellular events has been found to occur in a wide variety of regulated secretory cells throughout the vertebrates, but no concerted or methodical effort has been made to establish the precise extent of variation in the secretory process. Consequently, it is not known which elements, if any, of the process itself are pertinent to evolutionary change at the cellular level. As we have shown in previous studies, the exocrine cell product can vary greatly at the species level, but what other aspects of secretion vary? One unusual variation in the secretory process of salivary gland exocrine cells occurs in the seromucous cells of the accessory submandibular salivary gland in long-winged bats of the genus *Miniopterus* (Tandler et al., 1985). In these bats the accessory submandibular salivary gland consists of seromucous acini with serous demilunes. In the seromucous cells of *Miniopterus*, the apical cytoplasm typically is nearly filled with secretory granules of variable size (up to 2 μm in diameter) that are bounded by multiple layers of membrane (Fig. 14) rather than by the usual single limiting membrane (Tandler et al., 1985). Such multiple-membrane granules are unique, although a somewhat similar type of granule has been found in the gular gland of adult male free-tailed bats, *Tadarida brasiliensis* (Gutierrez and Aoki, 1974). In *Miniopterus*, the extra membranes apparently are added onto the developing secretory granule in the vicinity

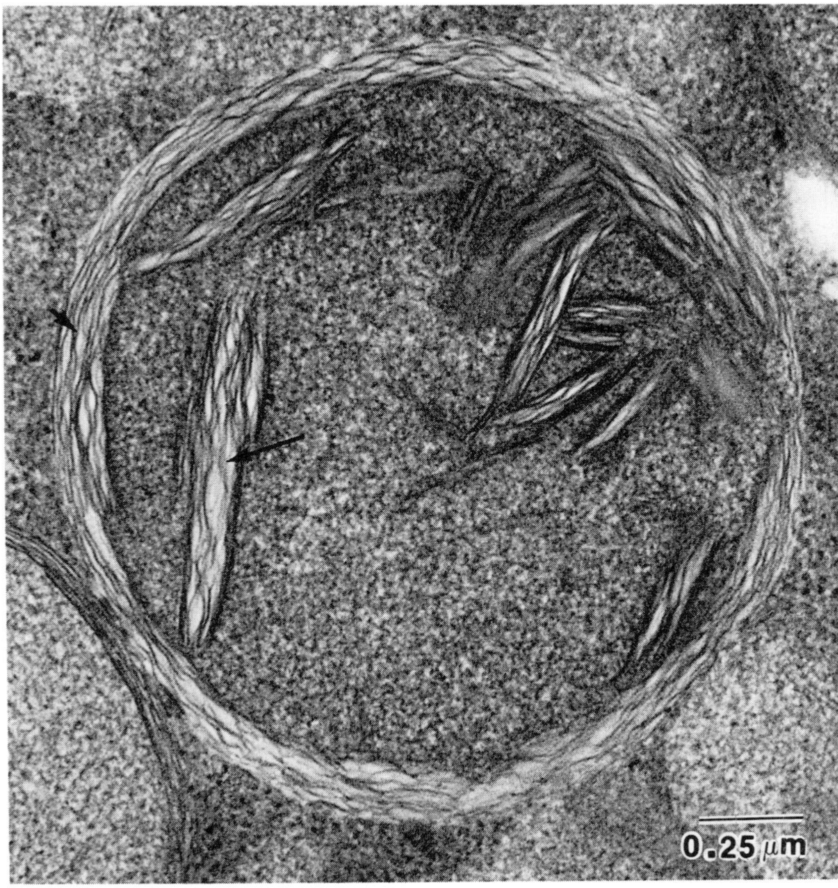

FIGURE 14. Unique type of mucous granule in acinar cell of the accessory submandibular salivary gland in an insectivorous vespertilionid bat, *Miniopterus schreibersi*. Note the many layers of membrane (arrows) surrounding and extending into this secretory granule. Specimen field-fixed in trialdehyde.

of the Golgi complex. When the secretory granules are exported from the cells, the outermost membrane of each granule fuses with the apical plasma membrane and both granule contents and extra membranes are released into the salivary gland lumen to become part of the initial saliva (Fig. 15).

The unexpected structure of the seromucous granules in *Miniopterus* constitutes an exception to an otherwise conservative process and also gives us an idea of what types of variation might be found in nature.

The occurrence of a seemingly unique process in these particular cells in *M. schreibersi* and in *M. magnator* raises questions about the nature of the secretory product and its role in these insectivorous Old World bats. The product of these cells is rich in phospholipids clearly because of the plasma membrane that is exported along with the other components of the secretory granules. This in itself could be significant because salivary lipid seems to facilitate penetration of the mucosa by lipophilic substances, affects the interactions between salivary proteins and oral bacteria, and possibly contributes to the development of certain oral diseases (Murty et al., 1984). Bats of the genus *Miniopterus* provide us with an example of variation in the secretory process through which

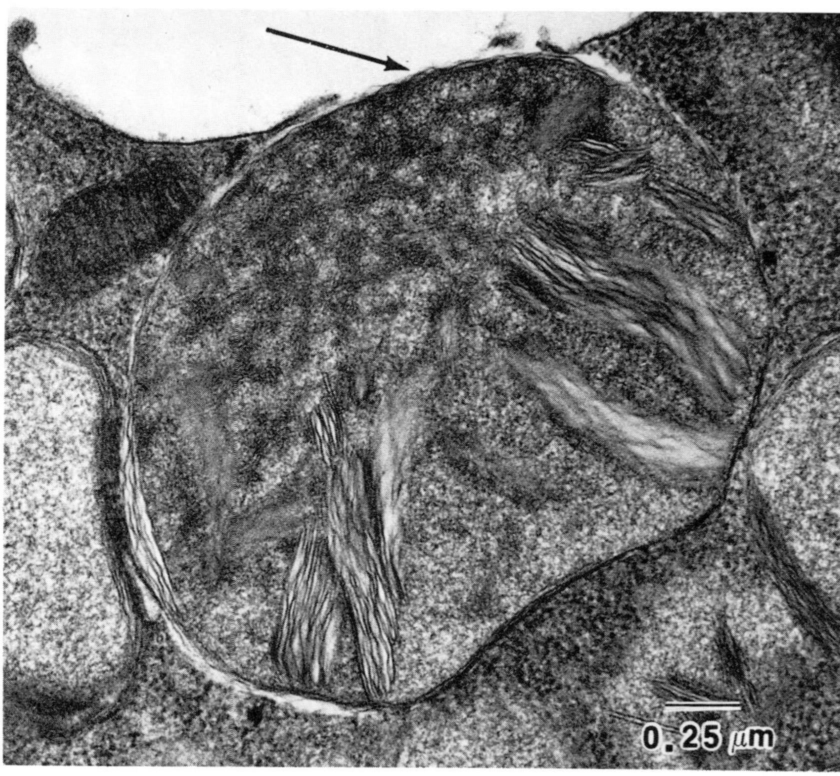

FIGURE 15. Exocytosis of a mucous granule from an acinar cell in the accessory submandibular salivary gland in *Miniopterus schreibersi*. Note that the secretory granule is still encased in multiple layers of membrane as it exits the cell (arrow). Field-fixed in trialdehyde.

interspecific differences in homologous cells can be appreciated, and eventually these bats may be useful for testing the functional significance of a specific cellular feature.

Salivary gland acinar cells can exhibit striking interspecific variation in ultrastructure, but these secretory cells have been modified even more substantially in the course of evolution among related species. They can undergo such dramatic evolutionary change in function that they nearly lose the various ultrastructural features normally associated with regulated, polarized secretory cells. Two examples are found in the fringe-lipped bat, *Trachops cirrhosus*, and in species of the false vampire bat, *Megaderma* (Phillips and Tandler, 1985).

Bats of the genera *Trachops* and *Megaderma* have different evolutionary histories. The fringe-lipped bat is a Neotropical phyllostomid species known to subsist primarily on frogs; based on the calls of frogs, these bats can distinguish edible frogs from species that are toxic (Ryan and Tuttle, 1983). These fringe-lipped bats apparently evolved from insectivorous or insectivorous/animalivorous ancestral phyllostomids (Smith, 1976). The false vampire bats are members of the Megadermatidae. They occur only in Old World tropical habitats, but like the fringe-lipped bat, they also feed on small vertebrates, including frogs (Lekagul and McNeely, 1977). *Trachops* and *Megaderma* share a histologically unique type of accessory submandibular salivary gland (Phillips and Tandler, 1985; Phillips et al., 1987a,b). Some of the "secretory" cells in these glands exhibit an ultrastructure that diverges greatly from the homologous exocrine cell found in the salivary glands of other species of mammals, including a large variety of other bats. The development of this structure, in conjunction with the exploitation of the same nutritional resource, is one of the most persuasive examples of convergent evolution ever found at the histological level.

The most radical departure from the usual acinar cell is found in *Trachops cirrhosus*, although the cells in *Megaderma* also are unusual. In *Trachops* these cells contain extensive RER and a few fairly typical serous-type granules (Fig. 16). However, it appears that the bulk of the material secreted is in the form of subluminal vermiculate structures that contain peculiar lucent ribbons (Phillips and Tandler, 1985). Numerous lipid droplets are also found in these cells; it appears that this lipid is released by an apocrine mechanism (Phillips and Tandler, 1985). Thus, in *Trachops*, and to some extent in *Megaderma*, one finds divergent exocrine cells that can be compared ultrastructurally, biochemically, and perhaps functionally to the homologous acinar cells in other species of bats. This type of comparison will provide a good idea of

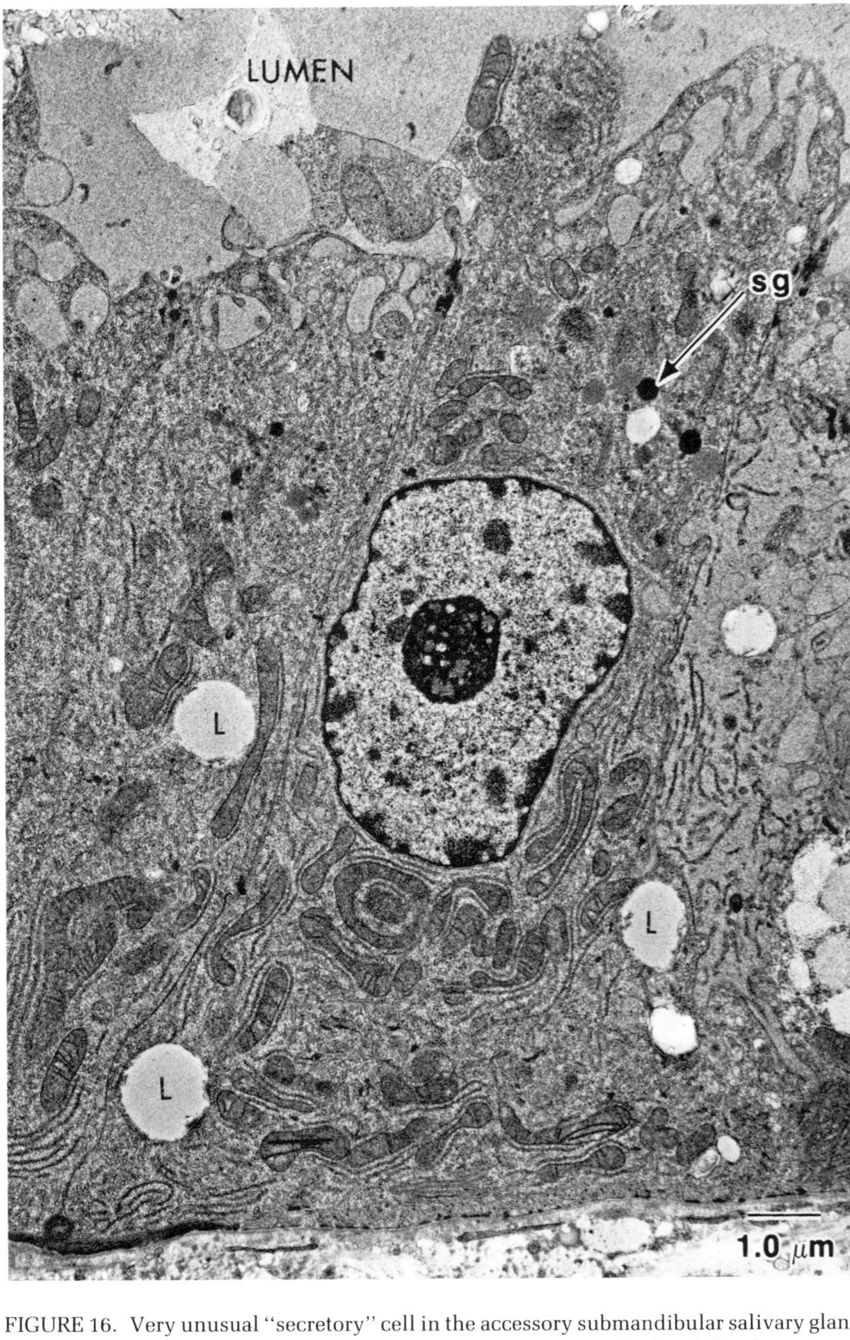

FIGURE 16. Very unusual "secretory" cell in the accessory submandibular salivary gland of the frog-eating phyllostomid bat, *Trachops cirrhosus*. Note the lipid droplets (L) and scattered electron-dense secretory granules (sg). Field-fixed in trialdehyde.

the different evolutionary changes that can occur in the function and ultrastructural appearance of exocrine cells.

4.4. From Ultrastructure to Organismal Biology: An Example

At the beginning of this chapter we stated that investigations at the cellular level might serve as a bridge between genes and organisms. We argued that this could be true because it is within the individual cell that the genome first is expressed. The following is offered as an final example of how ultrastructure ultimately may help us understand organisms in an evolutionary and ecological context.

One of the interesting ecological questions that can be posed about mammals in tropical environments is: "How do sympatric, ecologically similar, closely related species subdivide the habitat?" Habitats have been shared among species of small mammals in so many different ways that it seems likely that no single, simple answer will apply in each instance. Our example involves round-eared bats of the phyllostomid genus *Tonatia*. These bats occur from Mexico to the northern half of South America; six species are now recognized by systematists (Genoways and Williams, 1984). In Suriname two of the species, *Tonatia sylvicola* and *T. bidens,* have an overlapping distribution; specimens of each can be caught in the same mist nets set in patches of undisturbed tropical forest. Not only do these bats share the same habitat and flyways, but they also are similar in external appearance and size, apparently feed on a similar combination of fruit and insects, and differ more obviously from other species of *Tonatia* than they do from each other (Gardner, 1977; Genoways and Williams, 1984). In fact, the only notable external difference between the two is that *T. sylvicola* on the average is several millimeters larger than *T. bidens*. These species are a good example of two related and very similar mammals appearing to share an environment and perhaps even a nutrient resource.

In an investigation conducted with Toshikazu Nagato, we compared the ultrastructure of the salivary glands of *T. sylvicola* and *T. bidens* (Nagato et al., 1984) and found an unusual form of smooth endoplasmic reticulum in the submandibular acinar cells in male specimens of *T. sylvicola* but not in the females or in either sex of *T. bidens*. It consists of an array of membranous tubules organized in a hexagonal pattern with a latticework of 12-nm filaments running in three planes (Figs. 17 and 18). Although such an organelle has not been previously seen in any salivary gland acinar cells, it does resemble organelles found in other cell types that were known to be associated with production of steroids (Nagato et al., 1984). The presence of this unusual organelle

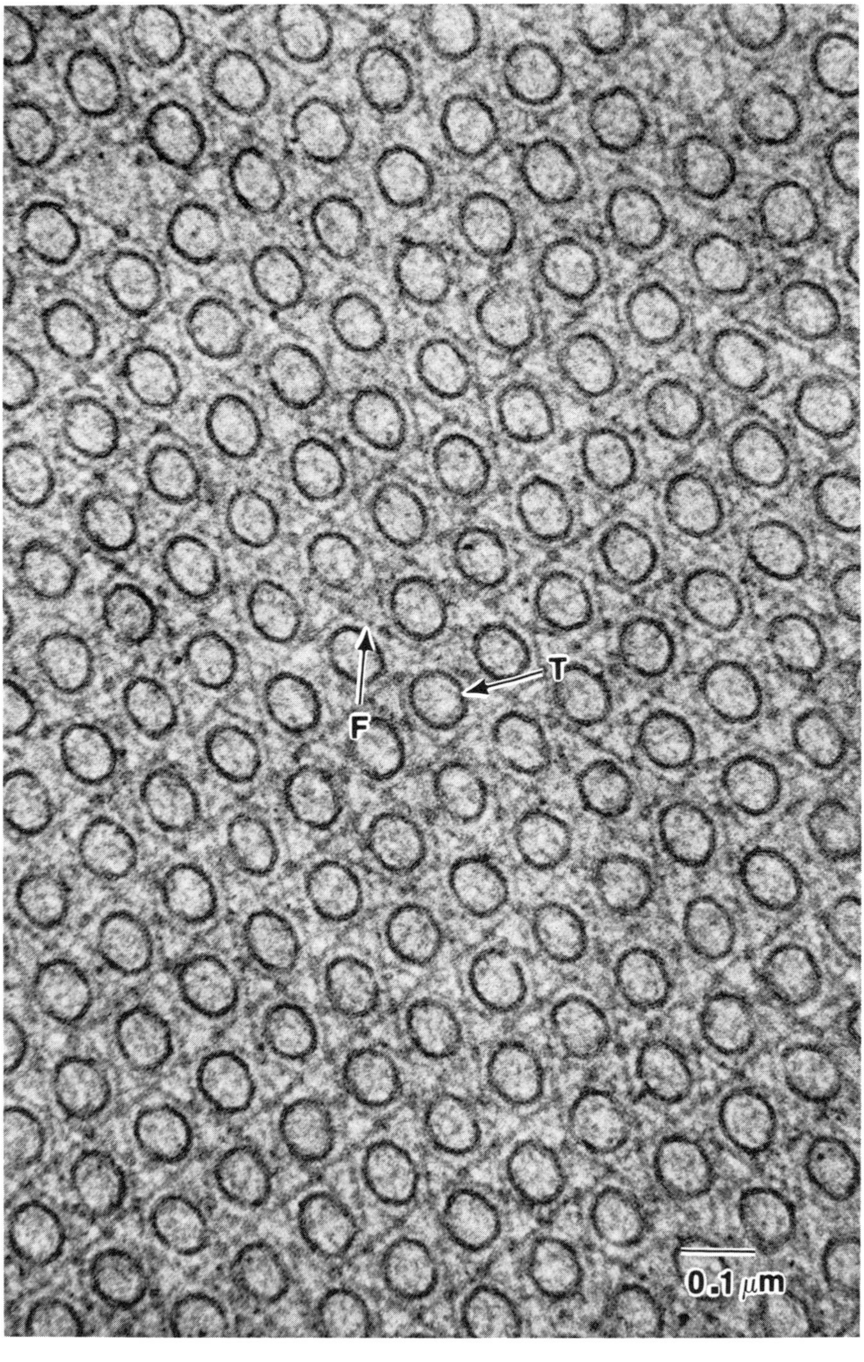
T
F
0.1 μm

in acinar cells in males of one species but not in males of the congeneric sympatric species suggests that male *T. sylvicola* add a special steroid product to their saliva that might play a role in communication, species recognition, isolation, or some combination of these.

The idea that saliva might contain pheromonelike molecules is fairly new, but has been substantiated in several animals (Müller-Schwarze, 1983). Saliva seems to play an important chemobehavioral role in marsupials, where it can elicit aggressive behavior or be used for identification of sex or individuals (Fadem, 1986), and may have similar effects in a wide variety of rodents from several families (Gray et al., 1984; Friedle and Fischer, 1984). 16-Androstene steroids found in high concentration in the submandibular saliva of boars act as pheromones that facilitate adoption of the mating stance by gilts (Perry et al., 1980; Booth, 1984). The occurrence of a steroid-producing type of organelle in acinar secretory cells in male *Tonatia sylvicola* is the first example of a qualitative sexual dimorphism in salivary glands and is in keeping with the conjectured role of saliva in reproductive behavior in this bat.

The production of a steroid component by the submandibular salivary gland in male *Tonatia sylvicola* but not by females or by either sex in *T. bidens* may be one of the factors that permits these animals to distinguish among themselves and reduce the possibility of nonproductive mating. In addition to the crystalloid SER in male *T. sylvicola*, it also is noteworthy that male *T. sylvicola* on the average are several millimeters larger than females of the same species, whereas both sexes are the same size in *T. bidens* (Genoways and Williams, 1984). Additionally, male *T. sylvicola* have a submetacentric X chromosome and a fundamental number of 60 ($2n = 34$), whereas male *T. bidens* have a metacentric X chromosome and a fundamental number of 20 ($2n = 16$) (Baker, 1979; Honeycutt et al., 1980).

In summary, species of the genus *Tonatia* offer yet another model for linking specific intracellular features to specific aspects of biology at the organismal level. The existence of a special organelle in the salivary gland acinar cells of these animals underscores the plasticity of these exocrine cells. When investigated with modern molecular tech-

FIGURE 17. Crystalloid smooth endoplasmic reticulum in a seromucous acinar cell in the submandibular salivary gland of a male specimen of *Tonatia sylvicola*. Note the membranous tubules in cross section (T) and the 12-nm filaments (F) that run in three planes; compare to Fig. 18 for a three-dimensional reconstruction. Specimen field-fixed in trialdehyde.

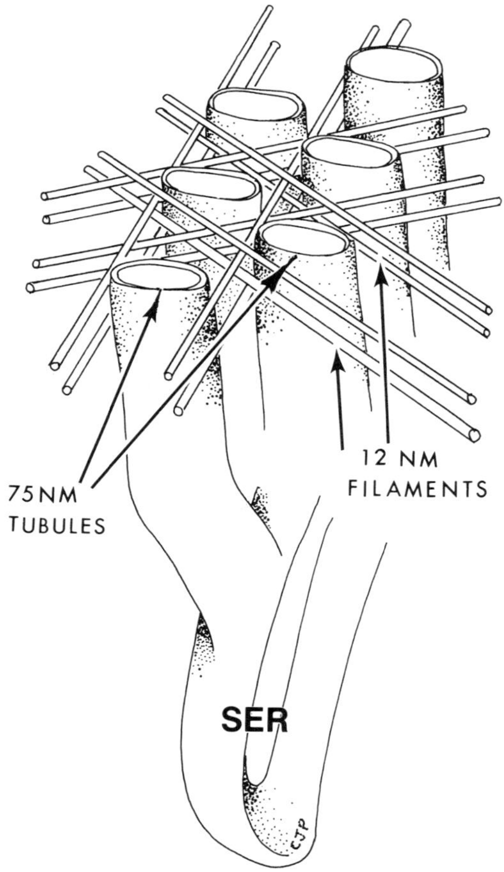

FIGURE 18. Three-dimensional representation of the crystalloid smooth endoplasmic reticulum found by Nagato et al. (1984) in seromucous cells of the submandibular salivary gland in male round-eared bats, *Tonatia sylvicola*. As shown here, the organelle consists of hexagonally stacked membrane tubules (75 nm) with a lattice-work of 12-nm filaments. The tubules apparently connect directly into conventional-appearing smooth endoplasmic reticulum (SER). Compare with Fig. 17.

niques, these cells should help us to understand better the evolutionary process.

Acknowledgments. We are pleased to acknowledge the support provided for our research by the Research Corporation (grants C-1251 and C-1855, to C.J.P.), NSF (grant CDP 801-8653, to C.J.P.), the Hofstra University HCLAS Executive Committee (grants to C.J.P.), an Exploration Grant from Penta Construction Corp. (to C.J.P.), and the NIH (grant AM-08305, to B.T.). The Carnegie Museum of Natural History directly or indirectly supported much of our field work. Many of our colleagues have been of great service to us over the years and have helped shape our ideas, helped us collect specimens, or have provided us with other types of support essential to our research. In particular, we would like to thank Hugh H. Genoways, Duane A. Schlitter, Dorothy E. Pumo, Keith M. Studholme, Carlin A. Pinkstaff, Toshikazu Nagato, Kuniaki Toyoshima, G. Lawrence Forman, Gary W. Grimes, and William F. Payson, Jr. Janice D. Tandler of Expletive Deleted, Lansing, Michigan, provided editorial assistance. Finally, our manuscript never would have been completed without the help of our friend Linda Cossen, Hofstra University Special Secretarial Services.

REFERENCES

Aherne, W. A., and Dunnill, M. S., 1982, *Morphometry*, Edward Arnold, London.

Alberch, P., 1980, Ontogenesis and morphological diversification, *Am. Zool.* **20**:653–667.

Azen, E. A., 1972, Genetic polymorphism of basic proteins from parotid saliva, *Science* **173**:673–674.

Azen, E. A., 1973, Properties of salivary basic proteins showing polymorphism, *Biochem. Genet.* **9**:69–85.

Azen, E. A., and Oppenheim, F. G., 1973, Genetic polymorphism of proline-rich human salivary proteins, *Science* **180**:1067–1069.

Baker, R. J., 1979, Karyology, in: *Biology of Bats of the New World Family Phyllostomatidae*, Part III (R. J. Baker, J. K. Jones, Jr., and D. C. Carter, eds.), Special Publication of the Museum, Texas Tech University, Lubbock, Texas, pp. 107–155.

Ball, W. D., 1974, Isolation of salivary ribonuclease, deoxyribonuclease and amylase from the parotid gland of the rat, *Rattus norvegicus albinus*, *Biochim. Biophys. Acta* **341**:305–317.

Baumgarten, H. G., Holstein, A. F., and Owman, C., 1970, Auerbach's plexus of mammals and man: Electron micrographic identification of three different types of neuronal processes in myenteric ganglia of the large intestine from rhesus monkeys, guinea-pigs and man, *Z. Zellforsch.* **106**:376–397.

Berezin, I., Sheppard, S., Daniel, E. E., and Yanaihara, N., 1985, Ultrastructural immunocytochemical distribution of VIP-like immunoreactivity in dog ileum, *Regul. Pept.* **11**:287–298.

Berzofsky, J. A., 1985, Intrinsic and extrinsic factors in protein antigenic structure, *Science* **229**:932–940.

Booth, W. D., 1984, Sexual dimorphism involving steroidal pheromones and their binding in the submaxillary salivary gland of the Gottingen miniature pig, *J. Endocrinol.* **100**:195–202.

Braniŝ, M., 1981, Morphology of the eye of shrews (Soricidae, Insectivora), *Acta Univ. Carol. Biol.* **1979**:409–445.

Calford, M. B., Graydon, M. L., Huerta, M. F., Kaas, J. H., and Pettigrew, J. D., 1985, A variant of the mammalian somatotropic map in a bat, *Nature* **313**:477–479.

Candipan, R. C., and Sjöstrand, F. S., 1984, An analysis of the contribution of the preparatory technique to the appearance of condensed and orthodox mitochondria, *J. Ultrastruct. Res.* **89**:281–294.

Castle, J. D., Jamieson, J. D., and Palade, G. E., 1975, Secretion granules in the rabbit parotid glands. Isolation, subfractionation, and characterization of the membrane and content subfractions, *J. Cell Biol.* **64**:182–210.

Cohen, A. I., 1964, Some observations on the fine structure of the retinal receptors of the American gray squirrel, *Invest. Ophthalmal.* **3**:198–216.

Cruz-Orive, L.-M., and Weibel, E. R., 1981, Sampling designs for stereology, *J. Microsc.* **122**:235–257.

Dawes, C., 1978, The chemistry and physiology of saliva, in: *Textbook of Oral Biology* (J. H. Shaw, E. A. Sweeney, C. C. Cappuccino, and S. M. Meller, eds.), Saunders, Philadelphia, pp. 593–629.

Dearden, L. C., 1966, Histology of the gastro-esophageal junction in certain rodents, *J. Mammal.* **47**:223–229.

Dearden, L. C., 1969, Stomach and pyloric sphincter histology in certain microtine rodents, *J. Mammal.* **50**:60–68.

DiSanto, P. E., 1960, Anatomy and histochemistry of the salivary glands of the vampire bat, *Desmodus rotundus murinus, J. Morphol.* **106**:301–335.

Dockray, G. J., 1981, Cholecystokinin, in: *Gut Hormones* (S. R. Bloom and J. A. Polak, eds.), Churchill Livingstone, New York, pp. 228–239.

Domeneghini, C., Castaldo, L., and Romanello, M. G., 1985, Ultrastructural study of the endocrine cells of the abomasum mucosa in pre-ruminant calves, *Basic Appl. Histochem.* **29**:49–80.

D'yachkova, L. N., 1984, Elements of the evolutionary morphology of the olfactory cerebral cortex in mammals (electron microscopic study), *Dokl. Akad. Nauk SSSR* **276**:1480–1483 [in Russian].

Eakin, R. M., 1963, Lines of evolution of photoreceptors, in: *General Physiology of Cell Specialization* (D. Mazia and A. Tyler, eds.), McGraw-Hill, New York, pp. 393–425.

Eakin, R. M., 1968, Evolution of photoreceptors, in: *Evolutionary Biology* (T. Dobzhansky, M. K. Hecht, and W. C. Steere, eds.), Appleton-Century-Crofts, New York, pp. 194–242.

Eakin, R. M., 1979, Evolutionary significance of photoreceptors: In retrospect, *Am. Zool.* **19**:647–653.

Eisenberg, J. F., 1981, *The Mammalian Radiations An Analysis of Trends in Evolution, Adaptation, and Behavior*, University of Chicago Press, Chicago.

Eckblad, E., Hakanson, R., and Sundler, F., 1984, VIP and PHI coexist with an NPY-like peptide in intramural neurones of the small intestine, *Regul. Pept.* **10**:47–55.

Fadem, B. H., 1986, Chemical communication in gray short-tailed opossums (*Monodelphis domestica*) with comparisons to other marsupials and with reference to monotremes, in:*Chemical Signals in Vertebrates, Vol. IV, Ecology, Evolution, and Comparative Biology* (D. Duvall, D. Müller-Schwarze, and R. M. Silverstein, eds.), Plenum Press, New York, pp. 587–607.

Feher, E., Sakharov, D. A., Salimova, N. B., and Vajda, J., 1980, Serotoninergic neurons in the vertebrate gut. An experimental and comparative study, in: *Neurotransmitters. Comparative Aspects* (J. Salanki and T. M. Turpaev, eds.), Akademiai Kiado, Budapest, pp. 101–122.

Feldman, J. L., and Phillips, C. J., 1984, Comparative retinal pigment epithelium and photoreceptor ultrastructure in nocturnal and fossorial rodents: The eastern woodrat, *Neotoma floridana*, and the plains pocket gopher, *Geomys bursarius, J. Mammal.* **65**:231–245.

Findley, J. S., and Wilson, D. E., 1982, Ecological significance of chiropteran morphology, in: *Ecology of Bats* (T. H. Kunz, ed.), Plenum Press, New York, pp. 243–260.

Fisher, S. K., Jacobs, G. H., Anderson, D. H., and Silverman, M. S., 1976, Rods in the antelope ground squirrel, *Vision Res.* **16**:875–877.

Forman, G. L., 1968, Comparative gross morphology of spermatozoa of two families of North American bats, *Univ. Kansas Sci. Bull.* **47**:901–928.

Forman, G. L., 1971, Histochemical differences in gastric mucus of bats, *J. Mammal.* **52**:191–193.

Forman, G. L., 1972, Comparative morphological and histochemical studies of stomachs of selected American bats, *Univ. Kans. Sci. Bull.* **59:**591–729.

Forman, G. L., and Genoways, H. H., 1979, Sperm morphology, in: *Biology of Bats of the New World Family Phyllostomatidae,* Part III (R. J. Baker, J. K. Jones, Jr., and D. C. Carter, eds.), Special Publication of the Museum, Texas Tech. Univ., Lubbock, Texas, pp. 177–204.

Forman, G. L., and Phillips, C. J., 1987a, Preparation and fixation of tissues for histological, immunohistological, and electron microscopic studies, in: *Ecological and Behavioral Methods for the Study of Bats* (T. H. Kunz, ed.), Smithsonian Institution Press, Washington, D.C., in press.

Forman, G. L., and Phillips, C. J., 1987b, Histological variation in the proximal colon of heteromyid and cricetid rodents, *J. Mammal.*, in press.

Forman, G. L., Baker, R. J., and Gerber, J. D., 1968, Comments on the systematic status of vampire bats (Family Desmodontidae), *Syst. Zool.* **17:**417–426.

Forman, G. L., Phillips, C. J., and Rouk, C. S., 1979, Alimentary tract, in: *Biology of Bats of the New World Family Phyllostomatidae,* Part III (R. J. Baker, J. K. Jones, Jr., and D. C. Carter, eds.), Special Publication of the Museum, Texas Tech University, Lubbock, Texas, pp. 205–227.

Friedle, R. E., and Fischer, R. B., 1984, Discrimination of salivary olfactants by male *Mesocricetus auratus, Psychol. Rep.* **55:**67–70.

Gans, C., and Northcutt, R. G., 1983, Neural crest and the origin of vertebrates: A new head, *Science* **220:**268–274.

Gardner, A. L., 1977, Feeding habits, in: *Biology of Bats of the New World Family Phyllostomatidae,* Part II (R. J. Baker, J. K. Jones, Jr., and D. C. Carter, eds.), Special Publication of the Museum, Texas Tech University, Lubbock, Texas, pp. 293–350.

Genoways, H. H., and Williams, S. L., 1984, Results of the Alcoa Foundation–Suriname Expeditions. IX. Bats of the genus *Tonatia* (Mammalia: Chiroptera) in Suriname, *Ann. Carnegie Mus. Nat. Hist.* **53:**327–346.

Gershon, M. D., 1981, The enteric nervous system, *Annu. Rev. Neurosci.* **4:**227–272.

Gershon, N. D., Porter, K. R., and Trus, B. L., 1985, The cytoplasmic matrix: Its volume and surface area and the diffusion of molecules through it, *Proc. Natl. Acad. Sci. USA* **82:**5030–5034.

Goldberg, M. E., 1985, The second translation of the genetic message: Protein folding and assembly, *Trends Biochem. Sci.* **10:**388–391.

Golenishchev, F. N., 1976, Structure and development of the teeth in great gerbil and red-tailed gerbil, in: Rodents Evolution and History of Their Recent Fauna, *Tr. Zool. Inst. Akad. Nauk SSSR* **66:**101–106 [in Russian].

Goodrich, J. T., Bernd, P., Sherman, D., and Gershon, M. D., 1980, Phylogeny of enteric serotinergic neurons, *J. Comp. Neurol.* **190:**15–28.

Gray, B., Fischer, R. B., and Meunier, G. F., 1984, Preferences for salivary odor cues by female hamsters, *Horm. Behav.* **18:**451–456.

Greenbaum, I. F., and Phillips, C. J., 1974, Comparative anatomy and general histology of tongues of long-nosed bats (*Leptonycteris sanborni* and *L. nivalis*) with reference to oral mites, *J. Mammal.* **55:**489–504.

Greenberg, G., Bringas, P., Jr., and Slavkin, H. C., 1984, The epithelial genotype controls the pattern of extracellular enamel prism formation, *Differentiation* **26:**32–43.

Grube, D., and Forssmann, W. G., 1979, Morphology and function of the entero-endocrine cells, *Horm. Metab. Res.* **1:**589–606.

Gutierrez, M., and Aoki, A., 1974, Ultrastructure of the gular gland of the free-tailed bat, *Tadarida brasiliensis, J. Morphol.* **141:**293–306.

Hand, A. R., 1980a, Cytochemical differentiation of the Golgi apparatus from GERL, *J. Histochem. Cytochem.* **28**:283–285.

Hand, A. R., 1980b, Salivary glands, in: *Orban's Oral Histology and Embryology* (S. N. Bhaskar, ed.), Mosby, St. Louis, pp. 336–370.

Hand, A. R., and Oliver, C., 1984, Effects of secretory stimulation on the Golgi apparatus and GERL of rat parotid acinar cells, *J. Histochem. Cytochem.* **32**:403–412.

Harding, H. R., Carrick, F. N., and Shorey, D. C., 1981, Marsupial phylogeny: New Indications from sperm ultrastructure and development in Tarsipes spenserae?, *Search* **12**:45–47.

Harding, H. R., Woolley, P. A., Shorey, D. C., and Carrick, F. N., 1982, Sperm ultrastructure, spermiogenesis, and epididymal sperm maturation in dasyurid marsupials: Phylogenetic implications, in: *Carnivorus Marsupials* (M. Archer, ed.), Royal Zoological Society, New South Wales, Sydney, Australia, pp. 659–673.

Hemon, D., Bourgeois, C. A., and Bouteille, M., 1981, Analysis of the spatial organization of the cell: A statistical method for revealing the non-random location of an organelle, *J. Microsc.* **121**:29–37.

Honeycutt, R. L., Baker, R. J., and Genoways, H. H., 1980, Results of the Alcoa Foundation–Suriname expeditions. III. Chromosomal data for bats (Mammalia: Chiroptera) from Suriname, *Ann. Carnegie Mus. Nat. Hist.* **49**:237–250.

Hood, C. S., and Smith, J. D., 1983, Histomorphology of the female reproductive tract in phyllostomid bats, *Occ. Pap. Mus. Texas Tech Univ.* **86**:1–38.

Horner, B. E., Taylor, M. J., and Padykula, H. A., 1964, Food habits and gastric morphology of the grasshopper mouse, *J. Mammal.* **45**:513–535.

Ikemoto, S., and Matsushima, Y., 1984, Sex and strain differences of mouse variant salivary proteins, *Differentiation* **26**:55–58.

Ishikawa, K., Matoba, M., Tanaka, H., and Ono, K., 1985, Anatomical study of the intestine of the insect-feeder bat, *Myotis frater kaguae*, *J. Anat.* **142**:141–150.

Jacobs, G. H., Fisher, S. K., Anderson, D. H., and Silverman, M. S., 1976, Scotopic and photopic vision in the California ground squirrel: Physiological and anatomical evidence, *J. Comp. Neurol.* **165**:209–228.

Jamieson, J. D., and Palade, G. E., 1971, Synthesis, intracellular transport, and discharge of secretory proteins in stimulated pancreatic exocrine cells, *J. Cell Biol.* **50**:135–158.

Jannett, F. J., Jr., 1975, "Hip glands" of *Microtus pennsylvanicus* and *M. longicaudus* (Rodentia: Mriudae), voles "without" hip glands, *Syst. Zool.* **24**:171–175.

Jerison, H. J., 1973, *Evolution of the Brain and Intelligence*, Academic Press, New York.

Junqueira, L. C. U., and Fava de Moraes, F., 1975, Comparative aspects of the vertebrate major salivary glands biology, in: *Funktionelle und Morphlogische Organisation der Zelle. Sekretion und Exkretion* (K. E. Wohlfarth-Bottermann, ed.), Springer, Berlin, pp. 36–48.

Junqueira, L. C. U., Fava de Moraes, F., and Toledo, A. M., 1967, Sialic acids in vertebrate salivary glands, saliva and pancreas, *Arch. Oral Biol.* **12**:151–157.

Junqueira, L. C. U., Toledo, A. M. S., and Doine, A. I., 1973, Digestive enzymes in the parotid and submandibular glands of mammals, *An. Acad. Bras. Cienc.* **45**:629–643.

Kalt, M. R., and Tandler, B., 1971, A study of fixation of early amphibian embryos for electron microscopy, *J. Ultrastruct. Res.* **36**:633–645.

Kamiya, T., and Pirlot, P., 1975, Comparative gastric morphology of Old World bats. I. In light microscopy, *J. Mammal. Soc. Japan* **6**:145–154.

Karasek, M., King, T. S., Hansen, J. T., and Reiter, R. J., 1982, Ultrastructure of the pineal gland of the eastern chipmunk (*Tamias striatus*), *J. Morphol.* **173**:73–86.

Kaung, H.-L., 1985, Electron microscopic immunocytochemical localization of glucagon and pancreatic polypeptide in rat pancreas: Characterization of a population of islet cells containing both peptides, *Anat. Rec.* **212:**292–300.

Keast, J. R., Furness, J. B., and Costa, M., 1985, Distribution of certain peptide-containing nerve fibers and endocrine cells in the gastrointestinal mucosa in five mammalian species, *J. Comp. Neurol.* **236:**403–422.

Kelly, R. B., 1985, Pathways of protein secretion in eukaryotes, *Science* **230:**25–32.

Koop, B. F., and Baker, R. J., 1983, Electrophoretic studies of relationships of six species of *Artibeus* (Chiroptera: Phyllostomidae), *Occ. Pap. Mus. Texas Tech Univ.* **83:**1–12.

Kousvelari, E. E., Oppenheim, F. G., and Cutler, L. S., 1982, Ultrastructural localization of salivary acidic proline-rich proteins from *Macaca fasicularis, J. Histochem. Cytochem.* **30:**274–278.

Kratzing, J. E., 1984, The structure and distribution of nasal glands in four marsupial species, *J. Anat.* **139:**553–564.

Kristensen, S. E., and Papadimitriou, J. M., 1981, The measurement of cell volume and surface area by SEM photogrammetry, *J. Microsc.* **124:**155–161.

Kuwabara, T., 1979, Species differences in the retinal pigment epithelium, in: *The Retinal Pigment Epithelium* (K. M. Zinn and M. F. Marmor, eds.), Harvard University Press, Cambridge, Massachusetts, pp. 58–82.

Larsson, L.-I., 1977, Ultrastructural localization of a new neuronal peptide (VIP), *Histochemistry* **54:**173–176.

Larsson, L.-I., and Rehfeld, J. F., 1977, Evidence for a common evolutionary origin of gastrin and cholecystokinin, *Nature* **269:**335–338.

Larsson, L.-I., Sundler, F., Hakanson, R., Pollack, H. G., and Kimmel, J. R., 1974, Localization of AAP, a postulated new hormone to a pancreatic endocrine cell type, *Histochemistry* **42:**377–383.

Larsson, L.-I., Sundler, F., and Hakanson, R., 1976, Pancreatic polypeptide—A postulated new hormone: Identification of its cellular storage site by light and electron microscopic immunocytochemistry, *Diabetologia* **12:**211–266.

Lee, R. M. K. W., McKenzie, R., Kobayashi, K., Garfield, R. E., Forrest, J. B., and Daniel, E. E., 1981, Effects of glutaraldehyde fixative osmolarities on smooth muscle cell volume, and osmotic reactivity of the cells after fixation, *J. Microsc.* **125:**77–88.

Lekagul, B., and McNeely, J. A., 1977, *Mammals of Thailand*, Association for the Conservation of Wildlife, Sahakarnbhat Company, Bangkok.

Leuenberger, P. M., and Novikoff, A. B., 1975, Studies on microperoxisomes. VII. Pigment epithelial cells and other cell types in the retina of rodents, *J. Cell Biol.* **65:**324–334.

Levine, J. S., 1985, The vertebrate eye, in: *Functional Vertebrate Morphology* (M. Hildebrand, D. M. Bramble, K. F. Liem, and D. B. Wake, eds.), Belknap Press of Harvard University Press, Cambridge, Massachusetts, pp. 317–337.

Liem, K. F., and Wake, D. B., 1985, Morphology: Current approaches and concepts, in: *Functional Vertebrate Morphology* (H. Hildebrand, D. M. Bramble, K. F. Liem, and D. B. Wake, eds.), Belknap Press of Harvard University Press, Cambridge, Massachusetts, pp. 366–377.

Loesch, A., and Burnstock, G., 1985, Ultrastructural identification of VIP-containing nerve fibers in the myenteric plexus of the rat ileum, *J. Neurocytol.* **14:**327–335.

Long, K. O., and Fisher, S. K., 1983, The distribution of photoreceptors and ganglion cells in the California ground squirrel, *Spermophilus beecheyi*, *J. Comp. Neurol.* **221:**329–340.

Lopez, L. C., Li, W.-H., Frazier, M. L., Luo, C.-C., and Saunders, G. F., 1984, Evolution of glucagon genes, *Mol. Biol. Evol.* **1:**335–344.

Machino, M., Morioka, H., Tachibana, M., Tsuruoko, T., and Mizukoshi, O., 1984, Studies

on the substructure of the lysozyme-rich secretory granule of the serous cell in the human nasal gland, *Arch. Histol. Jpn.* **47**:549–551.
Maderson, P. F. A., 1983, An evolutionary view of epithelial–mesenchymal interactions, in: *Epithelial–Mesenchymal Interactions in Development* (R. H. Sawyer and E. Mayr, eds.), Praeger, New York, pp. 215–242.
Maynard Smith, J., Burian, R., Kauffman, S., Alberch, P., Cambell, J., Goodwin, B., Lande, R., Raup, D., and Wolpert, L., 1985, Developmental constraints and evolution: A perspective from the Mountain Lake Conference on development and evolution, *Q. Rev. Biol.* **60**:265–287.
Mennone, A., Phillips, C. J., and Pumo, D. E., 1986, Evolutionary significance of interspecific differences in gastrin-like immunoreactivity in the pylorus of phyllostomid bats, *J. Mammal.* **67**:373–384.
Miller, R. J., 1984, The mucosa as a target tissue for gut neuropeptides, *Trends Neurosci.* **7**:2–3.
Minassian, H., and Huang, S.-N., 1979, Effect of sodium azide on the ultrastructural preservation of tissues, *J. Microsc.* **117**:243–253.
Müller-Schwarze, D., 1983, Scent glands in mammals and their functions, in: *Advances in the Study of Mammalian Behavior* (J. F. Eisenberg and D. G. Kleiman, eds.), American Society of Mammalogists, pp. 150–197.
Murphy, C. J., Howland, H. C., Kwiecinski, G. G., Kern, T., and Kallen, F., 1983, Visual accommodation in the flying fox (*Pteropus giganteus*), *Vision Res.* **23**:617–620.
Murty, V. L. N., Slomiany, B. L., Zdebska, E., Slomiany, A., Mandel, I. D., and Levy, B. M., 1984, Lipid composition of marmoset saliva, *Comp. Biochem. Physiol.* **79A**:41–44.
Nagato, T., Tandler, B., and Phillips, C. J., 1984, Unusual smooth endoplasmic reticulum in submandibular acinar cells of the male round-eared bat, *Tonatia sylvicola, J. Ultrastruct. Res.* **87**:275–284.
Naumova, E. I., 1981, *Functional Morphology of the Digestive System of Rodents and Lagomorphs*, Akademii Nauk SSSR, Moscow [in Russian].
Nielsen, J. T., and Sick, K., 1964, Genetics of amylase isozyme in the mouse, *Hereditas* **51**:291–296.
Novikoff, A. B., 1976, The endoplasmic reticulum: A cytochemist's view (a review), *Proc. Natl. Acad. Sci. USA* **73**:2781–2787.
Okon, E. E., 1977, Functional anatomy of the alimentary canal in the fruit bat, *Eidelon helvum*, and the insect bat, *Tadarida nigeriae, Acta Zool. (Stockh.)* **58**:83–93.
Oster, G. F., and Alberch, P., 1982, Evolution and bifurcation of developmental programs, *Evolution* **36**:444–459.
Oxberry, B. A., 1975, An anatomical, histochemical, and autoradiographic study of the ever-growing molar dentition of *Microtus* with comments on the role of structure in growth and eruption, *J. Morphol.* **147**:337–354.
Padykula, H. A., and Taylor, J. M., 1977, Uniqueness of the bandicoot chorioallantoic placenta (Marsupialia: Peramelidae) Cytological and evolutionary interpretations, in: *Reproduction and Evolution* (J. H. Calaby and C. H. Tyndale-Biscoe, eds.), Australian Academy of Science, Canberra, pp. 303–323.
Palade, G., 1975, Intracellular aspects of the process of protein synthesis, *Science* **189**:347–358.
Patel, Y. C., Zingg, H. H., Fitz-Patrick, D., and Srikant, C. B., 1981, Somatostatin: Some aspects of its physiology and pathophysiology, in: *Gut Hormones* (S. R. Bloom and J. M. Polak, eds.), Churchill Livingstone, New York, pp. 339–349.
Pedler, C., and Tilley, R., 1969, The retina of a fruit bat (*Pteropus giganteus brunnich*), *Vision Res.* **9**:909–922.

Perry, G. C., Patterson, R. L. S., MacFie, H. J. H., and Stinson, C. G., 1980, Pig courtship behavior; Pheromonal property of androstene steroids in male submaxillary secretion, *Anim. Prod.* **31**:191–199.

Pettigrew, J. D., 1986, Flying primates? Megabats have the advanced pathway from eye to midbrain, *Science* **231**:1304–1306.

Phillips, C. J., 1971, The dentition of glossophagine bats: Development, morphological variation, pathology, and evolution, *Univ. Kans. Publ. Misc. Ser. Mus. Nat. Hist.* **54**:1–136.

Phillips, C. J., 1976, Salivary glands: Comparative ultrastructure of secretory cells, *Proc. Nassau County Med. Cent.* **3**:162–167.

Phillips, C. J., 1985a, Field fixation and storage of museum tissue collections suitable for electron microscopy, *Acta Zool. Fenn.* **170**:87–90.

Phillips, C. J., 1985b, Microanatomy, in: *Biology of the New World Microtus* (R. H. Tamarin, ed.), American Society of Mammalogists, pp. 176–253.

Phillips, C. J., and Oxberry, B. A., 1972, Comparative histology of molar dentitions of *Microtus* and *Clethrionomys*, with comments on dental evolution in microtine rodents, *J. Mammal.* **53**:1–20.

Phillips, C. J., and Steinberg, B., 1976, Histological and scanning electron microscopic studies of tooth structure and thegosis in the common vampire bat, *Desmodus rotundus*, *Occ. Pap. Mus. Texas Tech Univ.* **41**:1–12.

Phillips, C. J., and Studholme, K. M., 1982, Comparative ultrastructure of gastric chief cells in carnivorous and frugivorous bats, *IRCS Med. Sci. Cell Memb. Biol.* **19**:573–574.

Phillips, C. J., and Tandler, B., 1985, Unique salivary gland structure in two genera of tropical bats, *Anat. Rec.* **211**:152A.

Phillips, C. J., Grimes, G. W., and Forman, G. L., 1977, Oral biology, in: *Biology of Bats of the New World Family Phyllostomatidae,* Part II (R. J. Baker, J. K. Jones, Jr., and D. C. Carter, eds.), Texas Tech University Press, Lubbock, Texas, pp. 121–246.

Phillips, C. J., Studholme, K. M., and Forman, G. L., 1984, Results of the Alcoa Foundation–Suriname expeditions. VIII. Comparative ultrastructure of gastric mucosae in four genera of bats (Mammalia: Chiroptera), with comments on gastric evolution, *Ann. Carnegie Mus. Nat. Hist.* **53**:71–117.

Phillips, C. J., Nagato, T., and Tandler, B., 1987a, Comparative ultrastructure and evolutionary patterns of acinar secretory product of the parotid salivary glands in Neotropical bats, in: *Studies in Neotropical Mammalogy: Essays in Honor of Philip Hershkovitz* (B. D. Patterson and R. M. Timm, eds.), Field Museum of Natural History, Chicago.

Phillips, C. J., Tandler, B., and Pinkstaff, C. A., 1987b, Unique salivary glands in two genera of tropical microchiropteran bats: An example of evolutionary convergence in histology and histochemistry, *J. Mammal.*, in press.

Pinkstaff, C. A., 1980, The cytology of salivary glands, *Int. Rev. Cytol.* **63**:141–261.

Pinkstaff, C. A., 1981, Histochemical characterization of salivary gland secretion, in: *Saliva and Salivation, Advances in Physiological Sciences,* Volume 28 (T. Zelles, ed.), Akademiai Kiado, Budapest, pp. 135–140.

Pinkstaff, C. A., Tandler, B., and Cohan, R. P., 1982, Histology and histochemistry of the parotid and principal and accessory submandibular glands of the little brown bat, *J. Morphol.* **172**:271–285.

Probert, L., De Mey, J., and Polak, J. M., 1981, Distinct subpopulations of enteric p-type neurones contain substance P and vasoactive intestinal polypeptide, *Nature* **294**:470–471.

Probert, L., De Mey, J., and Polak, J. M., 1983, Ultrastructural localization of four different neuropeptides within separate populations of p-type nerves in the guinea-pig colon, *Gastroenterology* **85**:1094–1104.

Purves, D., and Lichtman, J. W., 1985, Geometrical differences among homologous neurons in mammals, *Science* **228**:298–302.

Quay, W. B., 1954, The meibomian glands of voles and lemmings (Microtinae), *Misc. Publ. Mus. Zool. Univ. Michigan* **82**:1–17.

Quay, W. B., 1968, The specialized posterolateral sebaceous glandular regions in microtine rodents, *J. Mammal.* **49**:427–445.

Ralls, K., 1971, Mammalian scent marking, *Science* **171**:443–449.

Ravazzola, M., and Orci, L., 1980, Glucagon and glicentin immunoreactivity are topologically segregated in the α granule of the human pancreatic A-cell, *Nature* **284**:66–67.

Rieger, R., and Tyler, S., 1979, The homology theorem in ultrastructural research, *Am. Zool.* **19**:655–664.

Robinovitch, M. R., and Sreebny, L. M., 1969, Separation and identification of some of the protein components of rat parotid saliva, *Arch. Oral Biol.* **14**:935–949.

Rothman, J. E., 1986, Life without clathrin, *Nature* **319**:96–97.

Rouse, G. W., and Robson, S. K., 1986, An ultrastructural study of megachiropteran (Mammalia: Chiroptera) spermatozoa: Implications for chiropteran phylogeny, *J. Submicrosc. Cytol.* **18**:137–152.

Ryan, M. J., and Tuttle, M. D., 1983, The ability of the frog-eating bat to discriminate among novel and potentially poisonous frog species using acoustic cues, *Anim. Behav.* **31**:827–833.

Saki, T., and van Lennep, E. W., 1984, The harderian gland in Australian marsupials, *J. Mammal.* **65**:159–162.

Sands, M. W., Coppinger, R. P., and Phillips, C. J., 1977, Comparisons of thermal sweating and histology of sweat glands of selected canids, *J. Mammal.* **58**:74–78.

Schultzberg, M., Hökfelt, T., Nilsson, G., Terenius, L., Rehfeld, J. F., Brown, M., Elde, R., Goldstein, M., and Said, S., 1980, Distribution of peptide- and catecholamine-containing neurons in the gastro-intestinal tract of rat and guinea-pig: Immunohistochemical studies with antisera to substance P, vasoactive intestinal polypeptide, enkephalins, somatostatin, gastrin/cholecystokinin, neurotensin and dopamine β-hydroxylase, *Neuroscience* **5**:689–744.

Simson, J. A. V., Dom, R. M., Sannes, P. L., and Spicer, S. S., 1978, Morphology and cytochemistry of acinar secretory granules in normal and isoproterenol-treated rat submandibular glands, *J. Microsc.* **113**:185–203.

Smith, J. D., 1976, Chiropteran evolution, in: *Biology of Bats of the New World Family Phyllostomatidae*, Part I (R. J. Baker, J. K. Jones, Jr., and D. C. Carter, eds.), Special Publication of the Museum, Texas Tech University, Lubbock, Texas, pp. 49–69.

Smith, J. D., and Madkour, G., 1980, Penial morphology and the question of chiropteran phylogeny, in: *Proceedings Fifth International Bat Research Conference* (D. E. Wilson and A. L. Gardner, eds.), Texas Tech University Press, pp. 347–365.

Snyder, A. W., Bossomaier, T. R. J., and Hughes, A., 1986, Optical image quality and the cone mosaic, *Science* **231**:499–501.

Solcia, E., Capella, C., Buffa, R., Usellini, L., Fiocca, R., and Sessa, F., 1981, Endocrine cells of the digestive system, in: *Physiology of the Gastrointestinal Tract* (L. R. Johnson, ed.), Raven Press, New York, pp. 39–58.

Storch, V., 1979, Contributions of comparative ultrastructural research to problems of invertebrate evolution, *Am. Zool.* **19:**673–645.

Straney, D. O., 1984, The nasal bones of *Chiroderma* (Phyllostomidae), *J. Mammal.* **65:**163–165.

Studholme, K. M., Phillips, C. J., and Forman, G. L., 1986, Results of the Alcoa Foundation Suriname expeditions. Patterns of cellular divergence and evolution in the gastric mucosa of two genera of phyllostomid bats, *Trachops* and *Chiroderma*, *Ann. Carnegie Mus. Nat. Hist.*, **55:**211–239.

Studholme, K. M., Yazulla, S., and Phillips, C. J., 1987, Interspecific differences in immunohistochemical localization of retinal neurotransmitters in four species of bats, *Brain Behav. Evol.*, in press.

Studier, E. H., Wisniewski, S. J., Feldman, A. T., Dapson, R. W., and Wilson, D. E., 1983, Kidney structure in neotropical bats, *J. Mammal.* **64:**445–452.

Tandler, B., 1972, Microstructure of salivary glands, in: *A Symposium on Salivary Glands and Their Secretion* (N. H. Rowe, ed.), University of Michigan Press, Ann Arbor, Michigan, pp. 8–21.

Tandler, B., 1978, Salivary glands and the secretory process, in: *Textbook of Oral Biology* (J. H. Shaw, E. A. Sweeney, C. C. Cappuccino, and S. M. Meller, eds.), Saunders, Philadelphia, pp. 547–592.

Tandler, B., and Cohan, R. P., 1984, Ultrastructure of the parotid gland in the little brown bat, *Anat. Rec.* **210:**491–502.

Tandler, B., and Poulsen, J. H., 1976, Fusion of the envelope of mucous droplets with the luminal plasma membrane in acinar cells of the cat submandibular gland, *J. Cell Biol.* **68:**775–781.

Tandler, B., and Riva, A., 1986, Salivary glands, in: *Human Oral Embryology and Histology* (O. Fejerskov and I. A. Mjör, eds.), Munksgaard, Copenhagen, pp. 243–284.

Tandler, B., Nagato, T., and Phillips, C. J., 1985, Ultrastructure of the submandibular gland in four species of fruit bats (*Artibeus* sp.), *Anat. Rec.* **205:**198A.

Tandler, B., Nagato, T., and Phillips, C. J., 1986a, Systematic implications of comparative ultrastructure of secretory acini in the submandibular salivary gland in *Artibeus* (Chiroptera: Phyllostomidae), *J. Mammal.* **67:**81–90.

Tartakoff, A. M., and Vassalli, P., 1978, Comparative studies of intracellular transport of secretory proteins, *J. Cell Biol.* **79:**694–707.

Thomas, N. W., 1970, Morphology of endocrine cells in the islet tissue of the cod *Gadus callarias*, *Acta Endocrinol.* **63:**679–695.

Tyler, S., 1979, Contributions of electron microscopy to systematics and phylogeny: Introduction to the symposium, *Am. Zool.* **19:**541–543.

Ulinski, P. S., 1980, Functional morphology of the vertebrate visual system: An essay on the evolution of complex systems, *Am. Zool.* **20:**229–246.

Wallach, D., Tessler, R., and Schramm, M., 1975, The proteins of the content of the secretory granules of the rat parotid gland, *Biochim. Biophys. Acta* **382:**552–564.

Walls, G. L., 1942, *The Vertebrate Eye and Its Adaptive Radiation*, Hafner, New York.

Walsh, J. H., 1981, Gastrin, in: *Gut Hormones* (S. R. Bloom and J. M. Polak, eds.), Churchill Livingstone, New York, pp. 163–170.

West, R. W., and Dowling, J. E., 1975, Anatomical evidence for cone and rod receptors in the gray squirrel, ground squirrel, and prairie dog retinas, *J. Comp. Neurol.* **159:**439–460.

Williams, T. P., and Penn, J. S., 1985, Intracellular topography of rhodopsin regeneration in vertebrate rods, *J. Gen. Physiol.* **86:**413–422.

Wimsatt, W. A., 1956, Histological and histochemical observations of the parotid, submaxillary and sublingual glands of the tropical American fruit bat, *Artibeus jamaicensis* Leach, *J. Morphol.* **99:**169–209.

Wolff, J. O., and Johnson, M. F., 1979, Scent marking in the taiga voles, *Microtus xanthognathus*, *J. Mammal.* **60:**400–404.

Wong, Y. C., Breed, W. G., and Chow, P. H., 1985, Ultrastructure of the epithelial cells of the ventral prostate from the hopping mouse, *Notomys alexis*, *Acta Anat.* **121:**163–169.

Yamada, J., Campos, V. J. M., Kitamura, N., Pacheco, A. C., Yamashita, T., and Caramaschi, V., 1984, Immunohistochemical study of gastro-entero-pancreatic (GEP) endocrine cells in the vampire bat (*Desmodus rotundus*), *Gegenbaurs Morphol. Jahrb. Leipzig* **130:**845–856.

Young, J. A., and van Lennep, E. W., 1978, *The Morphology of Salivary Glands*, Academic Press, New York.

CHAPTER 2

ROLE OF CHROMOSOMAL BANDING PATTERNS IN UNDERSTANDING MAMMALIAN EVOLUTION

ROBERT J. BAKER, MAZIN B. QUMSIYEH, and CRAIG S. HOOD

1. INTRODUCTION

Understanding and elucidating the forces affecting chromosomal evolution is not a simple task and is by no means nearing completion. A major obstacle is documenting the exact time of fixation of a new chromosomal rearrangement in a natural population. Currently, it is not possible to determine at what point in its past a chromosomal rearrangement became established in the karyotype of a species. Therefore, we can only speculate on the circumstances surrounding fixation of a rearrangement in a single population that subsequently becomes characteristic of the entire species. An even more basic example is the lack of documentation that individuals with a new chromosomal rearrangement are more fit than individuals with the primitive condition. A list of the features of chromosomal evolution that have not been observed is long enough to suggest a huge gap between theoretical and empirical studies.

ROBERT J. BAKER, MAZIN B. QUMSIYEH, and CRAIG S. HOOD • Department of Biological Sciences and The Museum, Texas Tech University, Lubbock, Texas 79409.

Examples of theories proposing different primary forces regulating chromosomal evolution include what we call the deme size model and the adaptive karyotype model and are similar in principle to the neutralist–selectionist arguments associated with molecular variation. Although these models are not mutually exclusive, they clearly emphasize alternative primary forces in chromosomal evolution.

The deme size model has its roots in Wright's (1941) observation that for some types of chromosomal rearrangements, the chromosomal heterozygote for a new arrangement will be less fit than either the homozygotes for the original chromosomal condition or those for the rearrangement. This lower fitness results from impairment of the reduction division in meiosis, which causes the production of unbalanced gametes. Wright reasoned that if there is a large effective population size, a new chromosomal rearrangement has essentially no chance of increasing in frequency (and eventually becoming fixed), because of lowered fitness in heterozygous individuals. However, if the effective population size (deme) is small enough, then on some occasions a new rearrangement could become fixed through sampling error (genetic drift). Basically a new mutation has a finite probability of becoming fixed through genetic drift (Wright, 1941) in a natural population only when the size of the breeding population is small. This is especially true if the new mutation is a chromosomal rearrangement that causes its possessor to be less fertile than other individuals in the population that are homozygous for the old chromosomal condition (Lande, 1979; Chesser and Baker, 1986).

An extension of Wright's observations is that chromosomal evolution is restricted primarily to species with highly inbred demes or small founder populations. Lande (1979) defined several aspects of this model and presented formulas that would permit calculation of effective population size over the long term if certain assumptions are made concerning negative heterosis, mutation rate of chromosomal rearrangements, and the number of rearrangements that have become fixed in the history of a lineage.

An alternative theory is a more direct extension of neo-Darwinian evolution and natural selection (Darwin, 1859; Bickham and Baker, 1979). The adaptive karyotype theory is based on the hypothesis that the karyotype is adaptive and that under some circumstances natural selection will favor the incorporation of a new chromosomal rearrangement. This model will work only when (1) a new chromosomal rearrangement provides its possessor with a positive phenotypic or genetic benefit, (2) the benefit is present in the heterozygous as well as the homozygous condition, and (3) the magnitude of the benefit is greater than any reduced fitness resulting from meiotic malassortment in the heterozygote. In the adaptive karyotype model the effective population size is unimportant because the heterozygote for the new chromosomal rearrangement is more fit than the homozygote for the primitive state. The homozygote with the derived chromosomal condition would be more fit than the heterozygote because it would have the benefit of the new rearrangement, but no meiotic problems.

Given two populations of the same species with different chromosomal rearrangements, how can we determine which, if either, of the proposed models best explains the origin of the chromosomal differences? The truth is that there are so many variables associated with the factors potentially affecting chromosomal evolution that investigators do not agree on what kinds of data would result in the rejection of alternative theoretical explanations. Nonetheless, unless a theory is presented in a manner that permits falsification, it has limited value to science.

2. G-BANDS AS A MEASURE OF GENETIC HOMOLOGY AND TYPES OF CHROMOSOMAL CHANGE

The ability to determine accurately the type and number of rearrangements being studied is as critical to understanding chromosomal evolution as are theoretical considerations. Nondifferentially stained karyotypes can provide data on the number and size of chromosomes (diploid number) as well as general observations on centromere position. However, nondifferentially stained karyotypes provide little estimate of genetic homology of chromosomes. Some differential staining methods provide unique banding patterns over the entire karyotype and these patterns can be used to hypothesize homologous segments among individuals, populations, and even species. The most commonly used banding technique is G-banding. However, R-banding also has been used in a number of published reports (for example, Dutrillaux et al., 1981; Viegas-Piquignot et al., 1985). R-Bands produce the reverse patterns of those seen in G-banding. See Pathak (1976) for an explanation of the various types of banding techniques.

A critical question concerning any similarity measure of banding patterns used to indicate homology among individuals or species is, "does the similarity document genetic homology, that is, common evolutionary origin?"

Correspondence between G-band pattern and genetic homology of chromosomal segments has been demonstrated by numerous studies

(Francke and Taggart, 1980; Stubblefield, 1980; Creau-Goldberg et al., 1981; Ma et al., 1982; Stallings et al., 1985). Certain linkage groups have been conserved in mammals as evolutionarily divergent as *Mus musculus* and man (Lalley et al., 1978). Using gene mapping techniques, Francke and Taggart (1980) demonstrated that although the same loci were present, the order of the loci on the X chromosome of *Mus* and man are different, and that this difference corresponds to an inversion that was predicted by G-band analysis. Sawyer and Hotier (1986) also documented that high resolution of G-bands in *Mus* and humans predicted the specific location of homologous genes in the autosomes. Therefore, there are data documenting that similarity in G-band pattern can be valuable in identifying genetically homologous segments. Some chromosomes are so small as to have few bands, and in such cases there are no comparable patterns. Also, there is a certain amount of technical skill required to prepare comparable karyotypes. Finally, even for some species that are thought to be closely related, techniques presently available can detect little G-band similarity between karyotypes (Baker and Bickham, 1980, 1984).

Because there are many studies on the patterns, rates, and modes of chromosomal evolution based entirely on variation revealed by nondifferentially stained karyotypes (Wilson et al., 1974, 1975; Bush et al., 1977; Bengtsson, 1980; Imai and Crozier, 1980; Cothran and Smith, 1983; Imai, 1983; Imai et al., 1983; Bianchi and Merani, 1984; Larson et al., 1984), it is important to understand just how accurate such estimates are. This problem was addressed specifically in Haiduk et al. (1981) in a study of megachiropteran bats. Based on nondifferentially stained karyotypes, a minimum of seven rearrangements could explain the observed differences in their sample. However, a minimum of 34 rearrangements were required to explain the variation observed in G-band pattern. Additionally, in a study of phyllostomoid bats (Baker and Bickham, 1980), over five times as many rearrangements were observed in G-band analysis as were required to explain the variation in nondifferentially stained karyotypes. Other studies document that chromosomal variation has not been estimated correctly by nondifferentially stained karyotypes (felids, Wurster-Hill and Gray, 1973, 1975; equids, Ryder et al., 1978; primates, Dresser and Hamilton, 1979; and vespertilionid bats, Bickham and Baker, 1977, Bickham, 1979a).

On the other hand, there are examples where nondifferentially stained karyotypes were accurate in estimating the magnitude of chromosomal evolution (carnivores, Wurster-Hill and Gray, 1975; primates, Dutrillaux, 1979; phyllostomid bats, Baker and Bass, 1979; vespertilionid bats, Bickham, 1979b). If the magnitude of underestimation was relatively constant, one could simply correct by an appropriate factor, but, this does not appear to be the case.

Another significant problem in using nondifferentially stained karyotypes is that such data cannot be used to determine unequal amounts of chromosomal evolution when two taxa are being compared. Therefore, when taxa in such groups as bats or rodents are being compared, they are usually labeled as conservative or rapidly evolving. However, when the magnitude of chromosomal change is partitioned by a cladistic analysis of G-banded karyotypes, a different perspective is obtained. There are several examples where one species in a genus has retained the primitive karyotype for the genus (such as *Reithrodontomys fulvescens*), whereas other species in the genus have undergone 15 or 20 euchromatic rearrangements (such as *Reithrodontomys megalotis* and *R. montanus;* Hood et al., 1984). The data clearly show that within groups such as bats, horses, and rodents, some species have undergone extensive chromosomal evolution, whereas others have experienced little change (Baker and Bickham, 1980; Ryder et al., 1978; Hood et al., 1984). Therefore, to classify bats as having a slow rate of chromosomal evolution and rodents as having a rapid rate (for example, Bush et al., 1977; Bengtsson, 1980; Patton and Sherwood, 1983) is inaccurate. Clearly, some species of bats have experienced relatively rapid chromosomal evolution and some species of rodents have undergone a relatively conservative rate of chromosomal change. Any study designed to elucidate patterns of chromosomal evolution should be based on a more detailed examination of chromosomal evolution than is revealed by nondifferentially stained karyotypes.

3. NEGATIVE HETEROSIS AND ITS ROLE IN CHROMOSOMAL EVOLUTION

The idea that heterozygotes are less fit for certain rearrangements, known as negative heterosis in chromosomal evolution, has been well explored from a theoretical standpoint (Wright, 1941; Lande, 1979; Chesser and Baker, 1986). This aspect of heterozygosity for a chromosomal rearrangement potentially makes chromosomal evolution unique from neutral or adaptive genic evolution. If negative heterosis is significant, and the new rearrangement causes no selective advantage, then any individual heterozygous for a chromosomal rearrangement (assuming that new chromosomal mutations first appear in a population in a single individual in the heterozygous state) will be less fit than the homozygotes for the primitive chromosomal condition. The extent to which the heterozygous individual is less fit will equal the magnitude of meiotic malassortment resulting in the production of unbalanced gametes.

From the literature, it is obvious that some chromosomal rearrangements do cause meiotic problems in the heterozygous condition (White, 1978). However, there is a growing body of evidence that some species have found means of reducing meiotic problems for some types of rearrangements. Individuals of Peter's tent-making bat, *Uroderma bilobatum*, heterozygous for three rearrangements (a centric fusion, a translocation of an acrocentric to the telomere of a small biarmed element, and a tandem fusion) were fertile and played a significant role in the production of the next generation (Baker, 1979, 1981; Baker and Bickham, 1980).

In a sample of ten specimens of rice rats (genus *Oryzomys*) collected from a natural population on an isolated mountain top in Suriname, Koop et al. (1983) found nine different centric fusion/fission polymorphisms (Fig. 1) and only two of the ten individuals were homozygous for all chromosomal fusions. Three of the remaining eight individuals were heterozygous for a single fusion, two were heterozygous for two fusions, one was heterozygous for four fusions, and two were heterozygous for five fusions. It is difficult to explain how an isolated population could maintain such extensive polymorphism if each fusion heterozygote experienced significantly reduced fitness. The most viable explanation to us is that heterozygosity for centric fusions in this species of *Oryzomys* causes little or no negative heterosis.

The extent and nature of this polymorphic system could not have been determined if only nondifferentially stained karyotypes had been available. Eleven specimens of *Oryzomys* were collected and karyotypic preparations were made under field conditions (tents and no electricity) in Suriname, with G- and C-band preparation done subsequently at Texas Tech as described by Baker et al. (1982). Of the 11 animals obtained, analyzable G- and C-banded karyotypes were prepared from ten individuals. The lack of data for one individual was the result of a cryotube exploding upon being removed from liquid nitrogen. Figure 1 documents that analyzable G-bands can be prepared under typical field conditions.

Another example where a group of organisms appear to be experiencing reduced negative heterosis is in the peromyscine rodents. In *Peromyscus* two types of rearrangements (pericentric inversions and heterochromatic additions or deletions) have become fixed in the 30 species that have thus far been examined (Stangl and Baker, 1984). Within mammals, it is generally accepted that there is little or no neg-

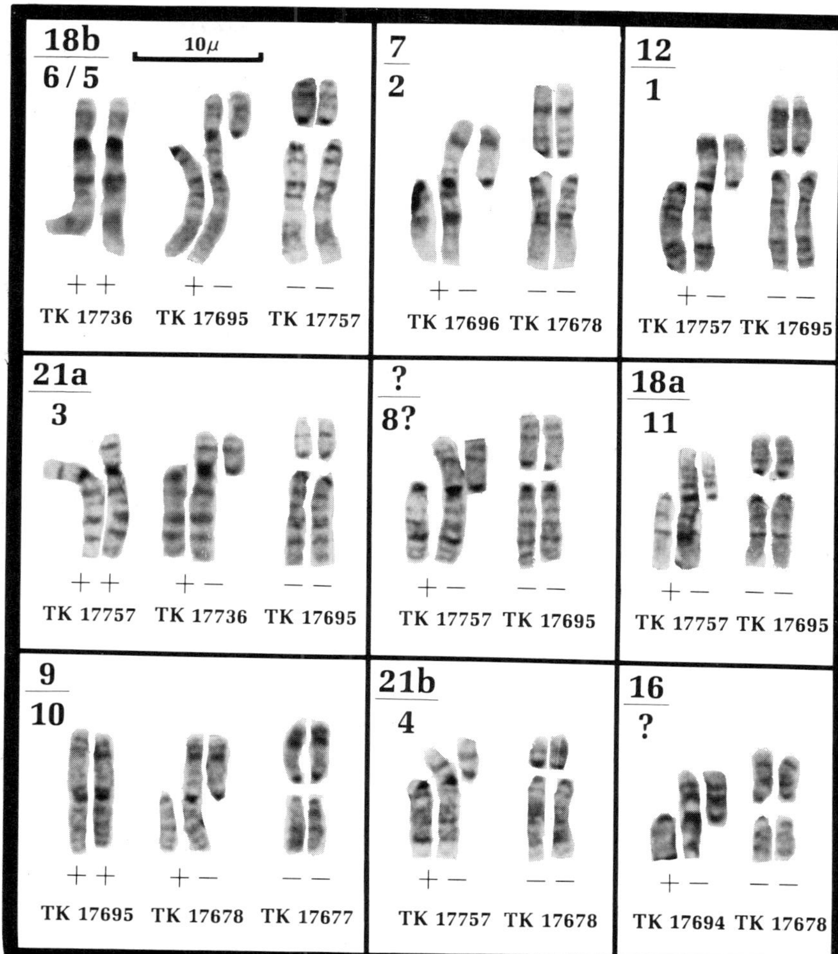

FIGURE 1. G-Banded chromosomes showing the nine polymorphic centric fusions/fissions observed in a population of *Oryzomys* (Koop et al., 1983). The numbering of the chromosomes refers to the proposed homology to those of *Peromyscus*; TK numbers are field numbers used to reference tissues and voucher specimens. (+) Fusion morph; (−) fission morph.

ative heterosis associated with the heterozygous condition for the presence or absence of heterochromatic short arms. However, for pericentric inversions the consensus is that most organisms incur an appreciable heterozygous disadvantage (White, 1978). Data from several sources suggest that pericentric inversions cause little or no heterozygous disadvantage in *Peromyscus*.

One such data set involves a G- and C-band study of a chromosomal hybrid zone in *P. leucopus* where three pericentric inversions distinguish the two parental races (Baker et al., 1983c; Stangl, 1986). An indirect evaluation of the magnitude of selection gradients, *b*, can be obtained from static cline models (Slatkin, 1973; May et al., 1975; Endler, 1977) as $b = l^2(1.66/w)^3$, where l is the gene flow parameter or estimate of average dispersal distance per year. The formula assumes homogeneous gene flow, large population size, and no dominance (Hafner, 1982). Dispersal distance l in *Peromyscus leucopus* averages 0.33 km/yr (Krohne et al., 1984). Results of a karyotypic study across this hybrid zone yielded an average zone width w of 30.6 km (Stangl, 1986). Thus, the intensity of selection b necessary to maintain the zone would be 1.7×10^{-5} km. Hence, little selection is necessary to maintain the zone, implying that there is essentially no negative heterosis associated with the three pericentric inversions. An extension of this observation is that the chromosomes are not functioning in a manner to facilitate speciation, or the zone would be much more narrow.

Another suggestion that meiotic problems associated with the presumed pericentric inversions in *Peromyscus* are minimal comes from the work on the synaptonemal complex by Ira Greenbaum and his students. Studies on individuals of *Peromyscus maniculatus* and *P. sitkensis* heterozygous for one or more pericentric inversions showed the pairing of the inverted region to be nonhomologous, and inversion loops were not formed (Greenbaum and Reed, 1984; Greenbaum et al., 1986). An analysis of diakinesis in these same species indicated that crossing over does not occur in the inverted regions (D. Hale, personal communication). This observation is important because if crossing over does not occur in the inverted chromosomal segments, then no unbalanced gametes are produced.

Frequencies of three pericentric inversion polymorphisms in populations of *Peromyscus boylii* from Mexico suggest little or no negative heterosis associated with these rearrangements. Davis et al. (1986) studied chromosomal morphs at a number of geographic localities and found that the frequencies did not vary significantly from those predicted by Hardy–Weinberg equilibrium. This finding is compatible with the hypothesis that pericentric inversions in *Peromyscus* cause little or no negative heterosis, because, if there was strong selection against the heterozygotes, one would predict that the frequencies would not be in Hardy–Weinberg equilibrium. At this point we might add a word of caution concerning the types of rearrangements found in *Peromyscus*. We are not convinced that it has been proven that the euchromatic rearrangements in peromyscine rodents are indeed pericentric inversions and not centric transpositions. Regardless of which type of euchromatic rearrangements prove to be characteristic of these rodents, the following statements on karyotypic orthoselection are still applicable.

These studies indicate that several species of *Peromyscus* have undergone karyotypic orthoselection (repeated occurrence of a single specific type of rearrangement in a species; White, 1978) for pericentric inversions which cause little or no negative heterosis in *Peromyscus*. A hypothesis to explain these data is that other types of rearrangements that have not become fixed in 30 species of *Peromyscus* cause severe negative heterosis, but pericentric inversions have become fixed in *Peromyscus* because they cause little or no negative heterosis. An extension of this hypothesis is that groups that have experienced karyotypic orthoselection for a given chromosomal rearrangement have evolved mechanisms by which negative heterosis for that type of rearrangement has been reduced or eliminated. A critical test of this hypothesis would involve examining mammalian taxa that exhibit karyotypic orthoselection for other types of rearrangements, to see if these rearrangements are likewise tolerated in the heterozygous condition with minimal meiotic problems. In *Aethomys* (Baker et al., 1986) and *Muntiacus* (Brinkley et al., 1984), it is possible that negative heterosis for tandem fusions has been reduced.

Finally, in computer simulations of fixation of newly formed chromosomal mutants, rearrangements that were assigned 0 and 10% reduction in fertility were essentially equal in frequency of fixation, whereas those mutants causing a greater loss of fertility (25 and 50%) rarely became fixed for the new rearrangement (Chesser and Baker, 1986). These data suggest that at higher levels of negative heterosis (25–50% loss of fitness), there is considerable stabilizing selection to maintain the original karotypic condition, but at lower levels of loss of fitness (10% or less) considerable chromosomal evolution may occur.

4. THE PROBLEM OF CONVERGENT CHROMOSOMAL EVOLUTION

A neutral hypothesis of random chromosomal breakage leading to a new rearrangement would predict that the probability of encountering convergent (independently derived) chromosomal rearrangements would be exceedingly low. If the fixation of chromosomal rearrangements within natural populations of mammals is governed by bottlenecks in small populations, then the probability of fixing exactly the same rearrangement in independent lineages would be even further reduced.

Cladistic analyses of G-band data for a variety of mammalian taxa have documented that convergence of chromosomal rearrangement events is a significant feature in the evolution of some groups of mammals. Within many species of peromyscine rodents, chromosome 6 has undergone pericentric inversion repeatedly (Robbins and Baker, 1981; Rogers, 1983; Rogers et al., 1984; Stangl and Baker, 1984; Hood et al., 1984). In the 30 species of *Peromyscus* examined to date, chromosome 6 has been inverted at least seven times (Fig. 2) (Stangl and Baker, 1984). Pericentric inversions of this chromosome are also found in the related genera *Reithrodontomys* and *Neotoma*, and in several cases the rearrangement breakpoints were located in the exact same G-band of the chromosome (Hood et al., 1984; Koop et al., 1984).

One explanation for these observations is that there are specific sites on some chromosomes that are susceptible to breakage at a higher frequency than would be expected due to chance alone. Examples of nonrandom breakage are well-documented in the human cytogenetic literature. An extreme example involves a reciprocal translocation between segments at specific G-band locations of chromosomes 11 and 22 in the human karyotype found in unrelated families (Fraccaro et al., 1980).

The significance of repeated chromosomal mutations resulting in karyotypic convergence has the following cytogenetic and evolutionary implications. First, the accumulation of chromosomal rearrangements in these cases is not random, and may involve factors that affect the rate of specific chromosomal mutations (King, 1982; Shaw et al., 1983). Second, if intrinsic rates of chromosomal mutation have not been affected, then it seems unlikely that population bottlenecks represent the sole driving force for fixation of new chromosomal rearrangements in independent lineages. Finally, the extent to which convergence and reversal (homoplasy) of chromosomal conditions occur reduces the resolving power of G-band chromosomal data for systematic studies (Baker et al., 1983b).

5. CANALIZATION AND DEME SIZE MODELS AS EXPLANATIONS OF VARIATION IN RATES AND TYPES OF CHROMOSOMAL EVOLUTION

The canalization model of chromosomal evolution (Bickham and Baker, 1979, p. 78) makes the assumption that the karyotype "contrib-

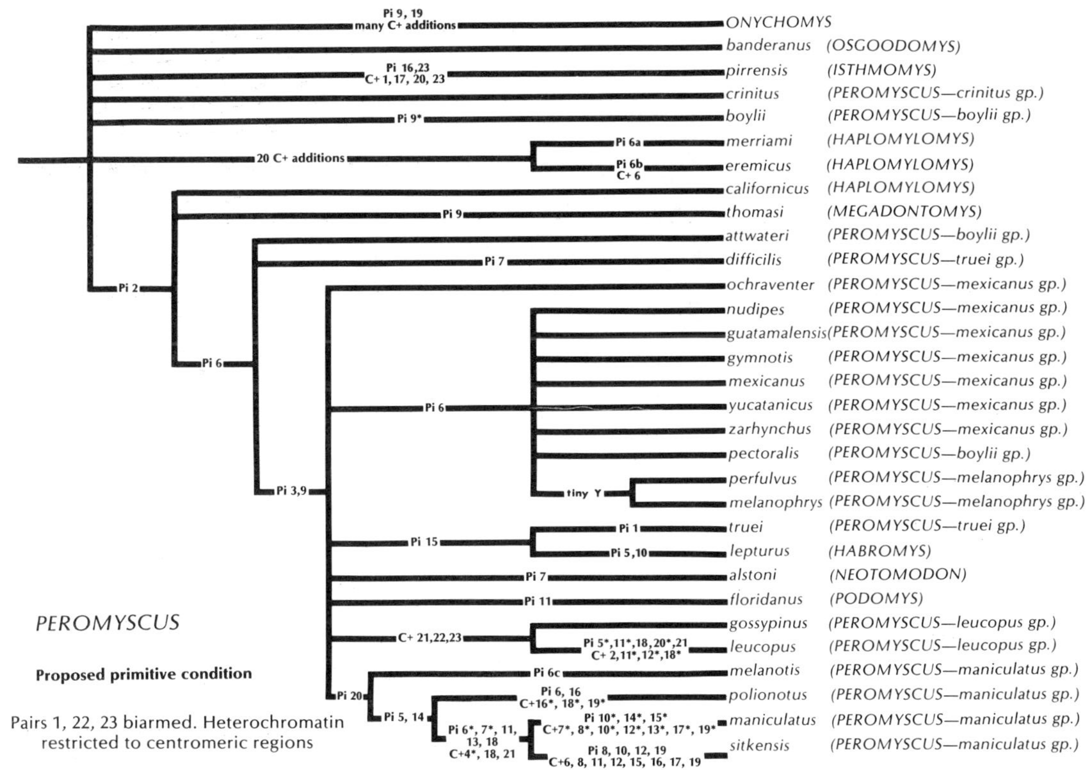

FIGURE 2. Arrangement of 30 species of *Peromyscus* and the genus *Onychomys*, derived from a cladistic analysis of chromosomal homology based on G- and C-banded chromosomes (Stangl and Baker, 1984). Numbers identify chromosomes to a standard, and letters following initiate a unique event; (PI) a pericentric inversion, (C+) addition of heterochromatin, (*) polymorphic rearrangements.

utes significantly to the fitness of the individual." This model was based on comparative studies of G-banded chromosomes of assemblages of turtles and bats. The patterns of chromosomal variation in these taxa indicate that chromosomal evolution occurs more rapidly "immediately after a lineage breaks into a new adaptive zone." In bats, most karyotypic divergence occurred simultaneously with the morphological evolution that distinguishes the modern families. This was followed by a slower rate of chromosomal evolution when the morphological characteristics of recent genera and species evolved.

The predictions from this model are that chromosomal evolution is more rapid as a lineage breaks into an adaptive zone and that once an optimum karyotype has developed for a lineage the rate of chromosomal evolution slows down. Additionally, the model predicts that when a group initially breaks into an adaptive zone the chromosomal rearrangements incorporated into the various lineages within the group will be more severe (from a standpoint of breaking up linkage groups; Bickham, 1981) than after the group has occupied the adaptive zone for a considerable time. The model received considerable comment and debate (Futuyma and Mayer, 1980; Bickham, 1981; Capanna, 1982; Shields, 1982; Sites, 1983; Tegelstrom et al., 1983; King, 1985). The model is based on the hypothesis that the karyotype can be adaptive, yet this is a difficult point to prove. The only example from the literature on mammals that demonstrates a statistically significance between a chromosomal rearrangement and a positive selective advantage involves a balanced polymorphism in pocket gophers of the genus *Geomys* (Baker et al., 1983a).

The model does not describe all data from chromosomal studies (Baker and Bickham, 1980; Baker et al., 1983b). The predictions of the canalization model do not seem to apply in cases of karyotypic megaevolution (Baker and Bickham, 1980; see Section 6) and in the chromosomal patterns observed in some muroid rodents (Baker et al., 1983b; Koop et al., 1984). Patterns of chromosomal evolution in rodents are sufficiently interesting to warrant further discussion here. In Fig. 3, we present the number of chromosomal rearrangements (in the 14 largest pairs of autosomes) that have accompanied the evolution of the morphological features on which 24 genera of muroid rodents are identified. The G-band data published for these genera were compared in an attempt to identify homologous sequences in the largest 14 autosomal pairs of the *Peromyscus* karyotype (Committee for Standardization of Chromosomes of *Peromyscus*, 1977). G-Bands of these genera were obtained from the following sources: *Peromyscus* (Stangl and Baker, 1984), *Onychomys* (Baker et al., 1979), *Baiomys* (Yates et al., 1979), *Reith-*

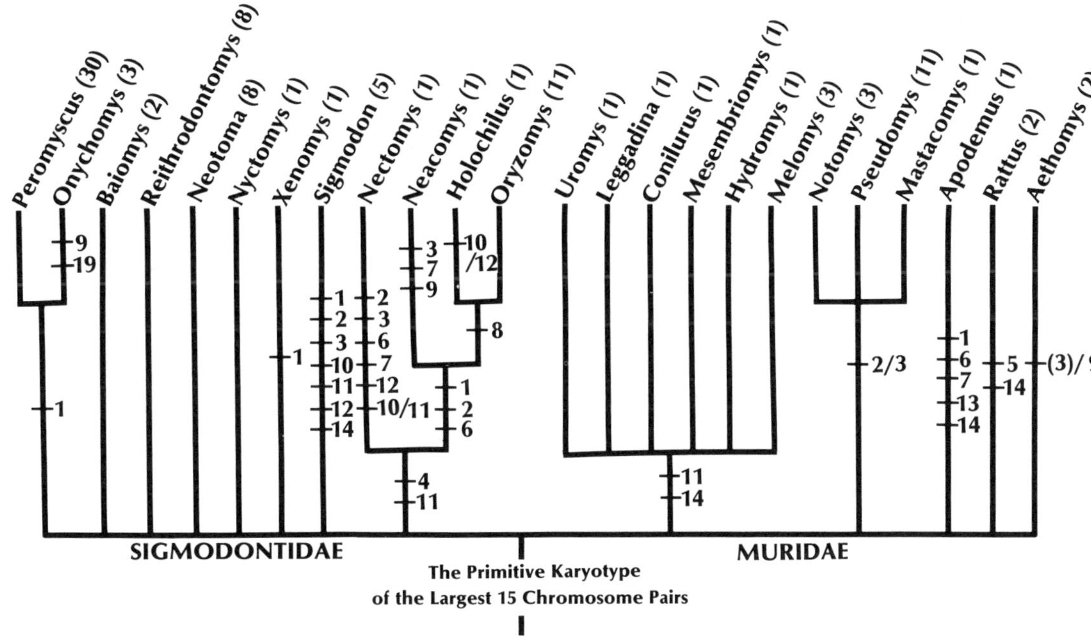

FIGURE 3. Cladistic relationships of the proposed primitive karyotypes for 24 genera of murid and sigmodontid rodents based on G-band analysis of the 14 largest autosomal pairs and the X chromosome. Numbers on branches refer to chromosomes that showed a derived character state (have undergone rearrangement) in the lineage indicated. Numbers after generic names indicate number of species examined in that genus.

rodontomys (Hood et al., 1984), *Neotoma* (Koop et al., 1985), *Nyctomys* and *Xenomys* (Haiduk et al., 1986), *Sigmodon* (Elder, 1980), *Holochilus, Neacomys, Nectomys,* and *Oryzomys* (Baker et al., 1983b), *Uromys, Leggadina, Conilurus, Mesembriomys, Hydromys, Notomys, Pseudomys, Mastacomys,* and *Melomys* (Baverstock et al., 1980, 1983), *Apodemus* (Koop et al., 1984), *Aethomys* (Baker et al., 1986), *Rattus* (Baverstock et al., 1982; and unpublished data from our laboratory).

Figure 3 shows the rearrangements in the largest 14 pairs of autosomes postulated to have occurred since genera diverged from the proposed primitive condition for the common ancestor of these genera and up to the establishment of the primitive karyotype of each genus. As can be seen from this figure, the proposed primitive condition for the 14 largest autosomes for the genus *Peromyscus* (Stangl and Baker, 1984) is divergent from that proposed as primitive for the Sigmodontidae by only one rearrangement, a biarmed chromosome 1, which is also shared with *Onychomys.* On the other hand, since the establishment of the primitive condition for *Peromyscus,* 43 pericentric inversions are postulated to have occurred in the 30 species thus far examined (Fig. 2) (Stangl and Baker, 1984). A similar pattern is observed in *Reithrodontomys, Neotoma,* and *Oryzomys,* where the primitive conditions for the 14 largest pairs were the same for a genus as they were for the family, yet considerable chromosomal variation was documented between species within a genus (for example, *Reithrodontomys,* Hood et al., 1984, and *Oryzomys,* Baker et al., 1983b). These patterns contradict the canalization model, which predicts that a greater amount of chromosomal rearrangement should have accompanied the establishment of muroid families and a lesser amount should have accompanied speciation within the various genera.

Several possible explanations for the muroid data are suggested. If muroids had large deme sizes earlier in their evolution, whereas modern species have smaller deme sizes (Lande, 1979), then these data are compatible with Wright's (1941) deme size model. This would imply that bats had small demes earlier in their evolution, whereas modern species are characterized by large demes. Alternatively, it could be that chromosomal evolution was slow or nonexistent earlier in muroid evolution and that subsequently chromosomal change accelerated, perhaps by an increased mutation rate. A third possibility is that there was an equal amount of chromosomal evolution through time, but the ancestors of modern genera were derived only from the chromosomally conservative species. The third explanation would suggest that the primitive karyotype may be valuable to a species in surviving a period of extinction compared to other species that possessed more derived karyotypes.

It must be emphasized that these ideas are based on the examination of the 14 largest pairs of autosomes. Additional variation likely exists in smaller chromosomes or small rearrangements not detectable from the quality of G-bands available for study of muroid rodents. Recent investigations of the muroid family Gerbillidae (Qumsiyeh, 1986a, 1986b), in which all the euchromatic segments can be identified across genera, indicate numerous rearrangements characterizing the family, genera, and species and that the rate of chromosomal evolution was relatively constant through time.

Test of the canalization model primarily include correlating age with chromosomal diversity, and comparing chromosomal banding patterns among higher categories to determine inter- versus intra-higher category variation. If there is no correlation between age and chromosomal diversity within many groups, this would represent a falsification of the model and would mean that the canalization model is not of general applicability (Bickham and Baker, 1979, p. 82). Data sets for bats and turtles (Bickham, 1981) are best explained by this model, but the data from muroid rodents do not fit the predictions of the canalization model.

An alternative explanation of why there is variation in rates of chromosomal evolution among mammalian taxa is that chromosomal evolution occurs in small inbred demes (Wilson et al., 1975; Bush et al., 1977; Larson et al., 1984). Therefore, these models conclude that the magnitude of chromosomal evolution is a reflection of long-term effective deme size that has been characteristic of a species (Lande, 1979). For a number of reasons we conclude that for mammals this explanation is not compatible with the available data. In computer simulations of factors affecting the fixation of chromosomal rearrangements, Chesser and Baker (1986) using data from *Myotis lucifugus*, concluded that if the effective population size was in excess of 20, there was little probability of a new chromosomal rearrangement becoming fixed through genetic drift. Additionally, if the deme size becomes lower than 20, there is a high probability that the population will become extinct. Therefore, demes of such small sizes have a lower probability of becoming the progenitors of a species than do demes composed of larger numbers of individuals.

Data from mammals suggest that long-term effective population sizes are considerably larger than the values shown to be critical in these calculations (Chesser and Baker, 1986). Wilson et al. (1985, p.

390) calculated that long-term effective populational sizes for ten mammalian species ranged from 6,000 to 2,900,000. Although we are of the opinion that their values are too high, clearly their values are several orders of magnitude above the populational sizes needed to produce an accelerated rate of chromosomal evolution. This is especially critical in the case of *Peromyscus maniculatus*, for which they calculate a long-term effection population size of 2,900,000 yet, chromosomally, *P. maniculatus* may well be the most actively evolving species in the genus *Peromyscus* (Stangl and Baker, 1984; Rogers et al., 1984).

In addition to an initial small deme size, genetic drift can be generated by inbreeding. However, Ralls et al., (1986) indicate that studies on natural populations of mammals fail to document the kind of inbreeding that would be required to fix chromosomal rearrangements, and Chesser and Ryman (1986) show that from a theoretical standpoint substantial dispersal among demes is required to avoid inbreeding depression. Even minimal dispersal among demes would negate the likelihood of fixation of chromosomal rearrangements.

If deme size is the most critical factor regulating the rate of fixation of chromosomal rearrangements (this would mean that mutation rate and magnitude of negative heterosis for a given type of chromosomal rearrangement is relatively constant), then rearrangements that cause the most severe negative heterosis would be expected to occur less frequently than rearrangements that cause little or no negative heterosis. Additionally, rearrangements that cause severe negative heterosis would be expected to occur as isolated events or if they occurred as clusters on branches of cladograms then the rearrangements that cause severe negative heterosis would be expected on the same branches of the cladogram to be outnumbered by rearrangements that cause little or no negative heterosis. However, the pattern that we observe in different groups of mammals is often an extension of White's (1978) karyotypic orthoselection. For example, *Mus* has undergone many centric fusions whereas peromyscine rodents have undergone many pericentric inversions (or centric transpositions) and no centric fusions. Another point to consider is the distribution of tandem fusions which are thought to cause severe negative heterosis. Tandem fusions rarely become fixed in mammalian populations, yet in several species of mammals, tandem fusions have become fixed almost to the exclusion of other types of rearrangements, for example in *Sigmodon* (Elder, 1980), *Muntiacus* (Shi et al., 1980), and *Aethomys* (Baker et al., 1986). It is difficult to explain such variations by simple manipulation of deme size to produce genetic drift. Clearly in these examples the most significant factors regulating the types of chromosomal rearrangements being incorporated must involve either variations in the rates of mutation for a given type of chromosomal rearrangement and/or variation in the magnitude of negative heterosis associated with a given type of chromosomal rearrangement.

We hypothesize that, species that have the fastest rate of chromosomal evolution are those that have: (1) evolved means of reducing negative heterosis (this would include but not be restricted to genetic or phenotypic benefits associated with a rearrangement that partially compensated for or even outweighed the negative heterosis associated with the rearrangement, Baker et al., 1983), for rearrangements that become fixed, and/or (2) experienced a greater mutation rate.

We think that this explanation is more compatible with the variations we see in rates (especially as related to different types of chromosomal rearrangement) of incorporation observed from recent studies based on cladistical analysis of G-banded chromosomes of various mammalian taxa as well as with data from studies on deme size and inbreeding in natural populations (Ralls et al., 1986).

6. KARYOTYPIC MEGAEVOLUTION

The phenomenon of karyotypic megaevolution was described by Baker and Bickham (1980) based on phylogenetic analysis of G-band patterns for 78 species of bats. Viewed in a cladistic framework, magnitude, type, and pattern of chromosomal change can be apportioned within and among taxa. For many lineages, G-band linkage groups are conserved among closely related taxa and chromosomal variation is limited to a few well-defined rearrangement events. However, Baker and Bickham (1980) documented examples of chromosomal change within lineages of closely related taxa (phyllostomid bats and a wide variety of other vertebrates) that present a fundamentally different pattern. There are cases where, within a lineage of closely related taxa, one taxon has undergone little or no chromosomal evolution, while a sister taxon has a G-band karyotype that has been radically reorganized. These taxa are considered karyotypically megaevolved.

In all examples of karyotypic megaevolution described to date (see also Baverstock et al., 1983), the following observations hold: (1) karyotypically megaevolved taxa have undergone extensive euchromatic rearrangement such that the G-band patterns of many, if not all, euchromatic arms have been altered (Baker and Bickham, 1984, referred to those arms as euchromatic linkage groups, but they should be referred to as G-band-defined arms); (2) little morphological and/or, in some cases, little genic divergence (as measured by starch gel electrophoresis; Nelson et al., 1984) has accompanied the extensive chromosomal divergence; (3) many different types of rearrangements are involved; and (4) these rearrangements include many that would be predicted to cause severe negative heterosis in the heterozygote (White, 1978) and therefore should rarely become fixed in natural populations.

The disparity in the amounts of morphological and genic change on the one hand and extensive chromosomal change on the other imply relatively rapid karyotypic divergence. Compared with the small amount of chromosomal variation that typically characterizes closely related mammalian taxa, this observation is striking.

Karyotypic megaevolution is not simply a fast rate of chromosomal evolution (Marks, 1983; Baker and Bickham, 1984). In a wide variety of mammalian examples, including *Mus* (Capanna et al., 1977; Capanna, 1982; Bickham and Baker, 1980), *Onychomys* (Baker et al., 1979; Baker and Barnett, 1981), *Sigmodon* (Elder, 1980), primates (Rumpler et al., 1983), and horses (Ryder et al., 1978), there is a rapid rate of chromosomal evolution. In many of these cases extensive chromosomal change occurred predominately through accumulation of a single type of rearrangement (for example, heterochromatic short-arm additions in *Onychomys* and centric fusions in *Mus*) that does not alter the internal structure of the euchromatic segments. Furthermore, in these examples of rapid chromosomal evolution, the types of rearrangements involved could occur in the heterozygous condition without causing severe negative heterosis. These patterns of chromosomal variation represent examples of karyotypic orthoselection (White, 1975).

We recently discovered an interesting example of an accelerated rate of chromosomal evolution in two species of the murid genus *Aethomys*. One species, *Aethomys chrysophilus* ($2n = 44$), has retained a karyotype that is very similar to that proposed as primitive for the family Muridae, whereas the other, *A. namaquensis* ($2n = 24$), has undergone eight tandem and five centric fusions. *Aethomys namaquensis* apparently has undergone karyotypic orthoselection for both tandem and centric fusions. This example possesses some characteristics of karyotypic megaevolution, such as a large number of rearrangements (tandem fusions) that would be predicted to encounter problems of negative heterosis have accumulated in one taxon, associated with little morphological evolution that distinguishes it from its congener. Nonetheless, *A. namaquensis* has not undergone as many chromosomal rearrangements as is generally characteristic of karyotypic megaevolution. Further study of examples such as that presented by *A. namaquensis* could contribute significantly to understanding the nature of accelerated rates of chromosomal evolution.

The pattern of karyotypic megaevolution has involved some cytogenetic mechanism that has allowed rapid and extensive repatterning of the euchromatic genome. Baker and Bickham (1980, 1984) suggested that karyotypically megaevolved species may have encountered one or more of the following: (1) periods of transposition activity (McClintock, 1978), (2) reduced meiotic constraints, and/or (3) radical changes in nuclear or cellular factors that normally stabilize chromosomal evolution (Bennett, 1982). Karyotypic megaevolution remains a phenomenon that is not easily explained by current theories of chromosomal evolution.

7. SYSTEMATIC VALUE OF G-BAND STUDIES

Identifying chromosomes by G-banding was first developed for studies of human chromosomes in the early 1970s. Since then, the use of G-banding has mushroomed for studies of higher vertebrates. G-Band sequences can be compared between species and even between higher taxonomic categories (Fig. 4) (Mascarello et al., 1974; Qumsiyeh and Baker, 1985; Sawyer and Hozier, 1986), and chromosomal rearrangement differences between taxa can be more accurately identified by these methods than by simple nondifferential-staining techniques.

Data from G-bands lend themselves to phylogenetic analysis by cladistic methods (Baker, 1979; Baker et al., 1983b). Variation in G-band patterns in different taxa can be treated as character states and the identification of primitive character states allows determination of kinds and number of rearrangements incorporated during the course of evolution. Character states that are derived (apomorphies) can be used to identify subsets of taxa as monophyletic. The resulting cladogram represents a hypothesis of phylogenetic relationship that tests previous hypotheses based on other data sets (for example, morphology, protein electrophoresis, immunology, or DNA analysis). Data from G-bands represent an independent data set for testing systematic relationships of taxa that have been traditionally grouped together on the basis of highly specialized morphological adaptations. Examples include studies of the nectar-feeding bats (Haiduk and Baker, 1982, 1984; Haiduk, 1983). Recently, data from G-bands have proved useful in identifying cryptic species of mammals (Capanna, 1982; Baker, 1984; Baker et al., 1985; Baker and Bickham, 1986).

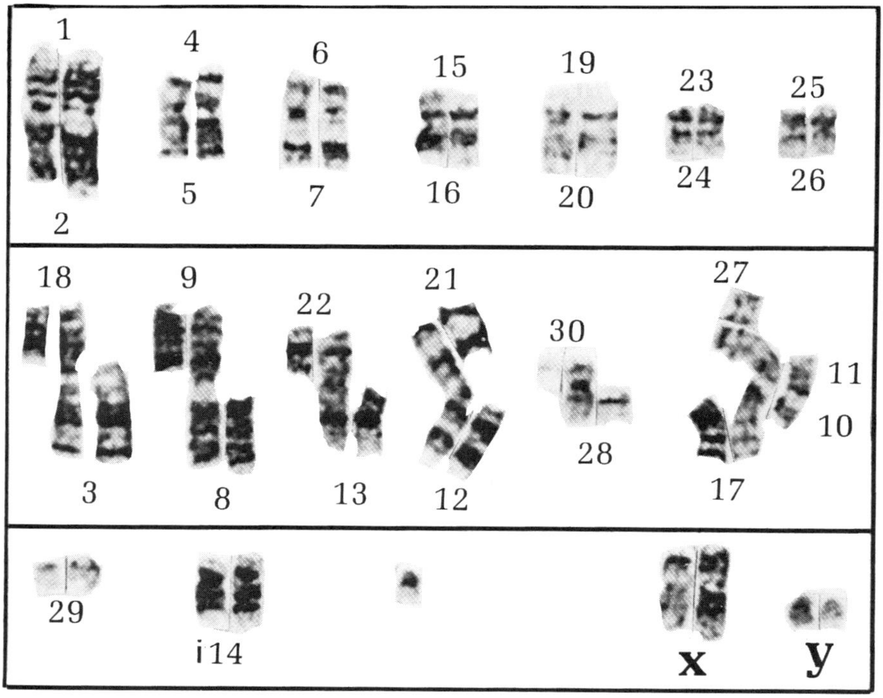

FIGURE 4. Comparison of G-banded haploid complements of the phyllostomid bats *Macrotus waterhousii* (Phyllostominae) and *Rhinophylla fisheae* (Carolliinae). Chromosomes in the top panel are completely homologous; those in the middle have undergone Robertsonian changes; and the bottom panel illustrates the sex chromosomes and autosomal elements. In all comparisons, the chromosome on the right (or center) is from *Macrotus*; those to the left (or left and right) are from *Rhinophylla*. Numbering system is that based on *Macrotus* (Patton and Baker, 1978).

Two problems that are not unique to G-bands arise in using these data for systematic studies. Chromosomal rearrangement characters are subject to reversals and convergence (see Section 4), and thus we are forced to choose the most parsimonious (least homoplastic) explanation of character change in the absence of additional data. A second problem is the fact that in many taxa, conservatism is the rule, and thus, shared primitive chromosomal conditions (symplesiomorphies) provide little resolution of relationships. An example of this conservatism was discussed for the muroid rodent genera. Other examples include some genera of vespertilionid bats (Bickham, 1979a).

In mammals, a multitude of G-band studies have been conducted to understand relationships of subspecies, species, genera, and even families. We will not attempt here to review all the literature in this area, but provide some examples of the utilization of G-banding for systematic studies in groups of mammals.

Order Marsupialia (Rofe and Hayman, 1985).

Order Insectivora (Searle, 1984).

Order Chiroptera. Family Pteropodidae (Haiduk et al., 1981; Haiduk, 1983), Rhinopomatidae (Qumsiyeh and Baker, 1985), Emballonuridae (Hood and Baker, 1986), Noctilionidae (Patton and Baker, 1978; Baker and Bickham, 1980), Phyllostomidae (Patton and Baker, 1978; Baker, 1979; Baker and Bass, 1979; Baker et al., 1982), Mormoopidae (Patton and Baker, 1978; Sites, et al., 1981), Vespertilionidae (Bickham, 1979a, 1979b; Baker et al., 1985).

Order Primates (Rumpler et al., 1983; Dutrillaux et al., 1981; Yunis and Prakash, 1982; Viegas-Pequignot et al., 1985).

Order Lagomorpha (Robinson et al., 1983, 1984).

Order Rodentia. Sigmodontidae (Mascarello et al., 1974; Mascarello and Hsu, 1976; Greenbaum and Baker, 1978; Bianchi et al., 1976; Rogers, 1983; Rogers et al., 1984; Stangl and Baker, 1984; Modi and Lee, 1985), Muridae (Baverstock et al., 1980, 1983), Gerbillidae (Qumsiyeh, 1986a, 1986b), Arvicolidae (Modi, 1986).

Order Carnivora (Wurster-Hill and Centerwall, 1982; Wurster-Hill and Bush, 1980).

Order Cetacea (Arnason, 1974).

8. C-BANDS AND GENETIC HOMOLOGY

Another banding technique produces bands referred to as C-bands, which indicate the presence of constitutive heterochromatin. This technique is distinctly different from those that produce a banding pattern throughout the entire karyotype. C-Bands are thought to appear where the sequence of nucleotides in the DNA is highly repetitive (often referred to as heterochromatin or satellite DNA, although the two terms are not exactly synonyms). Basically, the technique involves treatment of the chromosomes to preferentially extract non-C-band chromatin (Pathak, 1976). It is not well understood why the C-band positive regions remain unaltered while the euchromatic regions are extensively changed. It is clear that the C-band positive areas represent clusters of highly repetitive DNA sequences. Even though highly repetitive DNA would reassociate more rapidly than areas of single copy DNA, this simplistic mechanism has not been supported by other studies. This treatment is valuable to the study of chromosomal evolution because it distinguishes the highly repetitive regions of the chromosome from the region of the chromosome containing single-copy genes. Further, regions of highly repetitive DNA are thought to occur in the heterozygous condition with little or no negative heterosis.

Highly repetitive DNA varies quantitatively in mammalian genomes from approximately 50% to as little as 5% (Hsu and Arrighi, 1971; Miklos et al., 1980; Patton and Sherwood, 1982; Sherwood and Patton, 1982). The presence of blocks of heterochromatin does not necessarily indicate genetic homology, because several different highly repetitive DNA sequences may exist within uniformly C-banded chromosomal segments (John and Miklos, 1979). It is likely that all mammalian chromosomes have some C-band material associated with the centromere. C-Band material and/or highly repetitive DNA sequences have been suggested as important in a wide array of biological features, such as successful pairing of homologues, mutation rates, recombination rates, cell metabolism, and gene expression and speciation (John and Miklos, 1979; Brutlag, 1980; Singer, 1982).

9. SUMMARY AND CONCLUSIONS

1. There is a huge gap between theoretical considerations of chromosomal evolution and the available empirical data to test such theories.
2. G-Band variation provides a measure of genetic homology and allows for an assessment of type and magnitude of chromosomal variation in mammalian groups.
3. From a technical standpoint, it is now possible to prepare specimens for G- and C-banding as well as for a variety of other techniques under field conditions with adequate sample sizes to conduct evolutionary studies (Baker et al., 1982, 1983b,c). High-resolution G-bands (Yunis, 1980, 1981; Yunis and Prakash, 1982) and localization of DNA sequences within individual bands (in situ hybridization; Harper et al., 1981; Harper and Sanders, 1981) have been developed for humans and a few other mammalian taxa. These techniques promise to provide a precise measure of homology, and consequently of chromosomal change.
4. Nondifferentially stained karyotypes are undependable indicators of the magnitude of chromosomal evolution. To elucidate patterns of chromosomal evolution, studies should be designed to include more detailed examination than is revealed by nondifferentially stained karyotypes and should account for in-group variation in rates of evolution.

5. Data from several studies on mammalian chromosomes support the conclusion that "the negative heterosis assumption is an overstatement and leads one to question how important negative heterosis is in rearrangements which are successfully incorporated during chromosomal evolution" (Baker, 1981, p. 303). A possible explanation for some karyotypic orthoselection (White, 1978) is that negative heterosis for that particular type of rearrangement has been avoided.

6. Convergent evolution of chromosomal rearrangements appears to be a problem, at least with the level of G-bands available to workers doing comparative studies.

7. The canalization model is a testable hypothesis of chromosomal evolution that needs further empirical tests. The model explains patterns of chromosomal evolution in some groups, such as bats and turtles, but data from other groups, such as muroid rodents, do not fit the predictions of the model.

8. Models concerned with rate of fixation of chromosomal rearrangements that are based on variation in long-term effective population size have not been tested by empirical studies on natural populations. Such models do not appear to predict the observed patterns resulting from cladistical analyses of G-banded chromosomes.

9. Some species of mammals have undergone extensive reorganization of their euchromatic genome compared to closely related species. This phenomenon of karyotypic megaevolution, which may occur more widely among mammals than is currently documented, remains unexplained by current theories of chromosomal evolution.

ACKNOWLEDGMENTS. We thank Rodney Honeycutt, John Bickham, Don Wilson, Ben Koop, and Robert Bradley for critically evaluating our ideas. We thank Michael Haiduk for assistance in preparing Fig. 4. Supported in part by National Science Foundation grant No. DEB-83-00764.

REFERENCES

Arnason, V., 1974, Comparative chromosome studies in Cetacea, *Hereditas* **77**:1–36.

Baker, R. J., 1979, Karyology, in: *Biology of Bats of the New World Family Phyllostomatidae*, Vol. 16 (R. J. Baker, J. K. Jones, Jr., and D. C. Carter, eds.), Special Publication, The Museum, Texas Tech University, Lubbock, Texas, pp. 107–155.

Baker, R. J., 1981, Chromosome flow between chromosomally characterized taxa of a volant mammal, *Uroderma bilobatum*, *Evolution* **35**:296–305.

Baker, R. J., 1984, A sympatric cryptic species of mammal: A new species of *Rhogeessa* (Chiroptera: Vespertilionidae), *Syst. Zool.* **33**:178–183.

Baker, R. J., and Barnett, R. K., 1981, Karyotypic orthoselection for additions of heterochromatic short arms in grasshopper mice (*Onychomys:* Cricetidae), *Southwest. Nat.* **26**:125–131.

Baker, R. J., and Bass, R. A., 1979, Evolutionary relationships of the Brachyphyllinae to the glossophagine genera *Glossophaga* and *Monophyllus*, *J. Mammal.* **60**:364–372.

Baker, R. J., and Bickham, J. W., 1980, Karyotypic evolution in bats: Evidence of extensive and conservative chromosomal evolution in closely related taxa, *Syst. Zool.* **29**: 239–251.

Baker, R. J., and Bickham, J. W., 1984, Karyotypic megaevolution by any other name: A response to Marks, *Syst. Zool.* **33**:339–341.

Baker, R. J., and Bickham, J. W., 1986, Speciation by monobrachial centric fusions, *Proc. Natl. Acad. Sci. USA*, **83**:8245–8248.

Baker, R. J., Barnett, R. K., and Greenbaum, I. F., 1979, Chromosomal evolution in grasshopper mice (*Onychomys:* Cricetidae), *J. Mammal.* **60**:297–306.

Baker, R. J., Haiduk, M. W., Robbins, L. W., Cadena, A., and Koop, B. F., 1982, Chromosomal studies of South American bats and their systematic implications, in: *Mammalian Biology in South America*, Volume 4 (M. A. Mares and H. H. Genoways, eds.), Special Publication Series, Pymatuning Laboratory of Ecology, University of Pittsburgh, Linesville, Penn., pp. 303–327.

Baker, R. J., Chesser, R. K., Koop, B. F., and Hoyt, R. A., 1983a, Adaptive nature of chromosomal rearrangements: Differential fitness in pocket gophers, *Genetica* **61**: 161–164.

Baker, R. J., Koop, B. F., and Haiduk, M. W., 1983b, Resolving systematic relationships with G-bands: A study of five genera of South American cricetine rodents, *Syst. Zool.* **32**:403–416.

Baker, R. J., Robbins, L. W., Stangl, F. B., Jr., and Birney, E. C., 1983c, Chromosomal evidence for a major subdivision in *Peromyscus leucopus*, *J. Mammal.* **64**:356–359.

Baker, R. J., Bickham, J. W., and Arnold, M. L., 1985, Chromosomal evolution in *Rhogeessa* (Chiroptera: Vespertilionidae): Possible speciation by centric fusions, *Evolution* **39**: 233–243.

Baker, R. J., Qumsiyeh, M. B., and Rautenbach, J. L., 1986, Evidence for eight tandem fusions in the evolution of the karyotype of *Aethomys namequensis* A. Smith, *Genetica*, in press.

Baverstock, P. R., Watts, C. H. S., Adams, M., and Gelder, M., 1980, Chromosomal and electrophoretic studies of Australian *Melomys* (Rodentia: Muridae), *Aust. J. Zool.* **28**:553–574.

Baverstock, P. R., Gelder, M., and Jahnke, A., 1982, Chromosome evolution in Australian *Rattus*—G-banding and hybrid meiosis, *Genetica* **60**:93–103.

Baverstock, P. R., Watts, C. H. S., Gelder, M., and Jahnke, A., 1983, G-banding homologies of some Australian rodents, *Genetica* **60**:105–117.

Bengtsson, B. O., 1980, Rates of karyotype evolution in placental mammals, *Hereditas* **92**:37–47.

Bennett, M. D., 1982, Nucleotypic basis for the spacial ordering of chromosomes in eukaryotes and the implications of the order for genome evolution and phenotypic variation, in: *Genomic Evolution* (G. A. Dover and R. B. Flavell, eds.), Academic Press, London, pp. 239–261.

Bianchi, N. O., and Merani, S., 1984, Cytogenetics of South American akodont rodents (Cricetidae). X. Karyological distances at generic and intergeneric levels, *J. Mammal.* **65**:206–219.

Bianchi, N. O., Vidal Rioja, L., and Bianchi, M. S. A., 1976, Cytogenetics of the South

American akodont rodents (Cricetidae). II. Interspecific homology of G-banding patterns, *Cytologia* **41**:139–144.

Bickham, J. W., 1979a, Chromosomal variation and evolutionary relationships of vespertilionid bats, *J. Mammal.* **60**:350–363.

Bickham, J. W., 1979b, Banded karyotypes of 11 species of American bats (genus *Myotis*), *Cytologia* **44**:789–797.

Bickham, J. W., 1981, Two-hundred-million-year-old chromosomes: Deceleration of the rate of karotypic evolution in turtles, *Science* **212**:1291–1293.

Bickham, J. W., and Baker, R. J., 1977, Implications of chromosomal variation in *Rhogeessa* (Chiroptera: Vespertilionidae), *J. Mammal.* **58**:448–453.

Bickham, J. W., and Baker, R. J., 1979, Canalization model of chromosomal evolution, in: *Models and Methodologies in Evolutionary Theory, Bull. Carnegie Mus. Nat. Hist.* **13**:70–84.

Bickham, J. W., and Baker, R. J., 1980, Reassessment of the nature of chromosomal evolution in *Mus musculus, Syst. Zool.* **29**:159–162.

Brinkley, B. R., Valdivia, M. M., Toussom, A., and Brenner, S. L., 1984, Compound kinetochores of the Indian muntjac: Evolution by linear fusion of the unit kinetochores, *Chromosoma* **91**:1–11.

Brutlag, D. L., 1980, Molecular arrangement and evolution of heterochromatic DNA, *Annu. Rev. Genet.* **14**:121–144.

Bush, G. L., Case, S. M., Wilson, A. C., and Patton, J. L., 1977, Rapid speciation and chromosomal evolution in mammals, *Proc. Natl. Acad. Sci. USA* **74**:3942–3946.

Capanna, E., 1982, Robertsonian numerical variation in animal speciation: *Mus musculus*, an emblematic model, in: *Mechanisms of Speciation* (C. Baricozzi, ed.), Alan R. Liss, New York, pp. 155–177.

Capanna, E., Civitelli, M. V., and Cristaldi, M., 1977, Chromosomal rearrangements, reproductive isolation, and speciation in mammals. The case of *Mus musculus, Boll. Zool.* **44**:213–246.

Chesser, R. K., and Baker, R. J., 1986, On factors affecting the fixation of chromosomal rearrangements and neutral genes: Computer simulations, *Evolution* **40**:625–632.

Chesser, R. K., and Ryman, N., 1986, Inbreeding as a strategy in subdivided populations, *Evolution* **40**:616–624.

Committee for Standardization of Chromosomes of *Peromyscus*, 1977, Standardized karyotype of deer mice, *Peromyscus* (Rodentia), *Cytogenet. Cell Genet.* **19**:38–43.

Cothran, E. G., and Smith, M. H., 1983, Chromosomal and genic divergence in mammals, *Syst. Zool.* **32**:360–368.

Creau-Goldberg, N., Cochet, C., Turleau, C., and deGrouchy, J., 1981, Comparative gene mapping of man and *Cebus capucinus*: A study of 23 enzymatic markers, *Cytogenet. Cell Genet.* **31**:228–239.

Darwin, C., 1859, *The Origin of Species by Means of Natural Selection or the Preservation of Favoured Races in the Struggle for Life,* Murray, London.

Davis, K. M., Smith, S. A., and Greenbaum, I. S., 1986, Evolutionary implications of chromosomal polymorphism in *Peromyscus boylii* from southwestern Mexico, *Evolution* **40**:645–649.

Dresser, M. E., and Hamilton, A. E., 1979, Chromosomes of Lemuriformes. V. A comparison of the karyotypes of *Cheirogaleus medius* and *Lemur fulvus fulvus, Cytogenet. Cell Genet.* **24**:160–167.

Dutrillaux, B., 1979, Chromosomal evolution in primates: Tentative phylogeny from *Microcebus murinus* (prosimian) to man, *Hum. Genet.* **48**:251–314.

Dutrillaux, B., Couturier, J., and Viegas-Pequignot, E., 1981, Chromosomal evolution in primates, in: *Chromosomes Today*, Volume 7 (M. D. Bennett, M. Bobrow, and G. Hewitt, eds.), pp. 176–191.

Elder, F. F. B., 1980, Tandem fusion, centric fusions, and chromosomal evolution in the cotton rats, genus *Sigmodon*, *Cytogenet. Cell Genet.* **26:**199–210.

Endler, J. A., 1977, *Geographic Variation, Speciation and Clines*, Princeton University Press, Princeton, New Jersey.

Fraccaro, M., Lindstein, J., Ford, C. E., Iselius, L., Antonelli, A., Aula, P., Aurias, A., Bain, A. D., Bartsch-Sandhoff, M., Bernardi, F., Boud, E., Buchanan, L. F., Cameron, A. H., de la Chapelle, A., Ciuffa, G., Cuoco, C., Dutrillaux, B., Dutton, G., Ferguson-Smith, M. A., Francesconi, D., Gernaedts, J. P., Gimelli, G., Gueguen, J., Garsner, E., Hagemejer, A., Hansen, F. J., Holings, P. E., Hustinx, T. W. J., Kaakmen, A., van de Kamp, J. J. P., von Koskull, H., Lejeune, J., Lindenbaum, R. H., McCreanor, H. R., Mikkelsen, M., Mitelman, F., Nicoletti, B., Nilsby, I., Nilsson, A., Noel, B., Padavoni, E., Pasquali, F., de Pater, J., Pedersen, C., Petersen, F., Robson, E. B., Rotman, J., Ryynanen, M., Sachs, E., Salat, J., Smythe, R. H., Stabell, I., Subrt, I., Vampirelli, P., Wessner, G., Zergollern, L., and Zuffardi, O., 1980, The 11q;22q translocation: A European collaborative analysis of 43 cases, *Hum. Genet.* **56:**21–51.

Francke, U., and Taggart, R. T., 1980, Comparative gene mapping: Order of loci on the X chromosome is different in mice and humans, *Proc. Natl. Acad. Sci. USA* **77:** 3595–3599.

Futuyma, D. J., and Mayer, G. C., 1980, Non-allopatric speciation in animals, *Syst. Zool.* **29:**254–271.

Greenbaum, I. F., and Baker, R. J., 1978, Determination of the primitive karyotype for *Peromyscus*, *J. Mammal.* **59:**820–834.

Greenbaum, I. F., and Reed, M. J., 1984, Evidence of heterosynaptic pairing of the inverted segment in pericentric inversion heterozygotes of the deer mouse (*Peromyscus maniculatus*), *Cytogenet. Cell Genet.* **38:**106–111.

Greenbaum, I. F., Hale, D. W., and Fuxa, K. P., 1986, The mechanism of autosomal synapsis and the substaging of zygonema and pachynema from deer mouse spermatocytes, *Chromosoma* **93:**203–212.

Hafner, J. C., 1982, Genetic interactions between hybridizing cytotypes of the tent-making bat (*Uroderma bilobatum*), *Evolution* **35:**305–320.

Haiduk, M. W., 1983, Evolution in the family Pteropodidae (Chiroptera: Megachiroptera) as indicated by chromosomal and immunoelectrophoretic analyses, Ph. D. thesis, Texas Tech University, Lubbock, Texas.

Haiduk, M. W., and Baker, R. J., 1982, Cladistical analysis of G-banded chromosomes of nectar-feeding bats (Glossphaginae: Phyllostomidae), *Syst. Zool.* **31:**352–265.

Haiduk, M. W., and Baker, R. J., 1984, Scientific method, opinion, phylogenetic reconstruction and nectar-feeding bats: A response to Griffiths and Warner, *Syst. Zool.* **33:** 343–350.

Haiduk, M. W., Baker, R. J., Robbins, L. W., and Schlitter, D. A., 1981, Chromosomal evolution in African megachiroptera: G- and C-band assessment of the magnitude of change in similar standard karyotypes, *Cytogenet. Cell Genet.* **29:**221–232.

Haiduk, M. W., Sanchez Hernandez, C., and Baker, R. J., 1987, Relationships of *Nyctomys* and *Xenomys* to other cricetine genera based on data from G-banded chromosomes, *Southwest. Nat.*, in press.

Hall, W. P., 1983, Modes of speciation and evolution of the sceloporine iquanid lizards. I. Epistemology of comparative approach and introduction to the problem, in: *Advances in Herpetology and Evolutionary Biology*, Museum of Comparative Zoology, Harvard University, Cambridge, Massachusetts, pp. 643–679.

Harper, M. E., and Sanders, G. F., 1981, Localization of single copy DNA sequences on G-banded human chromosomes by *in situ* hybridization, *Chromosoma* **83**:431–439.

Harper, M. E., Ullrich, A., and Sanders, G. F., 1981, Localization of the human insulin gene to the distal end of the short arm of chromosome 11, *Proc. Natl. Acad. Sci. USA* **78**:4458–4460.

Hood, C. S., and Baker, R. J., 1986, G- and C-banding chromosomal studies of the bats of the family Emballonuridae, *J. Mammal.* **67**:705–711.

Hood, C. S., Robbins, L. W., Baker, R. J., and Shellhammer, H. S., 1984, Chromosomal studies and evolutionary relationships of an endangered species, *Reithrodontomys raviventris*, *J. Mammal.* **65**:655–667.

Hsu, T. C., and Arrighi, F. E., 1971, Distribution of constitutive heterochromatin in mammalian chromosomes, *Chromosoma* **34**:243–253.

Imai, H. T., 1983, Quantitative analysis of karyotype alteration and species differentiation in mammals, *Evolution* **37**:1154–1161.

Imai, H. T., and Crozier, R. H., 1980, Quantitative analysis of directionality of mammalian karyotype evolution, *Am. Nat.* **116**:537–569.

Imai, H. T., Maruyama, T., and Crozier, R. H., 1983, Rates of mammalian karyotype evolution by the karyograph method, *Am. Nat.* **121**:477–488.

John, B., and Miklos, G. L. A., 1979, Functional aspects of satellite DNA and heterochromatin, *Int. Rev. Cytol.* **58**:1–114.

King, M., 1982, A case for simultaneous multiple chromosome rearrangements, *Genetica* **59**:53–60.

King, M., 1985, The canalization model of chromosomal evolution: A critique, *Syst. Zool.* **34**:69–75.

Koop, B. F., Baker, R. J., and Genoways, H. H., 1983, Numerous chromosomal polymorphisms in a natural population of rice rats (*Oryzomys*: Cricetidae), *Cytogenet. Cell Genet.* **35**:131–135.

Koop, B. F., Baker, R. J., Haiduk, M. W., and Engstrom, M. D., 1984, Cladistical analysis of the primitive G-band sequences for the karyotype of the ancestor of the Cricetidae complex of rodents, *Genetica* **64**:199–208.

Koop, B. F., Baker, R. J., and Mascarello, J. T., 1985, Cladistical analysis of chromosomal evolution within the genus *Neotoma*, *Occ. Pap. Mus. Texas Tech Univ.* **96**:1–9.

Krohne, D. T., Dubbs, B. A., and Baccus, R., 1984, An analysis of dispersal of an unmanipulated population of *Peromyscus leucopus*, *Am. Midl. Nat.* **112**:146–156.

Lalley, P. A., Minna, J. D., and Francke, U., 1978, Conservation of autosomal gene synteny groups in mouse and man, *Nature* **174**:160–163.

Lande, R., 1979, Effective deme size during longterm evolution estimated from rates of chromosomal evolution, *Evolution* **33**:234–251.

Larson, A., Prager, E. M., and Wilson, A. C., 1984, Chromosomal evolution, speciation and morphological change in vertebrates: The role of social behavior, in: *Chromosomes Today*, Volume 8, (M. D. Bennett, A. Gropp, and U. Wolf, eds.), George Allen and Unwin, London, pp. 215–228.

Ma, N. S. F., Simeone, T., McLean, J., and Parham, P., 1982, Chromosomal localization and gene synteny of the major histocompatibility complex in the owl monkey *Aotus*, *Immunogenetics* **15**:1–16.

Marks, J., 1983, Rates of karyotypic evolution, *Syst. Zool.* **32**:207–209.

Mascarello, J. T., and Hsu, T. C., 1976, Chromosome evolution in woodrats, genus *Neotoma* (Rodentia: Cricetidae), *Evolution* **30**:152–169.

Mascarello, J. T., Stock, A. D., and Pathak, S., 1974, Conservatism in the arrangement of genetic material in rodents, *J. Mammal.* **55**:695–704.

May, R. M., Endler, J. A., and McMurtrie, R. E., 1975, Gene frequency clines in the presence of selection opposed by gene flow, *Am. Nat.* **109**:659–676.

McClintock, B., 1978, Mechanisms that rapidly reorganize the genome, *Stadler Genet. Symp. Univ. Missouri* **10**:25–47.

Miklos, G. L. G., Willcocks, D. A., and Baverstock, P. R., 1980, Restriction endonuclease and molecular analysis of three rat genomes with special reference to chromosomal rearrangement and speciation problems, *Chromosoma* **76**:339–363.

Modi, W. S., 1986, Karyotypic differentiation among two sibling species pairs of New World microtine rodents, *J. Mammal.* **67**:159–165.

Modi, W. S., and Lee, M. R., 1985, Systematic implications of chromosomal banding analyses of populations of *Peromyscus truei* (Rodentia: Muridae), *Proc. Biol. Soc. Wash.* **97**:716–723.

Nelson, K., Baker, R. J., Shellhammer, H. S., and Chesser, R. K., 1984, Test of alternative hypothesis concerning the origin of *Reithrodontomys raviventris:* Genetic analysis, *J. Mammal.* **65**:688–673.

Pathak, S., 1976, Chromosome banding techniques, *J. Reprod. Med.* **17**:15–18.

Patton, J. C., and Baker, R. J., 1978, Chromosomal homology and evolution in phyllostomatoid bats, *Syst. Zool.* **27**:449–462.

Patton, J. L., and Sherwood, S. W., 1982, Genome evolution in pocket gophers (genus *Thomomys*). I. Heterochromatin variation and speciation potential, *Chromosoma* **85**: 149.

Patton, J. L., and Sherwood, S. W., 1983, Chromosome evolution and speciation in rodents, *Annu. Rev. Ecol. Syst.* **14**:139–158.

Qumsiyeh, M. B., 1986a, Chromosomal evolution in the rodent family Gerbillidae, Ph. D. thesis, Texas Tech University, Lubbock, Texas.

Qumsiyeh, M. B., 1986b, Phylogenetic studies of the rodent family Gerbillidae: I. Chromosomal evolution in the southern African complex, *J. Mammal.* **67**:680–692.

Qumsiyeh, M. B., and Baker, R. J., 1985, G- and C-banded karyotypes of the Rhinopomatidae (Microchiroptera), *J. Mammal.* **66**:541–544.

Ralls, K., Harvey, P. H., and Lyles, A. M., 1986, Inbreeding in natural populations of birds and mammals, in: *Conservation Biology: The Science of Scarcity and Diversity.* (M. E. Soule, ed.), Shaver Associates, Inc., Sunderland, Mass., pp. 35–36.

Robinson, T. J., Elder, F. F. B., and Chapman, T. A., 1983, Evolution of chromosomal variation in cottontails, genus *Sylvilagus* (Mammalia: Lagomorpha). I. *Sylvilagus aquaticus, S. floridanus,* and *S. transitionalis, Cytogenet. Cell Genet.* **35**:216–222.

Robinson, T. J., Elder, F. F. B., and Chapman, T. A., 1984, Evolution of chromosomal variation in cottontails, genus *Sylvilagus* (Mammalia: Lagomorpha). II. *Sylvilagus audibonii, S. idahoensis, S. nuttallii,* and *S. palustris, Cytogenet. Cell Genet.* **38**: 282–289.

Robbins, L. W., and Baker, R. J., 1981, An assessment of the nature of chromosome rearrangements in 18 species of *Peromyscus* (Rodentia: Cricetidae), *Cytogenet. Cell Genet.* **31**:194–202.

Rofe, R., and Hayman, D., 1985, G-banding evidence for a conserved complement in the Marsupialia, *Cytogenet. Cell Genet.* **39**:40–50.

Rogers, D. S., 1983, Phylogenetic affinities of *Peromyscus* (*Megadontomys*) *thomasi:* Evidence from differentially stained chromosomes, *J. Mammal.* **64**:617–623.

Rogers, D. S., Greenbaum, I. F., Dunn, S. J., and Engstrom, M. D., 1984, Cytosystematic value of chromosomal inversion data in the genus *Peromyscus* (Rodentia: Cricetidae), *J. Mammal.* **65**:457–465.

Ryder, O., Epel, N., and Benirschke, K., 1978, Chromosome banding studies of the Equidae, *Cytogenet. Cell Genet.* **20:**323–350.

Rumpler, Y., Courturier, J., Warter, S., and Dutrillaux, B., 1983, Chromosomal evolution in Malagasy lemurs, *Cytogenet. Cell Genet.* **36:**542–546.

Sawyer, J. R., and Hozier, J. C., 1986, High resolution of mouse chromosomes: banding conservation between man and mouse, *Science* **232:**1632–1635.

Searle, J. B., 1984, Three new karyotypic races of the common shrew *Sorex araneus* (Mammalia: Insectivora) and a phylogeny, *Syst. Zool.* **33:**184–194.

Shaw, D. D., Wilkinson, P., and Coates, D. J., 1983, Increased chromosomal mutation rate after hybridization between two subspecies of grasshoppers, *Science* **220:** 1165–1167.

Sherwood, S. W., and Patton, J. L., 1982, Genome evolution in pocket gophers (genus *Thomomys*). II. Variation in cellular DNA content, *Chromosoma* **85:**163–179.

Shi, L. M., Ye, Y. Y., and Duan, X., 1980, Comparative cytogenetic studies on the red muntjac, chinese muntjac, and their F_1 hybrids, *Cytogenet. Cell Genet.* **96:**22–27.

Shields, G. F., 1982, Comparative avian cytogenetics, A review, *Condor* **84:**45–58.

Singer, M. F., 1982, Highly repetitive sequences in mammalian genomes, *Int. Rev. Cytol.* **76:**67–112.

Sites, J. W., Jr., 1983, Chromosome evolution in the iquanid lizard *Sceloporus grammicus*. I. Chromosome polymorphisms, *Evolution* **37:**38–53.

Sites, J. W., Jr., Bickham, J. W., and Haiduk, M. W., 1981, Conservative chromosomal change in the bat family Mormoopidae, *Can. J. Gen. Cytol.* **23:**459–467.

Slatkin, M., 1973, Gene flow and selection in a cline, *Genetics* **75:**733–756.

Stallings, R. L., Munk, A. C., Longmire, J. L., Jeff, J. H., Wilder, M. E., Sciliano, M. J., Adair, G. M., and Crawford, B. D., 1985, Oncogenes and linkage groups, Conservation during mammalian chromosome evolution, *Chromosoma* **92:**256–163.

Stangl, F. B., Jr., 1986, Dynamics of chromosomal variation between chromosomally characterized races of *Peromyscus leucopus* (Rodentia: Cricetidae), *J. Mammal.* **67:** 465–473.

Stangl, F. B., Jr., and Baker, R. J., 1984, Evolutionary relationships in *Peromyscus:* Congruence in chromosomal, genic, and classical data sets, *J. Mammal.* **65:**643–654.

Stubblefield, E., 1980, Chromosome bands and the subunit structure of Chinese hamster metaphase chromosomes, *Cytogenet. Cell Genet.* **26:**191–198.

Tegelstrom, H., Ebenhard, T., and Ryttman, H., 1983, Rate of karyotype evolution and speciation in birds, *Hereditas* **98:**235–239.

Viegas-Pequignot, E., Koiffman, C. D., and Dutrillaux, B., 1985, Chromosomal phylogeny of *Lagothrix, Brachyteles,* and *Cacajao, Cytogenet. Cell Genet.* **39:**99–104.

White, M. J. D., 1975, Chromosomal repatterning—Regularities and restrictions, *Genetics* **79:**63–72.

White, M. J. D., 1978, *Modes of Speciation,* Freeman, San Francisco.

Wilson, A. C., Maxson, L. R., and Sarich, V. M., 1974, The importance of gene rearrangement in evolution: Evidence from studies on rates of chromosomal, protein, and anatomical evolution, *Proc. Natl. Acad. Sci. USA* **71:**3028–3030.

Wilson, A. C., Bush, G. L., Case, S. M., and King, M. C., 1975, Social structuring of mammalian populations and rate of chromosomal evolution, *Proc. Natl. Acad. Sci. USA* **72:**5061–5065.

Wilson, A. C., Cann, R. L., Carr, S. M., George, M., Gyllensten, V. B., Helm-Bychouski, R. M., Higguchi, R. G., Palumbi, S. R., Prager, E. M., Sage, R. D., Stoneking, M., 1985, Mitochondrial DNA and two perspectives on evolutionary genetics, *Biol. J. Linnean Soc.* **26:**375–400.

Wright, S., 1941, On the probability of fixation of reciprocal translocations, *Am. Nat.* **75:** 513–522.

Wurster-Hill, D. H., and Bush, M., 1980, The interrelationships of chromosome banding patterns in the giant panda (*Ailuropoda melanoleuca*), hybrid bear (*Ursus middendorfi* × *Thalarctos maritimus*), and other carnivores, *Cytogenet. Cell Genet.* **27:** 147–154.

Wurster-Hill, D. H., and Centerwall, W. R., 1982, The interrelationships of chromosome banding patterns in canids, mustelids, hyena, and felids, *Cytogenet. Cell Genet.* **34:** 178–192.

Wurster-Hill, D. H., and Gray, C. W., 1973, Giemsa banding patterns in the chromosomes of twelve species of cats (Felidae), *Cytogenet. Cell Genet.* **12:**377–397.

Wurster-Hill, D. H., and Gray, C. W., 1975, The interrelationships of chromosome banding patterns in procyonids, viverrids, and felids, *Cytogenet. Cell Genet.* **15:**306–331.

Yates, T. L., Baker, R. J., and Barnett, R. K., 1979, Phylogenetic analysis of karyological variations in three genera of peromyscine rodents, *Syst. Zool.* **28:**40–48.

Yunis, J. J., 1980, Nomenclature for high resolution human chromosomes, *Cancer Genet. Cytogenet.* **2:**221–229.

Yunis, J. J., 1981, Mid-prophase human chromosomes. The attainment of 2000 bands, *Hum. Genet.* **56:**293–298.

Yunis, J. J., and Prakash, O. M., 1982, The origin of man: A chromosomal pictorial legacy, *Science* **215:**1505–1530.

CHAPTER 3

THE ORIGIN OF RODENTS AND LAGOMORPHS

LI CHUAN-KUEI, ROBERT W. WILSON, MARY R. DAWSON, and LEONARD KRISHTALKA

1. INTRODUCTION

Although the two major groups of gliriform mammals, the rodents and lagomorphs, are frequently grouped together by nonspecialists, the opinion of both paleontologists and neomammalogists has long been that the two orders are not closely related. Now, new evidence on this question has come from recent fossil discoveries in China and from reevaluation of other evidence, especially developmental. After a review of the past interpretations and new evidence, we return to the view that the Lagomorpha and Rodentia can be traced to a common ancestral group and that the Cohort Glires is a valid taxonomic unit.

2. RODENT ORIGINS

The origin of rodents has long been uncertain. Simpson (1945, p. 198) said, "There is not, even among fossils, any question as to whether

LI CHUAN-KUEI • Institute of Vertebrate Paleontology and Paleoanthropology, Academia Sinica, Beijing, China. ROBERT W. WILSON • Museum of Natural History, University of Kansas, Lawrence, Kansas 66045. MARY R. DAWSON and LEONARD KRISHTALKA • Section of Vertebrate Fossils, Carnegie Museum of Natural History, Pittsburgh, Pennsylvania 15213.

a given animal is or is not a rodent, however doubtful its position in the order may be." The mere fact that rodents appear full-blown at their earliest appearance suggests trouble in searching for a source of origin, whether by the fossil record or by systematic analysis. Most writers, if faced with the need for a statement, have simply written of the origin of rodents from primitive eutherians, from insectivores, or from proteutherians. Thus, for example, Olson (1971) concluded, "As currently viewed the Order Rodentia appears to be strongly isolated from all other placentals. The source is not certain, but presumably it lay among insectivores." McKenna (1975), in his cladistic classification of the Mammalia, placed the Rodentia as *incertae sedis* in the Cohort Epitheria. Szalay (1977) wrote, "Rodents are still very much an *incertae sedis* group compared to other Mammalia."

Older attempts to place the origin of rodents more precisely were mostly without much success. Osborn (1902) established the Proglires (with Family Mixodectidae), as the name indicates, to point to a more specific origin. Forsyth Major (1893), Hinton (1926), and Friant (1932) argued for multituberculate origin. Neither view is now supportable. More recently, primitive primates, or at least insectivores near the taxonomic border of primates, have been suggested as being ancestral to rodents. One may look to McKenna (1961), Van Valen (1966), and Lillegraven (1969) for at least tentative support for this idea. Although it was the popular view for a time, probably only Wood would wish to maintain this relationship at present. In 1977 his position on the origin of rodents was: "By elimination, southeastern North America becomes the most probable area, and the order may be presumed to have arisen (most probably from primates) in the forests of that area during the middle Paleocene" (Wood, 1977). As late as 1982 he (in Patterson and Wood, 1982, p. 454) repeated this view, and denied the affinity of the eurymylid *Heomys* with rodents.

The Eurymylidae, from the early Tertiary of Asia, are the latest candidates for being rodent ancestors, especially the middle and late Paleocene *Heomys*, the most rodentlike of the eurymylids. Li (1977), in describing this genus, recognized that it had rodent characteristics, but he was not sure whether or not the resemblance was phylogenetic or the result of convergence. Following his description of *Heomys*, a rodent relationship was suggested by Gingerich and Gunnell (1979), Gingerich and McKenna (1980), Hartenberger (1980), Dawson et al. (1984), and Korth (1984). However, even the most recent discussions of eurymylid–rodent affinities, found in a series of stimulating papers (Luckett and Hartenberger, 1985), do not reach a clear consensus, although several authors (Li and Ting, 1985; Luckett, 1985; Novacek, 1985) strongly support the concept.

3. LAGOMORPH ORIGINS

Authorities have been equally uncertain concerning lagomorph origins. Gidley (1906) and Ehik (1926) suggested an ancestry that ultimately involved the Mesozoic triconodonts. Hürzeler (1936) considered artiodactyl relationships as one result of his work on the European artiodactyl *Cainotherium*. Wood (1957) suggested a condylarth, especially periptychid, origin for lagomorphs. Some authorities now maintain that the lagomorphs should be included with the Asian Anagalida of Szalay and McKenna (1971), as later expanded by McKenna (1975). The only living mammals that have been suggested for inclusion in this group (other than the Rodentia and Lagomorpha) are the macroscelidids (elephant shrews), and this is far from certain (McKenna, 1975). According to Dene et al. (1982), the tree-shrew, *Tupaia*, shows a close relationship to living lagomorphs in myoglobin. In addition, both *Tupaia* and the lagomorphs have a prepenial scrotum, but because this is also a marsupial character, it seems to be a primitive feature. Much of the previously cited serological evidence for the origin and relationships of the Lagomorpha seems ambivalent.

When Matthew and Granger (1925) described *Eurymylus* (from the latest Paleocene of Mongolia) they referred it with question to the Plesiadapidae, an early Tertiary family of primates. At the same time they assigned another new Mongolian genus, *Baënomys*, to the Glires and noted similarities to the Lagomorpha. In 1929, Matthew et al. (1929) recognized that *Eurymylus* and *Baënomys* were synonyms and established the Family Eurymylidae. Wood (1942) concluded that the Eurymylidae should be recognized as an aberrant family of the Lagomorpha, where it remained until 1971, when Sych (1971) placed it in its own order, Mixodontia. Li (1977) described additional eurymyloids (= mimotonids, rhombomylids, and eurymylids of this paper), which, with few exceptions, show distinctive lagomorph and rodent trends, however they may be interpreted.

4. ADVOCATES OF ESSENTIALLY INDEPENDENT ORIGIN

In 1912, Gidley (1912) offered considerable evidence for the separation of rodents and lagomorphs, which convinced most North American workers that rodents and lagomorphs were simply convergent from a general eutherian base. Thus, Wilson (1949, p. 130) could say, "We now know that they (the lagomorphs) constitute an independent order, with little or perhaps nothing to unite them with the rodents," and "The Lagomorpha must be a very ancient order, distinct from other mammalian orders as far back as early Paleocene, or more probably Cretaceous times." Dawson (1967, p. 287) has written, "Though often united popularly with the other main group of recent gliriform mammals, the Order Rodentia, lagomorphs probably originated from a different ancestral stock and have had an entirely different pattern of development." McKenna (1975) thought he could position the lagomorphs among selected eutherian groups, but left the rodents *incertae sedis* in his Cohort Epitheria. Eastman and Bleefeld (1980) could find no synapomorphies among the Recent rodents and lagomorphs (but see Landry, 1974) and no real evidence to unite the two groups, but in the abstract cited they did not mention the eurymyloids.

5. ADVOCATES OF A SPECIAL RELATIONSHIP BETWEEN RODENTS AND LAGOMORPHS

The general opinion of 19th century naturalists was that rodents and lagomorphs shared a special relationship. This is seen in the usual classification as Glires with suborders Simplicidentata (rodents) and Duplicidentata (lagomorphs). Tullberg (1899), however, in his monograph was clearly doubtful about this relationship. He decided, finally, to keep them together on the basis of embryologic resemblances, and that they were descended from a common ancestor. Simpson (1945), in his classification of the Mammalia, grouped Rodentia and Lagomorpha as sole members of the Cohort Glires, an arrangement that, as he says, "is permitted by our ignorance rather than sustained by our knowledge" (p. 196). For many years, Stuart Landry, Jr. was the only active defender in North America of a special relationship. He listed six synapomorphies which he believed united the Lagomorpha and Rodentia (Landry, 1974): (1) a distinctive type of placentation, (2) movement of the jaw joint forward and upward away from the ear, (3) lip folds which push in behind the incisors (not found in other gliriform animals; for example, *Daubentonia*), (4) possession of a lateral alisphenoid canal for the anterior branch of the fifth nerve, (5) ability to pull a fold of skin to cover the rhinarium, and (6) coprophagy combined with a large cecum specialized for the production of night feces. Only points 2 and 4 are applicable to fossils, and do not seem to offer much in favor of union.

Fossil discoveries in China of the past 10 years (especially as described by Li) have made us rethink the relationship of rodents to lagomorphs, and these discoveries have for the first time furnished us with a reasonable and immediate ancestor for the two groups. These fossils—mimotonids, rhombomylids, and eurymylids (= the eurymyloids of Li, 1977)—are central Asiatic mammals ranging in age from middle or late Paleocene to ?Oligocene. They offer the best evidence among fossils for a special relationship between rodents and lagomorphs (Li, 1977; Hartenberger, 1977; Li and Yan, 1979; Li and Ting, 1985; Novacek, 1985).

6. MIMOTONIDS, RHOMBOMYLIDS, AND EURYMYLIDS

The history of discovery and knowledge of the eurymyloids, as previously mentioned, dates back to 1925. To repeat, at that time Matthew and Granger (1925) described upper jaw material as *Eurymylus*, and lower jaw specimens as *Baënomys* from a late Paleocene locality in Outer Mongolia. The names were later synonymized by Matthew et al. (1929) under *Eurymylus*, and a new family, Eurymylidae, was named. Tentative reference was made to ?Glires, an ordinal name including both rodents and lagomorphs. Wood (1942) referred the family to the Lagomorpha, as the oldest known but already aberrant lagomorphs. Bohlin (1951) described a second genus of possible eurymyloid as *Mimolagus*, referring the genus to the Duplicidentata (= Lagomorpha). In 1971, Sych (1971) established the new order Mixodontia for *Eurymylus* alone, maintaining that this genus was neither rodent nor lagomorph.

In the 1970s vertebrate paleontologists of the Institute of Vertebrate Paleontology and Paleoanthropology, Academia Sinica, including one of the present authors (L. C.-K.), conducted field work in China which resulted in recovery of a variety of eurymyloid specimens from several different geologic horizons. Eight genera have been named, although the validity of several may be questioned (*Mimotona* and *Gomphos* may be synonyms; *Hypsimylus* may be a lagomorph in the restricted sense). In general, the geologically younger genera tend to be more lagomorphlike (*Mimolagus*, possibly *Hypsimylus*) or even more specialized (*Rhombomylus*). Most interest attaches to the two oldest genera, *Heomys* and *Mimotona* of the late Paleocene, or perhaps middle Paleocene. Li (1977) named and described these genera, placing *Heomys* in the Eurymylidae and *Mimotona* in the new family, Mimotonidae. The former genus was thought to have greater affinities with the rodents, the latter with the lagomorphs.

Heomys (Fig. 1) is certainly the mammal closest to the ancestry of the Rodentia that is known, although it seems to be somewhat too specialized to be a direct ancestor of early rodents such as the Ischyromyidae and Ctenodactyloidea. This is seen, for example, in the degree

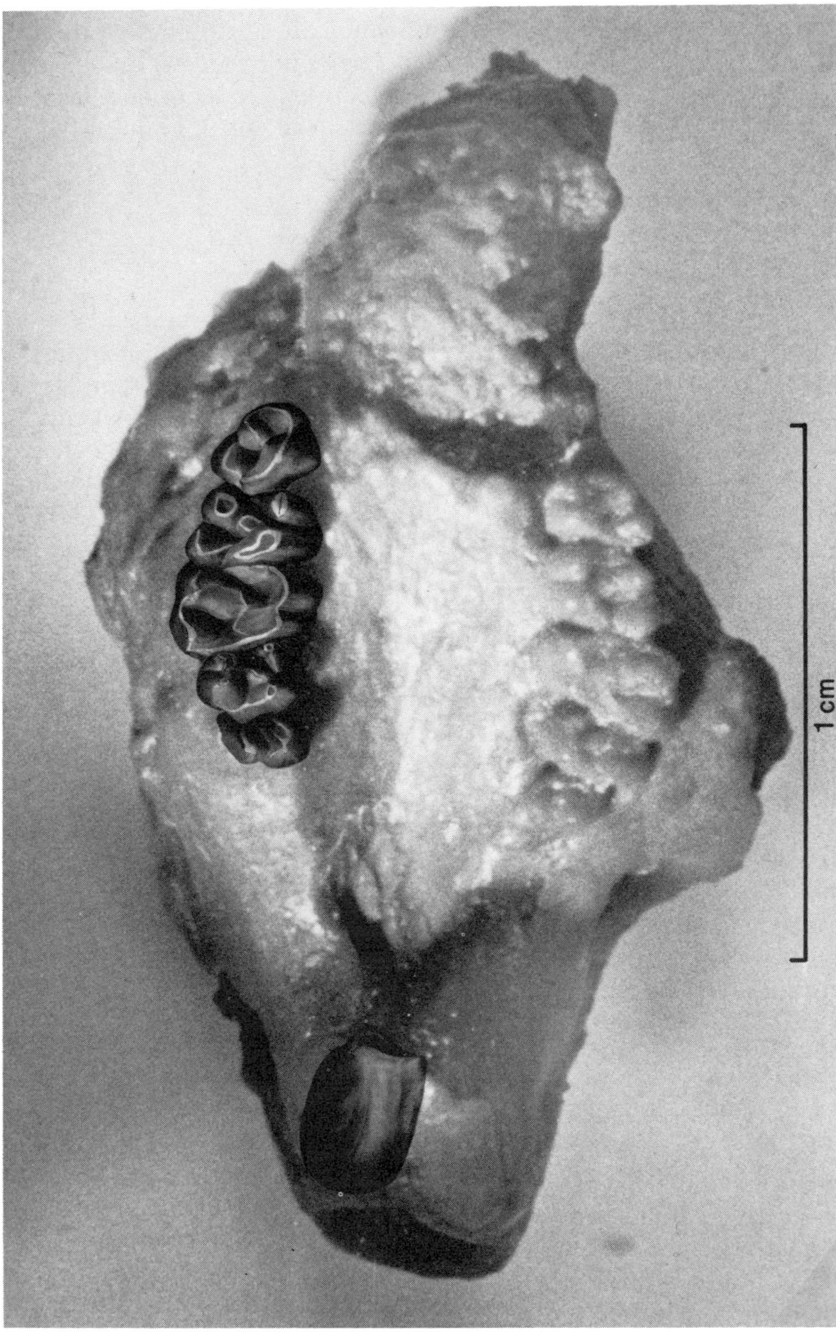

FIGURE 1. *Heomys orientalis* type, IVPP V4321. Palatal view of skull with left P^3–M^3 retouched (Li, 1977).

of anteroposterior compression of lower cheek-tooth trigonids, and in the somewhat too hypsodont crowns. The general trend in eurymylid evolution is also away from early rodents, as seen in: (1) transversely elongate upper-cheek teeth; (2) obscuring of conules (proto- and metaconules); (3) development of hypocone basins; (4) tooth wear indicating transverse jaw motion in all but the earliest taxa; and (5) inflected angular process—characters that are present in one or another of the later genera.

On the other hand, characters aligning *Heomys*, other eurymylids, and rhombomylids with rodents are several (see Fig. 2 and discussion below). In *Heomys* the patterns of P^4–M^1 are reasonably close to that in the earliest (early Eocene) ctenodactyloids. Several characters of *Heomys* are simply primitive (for example, large coronoid process of lower jaw, small infraorbital foramen), but do not disqualify *Heomys* from approximate rodent ancestry. Hence, although too specialized to be in the direct line of descent to the earliest rodents, *Heomys* is close to that position, and there is no reason not to think that as yet unknown members of the Eurymylidae actually are ancestral.

Mimotona, of the Family Mimotonidae, has a dental formula of 2⁄2, 0⁄0, 3⁄2, 3⁄3, and hence a formula which could be antecedent to that of the Lagomorpha (2⁄1, 0⁄0, 3⁄2, 3⁄3). *Mimotona* may well be antecedent to

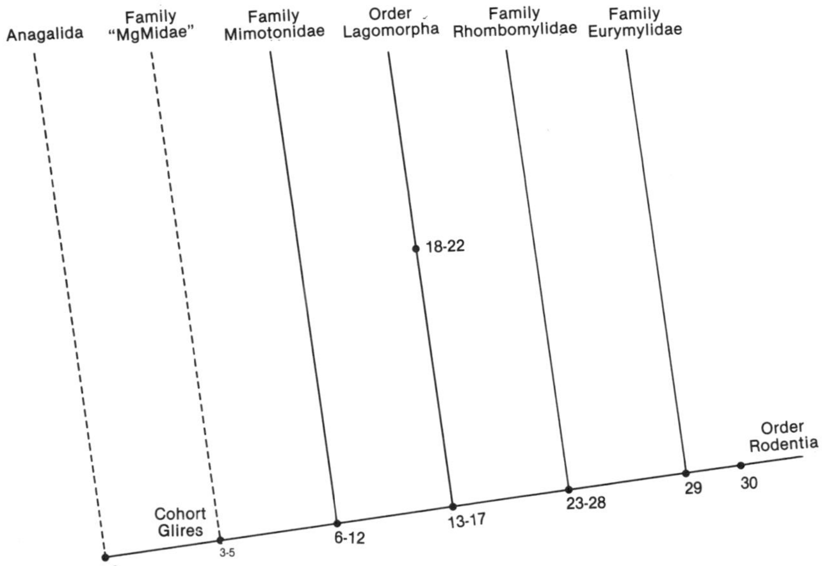

FIGURE 2. Hypothesized relationships among Anagalida and Glires. See text for discussion of characters.

Mimolagus, which possesses several characters of resemblance to the Lagomorpha other than dental formula—for example, enamel of incisors only imperfectly differentiated into two layers, wear on large upper incisors much like that in later lagomorphs in having a definite shelf (related to the presence of a second pair of incisors?), and premolars robust rather than relatively small as in *Eurymylus.* These features appear to be parallelisms among the Duplicidentata (Mimotonida and Lagomorpha).

7. RELATIONSHIPS

An attempt to organize the many relevant characters of the Anagalida, Eurymylidae, Mimotonidae, Lagomorpha, and Rodentia results in the cladogram shown in Fig. 2. The characters—many of them derived from papers in Luckett and Hartenberger (1985)—and suggested relationships are as follows: The entire clade (Anagalida plus Glires) appears to be united by (1) some unilateral hypsodonty and (2) bulla solely ectotympanic, features that will be tested as more anagalids are described. All Glires share (3) I1/1 and I2/2 (permanent) lost, (4) dI2/2 retained, enlarged and evergrowing, with dI_2 extending back to below M_2 or M_3 (extrapolation from living rodents and lagomorphs; see Luckett, 1985), and (5) molar trigonids compressed. One specimen assigned to *Barunlestes* (MgM I/135) may exhibit these conditions (Li and Ting, 1985). Remaining Glires are more derived in having (6) dI2/2 enamel restricted to the anterior face in the adult (Korventkontio, 1934; Grevstad and Selvig, 1986), (7) C1/1 and P1/1 lost (I3/3—P2/2 diastema), (8) enlarged incisive foramina, (9) enlarged premaxilla that makes contact with the frontal, (10) orbits shifted anteriorly relative to cheek teeth, (11) molar paraconids reduced to absent, and (12) loss of the calcaneofibular facet (and contact of the two bones)—conditions exhibited by the Mimotonidae (*Mimotona, Mimolagus, ?Hypsimylus*). Lagomorpha, Mixodontia, and Rodentia are more derived in that (13, 14) I_3 and P_2 are lost, (15) the lacrymal–palatine contact is absent, (16) the orbitosphenoid is large and located high in the orbit, and (17) the infraorbital canal is short. Lagomorphs possess these features, but are further specialized in having (18) a grooved dI^2, (19) a reduced temporal fossa and coronoid process, (20) a reduced or absent M_3, (21) a secondarily acquired calcaneofibular facet (Szalay, 1985, p. 120), and (22) a fused tibia and fibula. Remaining Glires—rhombomylids, eurymylids, and rodents—form the Simplicidentata, in which (23) I^3 and (24) P^2 are lost,

(25) the orbital fossa is shifted even more anteriorly, (26) the glenoid fossa is elongated longitudinally, (27) the orbital wing of the alisphenoid is large and elongated, and (28) the accessory foramen ovale is present—features preserved in *Rhombomylus*. Eurymylids and rodents are more derived in having (29) dI2/2 with two enamel layers, and rodents are further differentiated by (30) the absence of P_3, and many other features.

It should be noted that this analysis is based heavily on features shown by fossils. Other evidence in favor of the affinity between Lagomorpha and Rodentia, especially that from morphological characters not known from the fossil record and on developmental features (see, for example, Luckett, 1985, Novacek, 1985), was not utilized in this primarily paleontological analysis. The cladogram, however, indicates that rodents and lagomorphs share a close relationship, indeed a common ancestry—a view that had been all but abandoned in the past half century. The dichotomy would date at least from the Paleocene, and the long Eocene separation would be sufficient to establish the outstanding osteological differences between lagomorphs and rodents evident by the Oligocene in such leporids as *Palaeolagus haydeni*.

Presenting these affinities in a formal classification produces the acknowledged problems inherent in a cladistic classification. Also, although the cladogram depicts the differences between the Mimotonidae and the Eurymylidae, the two families are very similar in cheek-tooth pattern. *Rhombomylus* is isolated from eurymylids largely as a result of one character, its one-layered incisor enamel. The primitive or derived state of the layers of enamel is still not entirely clear (von Koenigswald, 1985), and further study may indicate that this character has a different meaning. Finally, *Heomys* has many rodentlike features in its dentition, but the character of retained P_3 is used here to indicate that its affinities are still within the Eurymylidae.

Our suggested classification once again returns the older concept "Glires" to respectability.

Cohort Glires
- Superorder Duplicidentata
  - Order Mimotonida (?late Cretaceous, late Paleocene–late Eocene, ?early Oligocene, Asia)
    - Family "MgMidae" (for MgM I/135, "Barunlestes")
    - Family Mimotonidae
      - *Mimotona* (? = *Gomphos*)
      - *Mimolagus*
      - *?Hypsimylus*

Order Lagomorpha (late Eocene, Asia, North America; Oligocene, Europe; Miocene, Africa; Recent, worldwide except Antarctica)

Superorder Simplicidentata

Order Mixodontia (mid-Paleocene–mid-Eocene, Asia)

Family Rhombomylidae

Rhombomylus

Matutinia

Family Eurymylidae

Eurymylus

Heomys

Suborder Rodentia (late Paleocene, North America; early Eocene, Eurasia; early Oligocene, Africa; Oligocene, South America; Recent, worldwide except Antarctica)

ACKNOWLEDGMENTS. We thank Nancy Perkins and Melinda McNaugher (Carnegie Museum of Natural History) for preparing Fig. 1, M. C. McKenna for noting that MgM I/135 is eurymyloidlike, Richard Stucky for his comments on the manuscript, and Hugh Genoways for a generous show of patience.

REFERENCES

Bohlin, B., 1951, Some mammalian remains from Shih-ehr-ma-cheng, Kui-hui-pu area, Western Kansu, *Rep. Sino-Swedish Expedition, Vertebr. Paleontol.* **5**:1–47.

Dawson, M. R., 1967, Lagomorph history and the stratigraphic record, in: *Essays in Paleontology and Stratigraphy, R. C. Moore Commemorative Volume* (C. Teichert and E. L. Yockelson, eds.), Department of Geology, University of Kansas Special Publication 2, University of Kansas Press, Lawrence, Kansas, pp. 287–316.

Dawson, M. R., Li, C.-K., and Qi, T., 1984, Eocene ctenodactyloid rodents (Mammalia) of Eastern and Central Asia, in: *Papers in Vertebrate Paleontology Honoring Robert Warren Wilson* (R. M. Mengel, ed.), Carnegie Museum of Natural History Special Publication 9, pp. 138–150.

Dene, H., Goodman, M., McKenna, M. C., and Romero-Herrera, A. E., 1982, *Ochotona princeps* (pika) myoglobin: An appraisal of lagomorph phylogeny, *Proc. Natl. Acad. Sci. USA* **79**:1917–1920.

Eastman, C. B., and Bleefeld, A. R., 1980, Lagomorph and rodent affinities reexamined (Abstract), in: *Program*, Annual Meeting of the American Society of Mammalogists, University of Rhode Island.

Ehik, J., 1926, The right interpretation of the cheekteeth tubercles of *Titanomys*, *Ann. Mus. Nat. Hung.* **23**:178–186.

Friant, M., 1932, Contribution à l'étude de la differentiation des dents jugales chez les mammifères, *Publ. Mus. Hist. Nat.* **1**:1–132.

Gidley, J. W., 1906, Evidence bearing on tooth-cusp development, *Proc. Wash. Acad. Sci.* **8**:91–110.

Gidley, J. W., 1912, The lagomorphs as an independent order, *Science* **36**(922):285–286.

Gingerich, P. D., and Gunnell, G. F., 1979, Systematics and evolution of the genus *Esthonyx* (Mammalia, Tillodontia) in the early Eocene of North America, *Contrib. Mus. Paleontol. Univ. Michigan* **25**:125–153.

Gingerich, P. D., and McKenna, M. C., 1980, Mammalian paleontology in China, *News Bull. Soc. Vertebr. Paleontol.* **118**:42–44.

Gregory, W. K., 1910, The Orders of Mammals, *Bull. Am. Mus. Nat. Hist.* **27**:1–524.

Grevstad, H. J., and Selvig, K. A., 1986, Ultrastructure of lingual enamel epithelium in rabbit permanent incisors, *Scand. J. Dent. Res.* **93**:409–420.

Hartenberger, J.-L., 1977, À propos de l'origine des rongeurs, *Géobios Mem. Spec.* **1**:183–193.

Hartenberger, J.-L., 1980, Données et hypothèses sur la radiation initiale des rongeurs, in: *Palaeovertebrata, Memoire Jubilaire en homage à René Lavocat,* (Jacques Michaux, ed.), Paleovertebrata Festschrift, pp. 285–301.

Hinton, M. A. C., 1926, *Monograph of Voles and Lemmings,* Vol. 1, British Museum (Natural History).

Hürzeler, J., 1936, Osteologie und Odontologie der Caenotheriden, *Abh. Schweiz Palaeontol. Ges.* **58**:1–89.

Korth, W. W., 1984, Earliest Tertiary evolution and radiation of rodents in North America, *Bull. Carnegie Mus. Nat. Hist.* **24**:1–71.

Korvenkontio, V. A., 1934, Mikroskopische Untersuchungen an Nagerincisiven unter Hinweis auf die Schmelzstruktur der Backenzahne, *Ann. Zool. Soc. Zool.-Bot. Fenn. Vanamo* **2**:1–274.

Landry, S. O., Jr., 1974, The fundamental relationship of the Lagomorpha and Rodentia (Abstract), in: *First International Theriological Congress Transactions,* Volume 2, N–Z, Moscow, pp. 202–203.

Li, C.-K., 1977, Paleocene eurymyloids (Anagalida, Mammalia) of Qianshan, Anhui, *Vertebr. PalAsiat.* **15**(2):103–118.

Li, C.-K., and Ting, S.-Y., 1985, Possible phylogenetic relationship of Asiatic eurymylids and rodents, with comments on mimotonids, in: *Evolutionary Relationships among Rodents* (W. P. Luckett and J.-L. Hartenberger, eds.), Plenum Press, New York, pp. 35–58.

Li, C.-K., and Yan, D.-F., 1979, The systematic position of eurymylids (Mammalia) and the origin of Rodentia, in: *Abstracts of Papers, 12th Annual Conference and 3rd National Congress Paleontological Society of China,* pp. 155–156.

Lillegraven, J. A., 1969, Latest Cretaceous mammals of upper part of Edmonton Formation of Alberta, Canada, and review of marsupial–placental dichotomy in mammalian evolution, *Univ. Kans. Paleontol. Contrib. Vertebr.* **12**:1–122.

Luckett, W. P., 1985, Superordinal and intraordinal affinities of rodents: Developmental evidence from the dentition and placentation, in: *Evolutionary Relationships among Rodents* (W. P. Luckett and J.-L. Hartenberger, eds.), Plenum Press, New York, pp. 227–276.

Luckett, W. P., and Hartenberger, J.-L. (eds.), 1985, *Evolutionary Relationships among Rodents,* Plenum Press, New York.

Major, C. J. F., 1893, On some Miocene squirrels, *Proc. Zool. Soc. Lond.* **1893**:179–215.

Matthew, W. D., and Granger, W., 1925, Fauna and correlation of the Gashato Formation of Mongolia, *Am. Mus. Novit.* **189**:1–12.

Matthew, W. D., Granger, W., and Simpson, G. G., 1929, Additions to the fauna of the Gashato Formation of Mongolia, *Am. Mus. Novit.* **376**:1–12.

McKenna, M. C., 1961, A note on the origin of rodents, *Am. Mus. Novit.* **2037**:1–5.

McKenna, M. C., 1975, Toward a phylogenetic classification of the Mammalia, in: *Phylogeny of the Primates* (W. P. Luckett and F. S. Szalay, eds.), Plenum Press, New York, pp. 21–46.

Novacek, M. J., 1985, Cranial evidence for rodent affinities, in: *Evolutionary Relationships among Rodents* (W. P. Luckett and J.-L. Hartenberger, eds.), Plenum Press, New York, pp. 59–81.

Olson, E. C., 1971, *Vertebrate Paleozoology*, Wiley-Interscience, New York.

Osborn, H. F., 1902, American Eocene primates, and the supposed rodent family Mixodectidae, *Bull. Am. Mus. Nat. Hist.* **17**:169–214.

Patterson, B., and Wood, A. E., 1982, Rodents from the Deseadan Oligocene of Bolivia and the relationships of the Caviomorpha, *Bull. Mus. Comp. Zool. Harv. Univ.* **149**:371–543.

Simpson, G. G., 1945, The principles of classification and a classification of Mammalia, *Bull. Am. Mus. Nat. Hist.* **85**:350.

Sych, L., 1971, Mixodontia, a new order of mammals from the Paleocene of Mongolia, in: Results of the Polish–Mongolian Paleontological Expeditions, Part III, *Palaeontol. Pol.* **25**:147–158.

Szalay, F. S., 1977, Phylogenetic relationships and a classification of the eutherian mammals, in: *Major Patterns in Vertebrate Evolution* (M. K. Hecht and P. C. Goody, eds.), Plenum Press, New York, pp. 315–374.

Szalay, F. S., 1985, Rodent and lagomorph morphotype adaptations, origins, and relationships: Some postcranial attributes analyzed, in: *Evolutionary Relationships among Rodents* (W. P. Luckett and J.-L. Hartenberger, eds.), Plenum Press, New York, pp. 83–132.

Szalay, F. S., and McKenna, M. C., 1971, Beginning of the age of mammals in Asia: The late Paleocene Gashato fauna, Mongolia, *Bull. Am. Mus. Nat. Hist.* **144**:269–318.

Tullberg, T., 1899, *Ueber das System der Nagethiere, eine phylogenetische Studie*, Akademischen Buchdruckerei, Upsala.

Van Valen, L., 1966, Deltatheridia, a new order of mammals, *Bull. Am. Mus. Nat. Hist.* **132**:1–126.

Von Koenigswald, W., 1985, Evolutionary trends in the enamel of rodent incisors, in: *Evolutionary Relationships among Rodents* (W. P. Luckett and J.-L. Hartenberger, eds.), Plenum Press, New York, pp. 403–422.

Wilson, R. W., 1949, Early Tertiary rodents of North America, *Carnegie Inst. Wash. Publ.* **584**:67–164.

Wood, A. E., 1942, Notes on the Paleocene lagomorph, *Eurymylus*, *Am. Mus. Novit.* **1162**:1–7.

Wood, A. E., 1957, What, if anything, is a rabbit?, *Evolution* **11**:417–425.

Wood, A. E., 1977, The Rodentia as clues to Cenozoic migrations between the Americas and Europe and Africa, in: *Paleontology and Plate Tectonics*, North American Paleontological Convention II, Lawrence, Milwaukee Public Museum, Special Publications in Biology and Geology, No. 2, pp. 95–109.

CHAPTER 4

PUNCTUATED EQUILIBRIUM AND PHYLETIC GRADUALISM

Some Facts from the Quaternary Mammalian Record

ANTHONY D. BARNOSKY

Many breaks in the fossil record are real; they express the way in which evolution occurs, not the fragments of an imperfect record.
Eldredge and Gould, 1972, p. 96

Absence of evidence is absence of evidence, not evidence of absence . . . . with respect to gaps in the evolutionary record.
Gingerich, 1984, p. 338

1. INTRODUCTION

Since Eldredge and Gould (1972) first wrote about "punctuated equilibrium" as a pattern for the history of life, three factions have appeared among evolutionary theorists. One faction favors the picture of punctuated equilibrium (Eldredge and Gould, 1972; Gould and Eldredge,

ANTHONY D. BARNOSKY • Section of Vertebrate Fossils, Carnegie Museum of Natural History, Pittsburgh, Pennsylvania 15213.

1977; Gould, 1982, 1985; Stanley, 1975, 1979, 1982). Another favors the traditional alternative of phyletic gradualism or at least gradual change (Gingerich, 1976, 1984, 1985; Charlesworth et al., 1982; Rose and Bown, 1984; Malmgren and Kennett, 1981, 1983; Ozawa, 1975). And the third faction suggests that the major differences between punctuated equilibrium and phyletic gradualism largely are semantic because evolution can and does produce both patterns, and moreover both were envisioned by the formulators of the modern evolutionary synthesis and even by Darwin (for example, Levinton and Simon, 1980; Stebbins and Ayala, 1981; Mayr, 1982; Hoffman, 1982; Johnson, 1982; Charlesworth and Lande, 1982; Ginzburg and Rost, 1982; Rhodes, 1983; Chaline, 1984; Newman et al., 1985). The differences in opinion persist largely because of a relative paucity of facts—facts about evolutionary relationships within lineages, about the rate of morphological change through time, about the distribution of the changes in space, and about the relationship between morphological change and speciation. The search for these facts has ranged with increasing success through several different geologic periods and kinds of organisms (for examples see recent summaries in Gingerich, 1985, Cope and Skelton, 1985, and Levinton, 1983), but only infrequently has it ventured into the realm of Quaternary mammals, that is, those that lived during the Pleistocene, which extends from about 1.8 million to 10,000 years ago, and the Holocene, which encompasses the last 10,000 years. In theory, however, Quaternary mammals should be able to provide information and precision about evolutionary tempos and modes that is not available from earlier parts of the geologic record. This is because the Quaternary is unique when compared to older time periods in at least five respects. It is a short period of geologic time, which provides the best glimpse of how evolution operates over a time scale of a few thousand up to 2 million years. Radiometric dating can be relatively precise, with uncertainties as little as 10^2–10^3 years rather than 10^4–10^6 years. Samples of fossil mammals commonly are larger and geographically more widespread than those from earlier times. Most of the genera and many of the species are still living or closely related to extant forms, so that we know more about their intraspecific morphological variation, their geographic distribution, their evolutionary relationships among species, and their habits and ecology. Finally, physical environmental changes are well enough documented for the Quaternary to aid us in assessing their impact on dispersal and evolution. What these special qualities of the Quaternary mammalian record add to the study of evolution is the subject of this chapter.

2. DEFINING THE MODELS

2.1. Punctuated Equilibrium

The picture of punctuated equilibrium clearly originated as an attempt to predict how allopatric speciation would appear in the fossil record. Eldredge (1971, p. 156) wrote, "I would suggest that the allopatric model (geographic speciation) be substituted in the minds of paleontologists for phyletic transformation as the dominant mechanism of the origin of new species in the fossil record." Eldredge and Gould (1972, p. 93) stated this theme more strongly: "We argue that . . . . the theory of allopatric speciation . . . . might reorient our picture for the origin of taxa." Eldredge and Gould (1972, p. 96) proposed that the fossil record could be used to test this assumption because in principle a certain pattern should result from the four "tenets and predictions" of allopatric speciation, namely:

> *Tenet 1*, New species arise by the splitting of lineages;
> *Tenet 2*, New species develop rapidly;
> *Tenet 3*, A small sub-population of the ancestral form gives rise to new species; and
> *Tenet 4*, The new species originates in a very small part of the ancestral species' geographic extent—in an isolated area at the periphery of the range.

From these tenets, Eldredge and Gould reasoned that we should not commonly see the transition of one species into another if we examine stratigraphically superposed fossils from a single geographic area, that is, in a local section. Rather, we should see a morphological break between ancestor and descendant. This is because the descendant evolved from a small, peripheral population that lived elsewhere and then migrated into the area of the local section. The four tenets *do not* mean that the descendant failed to change in sequential morphological steps from the ancestor in ecological time. They simply mean that we have a miniscule probability of finding the exact geographic region and slice of time where the change took place because that region is so small compared to the total range of the ancestral species, and because most morphological change is apt to be compressed into the short time when the ancestral population is still small and adapting to local conditions. Eldredge and Gould (1972, p. 96) clearly stated these points: "a sharp morphological break between the two forms . . . . *marks the migration of the descendent*, from the peripherally isolated area in which it developed. . . . Since speciation occurs *rapidly in small populations occupying small areas far from the center of ancestral abundance*, we will rarely discover the actual event in the fossil record'' [italics mine]. Punctuated equilibrium was the name they gave to the pattern that allopatric speciation should produce in the fossil record: relatively long stratigraphic intervals of morphological stasis punctuated by short stratigraphic intervals through which very different morphotypes appear. The new morphotypes appear suddenly because new species, evolved from peripheral populations, moved into the main part of the ancestral species' geographic range. Eldredge and Gould's null hypothesis became: The *pattern* of the fossil record frequently records stasis punctuated by rapid change. Reject the null hypothesis if gradual morphological change characterizes the fossils through most local sections. Do not reject it if punctuated equilibria prevail.

Eldredge and Gould (1972, pp. 108–110) then made a conceptual jump—they assumed punctuated equilibria *was* the dominant pattern in the history of life, which led to an important second-order hypothesis: Morphological trends in the fossil record result from selection at the species (or even higher) level, rather than at the level of organisms. The intent of their paper was to present the evidence and logic that led them to formulate the hypotheses, not the evidence to prove or disprove either one, although they openly admitted their bias to believe in punctuated equilibrium. Thus the stage was set for debate on two entirely different questions. Is punctuated equilibrium really the dominant pattern in the fossil record? And, assuming it is, what are the implications for macroevolution?

Gould and Eldredge (1977) more precisely drew the line between these questions. They devoted Section III of their paper to testing the hypothesis that punctuated equilibrium ruled in the fossil record, and with ten examples demonstrated that it did to their satisfaction (but certainly not to their opponents'!). Section IV concentrated on the second question in framing "punctuated equilibria as the basis for a theory of macroevolution" (Gould and Eldredge, 1977, p. 139). Because they accepted that punctuated equilibrium via allopatric speciation was real, it became logically possible for them to assert that natural selection operates at "hierarchical levels higher than the local population," notably at the species level (Gould and Eldredge, 1977, p. 139). Stanley (1975) termed this concept "species selection," and Gould and Eldredge (1977, p. 139) regarded it as the most exciting outgrowth of their 1972 paper, because it was a new idea about how evolution works, that is, about modes. In particular, Gould and Eldredge (1977, p. 139) invoked species selection to explain evolutionary trends and thereby "challenged the central assumption that secured the admission of paleontology into the modern synthesis of evolutionary theory (Simpson, 1944 and 1953): change in gene frequency within populations is the building block of major evolutionary events."

Vrba (1980, p. 80) took the concept one step further when she proposed the effect hypothesis, which interpreted "trends as effects .... of differential speciation rates, themselves effects of different degrees of environmental specialization between species, which in turn are effects of adaptation to local environments at the microevolutionary level." (She, later (Vrba, 1983, 1984a,b) was to point out that the effect hypothesis fundamentally differed from the concept of species selection, resembling it only in placing emphasis on selection at hierarchical levels higher than at the organismal level.) Whereas Gould and Eldredge (1977) restricted the term "punctuated equilibrium" to a model about tempos, and upon that model built species selection, the model about modes, Vrba (1980, p. 62) used "punctuated equilibrium" to encompass both tempo and mode.

Such a broad meaning became the common one when Gould (1982, p. 83) redefined punctuated equilibrium as

> a theory .... about the *tempo* and *mode* of evolution [italics mine]. It holds, speaking of mode, that significant evolutionary change arises in coincidence with events of branching speciation .... speaking of tempo, that the proper geologic scaling of speciation renders branching events as geologically instantaneous and that, following this rapid origin, most species fluctuate only mildly in morphology during a period of stasis that usually lasts for several million years .... 'geologically instantaneous' [is] .... defined as 1 percent or less of later existence in stasis.

Furthermore, Gould retracted the earlier emphasis that punctuated equilibrium was intimately tied to allopatric speciation, because almost any model of speciation might produce a punctuated pattern over a geologic time scale. (He now cites peripatric speciation as being most consistent with punctuated equilibrium; Gould, 1985, p. 5.) Species selection became an operational part of the definition because

> If the geometry of punctuated equilibrium prevails within the history of most clades ..., then traditional arguments for evolutionary trends [anagenetic trends within lineages] .... must fail. ... Trends must therefore be the product of a higher-order sorting that operates via the differential birth and death of species considered as entities ... (Gould, 1982, p. 101)

Species selection was restricted to mean sorting of species through differential origination rates.

The concept of punctuated equilibrium thereby matured from a geologic projection of allopatric speciation into a theory of macroevolution in its own right, a theory which proclaims: (1) morphology of a species remains relatively stable through most of its geologic range; (2)

most morphological change occurs at or near a cladogenetic speciation event, that is, during splitting of lineages; and (3) as a consequence, evolutionary trends result from the propensity of some kinds of organisms to speciate more frequently than others, that is, species selection sensu Gould (1982, p. 101). Punctuated equilibrium is a name for a theory about how evolution operates only within the second of Gould's (1985) three tiers of time, the first tier being ecological moments, the second being geologic time (millions of years), and the third being times of mass extinction.

2.2. Phyletic Gradualism

In contrast, the tiers of time are transcended by the most publicized alternative to punctuated equilibrium, phyletic gradualism. This is a theory rooted in the assumption that the morphological changes that evolve through geologic time—macroevolution—are the cumulative effects of the same processes that power evolution in ecological time—microevolution. Although the idea of such gradual, cumulative change stems from Darwin (1859) and permeates much of the literature about evolution, its implications for the fossil record were seldom articulated until 15 years ago. Eldredge and Gould (1972, p. 89) named their conception of gradual change "phyletic gradualism" and in emphasizing how it differed from punctuated equilibrium, defined its tenets as follows:

> *Tenet 1*, New species arise by the transformation of an ancestral population into its modified descendents;
>
> *Tenet 2*, Transformation is even and slow;
>
> *Tenet 3*, The transformation involves large numbers, usually the entire ancestral population; and
>
> *Tenet 4*, The transformation occurs over all or a large part of the species range.

Phyletic gradualism thus was cast as the opposite mode of speciation to allopatric as extended back in geologic time; it implied that "a long sequence of continuous, insensibly graded intermediate forms linking ancestor and descendent" should be found in most reasonably complete local sections (Eldredge and Gould, 1972, p. 89). Eldredge and Gould cited the frequent absence of such gradational links as evidence that phyletic gradualism did not occur often. But others envisioned that gradual change in the fossil record would result from less extreme modes of evolution than phyletic gradualism sensu Eldredge and Gould (1972), even from allopatric speciation, and that gradational links *were* found when data were examined properly. Gingerich (1976, p. 2), reporting on a stratigraphic sequence of Eocene mammals to make his case, summarized this view:

> General agreement exists regarding the usual *modes* of animal speciation, both phyletic and geographic. . . . Mutation and recombination provide genetic variability in the reproductively separate gene pool of each species, population, or deme. Effects of the natural environment tend to eliminate selectively individuals of a species that are least fit in each generation, and, if this selection is directional, small but usually significant changes in the genetic and phenotypic constitution of the entire species will result. New lineages arise when viable populations are separated geographically, and each develops its own evolutionary identity.
>
> The *tempo* of the speciation process in nature is less well documented. . . . If rates of divergence [in allopatric speciation] are relatively slow; that is, as slow as documented rates of phyletic evolution, the diversification of life is properly viewed as a gradual process, species are dynamic links in a great chain, and the traditional branching tree is an adequate representation.

In addition, Gingerich (1976, p. 25) suggested that character displacement was important in causing gradual morphological change, because when speciation was allopatric "differences in morphology postdate significantly the onset of genetic isolation, rather than predating genetic isolation as postulated by Eldredge and Gould (1972)."

Three pictures of evolution therefore competed—punctuated equilibrium, phyletic gradualism sensu Eldredge and Gould (1972), and gradualism as explained by Gingerich (1976). Vrba (1980, p. 62) applied the Eldredge–Gould term "phyletic gradualism" to a concept similar to the gradual evolution that Gingerich (1976) had written about. What she considered the most extreme versions of phyletic gradualism and punctuated equilibrium contrasted in eight ways (Table I).

With this contrast, the conception that both phyletic gradualism and punctuated equilibrium were names for models of macroevolution—not simply statements about tempo—became fixed. The points at which the models were not mutually exclusive—points 6–8 in Table I—revealed the potential for argument at the purely semantic level. Later Levinton (1983, p. 115) and Gingerich (1984, p. 335; 1985) explained a confusion about point 3 in Table I, concerning evolutionary rates. Although Vrba (1980, p. 62) said that under phyletic gradualism rates should be "uniformly low," the word "gradual" in fact connotes nothing about rates; the important concept is that change is continuous, advancing step by step in a single direction. Morphological change under phyletic gradualism therefore can be gradual over geologic time, yet still speed up and slow down. What point 3 *does* predict about phyletic gradualism is that, through time, within-species phenotypic

TABLE I
Contrasts between Phyletic Gradualism and Punctuated Equilibrium[a]

Real differences between the models

1. *Direction of phenotypic change within a species.* In phyletic gradualism change is unidirectional. In punctuated equilibrium change at most oscillates about a stable mean.
2. *Importance of splitting events (cladogenesis).* In phyletic gradualism, both anagenesis (transformation of the ancestor species into the descendent by phyletic evolution) and cladogenesis (splitting of lineages) are important mechanisms of speciation. In punctuated equilibrium, cladogenesis dominates to the virtual exclusion of anagenesis. (Vrba stated this as: Species arise "by phyletic speciation in sympatry, and by allopatric speciation in large or small populations" in the model of phyletic gradualism, and "generally only by allopatric speciation in small, isolated populations" in punctuated equilibrium.)
3. *Distribution of rate of phenotypic change.* In phyletic gradualism rates of phenotypic change do not increase significantly at the speciation event, be it cladogenetic or anagenetic. In punctuated equilibrium, most morphological change takes place at or near the cladogenetic speciation event, and not much occurs between speciation events. (Vrba's extreme version of phyletic gradualism also cited "uniformly low" rates of evolution.)
4. *Nature of species.* In phyletic gradualism anagenetic change can produce species that are arbitrary subdivisions of a lineage that continues through time. In punctuated equilibrium, species are real, discrete entities with beginnings and terminations.
5. *Macroevolutionary trends.* In phyletic gradualism, trends arise from additive phyletic evolution. In punctuated equilibrium, trends arise from sorting among species. (Vrba said from "nonrandom survival of stochastic speciation events," but elsewhere in her paper, pp. 80–81, she, like Gould, 1982, p. 10, stressed that it is birth rate of species, not survival, that results in species sorting.)

Points on which the models are not mutually exclusive

6. *Population size required for adaptive change to occur.* Small or large populations, or even whole species, can transform through phyletic gradualism. In punctuated equilibrium changes are confined to small, isolated populations.
7. *Dependence of rate of evolution on splitting events.* In phyletic gradualism, rates of evolution are independent of the number of splitting events (cladogenetic speciation events). In punctuated equilbrium, the number of lineage splitting events, and the amount of change that accompanies each one, control evolutionary rates.
8. *The role of environmental change.* In phyletic gradualism, evolution is accelerated by environmental change, but does not require it. Punctuated equilibrium requires environmental change. (Vrba's view here is not necessarily shared by all proponents of either model.)

[a]Modified from Vrba (1980, p. 62).

change should be nearly as pronounced as between-species change, and should proceed in the same direction.

It is only points 1–5 in Table I that define and contrast phyletic gradualism as presently understood. In those respects the differences between phyletic gradualism and punctuated equilibrium are very real—they cannot be dismissed as semantic.

2.3. Species Selection and the Effect Hypothesis

Although the importance of sorting among species is a principal difference between the two models, that concept itself is an ambiguous one. The concept commonly has been called "species selection," which means different things to different people. Vrba (1983, 1984a, p. 321, 1984b, p. 130) explained this in detail; the following discussion summarizes her review.

Prior to 1980, the term "species selection" referred to selection that operated by differential longevity, that is, survival or extinction, of species (Eldredge and Gould, 1972; Stanley, 1975, 1979; Gould and Eldredge, 1977; Eldredge and Cracraft, 1980). The process was conceived to be independent of selection that took place between organisms.

Vrba (1980) emphasized a different point of view, however, when she argued that birth rate of species—that is, speciation rate—also would tend to sort among species and might be more important than differential survival in causing macroevolutionary trends. Her view further differed in stating that the traits that caused sorting at the species level could result from traits that arose at the organismal or genomic level. Speciation rate, she posited, is the effect of microevolutionary processes, for example, adaptations that permit organisms to inhabit peripheral or restricted parts of a species range. Yet speciation rate per se is a trait that sorts among species over time, not among organisms. Kinds of species that speciate rapidly eventually will overwhelm the kinds that do not. Vrba (1980, p. 80) generalized that "Trends may be the *effects* of any factors which incidentally result in differential speciation rates" and called this idea the effect hypothesis.

Gould (1982) came to agree with Vrba's emphasis on speciation rate as an important cause of trends, but did not uphold the notion that speciation rate was linked to organism-level interactions. Accordingly, he retained the term species selection but recommended that it "be confined to . . . . claims for selection among species based on species level properties." He envisaged "true species selection . . . . is a dominant process when trends are produced by differential origination rather than by differential extinction" (Gould, 1982, p. 94).

But Vrba (1984a, p. 323) pointed out that high speciation rate in itself does not imply species selection if species selection is defined to avoid a tautology: "Species selection is that interaction between heritable, emergent character variation and the environment which causes differences in speciation and/or extinction rates among variant species within a monophyletic group." This means that if species selection is to be analogous to natural selection among individuals, then species, like individuals, must have certain traits that natural selection can act upon in a specified environment. The trait must originate (become emergent) at the level of the species, not at the level of the organism or population. It must vary among species within any given higher monophyletic group. And it must be possible for the ancestor species to pass the trait on to the descendant species. Examples of such traits might be (1) the characteristic population size, (2) the spatial and genetic separation between populations, or (3) characteristic dispersal rates. In practice, of course, it is very hard to demonstrate, even for the examples above, that a species-level trait is not an outgrowth of an organism-level interaction. Nevertheless, this perception of species selection is a theoretically sound one.

Therefore, as now understood, the concept of "sorting among species" that is specified in the punctuated equilibrium model includes at least two distinct processes—species selection and the effect hypothesis. The processes are similar in that both recognize species as "higher-order individuals," which are capable of interacting at a hierarchical level above that of organisms (Gould, 1985, p. 6). The principal differences are, first, that species selection requires the unit of selection to be above the organismal level; the effect hypothesis does not. Second, species selection requires that the trait being selected for is independent of characters that arise at the organismal level; the effect hypothesis requires that traits at the organismal level combine to form a character that becomes an active sorter only at the species level. Third, natural selection itself is the sorting mechanism in species selection; differences in speciation rates and extinction rates are the cause of sorting in the effect hypothesis (Vrba, 1983a, pp. 323–325).

3. CASE STUDIES OF QUATERNARY MAMMALS

We know many of the skeletal traits that characterize species of modern mammals. These traits in living species, or in extinct species that are closely related to living ones, can be traced through parts of the Quaternary record. In this respect Quaternary mammals are ideal for tracking evolutionary change within and between species, because we can analyze traits that are without doubt diagnostic of biological species.

In order to recognize evolutionary patterns, statistically large samples of fossils must be drawn from sequential small slices of time, and then the morphology at each slice of time compared. Such data are not commonly presented for Quaternary mammals. A computer search of the 1970–1984 BIOSIS and ZOOLOGICAL RECORD data bases found 732 articles that mention punctuated equilibria, or evolution and Pleistocene; approximately 30 more were found by sorting through a portion of the literature not included in the computerized files. Fewer than 25 (cited throughout the ensuing discussion) of these more than 760 articles gave the data necessary to support or reject conclusively some aspect of punctuated equilibrium or phyletic gradualism for Quaternary mammals.

3.1. Direction of Phenotypic Change within Species

The model of punctuated equilibrium predicts that morphology should not vary much from a stable mean during most of a species' existence, whereas phyletic gradualism predicts that the mean of the trait should shift unidirectionally.

3.1.1. Evidence for Stasis

3.1.1a. Short-Tailed Shrews. Blarina brevicauda, the short-tailed shrew, comprises two semispecies, *brevicauda* and *talpoides* (Jones et al., 1984). The semispecies are phena that have been distinct since at least 500,000 years ago (late Irvingtonian land-mammal age), but still hybridize where they are sympatric. *Blarina brevicauda brevicauda* is the only subspecies within the *brevicauda* semispecies. Eleven subspecies are included in the *talpoides* semispecies; one of these is *Blarina brevicauda kirtlandi*. A combination of 26 dental and dentary measurements subjected to canonical analysis reliably groups modern and fossil specimens of *B. brevicauda* into one or the other semispecies (Jones et al., 1984). Both semispecies are fairly common as Quaternary fossils.

At Baker Bluff Cave, Sullivan Co., Tennessee, specimens of *Blarina brevicauda* were recovered from seven superposed 1-ft thick stratigraphic levels (Guilday et al., 1978). Level A was 3–4 ft below the surface, Level B was 4–5 ft below the surface, and so on to Level G, which was 9–10 ft down (Fig. 1). Radiocarbon dates show that the sequence spans at least 8600 years, from 19,100 years at the bottom to about 10,500 years two-thirds of the way up the sequence, at Level C. A member of the *talpoides* semispecies, probably *B. brevicauda kirtlandi*, was present in all seven levels. The other semispecies, *B. brevicauda brevicauda*, appears only in the upper four levels (Guilday et al., 1978; Jones et al., 1984). The entrance of the *brevicauda* semispecies most likely was by immigration, because at the same level several boreal voles also make their first appearance.

Guilday et al. (1978) traced one taxonomically important character, length of the mandibular molar row (Jones et al., 1984; Graham and Semken, 1976), through all seven levels (Fig. 1). For the *talpoides* semi-

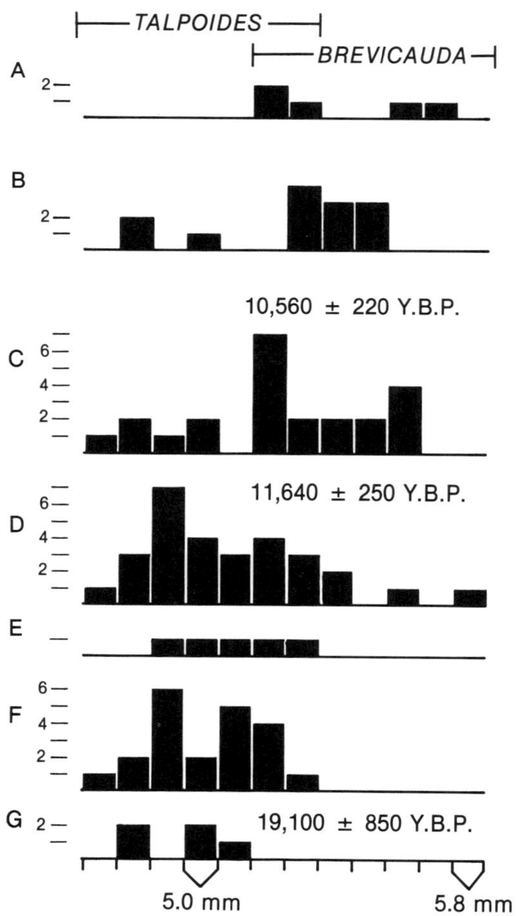

FIGURE 1. Length of M_1 to M_3 in the *talpoides* and *brevicauda* semispecies of the shrew *Blarina*. Specimens were excavated from Baker Bluff Cave, Tenessee. The histograms are arranged stratigraphically; each histogram plots specimens collected from a single 1-ft-thick stratigraphic level, with A being the youngest sample and G the oldest. Radiocarbon dates associated with the indicated levels provide the chronology. Y.B.P., Years before present. Tick marks at the left indicate the number of specimens in each frequency category. Data are from Guilday et al. (1978).

species, the character did not change much from Level G to Level B; the modal value remained near 4.9 mm. Likewise, molar-row length in the *brevicauda* semispecies remained relatively constant, near a mode of 5.4 mm through its stratigraphic range. This pattern indicates stasis through at least 8600 years, even through a time when the climate likely cooled coinciding with Level D (Guilday et al., 1978), and when an immigration brought the two closely related phena into sympatry. The example clearly rejects the model of phyletic gradualism at the ±10,000-year time scale, but supports punctuated equilibrium.

3.1.1b. Hamsters. Size is one criterion that differentiates modern hamster species, particularly within *Cricetulus* (Corbet, 1978, p. 90). Color of the fur, which is unknown for fossils, is the other taxonomically important trait.

Cricetulus bursae and *Rhinocricetus ehiki* are two hamsters that differ primarily in size, as reflected by the length of the mandibular tooth row (Kurtén, 1968, pp. 211–212). The two species coexisted in central and eastern Europe from at least 1 million years ago (Waalian warm stage; late Villafranchian land-mammal age) to about 700,000 years ago (second cold phase of the Gunz Glaciation; Biharian land-mammal age). The sites at which they coexist are not demonstrably superposed, but five of the localities were arranged by relative age through biostratigraphic comparisons (Kurtén, 1968, pp. 25, 241), which in Figure 2 are matched with appropriate parts of the radiometric scale via correlation charts compiled by Van Eysinga (1978) and Repenning (1983). The "absolute" dates therefore are very approximate and must be regarded as order-of-magnitude estimates that suggest the sequence spans in the neighborhood of 300,000 years.

During this time, tooth-row length in *Cricetulus bursae* fluctuated from a modal value of about 4.8 mm to about 5.1 mm (Beremend and Villany samples), then down to 4.7 mm (Episcopia sample), then back up to about 5.1 mm (Sackdilling sample), then back down to about 4.7 mm (Brasso sample) (Fig. 2). The modal values for *Rhinocricetus ehiki* also fluctuate back and forth, from about 5.7 mm at Beremend and Villany, to 5.4 mm at Episcopia, to 5.9 mm at Brasso. Size does not change significantly or unidirectionally for either species; instead the rather minor fluctuations are just what the model of punctuated equilibrium predicts. The fact that the samples by Quaternary standards are sparsely distributed through time does not weaken the case for punctuated equilibrium; discontinuous, but strongly unidirectional change would be predicted by phyletic gradualism in this case.

3.1.1c. Ground Squirrels. Spermophilus townsendii, Townsend's ground squirrel, is present today in southeastern Washington state, where its fossil history extends back at least 40,000 years. *Spermophilus*

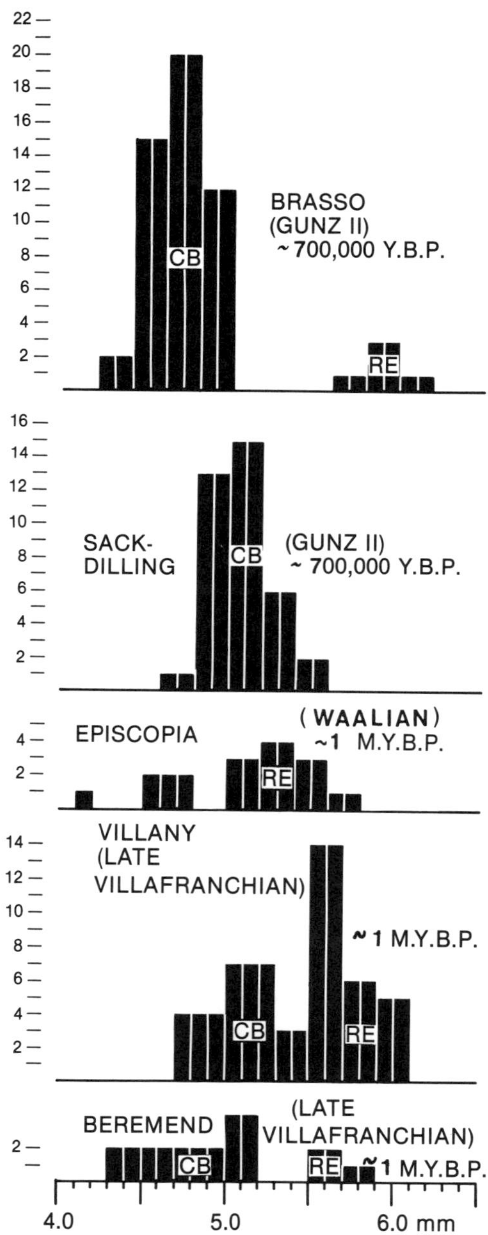

FIGURE 2. Length of mandibular tooth row in the hamsters *Cricetulus bursae* (CB) and *Rhinocricetus ehiki* (RE) from various localities in central and eastern Europe. Beremend and Villany are near the Yugoslavian border in south-central Hungary; Episcopia is near the Hungarian border in Romania; Brasso is also in Romania, but farther east in the Translyvania region; and Sackdilling is in the Oberpfalz of Germany. The histograms are arranged in stratigraphic order, with the oldest locality at the bottom. Radiometric dates are at best order-of-magnitude approximations, which are estimated from stepwise correlations with biostratigraphic and paleomagnetic sequences. (M.)Y.B.P., (Million) years before present. Tick marks at the left indicate the number of specimens in each frequency category. Data are from Kurtén (1968, Fig. 101).

townsendii is closely related to *S. washingtoni;* the two species are known to be separated chromosomally by only one pericentric inversion (Nadler et al., 1984). *Spermophilus townsendii* and *S. washingtoni* can be differentiated reliably by an osteological character, namely the length of the third upper molar relative to its width; the molars are relatively shorter in *S. townsendii* (J. E. Martin et al., 1983, p. 16; Rensberger et al., 1984, p. 30).

From data listed in Rensberger et al. (1984, pp. 5, 12–13, 17, 30), the ratio of width to length of *S. townsendii* third upper molars can be traced through 12 samples that are stratigraphically superposed in loess near Kennewick, Benton Co., Washington (Fig. 3). The oldest sample is between 40,000 and 320,000 years old, as estimated from the amount of time it must have taken for eight overlying calcretes (accumulations of calcium carbonate in fossil soils) to form (Gile et al., 1965, 1966; Warren, 1983; Pye, 1983; Wajtanowicz and Buranczynski, 1978; Tucholka, 1977). The youngest sample occurs above a tephra identified as the Mazama volcanic ash (Rensberger et al., 1984, p. 15), which is between 6600 and 7000 years old (Sarna-Wojcicki et al., 1983).

The width to length ratios fluctuate very slightly around a mean of about 1.00, never exceeding 0.98 or 1.15 as extremes (Fig. 3). The slight changes that do take place are not unidirectional. Therefore, proportions of the upper third molars in *S. townsendii* have remained static for at least 33,000 years, and possibly for as long as 320,000 years, depending on the age of the lowest sample.

3.1.2. Evidence for Unidirectional Change

3.1.2a. Sagebrush Voles. Complexity of the upper third molar is a diagnostic feature of species within *Lagurus*, the steppe and sagebrush voles. Triangles and reentrants are developed to varying degrees on the M^3, so that *Lagurus curtatus* and *Lagurus luteus* have only three labial and two lingual vertical ridges, whereas *Lagurus lagurus* has four labial and three lingual ridges.

Lagurus curtatus, the sagebrush vole, is found in the same eastern Washington loess deposits that are described above for *Spermophilus townsendii*. The fossils are grouped into seven stratigraphically superposed samples, A at the bottom through I at the top (Fig. 4) (Rensberger et al., 1984, p. 59). In the lowest two samples, all of the third upper molars are morphologically similar; they have a very shallow anterolabial reentrant. As a consequence, the labial margin of these M^3's is characterized by one poorly developed vertical ridge (the anterior one) and two prominent vertical ridges (the middle one and the posterior

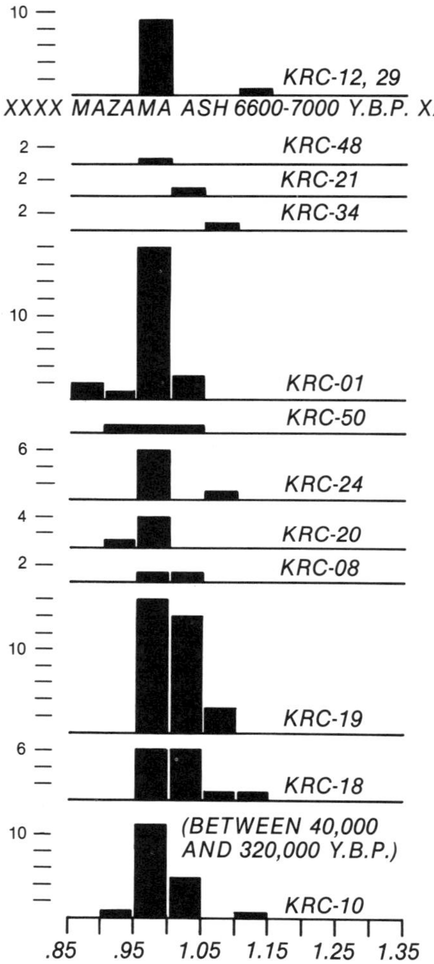

FIGURE 3. Ratio of the width to length of M^3 in *Spermophilus townsendii* from eastern Washington state. The histograms are arranged in stratigraphic order, with the oldest locality on the bottom. The upper date is based on the Mazama Ash, and the lower age range is estimated from the time required for eight calcretes to form. Y.B.P., Years before present. KRC followed by a number is the locality name. Tick marks at the left indicate the number of specimens in each frequency category. Data are from Rensberger et al. (1984).

one). In Level C a new morphotype of M^3 appears, in which the anterolabial reentrant is very deep. This makes the anterior and middle vertical ridges more prominent than the posterior one. The deep-reentrant morphotype becomes more abundant through time, comprising 25, 80, and finally 100% of the sample in Levels C and D, Level E, and Levels F and I, respectively (Fig. 4). Therefore, individuals with a deep reentrant composed increasingly more of the population, and at last completely replaced the shallow-reentrant forms over a time span of between 33,000 and 320,000 years (the dates explained above for *Spermophilus townsendii* also apply here). The same kind of unidirectional

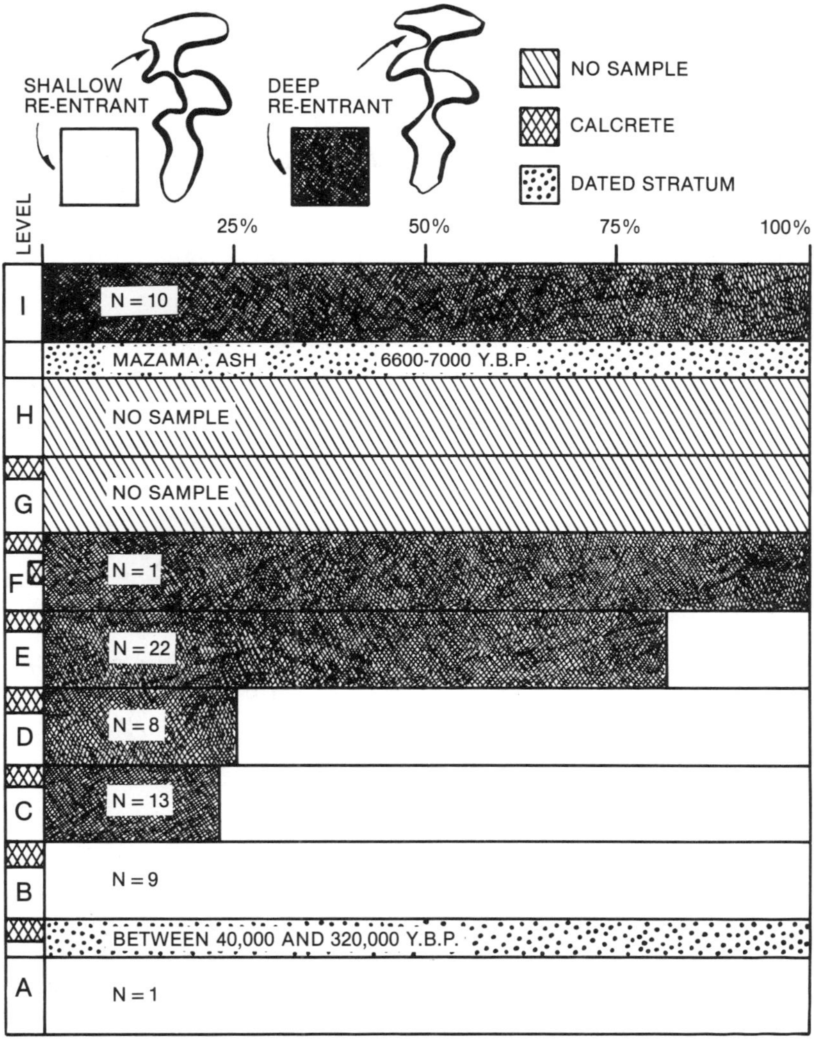

FIGURE 4. Bar diagram showing the sequentially changing percentages of M^3 morphotypes in *Lagurus curtatus* from eastern Washington state. Localities are arranged in stratigraphic order, with A the oldest and I the youngest. White bars illustrate the percentage of M^3's with a shallow anterolabial reentrant; black bars depict the percentage with a deep reentrant. The upper date is determined from the Mazama Ash; the lower age range is estimated from the time required for the eight calcretes to form. Y.B.P., Years before present. Data are from Rensberger et al. (1984).

increase takes place in the first lower molar, in which five-triangle forms replace four-triangle forms through the section (Rensberger et al., 1984, p. 59). The most parsimonious explanation is that the alleles that cause deep reentrants on M^3 and five triangles on M_1 spread through more and more of the eastern Washington population over time, until the traits finally became fixed. The other alternative—that a separate population with complex teeth gradually encroached into eastern Washington as the physical environment correspondingly altered—seems much less likely. Eight calcretes in the section argue for several dry–moist cycles that do not correlate well with the unidirectional changes in *Lagurus*.

3.1.2b. Water Voles. *Arvicola cantiana* lived in the British Isles from <350,000 years ago (Cromerian Interglacial) to between 70,000 and 50,000 years ago (early Devensian Glacial Stage, Chelford Interstadial). *Arvicola cantiana* was replaced stratigraphically by *Arvicola terrestris*, which first appeared between 50,000 and 30,000 years ago (Middle Devensian, Upton Warren Interstadial) and is still extant (Stuart, 1982, p. 168). Differences between these two species are evident in at least three characteristics of their molars—size, the ontogenetic age at which roots grow, and thickness of enamel (Stuart, 1982, p. 38). Molar size also distinguishes *A. terrestris* from the other living species, *A. sapidus* (Corbet, 1978, p. 105).

Stuart's (1982, p. 186) compilation of data shows that the length of M_1 in *A. cantiana* changes unidirectionally through seven stratigraphic intervals encompassing roughly 290,000 years. From the stratigraphically lowest to the highest sample, the mean lengths are 3.2, 3.3, 3.6, 3.7, 3.75, and 3.95 mm (Fig. 5).

The general trend for the ensuing species, *Arvicola terrestris*, likewise is toward increased size. Late Flandrian (<6000 years) and modern specimens are larger than late Devensian ones (10,000–26,000 years old), which in turn are larger than middle Devensian ones (26,000–50,000 years old) (Fig. 5). However, *A. terrestris* does not provide as conclusive evidence for unidirectional change as does *A. cantianus*, because the general size increase is interrupted by an apparent reversal or immigration at the Devensian/Flandrian boundary, and details are obfuscated partially by the uncertain stratigraphic position of the late Devensian samples (Stuart, 1982, p. 186).

3.1.2c. Elephants. The living species of elephants, *Loxodonta africana* and *Elephas maximus*, can be diagnosed by traits of their molars—the number of enamel plates, the frequency of plates per 10 cm of tooth (= lamellar frequency), thickness of plates, and crown height, among other traits. These characters change unidirectionally through

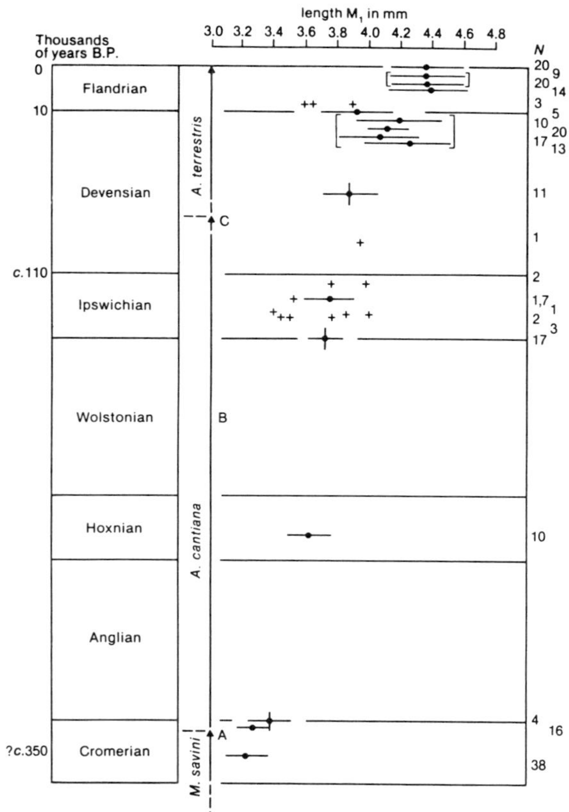

FIGURE 5. Preliminary diagram of size changes in the *Mimomys savini* to *Arvicola terrestris* lineage, relative timing of morphological changes, and suggested taxonomic divisions. Means of samples are indicated by dots, standard deviations by horizontal bars. Single measurements are indicated by crosses. Vertical bars denote some uncertainty in stratigraphic position. (A) Loss of roots and achievement of permanently growing molars; (B) last occurrence of "*Mimomys* fold"; (C) last occurrence of convex-thickened enamel. Note that the relative time scale is schematic only. (Reprinted from Stuart, 1982.)

time within some fossil species. A good example is *Elephas recki*, which lived in Africa from the late Pliocene to the middle Pleistocene, about 3.5 million to 500,000 years ago. Maglio (1973, p. 35) recognized "four evolutionary stages [within *E. recki*] based on morphological changes and stratigraphic succession (Maglio, 1970), but these grade into each other making the taxonomic designation impossible." From the earliest (Stage 1) to the latest (Stage 4) representatives, morphology of *E. recki* changed in the following ways: (1) increase in enamel plates, (2) increase in lamellar frequency, for example, in M_3, from 4.7–5.0 in Stage 1 up to 5.0–6.2 in Stage 4, (3) increase in enamel folding, (4) increase in crown height, for example, in M_3, from a hypsodonty index of 126.6–133.0 in Stage 1 up to 144.0–186.4 in Stage 4, and (5) decrease in enamel thickness, for example, an average thickness in M_3 from 3.0–3.1 mm in Stage 1 down to 2.5–3.6 mm in Stage 4 (Maglio, 1973, p. 36) (Fig. 6). Maglio (1973, p. 56) documented that molars of *Mammuthus meridionalis* underwent similar unidirectional modifications from about 3 million to 1 million years ago.

3.1.2d. Pigs. The third molars of living pigs are distinctive taxonomically (Harris and White, 1979, p. 7). *Mesochoerus limnetes* is a pig that frequently is represented by third molars in successive stratigraphic intervals of the Shungura Formation, Omo Basin, Ethiopia (Harris and White, 1979). The stratigraphic levels are identified from bottom to top as B, C, D, E, F, Gi, Gs, H, J, K, and L. Level B is thought to be between 2.6 and 3 million years old, Level G about 1.8–2 million years old, and the bottom of Level L about 1 million years old (Cooke, 1978, Fig. 2). The following aspects of *Mesochoerus limnetes* change gradually: (1) length of M^3 increasing through Levels B, C, E, F, G, H, and lumped K and L; (2) length of M_3 increasing through Levels B, C, D, Gi, Gs, H, J, K, and L; (3) ratio of M^3 length:trigon length increasing through Levels B, C, D, Gi, Gs, H, K, or L; (4) ratio of M^3 length:breadth increasing through Levels B, C, D, E, F, Gi, Gs, H, K, and L; and (5) ratio of M_3 length:breadth increasing through Levels B, C, D, Gi, Gs, H,

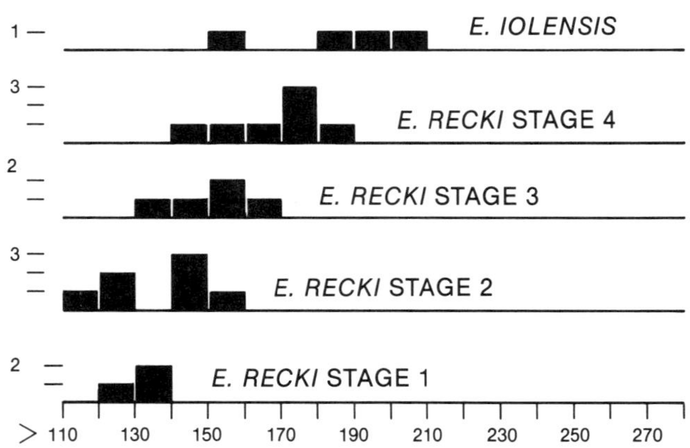

FIGURE 6. Successive stages of hypsodonty within the species *Elephas recki*, and the transition into *E. iolensis*. Histograms are arranged in stratigraphic order. Tick marks at the left indicate the number of specimens in each frequency category. Data are from Maglio (1973, Tables 13, 15).

J, and K (Harris and White, 1979, Figs. 80, 81, and 126–131). Morphology essentially remained static or fluctuated slightly between Levels C and F for M^3 length and the ratio M_3 length:trigon(id) length. The time spanned by this "stasis" was relatively short, perhaps 500,000 years, in comparison to the ± 1.5 million years over which unidirectional changes took place. The suite of characters for *Mesochoerus limnetes* therefore exemplifies unidirectional change in general, but at varying rates through time.

The same holds true for *Metridiochoerus andrewsi*, another pig from the Shungura Formation. The M^3 increases in length through levels B, C, and E, then remains static in Levels F, Gi, and Gs. The M_3 increases in length rather uniformly through Levels B, C, D, E, F, Gi, and Gs (Harris and White, 1979, Figs. 121 and 122). These characters and others led Harris and White (1979, p. 49) to recognize two distinct evolutionary stages within *Metridiochoerus andrewsi* in the Shungura Formation, and a third stage in the lower part of the upper member of the Koobi Fora Formation.

3.1.2e. Hominids. Wolpoff (1984) used 13 measurements to compare samples of *Homo erectus* from three stratigraphic intervals—the lower Pleistocene, the early-middle Pleistocene, and the later-middle Pleistocene, which total 1 million years. The measurements reflect cranial and mandibular size, both of which are used to distinguish various hominid species. Wolpoff's histograms conclusively show that seven of the nine cranial measurements increase unidirectionally through the three temporal levels, and three of the four mandibular measurements decrease (Fig. 7). Other studies, which claim stasis as the pattern for *Homo erectus*, do not present such conclusive data (Rightmire, 1981; see also Cronin et al., 1981).

3.2. Anagenesis and Cladogenesis

Phyletic gradualism recognizes *both* cladogenesis and anagenesis as important. Consequently, phyletic gradualism never can be rejected simply by presenting examples of cladogenesis, even though such examples abound in the fossil record—for instance, the Quaternary clades of African bovids discussed by Vrba (1984c), at least nine species of elephants (Maglio, 1973), and the three species within *Blarina* (Jones et al., 1984). However, the model of punctuated equilibrium denies that species arise by anagenesis. Therefore one *can* reject the model of punctuated equilibrium for a given lineage if anagenesis can be demonstrated.

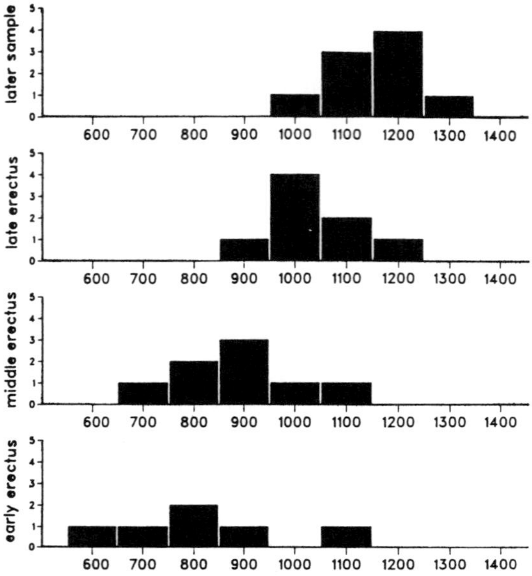

FIGURE 7. Distribution of cranial capacity in cubic centimeters for three *Homo erectus* samples and a sample of later hominids. Histograms are arranged in stratigraphic order. The tick marks at the left indicate the number of specimens in each frequency category. See Wolpoff (1984, Table 2) for means and other statistics. (Reprinted from Wolpoff, 1984.)

3.2.1. Examples of Anagenetic Change

To demonstrate anagenesis, one must show that the morphology of the ancestral species grades continuously into that of the descendant, until the descendant morphology completely replaces the ancestral morphology. By definition the ancestor and descendant cannot live in the same place at the same time—if they do, the lineage has split, not transformed. The following examples are consistent with the postulate of anagenetic transformation, but would be hard to reconcile with cladogenesis.

3.2.1a. Water Voles in the British Isles. Mimomys savini last appears in Cromerian Pollen Zone CrIIIb. In the overlying Pollen Zone CrIV, *Arvicola cantiana* replaces *M. savini*. The replacement spanned at most a few thousand years approximately 350,000 years ago. The generic distinction belies the extremely similar morphology of the two species where they are stratigraphically close. In *Mimomys savini*, roots appeared on the cheek teeth only in old animals. Stuart (1982, p. 185) noted that "the formation of roots was retarded until progressively later in life [throughout the history of *Mimomys*], until a point was reached where the cheek teeth ceased to root altogether, and had become permanently growing. This step is generally accepted as marking the boundary between the genera *Mimomys* and *Arvicola*." The other differences between *Mimomys savini* and *Arvicola cantiana* are in size, in the frequency of specimens with a crenulation on the anterior loop of M_1 (the "*Mimomys* fold"), and in enamel thickness. None of these differences are pronounced in stratigraphically adjacent samples of the two species. The mean size of M_1 is 3.25 mm in the highest sample of *M. savini*, and 3.3 mm in the lowest of *A. cantiana*. Later *A. cantiana* increase in size so that by the Hoxnian Interglacial they overlap little if any with *M. savini* (Fig. 5). The *Mimomys* fold is present in most of the unrooted teeth of *M. savini* (that is, teeth from young animals); the frequency of this character decreases to 30% in Cromerian *A. cantiana*, then becomes rarer yet in later samples, until it becomes "virtually absent" in the specimens from the Ipswichian Interglacial (somewhat earlier than 110,000 years ago) (Stuart, 1982, p. 185). Enamel in *M. savini* is thick on the convex angles of the cheekteeth; it is also thick in Cromerian and Hoxnian samples of *A. cantiana*. This trait becomes less frequent in Ipswichian and early Devensian *A. cantiana*.

By 50,000 years ago, *A. cantiana* was replaced by the modern *A. terrestris*. The morphological transformation was a smooth one. The species boundary between the two is drawn "for convenience" at the last appearance of thick enamel on the M_1 (Stuart, 1982, p. 185). Superposed samples on each side of this species boundary are otherwise nearly identical; subsequently *A. terrestris* became significantly larger than *A. cantiana* (Fig. 5).

3.2.1b. Elephants. Maglio (1973) provided convincing evidence that seven of the 17 Quaternary elephant species arose through anagenesis, and he suspected that at least one more did. Most of the morphological transformations are documented by stratigraphically sequential changes in the characters of the molars, such as those discussed above for *Elephas recki*. None of the descendent species coexist with their respective ancestors in time or space; rather, they replace them. The anagenetic species transitions are as follows:

Elephas ekorensis→E. recki→E. iolensis. There is progressive increase in the number of molar plates, in their relative height and spacing, and in folding of enamel, and progressive reduction in enamel thickness and in the median enamel loop. Maglio (1973) stated, "At present, a direct phyletic relationship between *E. ekorensis* and *E. recki* seems certain" (p. 81), and "*E. iolensis* appears to have been the direct descendent of *E. recki* and for this reason transitional specimens may be impossible to assign to one species unit or the other" (p. 38).

Elephas hysudricus→E. maximus. Elephas maximus continues the dental trends exhibited in *E. hysudricus*. Maglio (1973, p. 85) felt that retardation of ontogenetic development in *E. hysudricus* produced *E. maximus*.

Mammuthus meridionalis→M. armeniacus→M. primigenius. The progressive changes that merge one species into the next include increase in the number of molar enamel plates and in lamellar frequency and decrease in the median enamel loop and in enamel thickness. Maglio concluded (1973, p. 87) that *Mammuthus meridionalis* "passes both morphologically and stratigraphically into *M. armeniacus* and this into *M. primigenius*. The last two 'species' certainly represent arbitrary segments of a continuous lineage along which progressive molar and cranial changes occur through the middle and late Pleistocene."

Mammuthus imperator→M. columbi. Changes in this lineage parallel those described for the *M. primigenius* lineage (Maglio, 1973, p. 87).

Elephas planifrons→E. platycephalus. This is a lineage that Maglio (1973) speculates is an anagenetic one, but only a single specimen of *E. platycephalus* is known.

3.3. Distribution of Rate of Phenotypic Change

Under the model of punctuated equilibrium, most morphological change should take place at or near the speciation event. Speciation and the accompanying phenotypic changes should be accomplished within 1% of the time the species later exists in stasis (Gould, 1982, p. 83). Phyletic gradualism predicts that change throughout the life of the species should exceed the amount of change that takes place across a species boundary.

3.3.1. Morphological Change Confined to Speciation Events

3.3.1a. *Short-Tailed Shrews. Blarina brevicauda brevicauda* and *B. carolinensis* split off from the *B. brevicauda talpoides* semispecies in the late Irvingtonian. No transitional morphologies can be detected; the first representatives of the *B. carolinensis* and the *brevicauda* semispecies show all of the taxonomically important dental characters of modern representatives (Jones et al., 1984). Stasis prevailed for about 500,000 years. It is impossible to resolve for certain whether the new species actually originated within 5000 years—that is, 1% of the time that they later have existed in stasis—but they certainly must have undergone much more morphological change near the speciation event than they have subsequently. Speciation within short-tailed shrews undoubtedly has taken place in much less than 10,000 years, as shown by splitting of *B. hylophaga* from the *B. carolinensis* line between 9000 and 11,000 years ago, and the subsequent morphological stasis of *B. hylophaga* to the present (Jones et al., 1984, p. 135).

3.3.1b. African Bovids. Vrba (1984c) suggested that aspects of facial structure, particularly horn morphology, are important for specific-mate recognition for bovids. Therefore horn cores and cranial and mandibular characters are reliable for distinguishing between species. She subjected 58 mandibular and cranial characters to cladistic analysis to define fossil and recent species and recognize relationships among them. Within the tribe Alcelaphini, which includes the blesbucks, hartebeests, and wildebeests, 25 species lived during the Quaternary. The origin of all of these was apparently sudden, by cladogenesis in a minimum of 18 cases (Vrba, 1984c, pp. 65, 68). Ancestors with transitional morphologies are not known. The species-specific characters persisted approximately 250,000 years or somewhat less in seven species, 500,000 years in nine species, 1 million years in five species, 1.5 million years in two species, and 2 million years or more in two species. Through these times, none of the species underwent significant morphological changes; the changes that did occur were either oscillatory, or "gradual average modifications" that were minor compared to the morphological change that took place between species (Vrba, 1984c, p. 68). This pattern accords most closely with punctuated equilibrium, although data are not available to calculate whether most of the change indeed took place within the first 1% of the time the species existed.

3.3.1c. Elephants. Three species of elephants show very little change in their molar structure through stratigraphic ranges in excess of 750,000 years (Maglio, 1973). These include *Loxodonta adaurora* (ca. 2–4.5 million years ago), *L. africana* (ca. 750,000 years ago to present), and *Elaphus namadicus* (ca. 1 million years ago to late Pleistocene).

3.3.1d. Suggestive Data. In Europe, at least 111 modern species first appeared during the Pleistocene. A maximum of 25 of these can be linked to an ancestor by transitional morphologies (Kurtén, 1968, pp. 255, 258). That may suggest that the morphologies that characterize the other 86 modern species appeared relatively suddenly in the fossil record, and were followed by periods of stasis that range from 10,000 to 2 million years. Stanley (1978, p. 26) elaborated on these data by converting them to species-survivorship curves, which led him to conclude that "most net evolutionary change must have been associated with saltational speciation." Arguing for the validity of this approach is the fact that if the Pleistocene specimens were not morphologically identical to modern specimens, paleontologists probably would not refer them to the modern species. Arguing against it is that the morphology of stratigraphically superposed samples has not been compared adequately for the vast majority of the species in question.

3.3.2. Morphological Change Distributed through Stratigraphic Range

The examples discussed under Sections 3.1.2 and 3.2.1 illustrate cases in which less morphological change took place across species boundaries than in the succeeding life span of the species. Particularly convincing examples are the *Mimomys savini*→*Arvicola cantiana*→*A. terrestris* sequence (Fig. 5), the *Elephas recki*→*E. iolensis* sequence (Fig. 6), the *Mammuthus meridionalis*→*M. primigenius* line, and the transition through *Homo erectus* to post-*H. erectus* samples (Fig. 7) (Wolpoff, 1984).

3.4. Nature of Species

The previous examples also serve to illustrate that the following species emphatically *are* arbitrary subdivisions of a morphological continuum: *Mimomys savini, Arvicola cantiana, A. terrestris; Elephas ekorensis, E. recki, E. iolensis, E. hysudricus, E. maximus; Mammuthus meridionalis, M. armeniacus; Mammuthus primigenius, M. imperator,* and *M. columbi*. Kurtén (1968, p. 258) adds 22 more species for which transitions between pairs are "actually recorded": *Macaca florentina*→*M. sylvana, Felis issiodorensis*→*F.* cf. *pardina, Gulo schlosseri*→*G. gulo, Cuon majori*→*C. alpinus, Ursus minimus*→*U. etruscus, Ursus deningeri*→*U. spelaeus, Ursus etruscus*→*U. thibetanus, Dicerorhinus*→*D. etruscus, Dama clactonia*→*D. dama,* and *Lagurus pannonicus*→*L. lagurus*. Vrba (1984c, p. 65) suggests that the transformation of *Aepyceros* species no. 34 to *Aepyceros* species no. 35 was similarly continuous.

Other examples just as emphatically argue for species being discrete in space and time. Those discussed in previous parts of this paper include the short-tailed shrews *Blarina brevicauda, B. carolinensis,* and *B. hylophagus*, the hamsters *Cricetulus bursae* and *Rhinocricetus ehiki*, the squirrel *Spermophilus townsendii*, the elephants *Loxodonta adaurora, L. africana,* and *Elephas namadicus*, and the 25 species of alcelaphine bovids.

3.5. Macroevolutionary Trends

Some of the evolutionary trends exemplified by Quaternary mammals include the following:

1. Progressive development of dentine tracts, hypsodonty, and complexity in the anterior loop of M_1, along with increasing size in muskrats from the Pliocene through the Quaternary (Nelson and Semken, 1970; Zakrzewski, 1974; L. D. Martin, 1979, 1984).

2. Increasing crown height and dentine tracts in numerous arvicoline rodents, such as *Mimomys, Arvicola, Pliophenacomys,* and *Ophiomys,* through the Quaternary (Hibbard and Zakrzewski, 1967, 1972; Stuart, 1982; Barnosky, 1985).

3. Increasing crown height and addition of molar enamel plates in elephants from the Pliocene through the Quaternary (Maglio, 1973).

4. Increasing height and length of third molars in the *Nyanzachoerus–Notochoerus* and *Metridiochoerus* lineages of pigs from the late Pliocene through the middle Pleistocene (Harris and White, 1979).

5. An increase in size, enlargement of the carnassial tooth, reduction of the third premolar, development of an area of overlap between P_4 and M_1, lowering of the coronoid process, reduction of the anterior flange on the mandible, and increasingly inclined occiput region in dirk-toothed cats (Smilodontini) from the Pliocene through the late Pleistocene (L. D. Martin, 1984).

6. Increasing size in *Sigmodon* (cotton rats), complemented by increasing hypsodonty and increase in the roots of M_1 from two to four, from the latest Pliocene through the Pleistocene (R. A. Martin, 1979, 1984).

7. Increased complexity of horn-core and frontlet morphology, general increase in size, and reduction in premolar–molar ratio in alcelaphine bovids (the blesbucks, hartebeests, and wildebeests) through the Pliocene and particularly in the Quaternary (Vrba, 1984c, p. 64).

3.5.1. Additive Microevolution

The model of phyletic gradualism regards trends as the sum of many microevolutionary changes, initiated mainly by natural selection among phenotypes, that were produced through geologic time. To support this idea, the data used to recognize trends in the fossil record must have two qualities—successive (stratigraphically higher) populations or species should change only in the direction of the trend, and the evolving traits should be related clearly to some important function of the organism.

Trends 1–6 all fulfill the criterion of unidirectional change through successive stratigraphic intervals. These trends further are consistent with the model of phyletic gradualism because all can be adequately explained by the differential survival of organisms with functionally advantageous traits. In this sense, they parallel a very clear example of Vrba's (1984a, p. 324):

> If all organisms in a species are black, black colour is a species character. If black organisms are subsequently removed by selection, but those of different colour in a related species survive, this is differential species extinction. Selection of organisms is enough to account for events, and there is no need for concepts of species adaptation and selection.

In trends 1–4, which involve increasing crown height and related traits, low-crowned animals are analogous to the black ones in Vrba's example. Animals that eat abrasive foods, such as grass, wear down their teeth quickly. The more hypsodont and lophodont the tooth, the better an animal can cope with abrasive diets, as Rensberger (1973) and Maglio (1973) demonstrated by biomechanical analyses of mastication in rodents and elephants, respectively. Thus, natural selection among organisms would produce a trend by the following mechanism. In a given population of abrasive-food eaters, individuals with high-crowned teeth survive longer and produce more progeny than their low-crowned competitors. Generation by generation the average crown height thereby increases for the population as a whole. The process continues until teeth are so hypsodont that their abrasion no longer affects reproductive success. New, more hypsodont species originate either through anagenesis as the optimal crown height for a given food resource is approached, or through cladogenesis when a single population begins to utilize a new, more abrasive food.

Similar logic can be extended to the trends in the dirk-toothed cats (trend 5), because all of the modifications result in a more efficient predator (L. D. Martin, 1980; Emerson and Radinsky, 1980). Likewise, the trends for cotton rats (trend 6) are "directly coupled with the evolution of hypsodonty and the successful transition from browsing to grazing" (R. A. Martin, 1979, p. 1).

3.5.2. Sorting among Species

The model of punctuated equilibrium attributes trends to sorting among species, by the mechanisms of species selection or the effect hypothesis. Trends produced by these mechanisms should be apparent in the fossil record of a clade by, first, the presence of large numbers of relatively short-lived species, which would be necessary to supply a sufficient pool of variability, and second, by the random directions that speciation proceeded in with respect to the overall trend (analogous to the randomness of genetic mutations with respect to phenotypic trends at the organismal level) (Gould and Eldredge, 1977, p. 148).

The general trends in alcelaphine bovids—increasing complexity in horn morphology, increasing size, reduction of premolars—may provide an example. Twenty-five species comprising nine genera originate and/or become extinct at various times through the Quaternary (Vrba, 1984c, p. 65). In contrast, the trends cited above as fitting the model of phyletic gradualism spanned fewer species in the same amount of time—seven species in two genera of muskrats (trend 1), at most 12 known species of *Mimomys* and *Arvicola*, three species of *Pliophenacomys*, five species of *Ophiomys*, at most 15 species of *Sigmodon*, four species in two genera of dirk-tooth cats, and 17 species in three genera of elephants. Speciation of Quaternary alcelaphines was random with respect to the overall trends, because not *all* species evolve in the same morphological direction. For example, at least seven species proceed toward the overall trend for large individuals, but at least two other speciation events are toward smaller size (Vrba, 1984c, p. 66). Horn complexity increases in a minimum of eight species, but remains relatively simple in three others (see Fig. 2 in Vrba, 1984c). For most species the molar–premolar ratio decreases through time, but one of the latest appearing species, *Damaliscus dorcas*, has relatively long premolars (Vrba, 1980, p. 69).

4. DISCUSSION

Examples given in the last section comprise 98 species with well-documented data. The evidence one needs to test for phyletic gradualism or punctuated equilibrium falls into five data categories, which correspond to the five points of difference between the two models as explained in Table I. How the evidence distributes among the 98 Quaternary-mammal species and the five data categories is shown in Table II, where an E indicates that convincing data support punctuated equilibrium, and a G that convincing data support phyletic gradualism. These distributions can be examined both by taxon and by category to gain some idea of whether phyletic gradualism or punctuated equilibrium is supported more frequently.

4.1. Examination by Data Category

4.1.1. Category 1: Direction of Phenotypic Change within a Species

Only 14 of the 98 species provide convincing data about how phenotype changes through the life span of a species. Six of the species

TABLE II
Summary of Evidence Indicating Punctuated Equilibrium or Phyletic Gradualism for Species of Quaternary Mammals Discussed in the Text[a]

	1 Direction of phenotypic change within species	2 Anagenesis or cladogenesis	3 Morphological change at speciation	4 Nature of species	5 Cause of trends
Shrews					
Blarina brevicauda	E		E	E	
Blarina carolinensis	E		E	E	
Blarina hylophaga	E		E	E	
Hamsters					
Cricetulus bursae	E			E	
Rhinocricetus ehiki	E			E	
Cotton rats					
Sigmodon (15 species)					G ×15
Squirrels					
Spermophilus townsendii	E			E	
Voles					
Lagurus curtatus	G				
Mimomys savini		G	G	G	G
Arvicola cantiana	G	G	G	G	G
Arvicola terrestris	G	G	G	G	G
Ophiomys (five species)					G ×5
Pliophenacomys (three species)					G ×3

Muskrats					
Pliopotamys and *Ondatra* (seven species)					G ×7
Elephants					
Elephas ekorensis		G	G	G	G
Elephas recki	G	G	G	G	G
Elephas iolensis		G	G	G	G
Elephas hysudricus		G		G	G
Elephas maximus		G		G	G
Elephas namadicus			E	E	G
Mammuthus meridionalis	G	G	G	G	G
Mammuthus armeniacus		G	G	G	G
Mammuthus primigenius		G	G	G	G
Mammuthus imperator		G		G	G
Mammuthus columbi		G		G	G
Elephas planifrons					G
Elephas platycephalus					G
Loxodonta adaurora			E	E	G
Loxodonta africana			E	E	G
Pigs					
Nyanazachoerus tulotos					G
Nyanazachoerus kanamensis					G
Nyanazachoerus jaegeri					G

(Continued)

TABLE II (*Continued*)

	1 Direction of phenotypic change within species	2 Anagenesis or cladogenesis	3 Morphological change at speciation	4 Nature of species	5 Cause of trends
Pigs					
Notochoerus euilus					G
Notochoerus scotti					G
Mesochoerus limnetes	G				
Metridiochoerus andrewsi	G				G
Metridiochoerus modestus					G
Metridiochoerus hopwoodi					G
Metridiochoerus compactus					G
Bovids					
Alcelaphini (25 species)			E ×25	E ×25	E ×25
Aepycerotini (three species)		G ×2	G ×2, E ×1	G ×2, E ×1	
Hominids					
Homo erectus	G		G	G	
Dirk-toothed cats					
Smilodon and *Meganteron* (four species)					G ×4
Modern European species known as fossils[b]		G 25	G 25, E 86	G 25, E 86	
Total species 98					
Punctuated equilibrium	6		32	35	25
Phyletic gradualism	8	15	12	16	61
Cannot distinguish with available data	84	83	54	47	12

[a]E, Data provide convincing support for punctuated equilibrium; G, data provide convincing support for phyletic gradualism.
[b]Only suggestive (not firm) data are available; therefore these are not counted in the totals.

support punctuated equilibrium, and eight support phyletic gradualism. The relevant data are unknown for the remaining 84 species.

4.1.2. Category 2: Does Anagenesis Occur Frequently?

Anagenesis apparently took place in at least 15 species, which rejects punctuated equilibrium in those cases. The frequency of cladogenesis was not tallied, because it can occur in both punctuated equilibrium and phyletic gradualism. It is, however, common among the other 83 species. Data from this category only provide a minimum frequency for phyletic gradualism, because some of the cladogenetic events could also be encompassed in the model.

4.1.3. Category 3: Does Most Phenotypic Change Occur at Speciation?

Phenotypic change evidently was concentrated near the speciation event for 32 species and distributed throughout the stratigraphic range for 12 species. The prediction of punctuated equilibrium is fulfilled more frequently.

4.1.4. Category 4: Are Species Discrete Entities?

Species can be demarcated as discrete entities in 35 cases, as stipulated by punctuated equilibrium, but appear transitional in only 16 instances.

4.1.5. Category 5: Trends from Additive Microevolution or Species Sorting?

Most of the evolutionary trends are consistent with the model of phyletic gradualism. These include six different trends, which cumulatively involve 61 species. The fossil evidence suggests that only one trend, which involved 25 species, is consistent with what punctuated equilibrium predicts.

4.1.6. Quantitative versus Qualitative Morphological Traits

Mensural data support punctuated equilibrium in shrews and squirrels, but phyletic gradualism in voles, elephants, pigs, and *Homo erectus*. Qualitative characters, for example, the complexity of horn morphology and the presence or absence of a molar reentrant, suggest punctuated equilibrium in alcelaphine bovids, but phyletic gradualism in voles. Therefore, the correlation is at most weak between the kind of trait analyzed—qualitative or quantitative—and a particular model of evolution—punctuated equilibrium or phyletic gradualism. Nevertheless, it is important to recognize that qualitative traits are inherently more likely to show punctuated rather than gradual change, especially if only their presence or absence is the subject of analysis. One way to compensate for this bias is by tracing the frequency of a presence–absence trait among individuals of stratigraphically successive samples.

4.2. Examination by Taxon

4.2.1. Numbers of Species Supporting Each Model

Data in all five categories exist for only four species, all of which support phyletic gradualism. Data in three or four categories exists for 43 species; 31 of these argue for punctuated equilibrium. Therefore, of species that can provide data in at least three of the five data categories, 16 support phyletic gradualism, whereas 31 support punctuated equilibrium.

4.2.2. Correlation between Higher Taxon and Mode of Evolution

The distribution of species that support a given model is not random with respect to higher taxonomic group. Of the taxa for which evidence was available in at least three data categories, the shrews (soricids) and alcelephines (bovids) showed evolutionary patterns consistent with punctuated equilibrium. The voles (arvicolines), elephants (elephantids), and *Homo erectus* (hominid) strongly support phyletic gradualism; impalas (aepycerotines) also fit this model best. In taxa for which more limited data were available, two hamsters (cricetines) and a single squirrel (sciurid) suggest punctuated equilibrium, but muskrats (arvicolines), cotton rats (cricetines), pigs (suids), and dirk-tooth cats (felids) imply phyletic gradualism. Thus, evolution in three groups—soricids, alcelaphine bovids, and sciurids—seems inclined to punctuated equilibrium. In six others—arvicolines, elephants, pigs, impalas, *Homo erectus* and probably dirk-tooth cats—evolution tends toward phyletic gradualism. The cricetines may exhibit both patterns.

5. CONCLUSIONS

The Quaternary time scale comes into focus between the ecological moment and the usual conception of geologic time. That is, one can observe events that take place over thousands of years, rather than over the tens or hundreds that ecologists deal with, or the millions that studies of the pre-Quaternary encompass. The data presented here leave little doubt that, for Quaternary mammals, evolution proceeded within this intermediate time scale in ways consistent with the models of both phyletic gradualism and punctuated equilibrium. There is still considerable doubt, however, as to which mode of evolution was more frequent. An accurate assessment of this requires data from all five of the categories listed in Table II for more than just four species. Nevertheless, the patches of data that are available lead to some preliminary notions.

One or the other model seems to dominate within certain taxa. Punctuated equilibrium is evident for shrews, alcelaphine bovids, and probably squirrels and hamsters. It is worth noting that shrews and squirrels are morphologically conservative for most of their fossil history—stasis for such groups may go back a long way. Phyletic gradualism is apparently the norm for voles and muskrats, elephants, pigs, impalas, *Homo erectus*, and probably cotton rats and dirk-tooth cats. Therefore, if one considers only the higher taxa analyzed here, phyletic gradualism appears as the more frequent mode of evolution.

But if one considers instead only the total numbers of species consistent with a given model, punctuated equilibrium is supported twice as often as phyletic gradualism. The evidence leading to that conclusion is weighted heavily by a single taxonomic group of 25 alcelaphine species, and is found mainly in two of the five data categories (3 and 4, Table II)—the majority of species considered exhibit most of their morphological change near a speciation event, and most species seem to be discrete entities.

Punctuated equilibrium does not fare as well in the other two data categories that are able to test for it (1 and 5, Table II). It presently is impossible to tell if this is an artifact of the limited facts at hand or a real discrepancy. For example, if most of the alcelaphine bovids show stasis (which they may well, although I have not seen the data that would allow me to say for sure), category 1 undoubtedly would support the thesis of punctuated equilibrium. Category 5, on the other hand, presents a serious challenge to the model of punctuated equilibrium as now formulated—most of the evolutionary trends we have recognized in Quaternary mammals are not easily attributed to sorting among species.

Definitive resolution of such conflicting conclusions must of course await additional studies of how morphology changes within and among related species through well-dated stratigraphic intervals. Such paleontological studies are the best and most applicable tools we have to sculpt the models of phyletic gradualism and punctuated equilibrium into their finished form, or, if empiricism demands it, to disassemble them. The Quaternary mammalian record abounds in the information we need for refining and testing these two conceptions of evolution. Its relatively minor utilization already has shown that both punctuated equilibrium and phyletic gradualism can be corroborated over spans of time that are short in the geologic sense, but long in ecological terms—that is, through as little as 10,000–30,000 years, up to 2 million years. This portends an exciting possibility—bridging the temporal gap between studies of evolution at the ecological time scale and those at the geologic scale. We should not ignore the opportunity.

ACKNOWLEDGMENTS. I thank Mary R. Dawson, Leonard Krishtalka, Richard K. Stucky, and Carleton J. Phillips for reading an early draft of the manuscript and providing helpful comments and discussion; Nancy J. Perkins (Carnegie Museum Exhibits Division) for kindly drafting Figs. 1–4, and 6; Milford H. Wolpoff and J. John Sepkoski, Jr., for allowing reproduction of Fig. 7 from *Paleobiology;* Longman Group Ltd. for allowing reproduction of Fig. 5 from A. J. Stuart's book, *Pleistocene Vertebrates in the British Isles;* and Gerard McKiernan (Carnegie Museum Library) for help with the literature search.

REFERENCES

Barnosky, A. D., 1985, Late Blancan (Pliocene) microtine rodents from Jackson Hole, Wyoming: Biostratigraphy and biogeography, *J. Vertebr. Paleontol.* **5**(3):255–271.

Chaline, J., 1984, The polyphased evolution concept and its implications, *Geobios* **17**(6): 783–795.

Charlesworth, B., and Lande, R., 1982, Morphological stasis and developmental constraint: No problem for Neo-Darwinism, *Nature* **296**:610.

Charlesworth, B., Lande, R., and Slatkin, M., 1982, A Neo-Darwinian commentary on macroevolution, *Evolution* **36**(3):474–498.

Cooke, H. B. S., 1978, Suid evolution and correlation of African hominid localities: An alternate taxonomy, *Science* **201**:460–463.

Cope, J. C. W., and Skelton, P. W. (eds.), 1985, *Evolutionary Case Histories from the Fossil Record*, Special Papers in Palaeontology No. 33, Palaentological Association, London.

Corbet, G. B., 1978, *The Mammals of the Palaearctic Region—A Taxonomic Review*, British Museum (Natural History), London, and Cornell University Press, Ithaca, New York.

Cronin, J. E., Boaz, N. T., Stringer, C. B., and Yak, Y., 1981, Tempo and mode in hominid evolution, *Nature* **292**:113–122.

Darwin, C., 1859, *The Origin of Species*, Murray, London.

Eldredge, N., 1971, The allopatric model and phylogeny in Paleozoic invertebrates, *Evolution* **25**:156–167.

Eldredge, N., and Cracraft, J., 1980, *Phylogenetic Patterns and the Evolutionary Process*, Columbia University Press, New York.

Eldredge, N., and Gould, S. J., 1972, Punctuated equilibria: An alternative to phyletic gradualism, in: *Models in Paleobiology* (T. J. Schopf, ed.), Freeman, San Francisco, pp. 82–115.

Emerson, S. B., and Radinsky, L., 1980, Functional analysis of sabertooth cranial morphology, *Paleobiology* **6**(3):295–312.

Gile, L. H., Peterson, F. F., and Grossman, R. B., 1965, The K horizon: A master soil horizon of carbonate accumulation, *Soil Sci.* **99**:74–82.

Gile, L. H., Peterson, F. F., and Grossman, R. B., 1966, Morphological and genetic sequences of carbonate accumulations in desert soils, *Soil Sci.* **101**:347–360.

Gingerich, P. D., 1976, Paleontology and phylogeny: Patterns of evolution at the species level in early Tertiary mammals, *Am. J. Sci.* **276**:1–28.

Gingerich, P. D., 1984, Punctuated equilibria—Where is the evidence?, *Syst. Zool.* **33**(3):335–338.

Gingerich, P. D., 1985, Species in the fossil record: Concepts, trends, and transitions, *Paleobiology* **11**(1):27–41.

Ginzburg, L. R., and Rost, J. D., 1982, Are 'punctuations' artifacts of time-scales?, *Nature* **296**:610.

Gould, S. J., 1982, The meaning of punctuated equilibrium and its role in validating a hierarchical approach to macroevolution, in: *Perspectives on Evolution* (R. Milkman, ed.), Sinauer, Sunderland, Massachusetts, pp. 83–104.

Gould, S. J., 1985, The paradox of the first tier: An agenda for paleobiology, *Paleobiology* **11**(1):2–12.

Gould, S. J., and Eldredge, N., 1977, Punctuated equilibria: The tempo and mode of evolution reconsidered, *Paleobiology* **3**:115–151.

Graham, R. W., and Semken, H. A., 1976, Paleoecological significance of the short-tailed shrew (*Blarina*), with a systematic discussion of *Blarina ozarkensis*, *J. Mammal.* **57**:433–449.

Guilday, J. E., Hamilton, H. W., Anderson, E., and Parmalee, P. W., 1978, The Baker Bluff cave deposit, Tennessee, and the late Pleistocene faunal gradient, *Carnegie Mus. Nat. Hist. Bull.* **11**:1–67.

Harris, J. M., and White, T. D., 1979, Evolution of the Plio-Pleistocene African Suidae, *Trans. Am. Phil. Soc.* **69**(2):1–128.

Hibbard, C. W., and Zakrzewski, R. J., 1967, Phyletic trends in the late Cenozoic microtine, *Ophiomys* gen. nov., from Idaho, *Univ. Mich. Mus. Paleontol. Contrib.* **21**(12):255–271.

Hibbard, C. W., and Zakrzewski, R. J., 1972, A new species of microtine from the late Pliocene of Kansas, *J. Mammal.* **53**:834–839.

Hoffman, A., 1982, Punctuated versus gradual mode of evolution: A reconsideration, in: *Evólutionary Biology*, Vol. 15 (M. K. Hecht, B. Wallace, and G. T. Prance, eds.), Plenum Press, New York, pp. 411–436.

Johnson, J. G., 1982, Occurrence of phyletic gradualism and punctuated equilibria through geologic time, *J. Paleontol.* **56**(6):1329–1331.

Jones, C. A., Choate, J. R., and Genoways, H. H., 1984, Phylogeny and paleobiogeography of short-tailed shrews (genus *Blarina*), in: *Contributions in Quaternary Vertebrate Paleontology: A Volume in Memorial to John E. Guilday* (H. H. Genoways and M. R. Dawson, eds.), Carnegie Museum of Natural History Special Publication 8, Pittsburgh, pp. 56–148.

Kurtén, B., 1968, *Pleistocene Mammals of Europe*, Aldine, Chicago.

Levinton, J. S., 1983, Stasis in progress: The empirical basis of macroevolution, *Annu. Rev. Ecol. Syst.* **14**:103–137.

Levinton, J. S., and Simon, C. M., 1980, A critique of the punctuated equilibrium model and implications for the detection of speciation in the fossil record, *Syst. Zool.* **29**:130–142.

Maglio, V. J., 1970, Early Elephantidae of Africa and a tentative correlation of African Plio-Pleistocene deposits, *Nature* **225**:328–332.

Maglio, V. J., 1973, Origin and evolution of the Elephantidae, *Trans. Am. Phil. Soc.* **63**(3):1–149.

Malmgren, B. A., and Kennett, J. P., 1981, Phyletic gradualism in a late Cenozoic planktonic foraminiferal lineage: DSDP site 284, southwest Pacific, *Paleobiology* **7**:230–240.

Malmgren, B. A., and Kennett, J. P., 1983, Gradual evolution in a planktonic foraminiferal lineage reconsidered, phyletic gradualism in the *Globorotalia inflata* lineage vindicated, *Paleobiology* **9**:427–428.

Martin, J. E., Barnosky, A. D., and Barnosky, C. W., 1983, Fauna and flora associated with the West Richland mammoth from the Pleistocene Touchet beds in south-central Washington, *Thomas Burke Memorial Wash. State Mus. Res. Rep.* **3**:1–61.

Martin, L. D., 1979, The biostratigraphy of the arvicoline rodents in North America, *Trans. Nebr. Acad. Sci.* **7**:91–100.

Martin, L. D., 1980, Functional morphology and the evolution of cats, *Trans. Nebraska Acad. Sci.* **8**:141–154.

Martin, L. D., 1984, Phyletic trends and evolutionary rates, in: *Contributions in Quaternary Vertebrate Paleontology: A Volume in Memorial to John E. Guilday* (H. H. Genoways and M. R. Dawson, eds.), Carnegie Museum of Natural History Special Publication 8, Pittsburgh, pp. 526–538.

Martin, R. A., 1979, Fossil history of the rodent genus *Sigmodon*, *Evol. Monogr.* **2**:1–36.

Martin, R. A., 1984, The evolution of cotton rat body mass, in: *Contributions in Quaternary Vertebrate Paleontology: A Volume in Memorial to John E. Guilday* (H. H. Genoways and M. R. Dawson, eds.), Carnegie Museum of Natural History Special Publication 8, Pittsburgh, pp. 179–183.

Mayr, E., 1982, Speciation and macroevolution, *Evolution* **36**(6):1119–1131.

Nadler, C. F., Lyapunova, E. A., Hoffman, R. S., Vorontsov, N. N., Shaitarova, L. L., and Borisov, Y. M., 1984, Chromosomal evolution in Holarctic ground squirrels (*Spermophilus*) II. Giemsa-band homologies of chromosomes and the tempo of evolution, *Sonderdruck Z. Säugetierknd.* **49**:78–90.

Nelson, R. S., and Semken, H. A., 1970, Paleoecological and stratigraphic significance of the muskrat in Pleistocene deposits, *Geol. Soc. Am. Bull.* **81**:3733–3738.

Newman, C. M., Cohen, J. E., and Kipnis, C., 1985, Neo-Darwinian evolution implies punctuated equilibria, *Nature* **315**:400–401.

Ozawa, T., 1975, Evolution of *Lepidolina multiseptata* (Permian foraminifer) in East Asia, *Mem. Fac. Sci. Kyushu Univ. Ser. D Geol.* **23**:117–164.

Pye, K., 1983, Grain surface textures and carbonate content of late Pleistocene loess from West Germany and Poland, *J. Sediment. Petrol.* **53**:973–980.

Rensberger, J. M., 1973, An occlusion model for mastication and dental wear in herbivorous mammals, *J. Paleontol.* **47**(3):515–528.

Rensberger, J. M., Barnosky, A. D., and Spencer, P. K., 1984, Geology and Paleontology of a Pleistocene-to-Holocene Loess Succession, Benton County, Washington, Archaeological and Historical Services Eastern Washington University Reports in Archaeology and History 100-39.

Repenning, C. A., 1983, Quaternary rodent biochronology and its correlation with climatic and magnetic stratigraphies, Paper presented at Symposium on Correlation of Quaternary Chronologies, Atkinson College, York University, Toronto, Canada, May 1983.

Rhodes, H. T., 1983, Gradualism, punctuated equilibrium and the *Origin of Species*, *Nature* **305**:269–272.

Rightmire, G. P., 1981, Patterns in the evolution of *Homo erectus*, *Paleobiology* **7**(2):241–246.

Rose, K. D., and Bown, T. M., 1984, Gradual phyletic evolution at the generic level in early Eocene omomyid primates, *Nature* **309**:250–252.

Sarna-Wojcicki, A. M., Champion, D. E., and Davis, J. O., 1983, Holocene volcanism in the conterminous United States and the role of silicic volcanic ash layers in correlation of latest-Pleistocene and Holocene deposits, in: *Late-Quaternary Environments of the United States*, Volume 2, *The Holocene* (H. E. Wright, Jr., ed.), University of Minnesota Press, Minneapolis, pp. 52–77.

Simpson, G. G., 1944, *Tempo and Mode in Evolution*, Columbia University Press, New York.

Simpson, G. G., 1953, *The Major Features of Evolution*, Columbia University Press, New York.

Stanley, S. M., 1975, A theory of evolution above the species level, *Proc. Natl. Acad. Sci. USA* **72**:646–650.

Stanley, S. M., 1978, Chronospecies longevities, the origin of genera, and the punctuational model of evolution, *Paleobiology* **4**(1):26–40.

Stanley, S. M., 1979, *Macroevolution: Pattern and Process*, Freeman, San Francisco.

Stanley, S. M., 1982, Macroevolution and the fossil record, *Evolution* **36**:460–473.

Stebbins, G. L., and Ayala, F. J., 1981, Is a new evolutionary synthesis necessary?, *Science* **213**:967–971.

Stuart, A. J., 1982, *Pleistocene Vertebrates in the British Isles*, Longman, New York.

Tucholka, P., 1977, Magnetic polarity events in Polish loess profiles, *Biul. Inst. Geol.* **305**:117–123.

Van Eysinga, F. W. B., 1978, *Geological Time Table*, 3rd ed., Elsevier, Amsterdam.

Vrba, E. S., 1980, Evolution, species and fossils: How does life evolve?, *S. Afr. J. Sci.* **76**(2):61–84.

Vrba, E. S., 1983, Macroevolutionary trends: New perspectives on the roles of adaptation and incidental effect, *Science* **221**:387–389.

Vrba, E. S., 1984a, What is species selection?, *Syst. Zool.* **33**(3):318–327.

Vrba, E. S., 1984b, Patterns in the fossil record and evolutionary processes, in: *Beyond Neo-Darwinism: An Introduction to the New Evolutionary Paradigm* (M. W. Ho and P. T. Sanders, eds.), Academic, New York, pp. 115–142.

Vrba, E. S., 1984c, Evolutionary pattern and process in the sister-group Alcelaphini–Aepycerotini (Mammalia: Bovidae), in: *Living Fossils* (N. Eldredge and S. M. Stanley, eds.), Springer, New York, pp. 62–79.

Wajtanowicz, J., and Buraczynski, J., 1978, Materials to the absolute chronology of the loesses of Grzeda Skolska, *Ann. Univ. Mariae Curie-Sklodowska Sect. B* **30/31**:37–54.

Warren, J. K., 1983, Pedogenic calcrete as it occurs in calcareous dunes in coastal South Australia, *J. Sediment. Petrol.* **53**:787–796.

Wolpoff, M. H., 1984, Evolution in Homo erectus: The question of stasis, *Paleobiology* **10**(4):389–406.

Zakrzewski, R. J., 1974, Fossil Ondatrini from western North America, *J. Mammal.* **55**(2):284–292.

CHAPTER 5

MAMMALIAN REPRODUCTIVE PHYSIOLOGY

Adaptive Responses to Changing Environments

NORMAN C. NEGUS and PATRICIA J. BERGER

1. INTRODUCTION

Not many years ago, environments were regarded as being relatively stable and species were presumed to have evolved life histories in equilibrium with such environments. More recently, however, attention has been drawn to the fluctuating and uncertain nature of most environments and the necessity of organisms to adapt to this heterogeneity or perish (Levins, 1968). Biologists have begun seriously to consider life history strategies as phenotypic traits that are responding to sets of environmental variables. Today there is increasing awareness of the great complexity of selective forces acting on the evolution of life histories at both the genetic and phenotypic levels (Boyce, 1984; Emlen, 1984).

Because the ultimate "goal" of a life history strategy is to maximize fitness, it follows that much attention has been directed to the numerous parameters associated with reproductive success. However, the prob-

NORMAN C. NEGUS and PATRICIA J. BERGER • Department of Biology, University of Utah, Salt Lake City, Utah 84112.

lem is immensely more complex. On the one hand, the environment can range from highly predictable in space and time with high autocorrelation of events from year to year, to extreme temporal and spatial heterogeneity and great uncertainty. On the other hand, the organism is confronting its milieu and responding to it in such a way as to optimize survival and fitness. This dynamic interaction between environment and organism is the stuff of which life history strategies are made, and may involve selection for size, longevity, developmental rate, food habits, mating systems, and litter sizes, not to mention a host of other physiological parameters associated with energy allocation, reproduction, and reception of environmental information. Add to this the fact that the operational selective forces vary from one species to another and even within the same species in different environments, and one begins to glimpse the difficulty in developing robust theories concerning the evolution of life history strategies. Nonetheless, considerable progress from numerous vantage points has increased our understanding of the mechanistic aspects of reproductive strategies. Accordingly, in this paper we will attempt to summarize the patterns that are emerging in mammalian reproductive adaptations.

2. ENVIRONMENTAL CONSTRAINTS ON MAMMALIAN REPRODUCTION

Mammals are endotherms. As such, their homeostatic energy budgets are high as reflected in their basal metabolic rates (BMR). The problem of heat production and heat loss is a pervasive force affecting the energy budget of all mammals in a similar way regardless of size (Kleiber, 1932). Thus, BMR in mammals is proportional to mass raised to the ¾ power. It follows that with increase in size, the mass-specific metabolic rate decreases. Accordingly, a shrew-sized mammal has a high mass-specific metabolic rate, whereas a moose has a low mass-specific rate. This curious relationship exerts a potent influence on the allocation of energy to reproduction in mammalian species. In general, mammals with high metabolic rates have high reproductive rates, short generation times, and high growth rates. Why this should be so is not clear at present, although some insight may be gained by comparing closely related taxa. McNab (1980) has done much to elucidate the importance of such comparisons. Although the Kleiber curve reveals the general relation of size to metabolic rate in mammals, it gives no hint of the metabolic differences among closely related taxa. The difficulty lies in the fact that so few species have been studied with regard to their metabolic rates and reproductive parameters. Consider, for example, the microtine rodents. On the basis of metabolic determinations of *Microtus pennsylvanicus* and *montanus* (Bradley, 1976; V. Collins, unpublished), one is tempted to conclude that the genus exhibits high metabolic rates compared to similar-sized rodents. However, recent studies of the metabolic rate in *Microtus longicaudus* in our laboratory (V. Collins, unpublished) have demonstrated that this species has a significantly higher BMR than either *montanus* or *pennsylvanicus*. Similarly, the red-backed vole (*Clethrionomys gapperi*) has a higher BMR than either *M. pennsylvanicus* or *montanus*. However, the reproductive rate of *M. longicaudus* is lower than that of either *M. montanus* or *pennsylvanicus* (Conley, 1976) and populations of this species do not usually fluctuate greatly (Farris, 1971). *Microtus californicus* also has a metabolic rate similar to *M. montanus* and *pennsylvanicus* (Table I). The relationship that begins to emerge is that those microtines specializing primarily in monocotyledenous food resources (*M. californicus, pennsylvanicus, montanus*) have lower metabolic rates and higher reproductive rates than those species relying more on dicotyledenous plants and more varied diets. Thus, in microtines, high fecundity seems to be associated with low-quality food resources and low metabolic rates, in contrast to McNab's (1980) suggestion. This simply demonstrates how the scarcity of comparative data from related species can be quite misleading. It further points to the great need for comparative metabolic studies of related taxa, because there are strong correlations among metabolic rates, food habits, and reproductive strategies in mammals.

Not only is there a correlation with size and metabolic rate in mammals, but there is also one with longevity. In general, the larger

TABLE I
Standard Metabolic Rates (SMR) of Six Species of Microtine Rodents[a]

Species	*n*	SMR (cm^3 O_2/g per hr)	References
Microtus longicaudus	10	2.263 ± 0.439	V. Collins (unpublished)
Clethrionomys gapperi	10	2.01 ± 0.316	Bradley (1976)
Microtus pennsylvanicus	15	1.93 ± 0.387	Bradley (1976)
Microtus montanus	10	1.674 ± 0.186	V. Collins (unpublished)
Microtus californicus	8	1.43 ± 0.220	V. Collins (unpublished)
Microtus mexicanus	7	1.63 ± 0.185	Bradley (1976)

[a]Data from Bradley (1976) and V. Collins (unpublished). *Microtus longicaudus* and *Clethrionomys gapperi* have dietary specialization for dicotyledons during the breeding season. *Microtus pennsylvanicus, montanus, californicus,* and *mexicanus* specialize more on monocotyledons. SMR values are given as mean ± SD.

the mammal, the greater the longevity, with numerous exceptions among those forms that hibernate. Here again the reasons for the high correlation between metabolic rate and longevity are not clear, although it seems likely that rate of aging is associated with metabolic rate. In any event, many small mammals are faced with the problem of living 1 year or less. Clearly, such a constraint on life span imposes intense selective pressures on the evolution of reproductive strategies. The short-lived mammal must somehow manage to accumulate sufficient energy for growth, seasonal pelage molting, and reproductive effort, all of which require considerable energy above the homeostatic level. Environments that are temporally inhospitable and unpredictable pose immense problems to such species. Perhaps the most common solution to this problem by short-lived mammals is the selection of high-quality food resources that have maximal temporal predictability in a given environment. Accordingly, on a global basis, we would expect to find relatively few species of short-lived herbivorous mammals, and many more that are either seed-eaters or omnivorous in diet. Herbivory is a hazardous life style for a short-lived mammal, particularly if there is specialization for monocotyledenous plants, which are of low nutritional value on a gram-specific basis (Batzli and Cole, 1979). Microtine rodents seem to be the primary group of small, short-lived mammals that have successfully occupied this niche even in unpredictable environments. However, this requires extreme adaptation to temporal heterogeneity of resources (Negus and Berger, 1986), which we will consider in more detail later.

Irrespective of food habits, short longevity imposes severe constraints on life history evolution. In most terrestrial environments there are periods of scarcity and abundance of food items. If the autocorrelation of these events is high, then the availability of energy resources is predictable and a species may evolve a life history strategy with little or no phenotypic plasticity. Such species show little variation in the time of breeding each year and their reproductive strategy may be described as "obligate" (Negus and Berger, 1972). Under conditions of low autocorrelation of phenological events, short-lived species are faced with the problem of tracking the availability of sufficient energy resources to support reproduction. Under these circumstances there should be intense selection for mechanisms of information retrieval from the environment. The resulting "facultative" reproductive strategies will exhibit considerable phenotypic plasticity in some or all parameters of the life histories. Clearly, longevity and food habits are inseparably coupled to the energy budget of a species, which in large measure determine the selective forces that shape the evolution of life histories.

Given that the reproductive strategy peculiar to each species is a consequence of the selective forces associated with acquisition of sufficient energy from the environment, we can begin to dissect the problems confronting different kinds of mammals. From the energetic viewpoint, three aspects of the environment are of direct importance to the energy budget of a mammal: (1) food resources, (2) water, and (3) ambient temperature. In terrestrial environments these resources vary both temporally and spatially. Adaptive responses to their heterogeneity form the basis of reproductive strategies as we observe them. Superimposed on direct energy-related interactions are density-related factors such as competition and social behavior, but these are of indirect, secondary influence on reproductive strategies.

In temperate and polar habitats, virtually all mammalian food resources exhibit strong seasonal changes in quantity and quality. More than any other resource, the dynamics of food availability dictates the timing of mammalian reproduction (Bronson, 1985). Even in the aseasonal environments of the tropics many food resources fluctuate greatly in response to rainy and dry periods, and form the basis for the timing of reproduction in species that are not continuous breeders (McNab, 1980). Some environments express high autocorrelation of phenological events associated with food availability from year to year, whereas others exhibit various lesser degrees of autocorrelation. Mammals of differing sizes and longevities respond very differently to temporal heterogeneity of food resources. Large, mobile forms can move long distances to favorable food patches, or simply not breed at all and wait for a more favorable year. Small, short-lived mammals do not have such options and must somehow "make sense" out of their fluctuating environment. In either case, reproduction will always occur at the optimal time for offspring survival in terms of food resources.

Water as a resource does not impose the energetic constraints on mammalian reproduction that are so evident in food resources. Rather, the necessity of water for metabolic processes to occur represents the limiting aspect of water availability. In xeric environments that are either seasonally or always dry, water plays a crucial role in the regulation of mammalian reproduction (Beatley, 1969). Even in desert-adapted forms with highly efficient kidney function, the demands of lactation require a level of water intake that is frequently not available in desert habitats. Thus, the temporal availability of water becomes the regulator of reproductive timing in some desert mammals (Christian, 1979; Beatley, 1976; Van de Graaf and Balda, 1973).

The thermal environment represents an energetic resource in that it affects rates of heat loss and heat gain in the endothermic mammal.

As such, the thermal environment greatly influences mammalian homeostatic energy budgets. In large mammals this may have little effect on reproduction, but in small species with higher surface-to-volume ratios, the high energetic costs of homeostasis in unfavorable ambient temperatures may conflict with the energetic requirements for reproduction. In temperate and polar regions low ambient temperatures typically coincide with seasons of low food quality and availability. Accordingly, the thermal environment can have the effect of decreasing the time period in which there is acceptable energy flux for reproduction in many species of mammals.

3. INFORMATION CONTENT OF THE ENVIRONMENT

Up to this point we have simply described the interactions of mammalian size, longevity, and energy budgets with environmental resource heterogeneity. All organisms adapt in analogous ways to changing environments. This requires without exception that all organisms evolve mechanisms to receive relevant information from the environment and transduce the information content into useful responses. Viewed in this light, the information content of the environment assumes a role of primary importance to organisms. This is not a new idea. Levins (1968) emphasized the critical role of environmental information in the evolution of adaptive strategies.

Photoperiod is the most widely studied environmental information signal with regard to mammalian reproduction. In temperate and polar environments the information content of photoperiod is often high because day length accurately predicts forthcoming periods of favorable and unfavorable conditions. Many mammalian species, particularly those with relatively long life spans, respond to day length as the primary environmental signal for the timing of reproduction (Sadleir, 1972). However, in some environments, the information content of photoperiod may be very low with regard to optimal breeding time for a mammal. For example, in harsh deserts, where there may be little or no predictable seasonality to the rainfall patterns, photoperiod may convey no useful information regarding forthcoming food resources and water availability. Further, for short-lived species, there may be more noise than information in photoperiod signals. This results primarily from the variable lag time that is associated with day-length predictions and seasonal changes. Short-lived mammals have short generation times and cannot afford the luxury of variable lag times in their information retrieval systems, since this would almost certainly lead to extinction over time. In tropical latitudes as well, photoperiod may have low information content. Tropical day-length changes are subtle and are often poor predictors of the rainy and dry seasons that are the basis of ebb and flow of food resources in tropical environments (Bronson, 1985). In contrast, then, to Sadleir's (1972) suggestion that photoperiod is the only information by which most mammals cue initiation and cessation of breeding, it appears likely that a large number of the extant species of mammals find photoperiod information of little value in the timing of reproduction.

Ambient temperature, while of crucial importance to reproduction in ectothermic animals, seems to play a secondary role in mammals. The high variability of ambient temperature on a short-term basis means that the information content of the thermal environment is low. Unless a mammal can somehow accumulate information on averages from many data points, it appears that temperature offers little predictive information. On the other hand, the path between ambient temperature and energetic response in a mammal is very short. This means that while temperature is not of value for long-term predictions, it conveys immediate information concerning homeostatic requirements and can modify reproductive effort (Bronson and Pryor, 1983).

Food quality is of particular importance with regard to reproductive processes. The ability of mammals to obtain an adequate caloric intake to satisfy the needs of both pregnancy and lactation represents perhaps the primary constraint on the success of the breeding effort. Food restriction has been shown to have severe limitations on reproduction in wild house mice (Hamilton and Bronson, 1985) and rats (Glass and Swerdloff, 1980). Similarly, supplementing food to natural populations (Cole and Batzli, 1978; Taitt and Krebs, 1981) or to livestock maintained on poor pasture (Crichton et al., 1959) has been shown to enhance reproduction. These and other studies have concentrated primarily on the caloric and macronutrient content (protein, fats, vitamins, etc.) of the food and little attention has been given to the predictive value of the food resource itself. The use of cues directly associated with the food resource that serve as predictors of forthcoming abundance should be highly advantageous for animals living in unpredictable environments. Such cues would provide immediate information, regardless of other environmental parameters.

4. MAMMALIAN RESPONSES TO ENVIRONMENTAL INFORMATION

In the foregoing discussion we have attempted to portray the problems confronting various kinds of mammals in fluctuating environments as well as the potential information content of certain environmental parameters. Essentially, we have suggested that the energetic constraints in different kinds of mammals determine both how they view their environments as well as what information is essential to evolve an optimal reproductive strategy. In a more specific mechanistic sense, we can now ask, how do mammals make sense out of their environments? Obviously there is no simple model or answer to this question, but perhaps we can arrive at a few generalities.

Of paramount importance in the life history of any organism is the appropriate timing of reproduction. In mammals, with a very large range in generation time, the solution to this problem may be simple or complex. If the generation time of a species is long (>4 months), then the information content of photoperiod is usually high and can be effectively employed as a sufficient cue for reproduction. Although few species have been experimentally tested in this regard, it is apparent that many species with long generation times probably cue reproduction exclusively to photoperiod (Sadleir, 1972). Large mammals such as cervids and bovids with long gestation periods typically initiate breeding in response to decreasing photoperiod. The young are thus placed in the environment during the optimal time of food resource availability the following year. Even though there may be variation in the phenology from year to year, the strategy is successful statistically. In species with more intermediate generation times (that is, canids, felids, lagomorphs), breeding is typically initiated in response to increasing photoperiod, and the young enter the environment with favorable resources. In many of the temperate-latitude mustelids, photoperiod plays a dual role in the reproductive strategy. For example, *Mustela erminea* breeds in late summer in response to decreasing photoperiod. The blastocysts enter a diapause state and do not implant until late winter in response to increasing photoperiod (Wright, 1963). A normal gestation period follows and the young enter the environment at the optimal time of the year. The adaptive significance of delayed implantation in mustelids is unknown. Because this is a strategy on the part of the female, it seems possible that late summer breeding is designed to increase the probability of successful fertilization. Because mustelids are carnivores with relatively large home ranges and low densities, the probability of mating with a male is greatest at the end of summer, when densities are highest.

However, even for large mammals, if the environmental conditions are sufficiently enigmatic, photoperiod may not provide adequate information. Some of the macropod marsupials of Australia illustrate this point. The macropod generation time is long as a result of a short gestation period coupled with an extensive period of pouch life. In many parts of Australia, the temporal uncertainty of rainfall often far exceeds the generation time of even the largest macropods such as the red kangaroo (*Megaleia rufa*). Thus, photoperiod has no information content as a reproductive cue in most environments for these large herbivores. Under these conditions of uncertainty, the red kangaroo has become a continuous breeder (Sharman and Pilton, 1964). The female returns to estrus following parturition and mates. The resulting blastocyst is maintained in a state of diapause by lactation until the pouch young leaves the pouch or dies, at which point the blastocyst implants and the process is repeated. Under the stress conditions of prolonged drought, females enter anestrus following parturition. The cue to return to the estrus condition is apparently received via the vegetational food resource following a rainfall sufficient to elicit a flush of grass growth (Newsome, 1964a,b, 1965, 1966). In this case the temporal heterogeneity of the environment is so extreme that the only useful information to be gained is from the food resource itself.

Not all of the macropods exist in such unpredictable environments as the red kangaroo. In contrast, the tammar wallaby (*Macropus eugeneii*) inhabits the highly predictable seasonal environment of Kangaroo Island off the southeast coast of Victoria. The tammar has become a strict seasonal breeder, relying totally on photoperiod information to cue the timing of reproduction (Berger, 1970). Nulliparous females enter estrus within 10 days after the summer solstice in response to decreasing photoperiod and there is no delay in development. Following parturition, postpartum estrus and mating occurs, after which the resulting blastocyst enters an 11-month diapause (Berger, 1966). Decreasing photoperiod following the next summer solstice elicits blastocyst implantation and parturition. The rainy season occurs during the winter months on Kangaroo Island, and the pouch young are placed at foot during the optimal food resource conditions. Thus, in two closely related genera of macropods we see highly facultative phenotypic plasticity in the red kangaroo's timing of reproduction, and an obligate genetic response to photoperiod in the tammar wallaby. When tammar wallabies were transported across the equator to southern Louisiana in September 1966, within several weeks they cued to the decreasing photoperiod and became reproductively active (Fig. 1). Here, then, is an example of the differential information content of photoperiod under very different degrees of temporal heterogeneity in the environment.

In species of mammals with very short generation times, the problem of appropriately timing reproduction can be difficult. The problem is complicated by the fact that most species of mammals with short

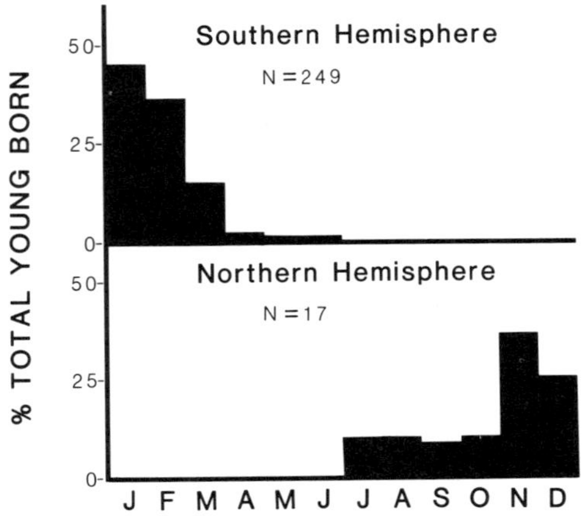

FIGURE 1. Breeding seasons of the tammar wallaby in the northern and southern hemispheres. Data for the southern hemisphere represents pouch young of females sampled from the wild populations on Kangaroo Island, South Australia, during 1965–1967. Date of birth was estimated from growth rates of pouch young in captivity. Data for the northern hemisphere represents pouch young of females in a captive colony at Tulane University, New Orleans, Louisiana, 1966–1968. (After Berger, 1970.)

generation times are very small. This means that energetic constraints are more severe than in larger forms (Bronson, 1985). Consequently, the food habits of a small mammal may have considerable influence on the type of reproductive strategy that evolves. In this sense it would seem most advantageous for a small mammal to be a generalist in food habits and switch from one food item to another as availability fluctuated. In fact, some of the most successful small rodents (for example, *Peromyscus leucopus, Peromyscus maniculatus, Mus musculus*) are extreme generalists (Bronson, 1979). As such, they have become highly opportunistic breeders, relying essentially on nutritional intake as the basic regulator of reproductive timing. However, competition and other factors seem to have forced many small mammals into more specialized food niches with attendant effects on their reproductive strategies (for example, heteromyid rodents, microtine rodents).

Definitive reproductive data on the seed-eating desert heteromyid rodents are essentially lacking. However, field studies have established that many species will breed at any time of the year if conditions are favorable (Kenagy and Bartholomew, 1985). Favorable conditions seem to be associated with sufficient rainfall to stimulate a wave of germination of desert annuals (Beatley, 1969; Reichman and Van de Graff, 1975). Photoperiod contains no useful information for many of the desert heteromyids, whereas the germination of desert annuals is a prediction of sufficient water for lactation. In at least one species of kangaroo rat, *Dipodomys microps*, the habit of peeling and eating saltbush (*Atriplex*) and blackbrush (*Coleogyne*) leaves (Kenagy, 1972; Csuti, 1979) seems to release this species from the constraints of water availability. *Dipodomys microps* is a seasonal breeder (Csuti, 1979; Quinn,

1983) and appears to use increasing photoperiod as the cue to begin breeding. In contrast, *D. merriami*, in the same habitat, breeds in response to germination of annuals and grasses. In this case congeners in the same location utilize different information from the environment to time their reproductive efforts.

Small mammals that specialize in herbivory confront formidable problems in many environments. Short generation times coupled with low-quality food resources that fluctuate drastically on a temporal basis demand that reproduction be timed with accuracy.

Microtine rodents are the most successful group of small mammals that have opted for a herbivorous life style. They deserve special consideration here for several reasons: (1) they are perhaps the most intensively studied group of wild mammals in the world, with literally thousands of publications on various aspects of their biology; (2) as small herbivorous mammals living primarily in strongly seasonal environments, they have been forced to adapt their reproductive strategies to a wide spectrum of environmental uncertainty; (3) it is primarily this group of mammals (aside from some lagomorphs and predator species) that has attracted widespread interest for many years concerning their enigmatic multiannual population fluctuations (Elton, 1942; Finerty, 1982); and (4) a number of microtine species have been successfully colonized in the laboratory, permitting the attainment of valuable experimental reproductive data.

At the outset, a consideration of reproductive strategies in microtine rodents requires that we recognize that microtines are a highly varied group. Microtines have been generally regarded as herbivores, but not all microtines are herbivores, and herbivory itself is a very general term (Batzli, 1985). Some species specialize primarily on grasses and sedges (monocotyledons), whereas others prefer herbaceous and shrubby plants (dicotyledons). Still others may include seeds and fruits as a large portion of their diet (Table II). These represent important differences in nutritional quality and seasonal predictability of food resources. On the basis of energetic constraints, we would expect such dietary differences to be reflected in the reproductive strategies of various species. Unfortunately, definitive data are lacking for many species. However, on the basis of present information, some patterns of microtine reproductive strategies are beginning to emerge. For this purpose, it seems useful to classify microtine species as to their basic dietary habits in the following manner: specialists on monocotyledons (M); specialists on dicotyledons (D); and more omnivorous in vegetative dietary habits (O). In Table II we attempt to classify a number of species based on information presently available.

TABLE II
A General Classification of the Dietary Habits of Some Microtine Species Based on Whether They Specialize for Monocotyledons (M) or Dicotyledons (D), or Are More Omnivorous in Vegetative Dietary Habits (O) during the Breeding Season

Species	Dietary habit
Dicrotonyx torquatus	D
Lemmus sibiricus	M
Synaptomys cooperi	M
Microtus pennsylvanicus	M
Microtus montanus	M
Microtus longicaudus	D
Microtus californicus	M
Microtus ochrogaster	O
Microtus oeconomus	M
Lagurus curtatus	D
Phenacomys intermedius	O
Arvicola richardsoni	O
Clethrionomys gapperi	D
Microtus mexicanus	M

One can now ask whether the dietary classification identifies any clear differences in reproductive strategies. The answer appears to be affirmative. Within the M group, we find many of the species that are known to have variable breeding periods, whereas the D and O groups contain species known to have more restricted breeding seasons (Keller, 1985; P. J. Berger and N. C. Negus, in preparation). Intuitively, this seems to make sense because both D and O groups should have food resources more temporally predictable and of higher quality than species in the M group. It is of interest also that the M group contains those species most noted for multiannual population fluctuations.

Despite the impressive volume of research that has focused on microtine rodents, the details of their reproductive strategies are poorly known for most species. Therefore, the concept we now present of microtine reproductive strategies is admittedly hypothetical to some extent, with the hope that more definitive data in the future will test the validity of the concept.

Unquestionably, the M-group microtines are typically living in temporally unpredictable environments. On a diet primarily of grasses and sedges, microtine reproduction can probably be sustained energetically only when the plants are in their vegetative growth stages. The germination and/or sprouting of grasses and sedges is dependent upon the thermal and moisture conditions in the soil. Thus, initiation and extent of growing seasons for such plants are often highly variable from year to year. As a consequence, M-group microtines do not find photoperiod to be a reliable cue for reproduction in most environments. Analogous to the strategy of the red kangaroo, a more immediate and reliable predictor of favorable energetic resources is in the food resources themselves. Considering the short generation time of M-group microtines (60 days), the response to such a cue must be very rapid to be effective. Such a cue was postulated many years ago (Bodenheimer and Sulman, 1946) for *Microtus guentheri* in Israel, and later for *Microtus montanus* (Pinter and Negus, 1965; Negus and Pinter, 1966). Subsequent experimental studies confirmed the presence of an active factor in young wheatgrass that was highly stimulatory to reproductive activity in *Microtus montanus* (Hinkley, 1966; Negus and Berger, 1971, 1977; Berger and Negus, 1974). Ultimately Sanders et al. (1981) isolated and identified the active compound, 6-methoxybenzoxazolinone (6-MBOA). Field experiments demonstrated that very small amounts of 6-MBOA (equivalent to the dosage from food plants) in the diet of *Microtus* were sufficient to trigger reproduction in nonbreeding winter populations (Berger et al., 1981). This was the first demonstration of a chemical in food resources acting as a cue for reproduction in mammals.

The information content of 6-MBOA is both high and precise. Because 6-MBOA is present primarily in young shoots of grasses and sedges (Figs. 2 and 3), it predicts a favorable food resource several weeks later. Employing this cue as the dominant environmental information, *Microtus montanus* tracks the condition of the food resources and breeds opportunistically when the cue is positive. It is probable that other species in the M group of microtines also cue reproduction to the presence of 6-MBOA or similar compounds.

As a test of the hypothesis that other M-group microtines also cue reproduction to chemical information from plant food resources, we undertook a field and laboratory study of lemming reproduction (*Dicrostonyx* and *Lemmus*) in high arctic tundra from 1980 to 1984 at Igloolik, Northwest Territory, Canada. At the outset of the investigation we hypothesized that *Dicrostonyx*, being a specialist on dicotyledons (D group), would not use plant chemical cues for reproductive timing, whereas *Lemmus*, a specialist on arctic grasses and sedges (M group), would synchronize breeding on the basis of 6-MBOA in the plant food resources. Our field data strongly supported the hypothesis (Tables III and IV; P. J. Berger and N. C. Negus, in preparation). *Dicrostonyx* typically started breeding under the snow shortly after the equinox in March and produced two to three litters prior to meltoff in June. In

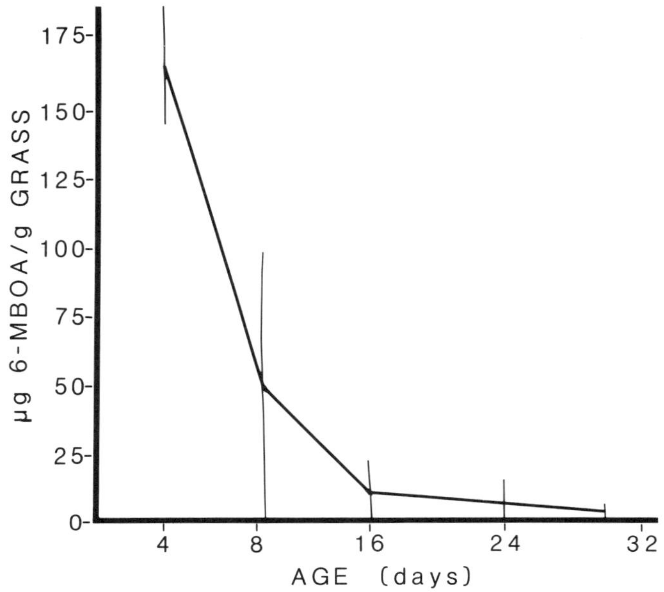

FIGURE 2. The concentration of 6-MBOA in winter wheat grass sampled at different ages after germination in a greenhouse ($n = 4$ for each data point). Vertical lines represent *SD*.

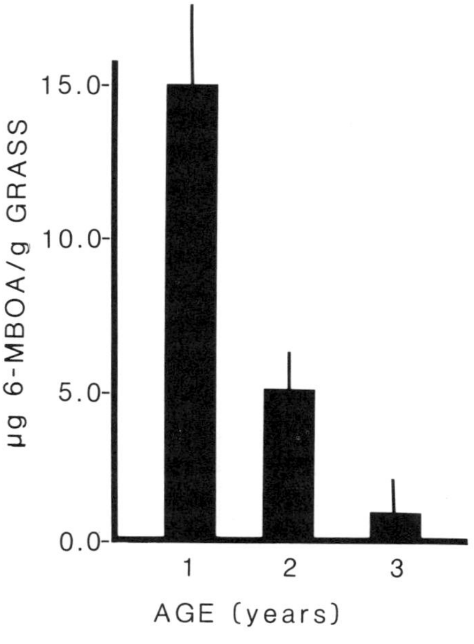

FIGURE 3. The concentration of 6-MBOA in samples of *Carex stans* tillers of different ages collected in July 1982 on Igloolik Island, Northwest Territory ($n = 3$–4 for each age of tillers sampled). Vertical lines represent *SD*.

TABLE III
Onset of Breeding in *Dicrostonyx torquatus* Based on Age Class and Estimated Date of Conception of Immature Animals Trapped between 31 May and 8 June 1983[a]

Age class (mean weight in g ± SD)	Weight range (g)	n	Estimated age (days)	Estimated date of conception
44.5 ± 3.4	40.0–49.0	7	65	16 March
34.0 ± 3.0	30.0–39.0	6	50	31 March
22.4 ± 2.3	20.0–27.0	12	35	15 April
16.1 ± 2.1	13.6–19.0	6	20	30 April

[a]For purposes of estimating the date of conception, all animals were considered to have been trapped on 8 June. The estimated age is based on weight data from known-age animals reared in the laboratory.

contrast, *Lemmus* initiated breeding in synchrony with meltoff when the first young sprouts of *Dupontia fisheri* and *Carex stans* appeared under the melting snow. Analysis of both *Carex* and *Dupontia* sprouts demonstrated high content of 6-MBOA (Table V and Fig. 3). Further, by mid-August, with the return of short periods of relative darkness, *Dicrostonyx* had ceased breeding completely, whereas *Lemmus* reproduction continued at a high level. Again, analysis of grass (*Dupontia*) samples revealed high 6-MBOA content in the young tillers (Table V). Laboratory tests gave further support to the hypothesis. Young female

TABLE IV
Comparison of Breeding Seasons for Female *Lemmus* and *Dicrostonyx*[a]

Date of capture	Species	n	Percent in first pregnancy	Percent pregnant and/or lactating (after first pregnancy)
1–23 June	*Lemmus*	16	69	0
	Dicrostonyx	24	12	75
7–31 July	*Lemmus*	20	35	65
	Dicrostonyx	23	13	87
17–24 August	*Lemmus*	13	8	85
	Dicrostonyx	18	0	6

[a]Only females of breeding age are included in the samples. Females were considered to be of breeding age based on weight and total length compared to laboratory-reared animals. Females were considered to be in their first pregnancy if they were not simultaneously lactating and histological examination demonstrated no corpora albicans. By these data, *Lemmus* appears to have initiated breeding in early June. During the 1–23 June trapping period, only one juvenile *Lemmus* was caught, as compared to 41 *Dicrostonyx* juveniles.

TABLE V
The 6-MBOA Content of *Dupontia fisheri* during the Growing Season on Igloolik Island in 1983[a]

Date	6-MBOA content (μg/g plant)		
	New shoots	Outer shoots	Secondary leaf
18 June	14.1 ± 4.4	—	—
1 July	—	3.8 ± 0.4	—
12 July	15.6 ± 5.5	7.3 ± 0.5	—
25 July	—	3.0 ± 0.8	2.0 ± 0.2
16 August	38.2 ± 6.8	9.3 ± 1.5	4.5 ± 0.5

[a]New shoots are the innermost, newly erupting shoots. The outer shoot is the leaf that encases the new shoot. The secondary leaf is the oldest leaf on the plant that grew in this season. It is quite probable that the new shoot of 12 July is subsequently sampled as the outer shoot on 25 July and as the secondary leaf on 16 August. Each data point represents the mean of 3–4 individual samples ± SD.

Lemmus showed a strong uterine weight response to 6-MBOA, but *Dicrostonyx* females showed no response whatever (Table VI; P. J. Berger and N. C. Negus, in preparation). Breeding experiments demonstrated that photoperiod manipulation caused initiation (long) or cessation (short) of breeding in *Dicrostonyx* (Table VII).

More data of the sort described above are needed for various species of microtines. However, a pattern seems to be emerging. Microtines that consume substantially monocotyledonous food resources cannot rely on the information content of photoperiod to cue reproduction. All such species probably employ plant chemical cues as the most accurate predictors of favorable resources for reproduction. There is now evidence that both *Microtus californicus* (Batzli and Pitelka, 1971) and *Microtus*

TABLE VI
Uterine Weight Response in 4-Week-Old *Lemmus* and *Dicrostonyx* to Three Daily I.P. Injections of 10 μg of 6-MBOA in 0.5 ml of 5% Propylene Glycol (PG)[a]

Species	Treatment	*n*	Mean body weight (g)	Mean uterine weight (mg)
Lemmus	Control	8	28.0 ± 2.3	13.0 ± 1.4
Lemmus	6-MBOA	8	27.7 ± 1.1	21.6 ± 3.5*
Dicrostonyx	Control	8	28.5 ± 2.4	12.5 ± 1.3
Dicrostonyx	6-MBOA	8	29.0 ± 2.0	12.2 ± 1.9

[a]Controls received 0.5 ml of 5% PG only. The data are expressed as mean ± SD. Asterisk indicates result significantly different from control at $P < 0.001$.

TABLE VII
Breeding Performance of *Dicrostonyx torquatus* under Different Photoperiod Regimes[a]

Regime	n	Mean number of litters	Mean litter size	Percent breeding at 60 days	Percent breeding at 100 days
24L:0D	17	3.8 ± 1.3*	3.2 ± 1.1**	94	71
14L:10D	14	1.9 ± 0.5	2.6 ± 1.1	50	7

[a]One group was reared and mated under 24 h of light (24L:0D) and the second group was reared under 24 h of light but transferred to 14L:10D on the day of mating. The pairs were mated for a total of 122 days. Only mated pairs that produced at least one litter were used in the data set. The initial litter is included in the data presented. The data are expressed as mean ± *SD*. *: $P < 0.001$ (Student's *t*-test). **: $P < 0.02$ (Student's *t*-test).

pennsylvanicus (Rose et al., 1982) cue reproduction to 6-MBOA in the food resource. We expect that as data accumulate, the following species may also be added to the list: *Microtus ochrogaster, M. mexicanus, M. agrestis, M. oeconomus, Clethrionomys glareolus,* and *Synaptomys cooperi*. Species with more varied food habits that may employ photoperiod as a sufficient cue for reproduction include *Microtus longicaudus, Phenacomys intermedius, Clethrionomys gapperi,* and *M. xanthognathus*. These lists are both tentative and incomplete, but are suggestive of two basic reproductive timing patterns in this interesting group of small herbivores.

The use of a plant chemical cue such as 6-MBOA for reproductive timing has further repercussions on the reproductive strategies of M-group microtines. First of all, it means that a species using such a cue is by definition an opportunistic breeder. As such, whenever the cue is present at stimulatory levels in the food resources, breeding should occur. Thus, if drought conditions occur in summer resulting in browning and cessation of grass growth, reproduction should cease; or if ambient thermal and moisture conditions in autumn or winter are favorable for sprouting of rhizomatous grasses and sedges, breeding should continue or be resumed. We have observed both circumstances in populations of *Microtus montanus* (Negus et al., 1977). Second, if 6-MBOA stimulates adult reproduction, it should have similar effects on the young members of a population. Negus and Berger (1986) have demonstrated that this is indeed the case and that growth of young *Microtus montanus* is significantly higher if they are receiving 6-MBOA in the diet. Thus, in species that cue reproduction to 6-MBOA we would expect to see variability in age at first reproduction from one cohort to the next as the breeding season progresses. This has been observed repeatedly in M-group microtines in the field (*M. montanus:* Negus et al., 1986; Negus and Berger, 1986; *M. californicus:* Lidicker, 1973; *M. pennsylvanicus:* Brown, 1973). In this case there is high phenotypic plasticity of developmental rates that respond appropriately to environmental information on food quality.

Our data on cohort analysis in *Microtus montanus* populations in a marsh environment exemplify the process (Negus et al., 1986; Negus and Berger, 1986). Cohort I animals, born in the spring (April–May), receive a strong 6-MBOA cue in their food and attain maturity at 3–4 weeks of age. Cohort II animals, born in June, receive a weaker cue and attain sexual maturity at 7–8 weeks of age. Cohort III animals, born in July–August, receive little or no plant cue and do not attain sexual maturity, but rather enter a growth diapause that extends through the winter (Fig. 4). These animals resume growth the following spring with the return of the 6-MBOA plant cue, and become the primary breeding stock for the reproductive season. In the ephemeral high-quality nutritional environment of *Microtus montanus,* the phenotypic responses of cohorts are the most appropriate compromises to attain maximal individual fitness in each generation.

The phenotypic cohort responses in *Microtus montanus* are doubtless mediated by a suite of environmental information: 6-MBOA, pho-

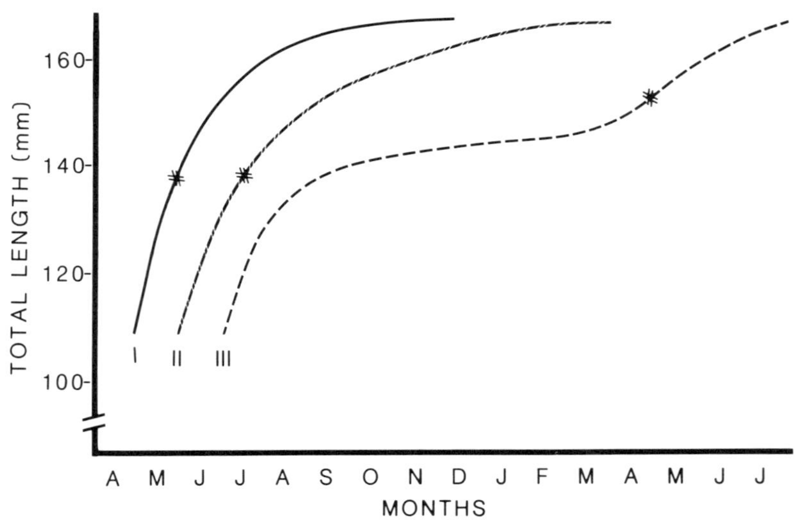

FIGURE 4. Representative individual growth rates of young *Microtus montanus* from cohorts I, II, and III at Timpie Springs, Utah. (#) Age at first reproduction. Growth rates were determined by recapturing marked animals in the field.

toperiod, and nutritional quality. Both Pinter (1968) and Horton (1984) have shown that photoperiod can cause an increase (long photoperiod) or decrease (short photoperiod) in growth rates of young *Microtus montanus*. In a general sense, we can envision each signal from the environment as conveying either positive or negative information. Positive signals would be conveyed by the presence of 6-MBOA, increasing photoperiod, and high nutritional quality. Conversely, negative signals would be represented by the absence of 6-MBOA, decreasing photoperiod, and low nutritional quality. In this perspective, cohort I animals receive all positive signals, cohort II animals receive two positive and one negative signal (photoperiod), whereas cohort III animals receive all negative signals.

From the viewpoint of environmental suites of information, it would be instructive to know if one signal can override another under natural conditions. For example, when autumn–winter breeding occurs in a microtine population, presumably negative photoperiod information is overridden by other cues. Some insight as to the extreme flexibility of this cueing system is gained from a population of *Microtus montanus* inhabiting a marsh formed by warm springs at Blue Lake on the Utah–Nevada border (Negus and Berger, 1986). Here, a number of ponds have been formed by the warm springs, with thousands of acres of surrounding marshland dominated by salt grass (*Distichlis stricta*). Samples taken in December–January in three successive winters demonstrated a high level of reproductive activity in animals inhabiting the narrow zone (15–20 ft wide) of actively growing vegetation bordering the ponds, whereas samples taken several hundred feet away from the pond edges showed typical winter inactivity and regression of reproductive tracts (Table VIII). Apparently, the positive nutritional level is sufficient to override the negative photoperiod information. Probably, *Microtus californicus* initiates reproduction in response to a similar suite of environmental information with the autumn–winter rainfall and growing season in California (Lidicker, 1973). With both opportunistic breeding and phenotypically plastic developmental rates, the life history and reproductive strategy of *Microtus montanus* is highly adaptive to temporally fluctuating environments. Based on the data of Brown (1973) for *Microtus pennsylvanicus* and Wiger (1982) for *Clethrionomys glareolus*, cohort developmental plasticity appears to be characteristic of M-group microtines.

The well-documented opportunistic reproductive strategy of *Mus musculus* (Bronson, 1979) is another striking example of extreme flexibility. However, this small muroid rodent is not a herbivore. Rather, wild *Mus* survive and breed on an impressive array of food items. The

TABLE VIII
Field Samples of *Microtus montanus* Trapped at Blue Lake, Utah[a]

A. Females

Location	n	Mean body weight (g)	Percent pregnant and/or lactating	Mean uterine weight (mg)
Pond edge	16	38.2 ± 7.4*	75	58.0 ± 23.3**
Marsh	15	30.7 ± 5.6	0	22.1 ± 8.0

B. Males

Location	n	Mean body weight (g)	Mean paired testes weight (mg)
Pond edge	7	52.8 ± 12.7	262.9 ± 148.9†
Marsh	6	40.8 ± 8.1	110.1 ± 44.1

[a]The data represent two different samples. The first was taken 24–25 January 1981, and the second was taken 12–13 January 1985. Animals trapped by the warm spring are designated as "Pond edge" and those trapped in the general marsh away from the pond edge are designated "Marsh." Data are expressed as mean ± *SD*. *: $P = 0.009$ (Student's *t*-test). **: $P = 0.001$ (Student's *t*-test); only uterine weights of nonpregnant animals are included. †: $P = 0.03$ (Student's *t*-test).

cue for reproduction in *Mus* appears to be based on caloric intake sufficient to support the energy demands of gestation and lactation in females (Bronson, 1985). Accordingly, wild *Mus* will breed at any season provided they have access to sufficient food resources. Supplemental feeding experiments with field populations demonstrate the reality of this response mechanism (Newsome, 1970). Tracking the environment on the basis of calories is a particularly adaptive strategy for mammals with extremely generalist food habits. *Mus* appears to be the ultimate example of this. *Peromyscus maniculatus* appears to employ this strategy in the lower latitudes of its distribution. Bronson (1985) has compiled data indicating that northern populations of this species may use photoperiod information to cue reproduction, whereas more southern populations do not. Desjardins (1981) also demonstrated greater photoperiodic sensitivity in northern *Peromyscus* than in southern individuals in the laboratory. Again the generalist food habits of *Peromyscus* permit caloric tracking at latitudes where the severity of winter conditions do not preclude reproductive success. One can envision that selection for heritable photoperiodic responsiveness becomes progressively less at lower latitudes.

An example such as *Peromyscus* demonstrates that even within a single species the reproductive strategy may become modified as a result of differing selective pressures in various parts of its range. The importation of the European rabbit (*Oryctolagus cuniculus*) to Australia lends further support to this concept. In Australia, *Oryctolagus* has become a totally opportunistic breeder, responding to rainfall-induced flushes of vegetative growth as the primary cue for reproduction in the unpredictable environments of that continent (Poole, 1960). The important point here seems to be that a species can shift from being a strongly seasonal breeder to an opportunistic breeder in response to environmental uncertainty.

Some mammals appear to have evolved endogenously programmed reproductive strategies, that is, genetically programmed reproductive cycles that require little or no environmental information to facilitate successful reproduction. Not surprisingly, hibernating species represent the primary examples of such strategies. The strategy of dormancy itself goes a long way towards rendering the environment predictable for a species. The typical hibernator emerges from torpor when the environmental resources have become favorable for reproduction. In such species, there is high adaptive value in breeding immediately following emergence so that the offspring will have sufficient time and resources for growth and sequestering of fat for torpor in the autumn. Intense selective pressure has obviously acted upon mechanisms for appropriate timing of torpor immergence and emergence. In several species of *Spermophilus* that have been studied under controlled laboratory conditions, the hibernation cycles as well as gonadal cycles are endogenous circannual rhythms (Pengelley and Asmundson, 1974). In a small hibernator such as *Zapus princeps*, endogenous emergence timing does not appear to be operative (Cranford, 1978). In years of late meltoff, ground squirrels emerge and burrow up through the snow with little difficulty. *Zapus*, however, must wait to emerge until meltoff is complete and vegetative growth has initiated. In this case, arousal is triggered by a thermal cue of about 8°C in the hibernaculum, which accurately signals the disappearance of snow cover and initiation of the growing season. The gonadal cycle, however, appears to be endogenous, since the animals emerge in a reproductively active state. Thus, the timing of reproduction in hibernators appears to be coupled to the timing mechanism controlling torpor duration.

In some desert-dwelling sciurids (for example, *Spermophilus tereticaudus* and *Spermophilus mollis*), postemergence environmental information appears to regulate litter size, which varies widely from year to year. In *S. mollis*, the litter size variation is immense (1–16!). In both species the mean litter size each year at a given locality is positively correlated with the amount of winter precipitation that has occurred.

Presumably, females upon emergence are able to assess the quality/quantity of the food resources and respond physiologically with regard to the number of young produced (Rickart, 1982). The presence of endogenous reproductive cycles in some mammals suggests the ultimate endpoint of genotypic control that can evolve by selection if the environment is highly predictable, minimizing selection for responses to environmental information.

5. CONCLUDING REMARKS

The primary aim of this chapter has been to describe patterns of mammalian reproductive strategies from a mechanistic viewpoint and to relate these to environmental heterogeneity. In the process we have attempted to show how size, longevity, food resources, metabolic rate, and environmental fluctuations interact to mold reproductive strategies in mammals. A clear dichotomy of strategies emerges. At the one extreme, with highly predictable temporal heterogeneity of resources, there is strong genotypic selection for reproductive traits, an obligate reproductive strategy with little or no phenotypic plasticity. The tammar wallaby exemplifies such evolutionary selection. At the other extreme are those environments that represent unpredictable temporal fluctuations of resources. For species inhabiting such environments, selection for genotypic reproductive traits cannot occur, but instead there is strong selection for information retrieval systems. High levels of phenotypic plasticity result. The M group of microtine rodents represent an extreme of this type of facultative reproductive strategy. Virtually all of the primary reproductive parameters, including age at first reproduction, are not genotypically programmed, and remain phenotypically responsive to environmental information. The reproductive strategies of most species fall somewhere between these extremes of adaptation, with various mixtures of genotypic selection and phenotypic plasticity.

The concept of environmental suites of information applies particularly to those species in which there is some degree of phenotypic plasticity. Thus, *Peromyscus* seems to employ both photoperiod and nutritional information to regulate breeding seasons in various parts of its distribution. It appears that environmental signals are viewed in a somewhat hierarchical manner by a given species such that a positive signal of high predictive value can override a negative signal of lesser value. The predictive value of 6-MBOA to an M-group microtine rodent must be very high, since it dominates over other, negative environmental signals. However, at present too little data are available to permit generalizations in this regard. Our understanding of mammalian reproductive strategies will be greatly enhanced in the future by those workers who choose to make careful comparative field and laboratory studies of related taxa that utilize different kinds of food resources. This much needed empirical data base is long overdue.

REFERENCES

Batzli, G. O., 1985, Nutrition, in: *Biology of New World Microtus* (R. H. Tamarin, ed.), American Society of Mammalogists, pp. 779–811.

Batzli, G. O., and Cole, F. R., 1979, Nutritional ecology of microtine rodents: Digestibility of forage, *J. Mammal.* **60:**740–750.

Batzli, G. O., and Pitelka, F. A., 1971, Condition and diet of cycling populations of the California vole, *Microtis californicus, J. Mammal.* **52:**141–163.

Beatley, J. C., 1969, Dependence of desert rodents on winter annuals and precipitation, *Ecology* **50:**721–24.

Beatley, J. C., 1976, Rainfall and fluctuating plant populations in relation to distributions and numbers of desert rodents in southern Nevada, *Oecologia* **24:**21–42.

Berger, P. J., 1966, Eleven-month "embryonic diapause" in a marsupial, *Nature* **211:**435.

Berger, P. J., 1970, The reproductive biology of the tammar wallaby, *Macropus eugenii* Desmarest (Marsupialia), Ph. D. thesis, Tulane University, New Orleans, Louisiana.

Berger, P. J., and Negus, N. C., 1974, Influence of dietary supplements of fresh lettuce on ovariectomized *Microtus montanus, J. Mammal.* **55:**747–750.

Berger, P. J., Negus, N. C., Sanders, E. H., and Gardner, P. D., 1981, Chemical triggering of reproduction in *Microtus montanus, Science* **214:**69–70.

Bodenheimer, F. S., and Sulman, F., 1946, The estrous cycle of *Microtus guentheri* and its ecological implications, *Ecology* **27:**255.

Boyce, M. S., 1984, Restitution of *r*- and *K*-selection as a model of density-dependent natural selection, *Annu. Rev. Ecol. Syst.* **15:**427–447.

Bradley, S. R., 1976, Temperature regulation and bioenergetics of some microtine rodents, Ph. D. thesis, Cornell University, Ithaca, New York.

Bronson, F. H., 1979, The reproductive ecology of the house mouse, *Q. Rev. Biol.* **54:**265–299.

Bronson, F. H., 1985, Mammalian reproduction: An ecological perspective, *Biol. Reprod.* **32:**1–26.

Bronson, F. H., and Pryor, S., 1983, Ambient temperature and reproductive success in rodents living at different latitudes, *Biol. Reprod.* **29:**72–80.

Brown, E. B., 1973, Changes in patterns of seasonal growth in *Microtus pennsylvanicus, Ecology* **54:**1103–1110.

Christian, D. P., 1979, Comparative demography of three Namib Desert rodents: Responses to the provision of supplementary water, *J. Mammal.* **60:**679–690.

Cole, F. R., and Batzli, G. O., 1978, The influence of supplemental feeding on a population of the prairie vole *Microtus ochrogaster, J. Mammal.* **59:**809–819.

Conley, W., 1976, Competition between *Microtus:* A behavioral hypothesis, *Ecology* **57:**224–37.

Cranford, J. A., 1978, Hibernation in the western jumping mouse (*Zapus princeps*), *J. Mammal.* **59:**496–509.

Crichton, J. A., Aitken, J. N., and Boyne, A. W., 1959, The effect of the plane of nutrition during rearing on growth, production, reproduction and health of dairy cattle, *Anim. Prod.* **1**:145–162.

Csuti, B. A., 1979, Patterns of adaptation and variation in the Great Basin kangaroo rat (*Dipodomys microps*), Univ. Calif. Publ. Zool. **111**:1–69.

Desjardins, C., 1981, Latitudinal gradients in the responsiveness of the rodent reproductive system to photic stimuli, *Biol. Reprod.* **24**:23A.

Elton, C. S., 1942, *Voles, Mice and Lemmings*, Clarendon Press, Oxford.

Emlen, J. M., 1984, *Population Biology. The Coevolution of Population Dynamics and Behavior*, Macmillan, New York.

Farris, A. L., 1971, Population dynamics and habitat distribution of *Microtus longicaudus*, *Microtus montanus* and *Peromyscus maniculatus* in southeastern Washington, Ph. D. thesis, Washington State University, Pullman, Washington.

Finerty, P. F., 1982, *The Population Ecology of Cycles in Small Mammals*, Yale University Press, New Haven.

Glass, A. R., and Swerdloff, R. S., 1980, Nutritional influences on sexual maturation in the rat, *Fed. Proc.* **39**:2360–2364.

Hamilton, G. D., and Bronson, F. H., 1985, Food restriction and reproductive development in house mice, *Biol. Reprod.* **32**:773–779.

Hinkley, R., Jr., 1966, Effects of plant extracts in the diet of male *Microtus montanus* on cell types in the anterior pituitary, *J. Mammal.* **47**:396–400.

Horton, T. H., 1984, Growth and maturation in *Microtus montanus:* modification by photoperiods before and after weaning, *Can. J. Zool.* **62**:741–746.

Keller, B. L., 1985, Reproductive patterns, in: *Biology of New World Microtus* (R. H. Tamarin, ed.), American Society of Mammalogists, pp. 725–778.

Kenagy, G. J., 1972, Saltbush leaves: Excision of hypersaline tissue by a kangaroo rat, *Science* **178**:1094–1096.

Kenagy, G. J., and Bartholomew, G. A., 1985, Seasonal reproductive patterns in five coexisting California desert rodent species, *Ecol. Mongr.* **55**:371–397.

Kleiber, M., 1932, Body size and metabolic rate, *Physiol. Rev.* **27**:511–41.

Levins, R., 1968, *Evolution in Changing Environments*, Princeton University Press, Princeton, New Jersey.

Lidicker, W. Z., Jr., 1973, Regulation of numbers in an island population of the California vole, a problem in community dynamics, *Ecol. Monogr.* **43**:271–302.

McNab, B. K., 1980, Food habits, energetics and the population biology of mammals, *Am. Nat.* **116**:106–24.

Negus, N. C., and Berger, P. J., 1971, Pineal weight response to a dietary variable in *Microtus montanus*, *Experientia* **27**:215–216.

Negus, N. C., and Berger, P. J., 1972, Environmental factors and reproductive processes in mammalian populations, in: *Biology of Reproduction: Basic and Clinical Studies* (J. T. Velardo and B. A. Kasprow, eds.), Third Pan American Congress on Anatomy, New Orleans, Louisiana, pp. 89–98.

Negus, N. C., and Berger, P. J., 1977, Experimental triggering of reproduction in a natural population of *Microtus montanus*, *Science* **196**:1230–1231.

Negus, N. C., and Berger, P. J., 1986, Cohort analysis: Environmental cues and "diapause" in microtine rodents, in: *Symposium on Population Biology and Life History Evolution*, Third International Theriological Congress, Edmonton, in press.

Negus, N. C., and Pinter, A. J., 1966, Reproductive responses of *Microtus montanus* to plants and plant extracts in the diet, *J. Mammal.* **47**:596–601.

Negus, N. C., Berger, P. J., and Forslund, L. G., 1977, Reproductive strategy of *Microtus montanus*, *J. Mammal.* **58**:347–353.

Negus, N. C., Berger, P. J., and Brown, B. W., 1986, Microtine population dynamics in a predictable environment, *Can. J. Zool.* **64:**785–792.

Newsome, A. E., 1964a, Anoestrus in the red kangaroo, *Megaleia rufa* (Desmarest), *Aust. J. Zool.* **12:**9.

Newsome, A. E., 1964b, Oestrus in the lactating kangaroo, *Megaleia rufa* (Desmarest), *Aust. J. Zool.* **12:**315.

Newsome, A. E., 1965, Reproduction in natural populations of the red kangaroo, *Megaleia rufa* (Desmarest) in central Australia, *Aust. J. Zool.* **13:**735.

Newsome, A. E., 1966, The influence of food on breeding in the red kangaroos in central Australia, *CSIRO. Wildl. Res.* **11:**187.

Newsome, A. E., 1970, An experimental attempt to produce a mouse plague, *J. Anim. Ecol.* **39:**299–311.

Pengelley, E. T., and Asmundson, S. J., 1974, Circannual rhythmicity in hibernating mammals, in: *Circannual Clocks Annual Biological Rhythms* (E. T. Pengelley, ed.), Academic Press, New York, pp. 95–101.

Pinter, A. J., 1968, Effects of diet and light on growth, maturation and adrenal size of *Microtus montanus, Am. J. Physiol.* **215:**461–466.

Pinter, A. J., and Negus, N. C., 1965, Effects of nutrition and photoperiod on reproductive physiology of *Microtus montanus, Am. J. Physiol.* **208:**633–638.

Poole, W. E., 1960, Breeding of the wild rabbit, *Oryctolagus cuniculus* (L.), in relation to the environment, *CSIRO Wildl. Res.* **5:**21.

Quinn, J. B., 1983, The ecology of two kangaroo rat species (genus *Dipodomys*) in southwestern Utah, M. S. thesis, University of Utah, Salt Lake City, Utah.

Reichman, O. J., and Van de Graff, K. M., 1975, Association between ingestion of green vegetation and desert rodent reproduction, *J. Mammal.* **56:**503–506.

Rickart, E. A., 1982, Ecology of Townsend's ground squirrel, Ph. D. thesis, University of Utah, Salt Lake City, Utah.

Rose, R. K., Everton, R. K., and Glass, G. G., 1982, Experimentally induced winter breeding in populations of small mammals, in: *Third International Theriological Congress, Abstracts of Papers,* p. 208.

Sadleir, R. M. F. S., 1972, Environmental effects, in: *Reproduction in Mammals, Book 4 Reproductive Patterns* (C. R. Austin and R. V. Short, eds.), Cambridge University Press, Cambridge, pp. 69–94.

Sanders, E. H., Gardner, P. D., Berger, P. J., and Negus, N. C., 1981, 6-Methoxybenzoxazolinone: A plant derivative that stimulates reproduction in *Microtus montanus, Science* **214:**67–69.

Sharman, G. B., and Pilton, P. E., 1964, The life history and reproduction of the red kangaroo (*Megaleia rufa*), *Proc. Zool. Soc. Lond.* **142:**29.

Taitt, M. J., and Krebs, C. J., 1981, The effect of extra food on small rodent populations: II. Voles (*Microtus townsendii*), *J. Anim. Ecol.* **51:**681–690.

Van de Graaf, K. M., and Balda, R. P., 1973, Importance of green vegetation for reproduction in the kangaroo rat, *Dipodomys merriami, J. Mammal.* **54:**509–512.

Wiger, R., 1982, Roles of self regulatory mechanisms in cyclic populations of *Clethrionomys* with special reference to *C. glareolus:* A hypothesis, *Oikos* **38:**60–71.

Wright, P. L., 1963, Variation in reproductive cycles in North American mustelids, in: *Delayed Implantation* (A. C. Enders, ed.), University of Chicago Press, Chicago.

CHAPTER 6

THE BEHAVIOR, PHYSIOLOGY, AND ANATOMY OF LACTATION IN THE PINNIPEDIA

OLAV T. OFTEDAL, DARYL J. BONESS, and RAYMOND A. TEDMAN

1. INTRODUCTION

Lactation represents a major proportion of maternal reproductive effort. In ungulates the energetic costs of lactation comprise 75–80% of the total energetic costs of reproduction (Oftedal, 1985). Although pinnipeds feed at sea, they return to land or ice to give birth and to nurse their young (Fig. 1) (Bonner, 1984). Thus, the bulk of a female's reproductive expenditure is in an environment in which she cannot feed. The temporal separation of feeding and lactation is achieved by a wide range of behavioral, physiological, and anatomic adaptations.

The three pinniped families have adopted different lactation strategies. The phocids (true seals) accumulate energy and nutrient reserves prior to parturition. Lactation is a relatively brief period of 4–50 days characterized by maternal fasting, rapid weight gain by the pup, and abrupt weaning when the mother departs to feed at sea. The otariids (fur seals and sea lions) alternate brief periods of nursing with foraging

OLAV T. OFTEDAL and DARYL J. BONESS • Department of Zoological Research, National Zoological Park, Smithsonian Institution, Washington, D.C. 20008. RAYMOND A. TEDMAN • Department of Anatomy and Histology, University of Adelaide, Adelaide, South Australia 5001, Australia.

trips at sea. The ability to replenish maternal reserves during lactation permits an extended period of care (typically 4–12 months) during which pups grow slowly. The sole surviving odobenid (the walrus) deviates from the pinniped norm in that the pup accompanies the mother and suckles at sea after leaving the natal site. Concurrent feeding and lactation after departure enables weaning to be gradual and lactation to persist for up to 2 years (Fay, 1982).

Lactation patterns have received much attention in pinnipeds, in part because many species are accessible and easily observed during this period. The predictability of breeding site and time, gregariousness, relative immobility on solid substrate, and site tenacity of most pinniped species facilitate studies in which extended observation, marking of individuals, or repeated captures are required.

In this paper we review data on behavioral, physiological, and

FIGURE 1. Lactating southern elephant seal with her pup on Macquarie Island. Note the unusual circumstance of milk squirting from the lower teat while the pup pauses during suckling. (Photograph courtesy of M. M. Bryden.)

anatomic aspects of lactation in pinnipeds. Although behavior has historically received more attention than other aspects of the lactation process, quantitative measurement of milk production and lactation energetics is an area of much recent activity. Anatomic and secretory adaptations associated with the diverse lactation patterns of pinnipeds are of particular interest; unfortunately, data are limited to a few species. A listing of the common and scientific names of the 33 species of extant pinnipeds is provided in Table I. For the sake of simplicity, species will be referred to by common names in this paper.

2. BEHAVIOR AND ECOLOGY OF LACTATION

2.1. Distribution and Breeding Habitat

Pinnipeds are widespread in distribution, ranging from the Arctic to the Antarctic. Otariids are found predominantly in subtropical to temperate regions, whereas phocids tend to inhabit temperate to polar latitudes (Table I). One otariid, the Antarctic fur seal, ranges into polar regions, and another, the Galapagos fur seal, is the only extant pinniped with a range that is exclusively tropical. The walrus has an Arctic and subarctic distribution. Differences among the three pinniped families in current patterns of distribution and whelping habitat can be traced to paleoecological differences in evolutionary origin and subsequent radiation (Ray, 1976; Repenning, 1976; Repenning et al., 1979).

All otariids whelp and rear young on land, especially on remote islands or regions of mainland inaccessible to humans. Otariids that inhabit areas that may be covered seasonally by pack ice (for example, Steller sea lion, northern fur seal, subantarctic fur seal, Cape fur seal) give birth in late spring or summer when the areas are free of ice. By contrast, phocids use diverse substrates for whelping and rearing of young. In temperate regions phocids may give birth on remote islands, on secluded mainland beaches, or on islands now inhabited by people (Bonner and Witthames, 1974; Scolaro, 1976; LeBoeuf and Panken, 1977; Mansfield and Beck, 1977; Renouf et al., 1981; Alcorn, 1984). In the Arctic, Antarctic, and extreme temperate latitudes, phocids whelp on ice. Ice can be categorized as pack ice (floating ice shelves), which is structurally unstable and variable in distribution, or as fast ice (grounded ice sheets, sometimes referred to as sea ice), which is very stable and is thus more akin to land (Bunt, 1968; Knox, 1970). Ten species are known to whelp on pack ice and three whelp on fast ice (Table I). The walrus gives birth on the pack ice of the Arctic.

TABLE I
Geographic Distribution of Pinnipeds in the Northern (N) and Southern (S) Hemispheres[a]

Taxa	Tropical (0–20°)	Subtropical (20–40°)	Temperate (40–60°)	Polar (60–80°)
		Phocidae		
		Pack-ice whelping		
Bearded seal, *Erignathus barbatus*			N	N
Caspian seal, *Phoca caspica*			N	
Crabeater seal, *Lobodon carcinophagus*			S	S
Grey seal, *Halichoerus grypus*			N	N
Harp seal, *Phoca groenlandica*			N	N
Hooded seal, *Cystophora cristata*			N	N
Leopard seal, *Hydrurga leptonyx*			S	S
Ribbon seal, *Phoca fasciata*			N	N
Ross seal, *Ommataphoca rossi*				S
Spotted seal, *Phoca largha*			N	N
		Fast-ice whelping		
Baikal seal, *Phoca siberica*			N	
Ringed seal, *Phoca hispida*			N	N
Weddell seal, *Leptonychotes weddellii*			S	S
		Land whelping		
Hawaiian monk seal, *Monachus schauinslandi*		N		
Harbor seal, *Phoca vitulina*		N	N	
Mediterranean monk seal, *Monachus monachus*		N		
Northern elephant seal, *Mirounga angustirostris*		N	N	
Southern elephant seal, *Mirounga leonina*		S	S	

TABLE I (*Continued*)

Taxa	Tropical (0–20°)	Subtropical (20–40°)	Temperate (40–60°)	Polar (60–80°)
		Otariidae		
Antarctic fur seal, *Arctocephalus gazella*			S	S
Australian sea lion, *Neophoca cinerea*		S		
California sea lion, *Zalophus californianus*	S	N		
Cape fur seal, *Arctocephalus pusillus*		S	S	
Galapagos fur seal, *Arctocephalus galapagoensis*	S			
Guadulupe fur seal, *Arctocephalus townsendi*		N		
Hooker's sea lion, *Phoca hookeri*			S	
Juan Fernandez fur seal, *Arctocephalus philippii*		S		
New Zealand fur seal, *Arctocephalus forsteri*		S	S	
Northern fur seal, *Callorhinus ursinus*		N	N	
South American fur seal, *Arctocephalus australis*	S	S		
Southern sea lion, *Otaria byronia*	S	S	S	
Steller sea lion, *Eumetopias jubatus*		N	N	
Subantarctic fur seal, *Arctocephalus tropicalis*		S	S	
		Obodenidae		
Walrus, *Odobenus rosmarus*			N	N

[a]Data are from FAO (1982) and Harrison and Ridgway (1981). Zones of latitude are taken from Kormondy (1969).

Variation in whelping habitat occurs within species in some phocids. The Weddell seal, normally a fast-ice species, uses beaches on nearby islands in years of bad ice (E. A. Smith and Burton, 1970; Popov and Krylov, 1978). The southern elephant seal whelps on ice rather than on land in years when the ice does not recede from around Signy

TABLE II
Duration of Pupping Season and Lactation Period in Pinnipeds[a]

Species	Pupping duration (days)	Lactation duration (days)	Data quality[b]	References
		Phocids		
		Pack-ice whelping		
Hooded seal	15	4	$n = 15$	Bowen et al. (1985, 1986a)
Harp seal	10	12	$n = 33$	Bowen and Sergeant (1983), Kovacs (1986)
Bearded seal	24	12–18	nd	Burns (1981a)
Grey seal	36	17[c]	$n = 38$	Boness (1979)
Caspian seal	—	21	nd	Ivashin et al. (1972)
Ribbon seal	—	21–28	nd	Burns (1981b)
Spotted seal	—	35	nd	Golt'sev and Fedoseev (1970)
	15	14–21	est	Naito and Nishiwaki (1972)
Crabeater seal	—	35–42	est	Laws (1958)
	—	28	nd	Siniff et al. (1979)
Leopard seal	—	~30	nd	Tikhomirov (1975)
		Fast-ice whelping		
Weddell seal	40	33–44	$n = 10$	Kaufman et al. (1975)
	—	50	$n = 50$	Lindsey (1937)
Ringed seal	30	~60	est	McLaren (1958)
Baikal seal	~60	60	nd	Pastukhov (1975a)
		Land whelping		
Southern elephant seal	45	23	$n = 60$	Carrick et al. (1962)
	—	23	$n = 28$	Condy (1980)
	—	24	$n = 12$	McCann (1982)
Northern elephant seal	36	27	$n = 44$	Reiter et al. (1978)
Harbor seal (*concolor*)	32	33	$n = 4$	Boulva and McLaren (1979), J. W. Lawson (personal communication)
Harbor seal (*richardsii*)	—	21–35	nd	Bishop (1967)
	75	35–42	est	Bigg (1969)

Harbor seal (*stejnegeri*)	—	90	nd	Belkin et al. (1969)
Harbor seal (*vitulina*)	21	28	nd	Venables and Venables (1955)
Hawaiian monk seal	135	39	n = 16	Alcorn (1984)
Mediterranean monk seal	—	42–49	nd	Boulva (1979)
		Otariids		
Antarctic fur seal	30	117	n = 8	Payne (1979), Kerley (1983), Doidge et al. (1986)
Northern fur seal	60	120	est	Peterson (1965)
Steller sea lion	60	120–365[d]	est	Gentry (1970), Edie (1977), Pitcher and Calkins (1981)
Subantarctic fur seal	45	300–330	est	Bester (1981), Kerley (1983), Roux and Hess (1984)
New Zealand fur seal	18	300–365	est	Stirling (1971), Miller (1975)
Cape fur seal	30	300–330	nd	Rand (1955), Stirling (1970)
	—	≤ 365[d]	nd	Stirling and Warneke (1971)
California sea lion	45	≤ 365[d]	nd	Peterson and Bartholomew (1967), B. S. Stewart and Yochem (1984)
Australian sea lion	180	~365	nd	Marlow (1975), Ling and Walker (1976)
Southern sea lion	60	≤ 365[d]	nd	Hamilton (1934)
Hooker's sea lion	30	~365	nd	Marlow (1975)
	60	—	—	Best (1974)
Galapagos sea lion	~180	≤ 365[d]	nd	Trillmich (1979)
South American fur seal	~90[e]	365–730	7% of 41[f]	Trillmich and Majluf (1981), Trillmich et al. (1986)
	—	270–365	nd	Vaz Ferreira (1956)
Galapagos fur seal	~90	~730	nd	Trillmich (1979)
		Odobenids		
Walrus	~60	730–1095	est	Fay (1982)

[a]Data converted to days on the assumption that 1 week equals 7 days, 1 month equals 30 days, and 1 year equals 365 days.
[b]Data refer to lactation duration; est, indirect estimate; nd, no data; n, number of individuals followed to weaning.
[c]Data are from a land-breeding colony of grey seals.
[d]Yearlings have been observed suckling, but the percentage doing this is unknown if the quality of data is shown as nd.
[e]Recent unpublished observations by P. Majluf reveal that this value should be about 45–60 days.
[f]Out of 41 females at a rookery, 7% were still nursing yearlings as well.

Island prior to the onset of parturition (Laws, 1956). The grey seal is the only species with some colonies that produce pups on land (Davies, 1949; Hewer, 1974; Anderson et al., 1975; Boness and James, 1979) and others that typically give birth on pack ice (Curry-Lindahl, 1975; Hook and Johnels, 1972). It has been suggested that the grey seal originated as a pack-ice species and has secondarily taken up land-whelping (McLaren, 1960; Boness, 1979).

Migratory behavior of pinnipeds may be related to whelping habitat. Most pack-ice phocids migrate at the end of the pupping season, whereas fast-ice and terrestrial phocids appear to be nonmigratory. The migration of pack-ice species may reflect a lack of adequate food supplies in the vicinity of the ice at the end of lactation (Bowen et al., 1985). Although nonmigratory, the fast-ice and terrestrial phocids radiate for distances of several hundred km to more than 1000 km from the whelping grounds to feed after lactation (Hewer, 1974; Condit and LeBoeuf, 1984).

Otariid females are nonmigratory, with the exception of northern fur seal females (Lander, 1979), which migrate south after the relatively short lactation period (Table II). In the Steller and California sea lions, males migrate north after mating with females on the rookeries (Mate and Gentry, 1979; Bigg, 1973). Otariid males are not constrained to forage near the rookery, since they play no role in caring for the young.

The walrus gives birth on pack ice, but in contrast to phocids, appears to use the floating ice as a means of conveyance to new foraging areas for benthic invertebrates on the outer banks of the Bering Sea (Fay, 1982). Like the phocids of the pack ice, the walrus is migratory. In the spring, as temperatures rise and the ice breaks up, animals move northward, where new ice is beginning to form (Fay, 1982).

2.2. Timing of Whelping

All pinnipeds for which information is available give birth seasonally. Both sexes aggregate when pupping begins, and mating occurs a few days to a few weeks postpartum. For terrestrial and fast-ice species, the sites where seals aggregate are very precise from year to year because of the long-term stability of these substrates (for example, McLaren, 1958; Hewer, 1960; Peterson, 1965; Gentry, 1970; T. G. Smith, 1973; LeBoeuf, 1974; Kaufman et al., 1975; Boness, 1979; Campagna, 1985). For pack-ice species, the location of whelping aggregations is less precise, but normally occurs in the same region from year to year (Lavigne, 1976; Burns and Frost, 1979).

The duration of the pupping season varies across species, ranging from about 10 to approximately 180 days and is related to whelping habitat (Table II). Pack-ice phocids tend to have the shortest pupping seasons, 10–36 days. Fast-ice and terrestrial phocids and most otariids produce pups over a period of about 30–60 days. The periods of pupping of two otariids (Galapagos sea lion and Australian sea lion) and one phocid (Hawaiian monk seal) are markedly longer than those of other species in their respective families (Table II). All three species occur at latitudes that are among the lowest of their respective families, but not all low-latitude species have extended pupping seasons. Ling and Walker (1976) postulate that the long season in the Australian sea lion results from a lack of the constraint that is placed on other species by extremely harsh seasonal conditions. Consistent with this, Gentry et al. (1986) argue that short pupping periods in the Antarctic and northern fur seals are related to seasonal changes in climate and food supply. However, if seasonality alone was the major factor controlling variation in the length of pupping periods, we would expect more temperate and subtropical species to have longer periods.

Other factors of possible importance include the instability of pack ice and ability of males to remain at the breeding grounds. Pack ice can break up and shift several kilometers overnight (Hook and Johnels, 1972) and the likelihood of this increases as spring approaches. These conditions should result in an increased synchronization of pupping. Anderson and Fedak (1985) present data suggesting that the energetic demands of fasting and conflict between males force some grey seal males to depart from the rookery before all females are inseminated. Females that pup outside the primary period of male aggregation may be at a selective disadvantage in that prime males may have already departed to replenish their body stores.

2.3. Maternal Behavior and Suckling Patterns

2.3.1. Perinatal Behavior

Pinniped females haul out at parturition sites shortly before giving birth. For example, harp seals typically come onto the ice the day they give birth, whereas Steller sea lions haul out an average of 4 days prior to birth (Sandegren, 1970; R. E. A. Stewart et al., 1981).

A bond between the mother and neonate is formed immediately after parturition through frequent nosing and vocalizations, which provide an opportunity for recognition of individual odors and vocal traits. Among phocids, the interactions typically take the form of the female sniffing the pup, although pups may vocalize (Fogden, 1971; LeBoeuf et al., 1972; Burton et al., 1975). Among otariids, females vocalize to their pups with special "pup attraction" calls and nose them frequently and pups respond to their mothers' calls (Bartholomew, 1959; Sandegren, 1970; McNab and Crawley, 1975; Trillmich, 1981). The combined auditory and olfactory exchanges between otariid females and pups are associated with few errors in recognition during subsequent suckling events. Individual recognition is important for otariids because females leave the rookery for several days at a time and upon return must relocate their pups among many others. By contrast, although phocid females tend to stay in close proximity to their pups throughout lactation, nursing of foster and unrelated pups by females is much more common among phocids (Fogden, 1971; Stirling, 1975; Alcorn, 1984). This may be associated with overcrowding or other factors that cause females to move into the water between periods of nursing (for example, Fogden, 1971).

Females of all species are protective of their pups, particularly just after parturition. Threats involving visual and vocal components are used by females against conspecifics of either sex that approach too closely (for example, Boness et al., 1982). In the northern elephant seal, there is lower mortality among pups whose mothers are relatively aggressive compared to pups of less aggressive mothers (Christenson and LeBoeuf, 1978). On the whole, otariid and odobenid females seem to be more tolerant of conspecifics near their pups than phocids, but this has not been addressed quantitatively in many species.

2.3.2. Problems in Quantifying Suckling

Not all quantitative studies of suckling in free-ranging pinnipeds can be compared directly, because definitions and exactness in measurements have differed among studies. Pinniped pups do not remain attached to the teat until completion of the suckling session, unlike the young of some mammalian species (Drewett, 1983; Hall, 1983). A pup usually sucks on a teat for a short period, breaks contact, and then returns either to the same or a different teat. Detailed examination of suckling sessions in three species indicates variability in the pattern of periods on and off teats. Grey seals in the Orkney Islands exhibit a pattern of sucking for about 6 min with very few breaks or shifts from one teat to another, followed by a secondary period of up to 15–20 min in which there are frequent teat changes (Fogden, 1971). Weddell seal sessions involve either frequent teat changes or relatively unbroken sucking rather than a combination of both within a single session (Tedman and Bryden, 1979). Suckling sessions in captive Cape fur seals are more uniform, with on-teat durations of about 24 sec and off-teat intervals averaging 36 sec (Carnio, 1982).

Suckling durations reported in the literature may refer only to on-teat duration or may include both on-teat and off-teat intervals (Table III). In some cases, insufficient detail is given to determine how duration has been defined. These methodological differences can result in spurious differences when comparing species. For example, on-teat time comprises 68, 45, and 40% of total session duration (including off-teat intervals) in northern elephant seals (LeBoeuf et al., 1972), Cape fur seals (Carnio, 1982), and California sea lions (D. J. Boness, K. A. Ono, and O. T. Oftedal, unpublished), respectively.

Cumulative on-teat duration would seem the most useful measure of suckling if the intent is to correlate suckling behavior to milk ingestion. Although there is no certainty that any particular sucking action results in milk flow, inclusion of off-teat intervals in measuring suckling duration only adds an additional source of variation. For present purposes a "suckling session" consists of both on-teat periods and short breaks between them. A long break or transition to other activity terminates a session. Conventions adopted in studies of suckling behavior should be well defined such that meaningful interpretation of results is possible.

2.3.3. Suckling Behavior

Suckling may be initiated by the mother or the pup. Weddell, northern elephant, and harbor seal pups usually initiate suckling by approaching, vocalizing to, and nuzzling the female (LeBoeuf et al., 1972; Tedman and Bryden, 1979; Renouf and Diemand, 1984). In the northern fur seal, the pup initiates suckling about 60% of the time and most of the other suckles are begun through a mutual process; females rarely are solely responsible for initiating suckles (Macy, 1982). Sandegren (1970) reports that Steller sea lion females initiate suckling in the first few days following birth, but does not indicate whether this behavior changes with pup age.

A few studies have examined whether the female or the pup terminates suckling sessions. It is usually reported that pups terminate sessions (Fogden, 1971; Macy, 1982; Carnio, 1982). Harbor seal females initially terminate about half of the sessions and the proportion increases as pups get older (Renouf and Diemand, 1984).

Temporal parameters of suckling have been measured in several phocids and otariids (Table III). Data representing cumulative on-teat time within suckling sessions are available for three of the phocids and

TABLE III
Temporal Parameters of Suckling in Pinnipeds

Species	Suckling duration (min)	Session frequency (h^{-1})	Intersession interval (h)	Daily duration[e] (h/24 h)	Pup age-related changes[d] Duration	Frequency	Daily duration	References
				Phocids				
Weddell seal	30.0[a]	0.34	2.9	4.1	–	–	–	Tedman and Bryden (1979)
Grey seal	10.4[b]	0.33	3.0	1.4	+	0	+	D. J. Boness (unpublished)
Harp seal	7.3[b]	0.40	2.5	1.2	+	–	0	R. E. A. Stewart (1983)
Northern elephant seal	6.8[b]	0.31	3.2	0.9	+	+	+	LeBoeuf et al. (1972)
Southern elephant seal	5.8[c]	0.56	1.8	1.3	+	+	+	Bryden (1968)
				Otariids				
Subantarctic fur seal	21.5[b]	0.83	1.2	7.2	nd	nd	nd	Bester (1977)
Guadalupe fur seal	18.5[c]	0.20	~5.0	1.5	nd	nd	nd	Pierson (1978)
Steller sea lion	16.2[b]	0.37	2.7	2.4	+[f]	–	–	Sandegren (1970), Higgins (1984)
California sea lion	10.7[b]	0.41	2.4	1.8	nd	nd	nd	D. J. Boness (unpublished)
Northern fur seal	8.5[c]	—	—	—	nd	nd	nd	Macy (1982)

[a]Suckling duration includes off-test intervals.
[b]Suckling duration is cumulative on-teat time.
[c]Not enough information is given in the study to discern what the duration includes.
[d]+, Increase; –, decrease; 0, no change; nd, no data available.
[e]Data are expressed relative to time mother is present.
[f]These changes in suckling parameters cover only the first one-tenth of the lactation period.

three of the otariids in Table III. Among these species, the average duration of suckling sessions in otariids is about twice as long as that of phocids (mean ± standard deviation of 16.1 ± 5.40 as compared to 8.2 ± 1.95 min). The otariids also suckle more frequently (when the mother is present at the rookery) than do the phocids, but the difference is of smaller magnitude than the difference in duration (0.53 ± 0.25 as compared to 0.35 ± 0.10 suckle/h). These patterns of frequency and duration of suckling sessions yield calculated mean daily suckling durations of 3.8 h per 24 h (SD = 2.95) for the otariids and 1.2 h per 24 h (SD = 0.25) for the phocids. The small number of species and differences in variances preclude statistical analysis of these data. The variance in all measures is substantially greater in the otariids than in the phocids. This is predominantly due to the greater average duration and frequency of suckling in the subantarctic fur seal than in the other two otariids. The long daily duration of suckling in the Weddell seal compared to other phocids may simply reflect relatively frequent or long off-teat intervals, since this estimate includes these intervals.

Changes in suckling pattern over the course of lactation differ among species. Daily suckling duration increases with pup age in the grey seal and both elephant seals, decreases with pup age in the Weddell seal, and remains relatively unchanged in the harp seal (Fig. 2). The increasing trend in northern and southern elephant seals is the result of an increase in both session duration and frequency (Bryden, 1968; LeBoeuf et al., 1972). In the grey seal, session duration increases, while frequencies remain the same (D. J. Boness, unpublished). The decrease in daily suckling duration with age in the Weddell seal results from a decrease in both session frequency and duration (Tedman and Bryden, 1979). In harp seals a slight increase in session duration is counteracted by a decrease in session frequency such that daily duration is unchanged (R. E. A. Stewart, 1983).

Changes in suckling behavior over the entire course of lactation have not been published for any otariid, although South American fur seal pups suckle for no more than 1 h per session when they are less than 3 months old, but as yearlings, before they are weaned, suckling durations may exceed two hours (P. Majluf, personal communication). In the Steller sea lion an increase in session duration over the first month after birth is more than compensated by a decrease in frequency such that daily suckling duration decreases (Sandegren, 1970; Higgins, 1984). Nothing is known of subsequent patterns, however. In the absence of longitudinal studies of the behavior of the walrus, little is known about its maternal behavior or suckling patterns. Suckling sessions appear to be intermediate in duration to those of phocids and

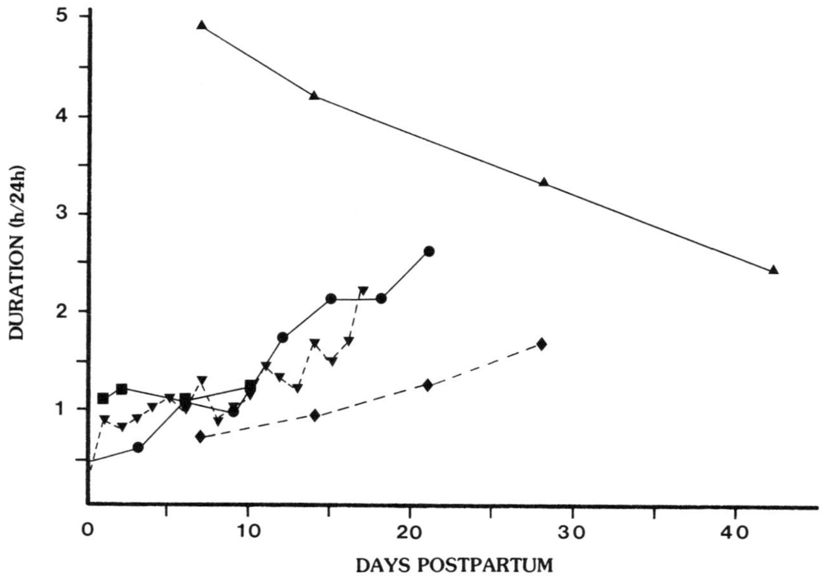

FIGURE 2. The relationship of mean daily suckling duration to pup age in phocids. (▲) Weddell seal, (●) southern elephant seal, (◆) northern elephant seal, (▼) grey seal, (■) harp seal.

otariids, but aquatic suckles are shorter on average than terrestrial ones (11.7 as compared to 16.0 min) (Miller and Boness, 1983; Salter, 1978). There are no data on the frequency of suckling.

2.3.4. Patterns of Maternal Attendance

Phocid females remain in the vicinity of the pupping site throughout lactation. Females of some species stay in close proximity to their pups, whereas females of others leave their pups and go into the water, but apparently remain nearby (Table IV). Most pack-ice species enter the water during the lactation period, although the hooded seal remains with its pup on the ice. This may relate to the extremely short lactation period observed in this species (see Section 2.4). Among those species in which females spend time in the water, as much as 70% of their time may be spent there, as is the case in harp seals (R. E. A. Stewart, 1983). In four of these species, pups follow their mothers, although this usually does not occur immediately after birth. The grey seal exhibits both types of attendance patterns when pupping on land, but not enough information is available to know if this is true of ice-pupping populations as well.

One obvious reason for females to spend time in the water is to feed on prey available in the immediate vicinity. Stomach samples taken from females at pupping sites contained food in about half the species in which females enter the water during lactation (Table IV). The amount consumed by lactating females is markedly reduced in comparison to the amount ingested outside the lactation period, however. The lack of evidence of feeding in some species in which females spend time in water could reflect insufficient data or may indicate that water entry may be related to factors other than food. Among factors that might be important are avoidance of thermal stress (Pierotti and Pierotti, 1980; R. E. A. Stewart, 1983), avoidance of either land- or ice-based predators (Table IV), and mating activities (Stirling, 1983). Animal movements may also be influenced by population density at pupping sites.

Otariid females attend their young continuously for a short period after birth, but then alternate between periods of foraging at sea and nursing on land ("feeding cycles"). Intraspecific variation is apparent in feeding cycles in at least three species in Table V. Part of this variation might be attributed to varying levels of accuracy of the data, but differences also relate to pupping sites and year of study. For example, northern fur seal females at St. Paul Island stayed at sea longer than females at St. George Island (Gentry and Holt, 1986). St. Paul is further from the continental shelf where fur seals feed than is St. George. Likewise, California sea lion females at San Nicolas Island had longer foraging trips in 1983 and 1984, during a food shortage caused by El Niño, compared to 1982 and 1985 (mean ± *SD*: 1982, 1.4 ± 0.32; 1983, 2.0 ± 0.78; 1984, 2.5 ± 0.72; 1985, 1.7 ± 0.54; Scheffe's range test indicated 1982 and 1985 < 1984, $P < 0.01$) (K. A., Ono, D. J. Boness, and O. T. Oftedal, unpublished).

Substantial interspecific variation occurs in duration of foraging periods (1–10 days; Table V), whereas there is less variation in the length of time to first departure (4–10 days) and the duration of periods of presence at the rookery (0.5–3 days). The variation in duration of foraging periods might reflect differences in type and accessibility of food. The species with the longest foraging periods, the northern fur seal, is a pelagic feeder, traveling as much as 250 km to feed (Scheffer, 1970; Lander and Kajimura, 1982; Stirling, 1983). Similarly, Antarctic fur seal females forage at sea for over 4 days on average and are known to travel approximately 200 km to feed (Croxall et al., 1985). In contrast, California sea lion and Cape fur seal females are absent on average for only 2.5 and 2.9 days, respectively, while feeding in shallow water less

TABLE IV
Aquatic Behavior and the Occurrence of Predators at Phocid Pupping Sites

Species	Time spent in water by females	Do pups accompany females in water?	Evidence of feeding by females	Predators at pupping site	References
			Pack-ice		
Bearded seal	Some	Yes	None[a]	Infrequent, polar bears	Burns (1970, 1981a)
Caspian seal	Large proportion	No	None	Wolves, large eagles	Ivashin et al. (1972), Popov (1979)
Grey seal	≤49%[b]	No[b]	None	Sharks	Brodie and Beck (1983), Coulson and Hickling (1964), Fogden (1971), D. J. Boness (unpublished), Anderson and Harwood (1985)
Harp seal	≤70%	No	Yes	Polar bears	Sivertsen (1941), R. E. A. Stewart (1983), R. E. A. Stewart and Murie (1986)
Hooded seal	None	No	None	Polar bears	Shepeleva (1971), Bowen et al. (1985)
Ribbon seal	Large proportion	No	None	None	Burns (1981b), Frost and Lowery (1980)

Spotted seal	Some	No	Yes	Polar bears, wolves	Burns et al. (1972), Bonner (1979a)
			Fast-ice		
Baikal seal	Some	No	Yes	Infrequent, brown bears	Pastukhov (1975b), Thomas et al. (1982)
Ringed seal	Some	Yes	Yes[a]	Frequent, polar bears	McLaren (1958), Stirling and McEwan (1975), Smith and Hammil (1981)
Weddell seal	≤56%	Yes	None	None	Kaufman et al. (1975), Demaster (1979)
			Terrestrial		
Harbor seal	Large proportion	Yes	Yes	Infrequent, eagles, sharks	Bonner (1979b), Boulva and McLaren (1979), Renouf (1984)
Hawaiian monk seal	Some	Yes	None	Sharks	Kenyon and Rice (1959), Alcorn (1984)
Northern elephant seal	None	No	None	None	LeBoeuf et al. (1972)
Southern elephant seal	None	No	None	None	Laws (1956), McCann (1982)

[a]Evidence indicates that pups may begin feeding on solid food before being weaned.
[b]In some colonies females do not go into the water nor do the pups.

TABLE V
Temporal Parameters of Feeding Cycles in Otariids[a]

Species	Time to first departure (days)	Absence duration (days)	Presence duration (days)	Percent time absent	Reference
Cape fur seal	4.3	2.9	2.4	54.4	David and Rand (1986)
Antarctic fur seal	6.9	4.3	2.1	64.9	Doidge et al. (1986)
Northern fur seal	8.0	9.4	2.0	—	Peterson (1968)
	~4–7	7.7	1.5	—	Bartholomew and Hoel (1953)
	7.4	5.9	2.2	73.0	Gentry and Holt (1986)
Galapagos fur seal	7.3	1.3[b]	1.0[b]	55.3	Trillmich (1986a)
California sea lion	7.9	2.5	1.4	52.3	D. J. Boness (unpublished)
	6.8	0.8	0.5	52.0	Trillmich (1986b)
Steller sea lion	9.0	0.7	0.7	—	Sandegren (1970)
	6.7[c]	~2[c]	~1[c]	50.0[c]	Higgins (1984)
Guadalupe fur seal	8.8	—	—	—	Pierson (1978)
New Zealand fur seal	~7	5.0	1.2	—	Stirling (1970)
	8.5	~1–2	—	—	McNab and Crawley (1975)
	10.0	~3–5	~2–4	—	Miller (1971)
South American fur seal	—	4.6[c]	1.3[c]	76.5[c]	Trillmich et al. (1986)

[a]Values are means unless otherwise stated and may reflect observations from more than one rookery or over more than one season.
[b]Values are medians; females with 1- to 2-year-old young have slightly different values for absence (0.9 day) and presence (0.5 day).
[c]Values obtained in an El Niño year and therefore may not reflect normal years.

than 100 km from the rookery (Fiscus and Baines, 1966; Costa et al., 1985; David and Rand, 1986). Steller sea lions and Galapagos fur seals, known to feed in close proximity to the rookeries, spend only about 1 day on foraging trips (Fiscus and Baines, 1966; Trillmich, 1986).

Feeding cycles change over the course of lactation, but not in a consistent pattern for all species. Maternal foraging trips increase in duration as pups get older in the Cape fur seal, Steller sea lion, and South American fur seal (David and Rand, 1986; Gentry and Holt, 1986; Higgins, 1984; P. Majluf, personal communication) and in one study of northern fur seals (Bartholomew and Hoel, 1953). However, the relationship between age and trip duration is more complex in the California sea lion and a second study of the northern fur seal. Trip duration in California sea lions decreases during the third month postpartum following an increase during the first 2 months. There is a subsequent increase in the length of foraging trips from the fourth to the sixth month postpartum (Boness et al., 1985a). Peterson (1965) found that northern fur seals initially increase the duration of trips to sea, but then decrease the length of time away as their pups get older. It is not clear why the two studies of northern fur seals yield different results. Changes in the duration of foraging trips as pups get older may indicate that the nutritional demands of lactation increase as pups get older (see Section 5.1) and/or that changes occur in prey abundance or composition. Longitudinal data on the duration of periods of presence indicate relatively little variation, although both Cape fur seal and California sea lion females show an initial decrease in duration, after which the duration becomes constant (Peterson, 1965; Boness et al., 1985a; David and Rand, 1986; Gentry and Holt, 1986).

The period of fasting and lactation prior to first departure entails a progressive depletion of female reserves. Interspecific comparisons indicate that neither absolute nor relative female size correlates with length of time to first departure ($r = 0.190$, $n = 9$, $P > 0.05$ and $r = -0.117$, $n = 9$, $P > 0.05$, respectively). However, in the South American fur seal, the length of time a female remains in attendance of her pup is significantly correlated with maternal weight (P. Majluf, personal communication). In the California sea lion, the duration of this period may correlate with the magnitude of female reserves, as evidenced by a significantly shorter initial period of attendance during a severe El Niño that affected food supply (Ono et al., 1985).

In comparison to phocids, otariids have been shown to exhibit a higher frequency and longer duration of suckling during periods of maternal presence. Apparently otariid pups must compensate for periods of maternal absence by increased suckling when mothers are available. We predict that those otariids with relatively long maternal foraging periods (for example, northern fur seal, Antarctic fur seal) will have relatively long daily suckling durations. Available data do not permit this hypothesis to be tested.

2.4. Duration of Lactation

Lactation duration also varies markedly among and within pinniped families. The phocids typically have short lactation periods, whereas otariids have relatively long lactation periods (Table II). Walruses, which take their pups with them and nurse aquatically, have the longest lactation period of any pinniped (Mansfield, 1958; Loughrey, 1959; Miller and Boness, 1983; Fay, 1982).

The quality of data available on the duration of lactation is quite variable. Known individuals have been followed from parturition to weaning in only a few phocids and one otariid. Many studies have estimated lactation length from census data, and others provide only a qualitative statement without further explanation. The problem is clearly greater in otariids than in phocids, and reflects the difficulty of following individuals over an extended period. In many otariids, lactation length is described as about 1 year because some yearlings are observed suckling (Table II). Few studies, however, have attempted to determine the proportion of the yearling cohort that is still suckling at a given age (Trillmich and Majluf, 1981; Boness et al., 1985b). Construction of weaning curves (for example, Bowen et al., 1985) would permit more meaningful interpretation of the duration of lactation in otariids.

Despite the need for better quantitative data on the duration of lactation, it is obvious that there is substantial variation within each family. Among phocids, the range in duration of lactation is at least in part associated with the stability and harshness of the breeding habitat ($P < 0.05$; Scheffe's range test). Fast-ice species have significantly longer periods of lactation (55.6 ± 7.51 d, $n = 3$) than either pack-ice species (20.3 ± 9.98 d, $n = 9$) or land-whelping species (30.5 ± 7.0 d, $n = 4$). One pack-ice species, the hooded seal, has the shortest lactation period (4 days) known for any mammal (Bowen et al., 1985). Instability of ice due to storms during the lactation period and the potential for polar bear predation may contribute to the selective advantage of a shortened lactation period.

The finding that fast-ice species have longer periods of care than terrestrial species may seem surprising, particularly because ambient conditions on fast ice are typically harsher than on land and, in the northern hemisphere, polar bears are a known predator on the fast ice. The two northern fast-ice species (ringed seal and Baikal seal) build birth lairs under ice hummocks or snow; the southern species (Weddell seal) does not. These dens can serve both to hide the female and her young from predators and to provide protection against severe surface weather (McLaren, 1958; Pastukhov, 1975b; T. G. Smith and Stirling, 1978). Thus, there could be some advantage to remaining in lairs as long as possible. The relatively long lactation period of the Weddell seal in the southern hemisphere might relate to aquatic predation. The leopard seal preys upon pups that have left the ice (Demaster, 1979).

It has been suggested that variation in lactation duration among fur seals may be related to differences in environmental predictability and climatic harshness associated with latitude. According to Gentry et al. (1986), high latitudes have strong seasonality with severe weather, but conditions are relatively similar from year to year, whereas low latitudes show less seasonal change but are more prone to unpredictable environmental oscillations such as El Niño. They hypothesize that high-latitude conditions favor short lactation periods, whereas low-latitude conditions favor longer periods. The significant negative correlation between duration of lactation and latitude among otariids ($r = -0.73$, $n = 13$, $P < 0.01$) is consistent with this hypothesis (Fig. 3).

The comparison of lactation duration to latitude is confounded with accessibility of food resources, however. The two subpolar species that have the shortest lactation periods, the Antarctic fur seal and the northern fur seal, also feed furthest from the rookery, whereas the Galapagos fur seal feeds in close proximity (see Section 2.3.4). The energetic costs to females associated with returning to the rookery to nurse young will be higher in the two subpolar species and may place a premium on reduction in lactation length. The converse may apply to the Galapagos fur seal. There is no evident relationship between lactation duration and latitude among the nine remaining species in Fig. 3 despite a distributional range from 10° to 50° from the equator. Intraspecific studies of species whose whelping distributions cover a large range (for example, Steller sea lion, southern sea lion, California sea lion, South American fur seal) could help refine our understanding of the relationship of lactation duration to latitude in otariids.

2.5. The Process of Weaning

Weaning tends to be abrupt in phocids. Phocid females simply depart from the rookery at the end of lactation, either just after or just before copulation, and do not return to their offspring (Bryden, 1968;

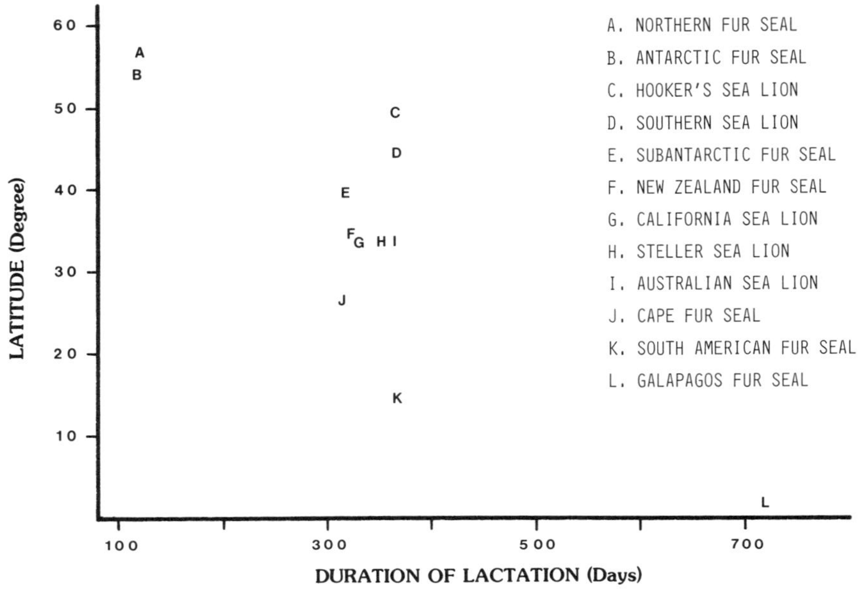

FIGURE 3. The relationship between duration of lactation and latitudinal distribution of otariids. Data are from studies referred to in Table II.

Golt'sev, 1971; Reiter et al., 1978; Boness and James, 1979; Condy, 1979; R. E. A. Stewart, 1983; Bowen et al., 1985). In some species weaned pups remain on the rookery and fast for several days or weeks, but in others the pups depart almost immediately after weaning (Coulson and Hickling, 1964; Bryden, 1968; Ortiz et al., 1978; Condy, 1980; Worthy and Lavigne, 1983a). A more gradual weaning may occur in those phocids in which pups spend time in the water with females (Table IV). Evidence of both milk and solid foods has been found in the stomachs of a few bearded and ringed seal pups (McLaren, 1958; Burns, 1970).

A gradual weaning process probably occurs in most otariids, although considerably more information is needed to confirm this hypothesis. Pups of some species are known to consume both milk and solid food (Hamilton, 1934; Rand, 1959; Trillmich, 1979). In other otariids, pups swim off the rookery during periods when their mothers are absent, but stomach contents have not been examined to determine if feeding occurs. Some evidence of abrupt weaning has been obtained for northern fur seals: 77% of observed pups departed from the rookery before their mothers (Macy, 1982), and females and pups are never sighted together at sea (Gentry and Holt, 1986).

A dichotomy in patterns of behavior emerges from studies of phocids and otariids. The phocids have relatively short lactation periods associated with continuous attendance of pups and fasting by females, whereas the otariids have relatively long periods of lactation, with females alternating between feeding trips at sea and attending their pups at the rookery. Both patterns are different from maternal behavior in most mammals. The accompanying daily suckling durations are not substantially higher than those of terrestrial mammals (Berger, 1979; Reinhardt and Reinhardt, 1981; Whitworth, 1984) despite short lactation periods (phocids) and intermittent suckling (otariids). The explanation of this paradox may be found in the ability of pinnipeds to secrete concentrated milks that permit rapid transfer of energy to the young.

3. THE COMPOSITION OF PINNIPED MILKS

3.1. Factors Affecting Gross Composition

3.1.1. Milk Sampling Procedures

Milk samples have been collected and analayzed for 16 of the 33 extant species of pinnipeds (Table VI). Relative to other mammalian orders this represents a high proportion of species (48%) (Oftedal, 1984a). The data available for pinnipeds vary in quality, however. Many reports are based on only one or a few milk samples obtained opportunistically. Milk has been collected by various sampling methods and at different lactation stages (Table VI). Prior to review of interspecies trends, the possible effects of these variables need to be examined.

The first analysis of a pinniped milk was based on gastric contents of a suckling California sea lion pup (Schroeder and Wegeforth, 1935). More recently, compositional data have been reported for gastric samples from five otariids, four phocids, and the walrus (Table VI). Gastric sampling is an attractive procedure due to the ease of capture and intubation of pups, with minimal disturbance to breeding groups. Unfortunately, gastric samples are apt to be contaminated by salivary and gastric secretions, and are subject to proteolytic and lipolytic enzymes. Milk constituents pass from the stomach at different rates as a consequence of milk curd formation and erosion (Hill et al., 1970). Thus, gastric samples may differ in composition from milk collected directly from mammary glands (Oftedal, 1984a).

In pinnipeds, gastric milk samples are consistently lower in dry matter, fat, and perhaps protein than are mammary samples at the same lactation stage (Table VII). In harp seals, samples obtained by intubation

TABLE VI
Sampling and Analytical Methods Applied in Studies of Milk Composition in Pinnipeds

Species	Sample type[a]	n	Lactation stage[b]	Analysis type[c]	References
			Phocidae		
Bearded seal	G	1	L(14–21)	P	Fay (1982)
Grey seal	M?	1	?	P	Amoroso et al. (1951)
	M?	1	?	F	Meara (1952)
	G	1	L(14)	P,A	Bock et al. (1968)
	X?	1	?	F	Ackman and Burgher (1963)
Harbor seal	?	?	?	(P)	Harrison (1960)
Harp seal	X	2	?	P,M	Sivertsen (1941)
	X?	?	?	(P),F	Jangaard and Ke (1968)
	M+	3	M?(<7)	P,M,F	Cook and Baker (1969)
	?	1	?	P	Jenness and Sloan (1970)
	M+	5	E–L(2–16)	P,M,F	Van Horn and Baker (1971)
	G	26	E–L(0–9)	P	Lavigne et al. (1982)
	G	22	E–L(0–9)	P	R. E. A. Stewart et al. (1983)
	X,M+,G	36	E–L(0–11)	P	O. T. Oftedal, D. J. Boness, and W. D. Bowan, (unpublished)
Hooded seal	X	1	?	P,M	Sivertsen (1941)
	X,G	16	?	P	Shergin et al. (1969)
	X?	?	?	F	Jangaard and Ke (1968)
	X,M+,G	22	E–L(0–4)	P	O. T. Oftedal, D. J. Boness, and W. D. Bowen, (unpublished)
Northern elephant seal	?	7	?	P	Jenness and Sloan (1970)
	M+	9	L(21–28)	P,(M)	LeBoeuf and Ortiz (1977)
	M	48	E–L(0–30)	P,(M),F	Riedman and Ortiz (1979)

TABLE VI (*Continued*)

Species	Sample type[a]	n	Lactation stage[b]	Analysis type[c]	References
Southern elephant seal	M+	7	E–L(0–21)	P	Bryden (1968)
	M	?	?	—	Shaughnessy (1974)
	M	12	E–L(0–26)	(P,M)	Peaker and Goode (1978)[d]
Weddell seal	M	7	E–M(1–18)	(P),F	Stull et al. (1967)
	M	39	E–L(0–35)	(P)	Kooyman and Drabek (1968)
	?	8	?	P	Jenness and Sloan (1970)
	M+	11	E–L(0–45)	P,(M),F	Tedman (1980)
Otariidae					
Antarctic fur seal	G	1	?	P	Bonner (1968)
California sea lion	G	1	?	P	Schroeder and Wegeforth (1935)
	X	3	E–L(<30–240)	P	Pilson and Kelly (1962)
	M+	28	E–M(0–160)	P	Oftedal et al. (1983)
	G,M+	10	E(2–60)	(P,M)	Trillmich and Lechner (1986)
	M+,G	64	E–M(0–160)	P,M,F	O. T. Oftedal, D. J. Boness, S. J. Iverson, M. E. Allen, and D. L. Wood (unpublished)
Cape fur seal	X	1	?	(P)	Rand (1955)
	X	1	E(<30)	P	Kerry and Messer (1968)
Galapagos fur seal	M+,G	27	E(1–30),L(360)	(P,M)	Trillmich and Lechner (1986)
Northern fur seal	M+	5	E–L(0–120)	P,(M),F	Ashworth et al. (1966)
	M	5	E(0–>10)	P	Fay (1982; personal communication)
	X	?	M(30–60)	P,M,F,A	Dosako et al. (1983)
	M+	58	E–M(0–60)	(P)	Costa and Gentry (1986)

(*Continued*)

TABLE VI (*Continued*)

Species	Sample type[a]	n	Lactation stage[b]	Analysis type[c]	References
South American fur seal	X,G	13	E–L(0–330)	P	Ponce de Leon (1984)
Steller sea lion	?	2	?	P,(M)	Poulter et al. (1965)
			Odobenidae		
Walrus	X,G	4	E(0–30)	P	Fay (1982)

[a]Samples were collected from gastric contents of pups (G), by manual expression from live animals with (M+) or without (M) oxytocin, or immediately postmortem by expression or incision (X).
[b]Stage of lactation characterized as early (E), mid (M), or late (L); range of days after birth indicated inside parentheses.
[c]Analyses reported in source are characterized as proximate analysis (P), mineral analysis (M), fatty acid analysis (F), and amino acid analysis (A); incomplete analysis indicated by parentheses around letter designation.
[d]Species incorrectly identified as Antarctic fur seal (see Bonner, 1984; M. Peaker, personal communication).

of pups are 9.8 ± 2.21 *SE* percentage points lower in dry matter and 10.2 ± 1.66 *SE* percentage points lower in fat than milk samples collected from the mothers of these pups ($n = 8$) (O. T. Oftedal, D. J. Boness, and W. D. Bowen, unpublished observations). The data in Table VII indicate that gastric samples contain about 74, 78, and 83% of the calculated energy content of mammary milk in California sea lions, harp seals, and Galapagos fur seals, respectively. Mean levels of dry matter and fat in gastric samples tend to parallel changes in mammary milk composition over lactation (Table VII) such that gastric samples may reveal relative trends in milk composition even if the absolute values are in error. Thus, Lavigne et al. (1982) were able to demonstrate an age-related increase in fat content of harp seal milk on the basis of gastric samples.

Mammary milk samples have been collected by manual expression from live animals after physical or chemical restraint, or by expression or incision of recently killed animals. In some large phocids, it is possible to collect milk surreptitiously from one nipple while the pup suckles on the other (Stull et al., 1967; Kooyman and Drabek, 1968; Riedman and Ortiz, 1979). Disadvantages to various procedures vary, and include potential postmortem changes due to loss of mammary metabolic function, contamination with blood and lipids of blubber origin during incision, contamination with dried salts or dirt on the

TABLE VII

Comparison of Mammary Milk Samples (M) to Gastric Contents (G) of Suckling Pups in Three Pinniped Species[a]

Lactation stage (days PP)	Samples analyzed		Dry matter (%)		Fat (%)		Crude protein (%)		Gross energy[b] (kcal/g)	
	M	G	M	G	M	G	M	G	M	G
California sea lion[c]										
0–60	21	27	41.1	33.6	31.7	21.9	9.8	8.9	3.46	2.52
90–120	12	9	54.2	43.5	43.7	32.4	9.6	8.9	4.54	3.47
Galapagos fur seal[d]										
1–30	19	8	—	—	29.4	23.8	12.1	10.9	3.39	2.81
Harp seal[e]										
~3	5	6	56.3	44.8	44.7	34.2	8.1	6.8	4.55	3.51
~8	6	6	61.5	51.4	49.2	39.3	7.9	6.4	4.94	3.96
~11	3	3	67.6	57.8	57.1	42.9	7.7	6.9	5.65	4.31

[a]Data are mean values; days PP, days postpartum.

[b]Gross energy calculated as (fat × 9.11 + protein × 5.86)/100 (Oftedal, 1984a).

[c]O. T. Oftedal, D. J. Boness, and S. J. Iverson (unpublished); milk samples from immobilized females; gastric samples collected directly after observed suckling.

[d]Trillmich and Lechner (1986).

[e]O. T. Oftedal, D. J. Boness, and W. D. Bowen (unpublished); milk samples from immobilized and shot females; gastric samples from suckling pups.

nipple or surrounding pelage, and inability to obtain large samples due to time constraints associated with immobilization procedures.

Among many domestic and terrestrial mammals, the fat content of milk rises progressively over the course of mammary evacuation such that the final fraction may be severalfold higher in fat content than the first fraction obtained (Oftedal, 1984a). Data for hooded seals, harp seals, and California sea lions do not show a significant increase in the fat content of sequential samples other than a small rise (from 32 to 34% fat) in the California sea lion in the first week postpartum (O. T. Oftedal, D. J. Boness, and W. D. Bowen, unpublished; O. T. Oftedal, D. J. Boness, and S. J. Iverson, unpublished). Nonetheless, it is probably prudent during milk collection to evacuate the mammary gland as completely as conditions permit.

In many pinniped studies, only one or a few samples of milk have been analyzed (Table VI), yet a few samples may not be representative. In California sea lions considerable variation can occur among samples collected from different teats of the same animal. The great differences in levels of dry matter (32–54%) and fat (19–44%) among teats of an animal that had recently given birth may reflect which teats had been suckled and to what extent (O. T. Oftedal, D. J. Boness, and S. J. Iverson, unpublished). Later in lactation interteat variation appears to be less, although aberrant samples may result from mastitis or involution processes in a gland that fails to be suckled.

The use of exogenous oxytocin facilitates milk collection in pinnipeds, presumably by initiating the milk ejection reflex. In some domestic animals the use of repeated, large doses or continuous infusion of oxytocin may reduce milk lactose content and alter electrolyte ratios, probably through effects on intercellular junctions in the mammary alveoli (Linzell et al., 1975). Because alveolar milk is apt to constitute but a small fraction of the milk collected in pinniped studies, the effects on milk composition of single oxytocin doses are probably negligible. In hooded and harp seals, milk samples obtained after oxytocin administration are similar in gross composition to samples obtained immediately postmortem (O. T. Oftedal, D. J. Boness, and W. D. Bowen, unpublished).

3.1.2. Lactation Stage

Milk composition changes markedly through the course of lactation in some pinniped species. Most notable is the enormous increase in fat content from about 10% at birth to about 50% at 2 weeks in southern (Bryden, 1968; Peaker and Goode, 1978) and northern elephant seals

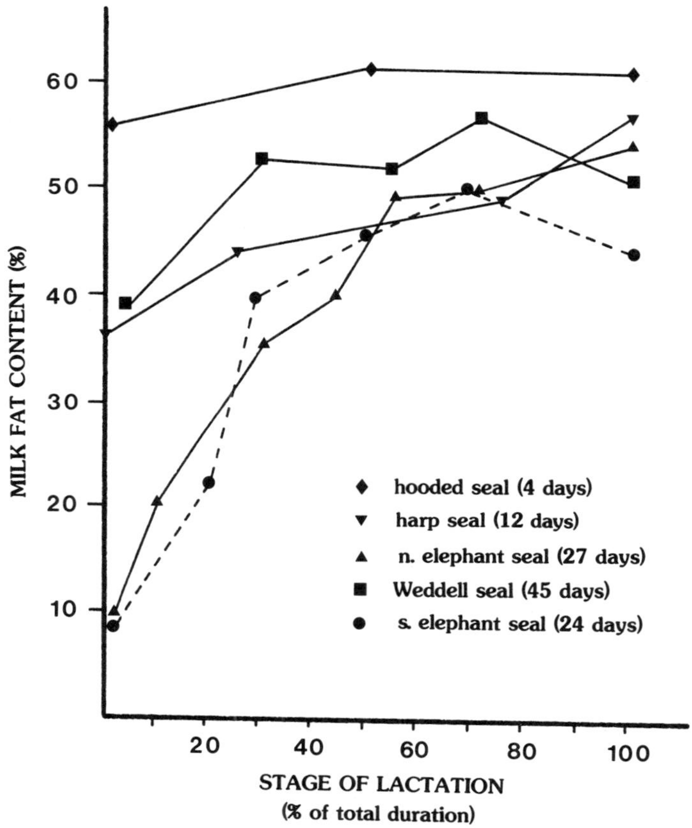

FIGURE 4. The relationship of milk fat content to stage of lactation in phocids. Each point represents a mean value at that stage. The number of samples analyzed for each species are as follows: hooded seal, $n = 22$; harp seal, $n = 19$; southern elephant seal, $n = 12$; northern elephant seal, $n = 44$; Weddell seal, $n = 11$. References as in Table VIII.

(Riedman and Ortiz, 1979) (Fig. 4). Protein content shows no change or a slight decline in this period. Thereafter samples exhibit variability in composition, but no consistent trends in either fat or protein content. The decline in fat and protein content beyond 2 weeks postpartum observed by Bryden (1968) was probably an artifact of the small number of samples ($n = 2$). Peaker and Goode (1978) did not confirm this trend (Fig. 4).

A similar but less pronounced rise in fat content in the first few weeks postpartum is found in both Weddell seals (Stull et al., 1967;

Kooyman and Drabek, 1968; Tedman, 1980) and harp seals (O. T. Oftedal, D. J. Boness, and W. D. Bowen, unpublished). (Fig. 4). In these species milk dry matter content tracks fat content, but is about 10–15% higher. In harp seals crude protein content of milk is significantly higher at birth (10.3%, $n = 5$) than at subsequent sampling times (7.9%, $n = 14$) (O. T. Oftedal, D. J. Boness, and W. D. Bowen, unpublished), but the relatively few samples of Weddell seal milk analyzed reveal no consistent trend (Tedman, 1980). The hooded seal exhibits little change in milk composition during its 4-day lactation period ($n = 22$ samples) (Fig. 4). Crude protein content declines from 6.2% at birth to 4.9% at 2–4 days, but no trends are apparent in dry matter or fat (O. T. Oftedal, D. J. Boness, and W. D. Bowen, unpublished).

Compositional trends during lactation are not as well documented for otariids. Milk of the South American fur seal increases in fat content from 31% in the first month postpartum to 44% in the fifth month and further to 55% in the 10–11th months. Dry matter follows a parallel trend (46, 54, and 68% in the first, fifth, and 10–11th months, respectively), but protein level averages 10–12% throughout (Ponce de Leon, 1984). Trends in California sea lions are remarkably similar in the first half of lactation. Fat increases from 32% at 0–2 months to 44% by the fourth month after birth, whereas dry matter rises from 41 to 54%. Protein content remains at about 9–10% over this period (Oftedal et al., 1983).

In the northern fur seal, it appears that milk fat content decreases somewhat during the initial postpartum period before females depart on feeding trips. Costa and Gentry (1986) report a decline in fat content from 46 to 40% in the first 8 days after birth. Data available from other laboratories are consistent with this trend, but the absolute fat values on the day of birth (58.7%, $n = 2$), at 1 day postpartum (52.9%, $n = 2$), and for the period of 10 days to 2 months (48.8%, $n = 6$) are considerably higher (Ashworth et al., 1966; Fay, 1982, personal communication; Dosako et al., 1983). According to Costa and Gentry (1986), the initial fat decline is followed by a gradual increase in average fat level such that by 2 months the fat content of northern fur seal milk is comparable to that at birth. California sea lion milk at 2 months postpartum is not significantly different in fat from that at birth (Oftedal et al., 1983), but trends in the interim have not been examined. In the Galapagos fur seal, milk fat content is reported to decrease by 7 percentage points during the first month, whereas crude protein remains unchanged (Trillmich and Lechner, 1986).

In otariids, time of sampling relative to female arrival and departure on feeding cycles may have an effect on milk fat content. Milk obtained from northern fur seals on arrival is on average 6 percentage points higher in fat content than milk collected just prior to departure (Costa and Gentry, 1986). Comparable data are not available for other species. Northern fur seals may represent an extreme case, since periods of absence are longer than in other otariids (Table V). Some of the discrepancy in reported fat values for northern fur seal milk may stem from sampling procedures; incomplete fat extraction due to problems with analytical methodogy may also be involved.

Data on milk composition in the last half of lactation are scanty for most otariids. Single samples of late-lactation milk have been assayed for both the California sea lion (Pilson and Kelly, 1962) and northern fur seal (Ashworth et al., 1966). The variability of fat content among samples implies that single samples may not be good indicators of trends.

3.1.3. Interspecies Variation at Mid- to Late Lactation

Given the effects of lactation stage on milk composition, direct interspecies comparisons are valid only at equivalent lactation stages. Comparisons among terrestrial mammals usually focus on midlactation, which may be defined as the period about peak lactation during which milk composition is relatively constant (Oftedal, 1984a). The timing of peak lactation in pinnipeds is not known, but constancy in milk composition is typically reached in phocids in the middle or latter part of lactation (Fig. 4). Milk composition data for this time period are available for two otariids and five phocids (Table VIII).

At this lactation stage the milks of all studied pinnipeds are remarkably similar in dry matter, fat, protein, sugar, and gross energy (Table VIII). Five species (northern fur, South American fur, harp, Weddell, and northern elephant seals) produce milks that are very similar: 63–68% dry matter, 51–55% fat, 8–11% protein, trace levels of sugar, and 5.2–5.7 kcal gross energy per gram. The lower values of dry matter, fat, and energy in the California sea lion (Table VIII) are for samples from the first half of lactation, and may not be representative of subsequent secretion. The similarity in compositional changes for milks of South American fur seals and California sea lions (see Section 3.1.2) suggest that late-lactation milk of California sea lions may be comparable to the five species listed above. The difference between northern and southern elephant seals may be an artifact of small sample numbers, because the trends in milk composition over the first few weeks are so similar (Fig. 4).

TABLE VIII
Milk Composition at Mid- to Late Lactation among Pinnipeds[a]

Species (lactation duration, days)	Lactation stage: Days PP	Lactation stage: n	Dry matter (%)	Fat (%)	Crude protein (%)	Sugar (%)	Gross energy[b] (kcal/g)	References
Phocids								
Hooded seal (4)	2–4	15	69.7	61.4	4.9	—	5.88	O. T. Oftedal, D. J. Boness, and W. D. Bowen (unpublished)
Harp seal (12)	~8–12	9	63.5	51.8	7.8	—	5.18	O. T. Oftedal, D. J. Boness, and W. D. Bowen (unpublished)
Southern elephant seal (23)	11–26	5	—	46.9	7.4	0.03	4.71	Peaker and Goode (1978)[d]
Northern elephant seal (27)	14–28	9	67.2	54.4	9.0	<0.25	5.48	Riedman and Ortiz (1979)
Weddell seal (45)	10–43	7	66.2	53.6	8.9	0.0	5.40	Tedman (1980)
Otariids								
Northern fur seal (120)	30–120	5	63.3	50.7	10.3	0.1	5.20	Ashworth et al. (1966), Dosako et al. (1983)
California sea lion[c] (~300)	90–120	12	54.2	43.7	8.9	0.6	4.50	O. T. Oftedal, D. J. Boness, and S. J. Iverson (unpublished)
South American fur seal (~365+)	270–330	5	67.9	55.2	11.6	0.0	5.71	Ponce de Leon (1984)

[a]Data are mean values.
[b]Gross energy calculated as (fat% × 9.11 + protein% × 5.86)/100 (Oftedal, 1984a).
[c]Results probably do not represent late lactation (see text).
[d]Species incorrectly identified as Antarctic fur seal (see Bonner, 1984; M. Peaker, personal communication).

It has been suggested that otariids produce more dilute milk than phocids (Laws, 1959; Bonner, 1984) and that the fat content of otariid milk is correlated to the length of time the mother is away on feeding trips (Gentry et al., 1986; Trillmich and Lechner, 1986). While many reports for otariids do indicate rather low dry matter and fat values, most of these reports are based on milk samples taken in the first half of lactation (for example, Pilson and Kelly 1962; Fay, 1982; Oftedal et al., 1983; Costa and Gentry, 1986; Trillmich and Lechner, 1986) or gastric samples (Rand, 1955; Bonner, 1968). Available information of milk samples from mid–late lactation (Table VIII) do not support hypotheses of a difference between otariids and phocids, or of a correlation to feeding cycle duration in otariids. As more complete information is obtained for additional species, it may become possible to correlate the profound differences in fat content in early lactation to behavioral or physiological parameters. It is intriguing, for example, that both the phocid (hooded seal) and the otariid (northern fur seal) with the highest known fat content in early lactation are species with especially short lactation periods.

Relative to other mammals, pinnipeds secrete milk that is very energy-dense (4.5–5.9 kcal/g at mid- to late lactation). Energy densities above 3.5 kcal/g are found only in some bears in late lactation (Jenness et al., 1972) and in some cetaceans (Jenness and Sloan, 1970). Hooded seals produce the most highly energy-rich milk that has been reported for any mammalian species (Table VIII).

Various hypotheses have been put forth to explain the adaptive significance of the characteristically high fat and energy content of pinniped milks. High fat content is accompanied by low water content. In phocids that fast during relatively long lactation periods (for example, elephant and Weddell seals, Table II), low milk water content may be important to maternal water balance. In species with a very abbreviated lactation period (for example, the hooded seal, Table II), high fat content may facilitate the rapid transfer of fat and energy to pups as needed for rapid postnatal blubber deposition. In otariids, high-energy-density milks are associated with relatively low volumes of secretion (see Table XIII) and may relate to the finite capacity of the mammary glands as storage organs during foraging trips. Mammals with long intersuckling intervals typically produce concentrated milks (Oftedal, 1984a).

Protein accounts for about 12% of milk energy in the California sea lion, South American fur seal, and northern fur seal, 9–10% in the harp seal, Weddell seal, and the two elephant seals, and 5% in the hooded seal. These values are all low relative to typical values of 20–35%

for most terrestrial mammals, but are comparable to primate values of 7–17% (Oftedal, 1984a). Otariids, like primates (Payne and Wheeler, 1968), have slow growth rates relative to metabolic size; protein requirements should be low relative to energy needs. Although phocids gain mass rapidly after birth, much of this gain is energy-rich blubber rather than protein-containing tissue (Worthy and Lavigne, 1983b; Bowen et al., 1986b). Phocid pups have high fat needs rather than low protein requirements. The fact that the species with the greatest postnatal growth rate (hooded seal; Bowen et al., 1985) has the lowest percentage of milk energy as protein is consistent with this hypothesis.

3.2. Detailed Milk Constituents

3.2.1. Proteins

In most pinniped studies milk protein is calculated as total nitrogen × 6.38. Because nonprotein nitrogen comprises about 3–7% of total nitrogen in milk of California sea lions, northern fur seals, and Galapagos fur seals (Ashworth et al., 1966; Oftedal, 1984a; Trillmich and Lechner, 1986), this calculation probably overestimates milk protein content by a small amount.

Milk proteins are characterized as caseins or whey proteins, depending on solubility at acid pH. Upon exposure to gastric acidity and proteases, casein micelles coallesce to form curds that erode slowly, releasing protein and peptides for further digestion in the duodenum. Casein micelles are considerably larger in northern fur seal milk than in bovine milk, and include five major casein components (Dosako et al., 1983). Caseins are reported to account for 44–72% of total protein in several species of phocids (Jenness and Sloan, 1970), and for 52% in the northern fur seal (Ashworth et al., 1966).

Little work has been done on whey proteins of pinniped milks. Gel electrophoresis of whey proteins from northern fur seal milk revealed a prominent band corresponding in mobility to β-lactoglobulin; other minor bands were also observed (Dosako et al., 1983). Southern elephant seal milk contains a whey protein with the electrophoretic mobility of serum albumin (Shaughnessy, 1974). α-Lactalbumin levels are very low or undetectable in California sea lion and northern fur seal milks (Schmidt et al., 1971; Johnson et al., 1972). Pinniped milks undoubtedly contain enzymes and other proteins found in milks of other mammals. High activity of bile salt-stimulated lipase has recently been demonstrated in hooded seal milk (Freed et al., 1986).

Amino acid analyses of total protein, casein, and/or whey protein fractions have been published for northern fur seal and harp seal milks (Ashworth et al., 1966; Lauer and Baker, 1977; Dosako et al., 1983).

3.2.2. Lipid Constituents

Milk fat consists of various lipid classes. Triglycerides account for about 98–99% of the lipids in California sea lion milk (J. Bitman, D. L. Wood, S. J. Iverson, and O. T. Oftedal, unpublished). Other neutral lipid classes include diglycerides (1% or less of total lipids), monoglycerides (0.1–0.2%), free fatty acids (0.2–0.6%), and cholesterol (0.3%). Phospholipids account for about 0.5% of the total lipids. The patterns in other pinnipeds are probably similar.

Fatty acid constituents of the neutral lipids have been determined for seven species of pinnipeds (Tables IX and X). Methods for chromatographic isolation and identification of medium to moderately long-chain fatty acids (C_{14}–C_{18}) are well established. Oleic (18:1), palmitic (16:0), and palmitoleic (16:1) are predominant, although their relative contributions vary among species (Table IX). Pinniped milks are also very rich in long-chain, polyunsaturated fatty acids that exhibit prolonged retention times on gas-liquid chromatography (GLC) columns (Table X). Fatty acids with retentions equivalent to or greater than linolenic acid (18:3) account for 20–42% by weight of the total fatty acids. Investigators have assigned different identities to the major GLC peaks, even for the same species (Table X). Both the resolution and the assigned identities of these peaks may be questioned, given the limitations of methodology in the 1960s, when most of these analyses were conducted. Further investigation and identification of the very long-chain fatty acids of pinniped milks are needed using currently available capillary column methods and appropriate standards.

In pinnipeds, milk fatty acids are probably derived with little modification from dietary organisms such as plankton and fish (Ashworth et al., 1966; Stull et al., 1967; Van Horn and Baker, 1971) and may therefore reflect differences in fatty acid composition of diets. The proportions of component fatty acids of blubber vary among species, locality, stage of lactation, fasting state, and location on the body (Meara, 1952; Bryden and Stokes, 1969; Jangaard and Ke, 1968). Some of these factors may also influence milk fatty acid patterns, especially when milk lipids must be mobilized from blubber stores.

3.2.3. Carbohydrates

Pinniped milks contain very low levels of carbohydrate (Table VIII). Although lactose is the predominant sugar in milks of many mammals,

TABLE IX
Predominant Medium- to Long-Chain Fatty Acids in Pinniped Milks

Species	n	Percentage by weight[a]						References
		14:0	16:0	16:1	18:0	18:1	18:2	
Phocids								
Grey seal	1–2	3.0	17.2	14.3	1.8	30.6	0.6	Meara (1952), Ackman and Burgher (1963)
Harp seal	10	4.9	11.8	19.9	4.2	23.5	1.7	Jangaard and Ke (1968), Glass and Jenness (1971), Cook and Baker (1969), Van Horn and Baker (1971)
Hooded seal	1	3.6	9.5	13.5	2.8	27.2	1.3	Jangaard and Ke (1968)
Northern elephant seal	30	2.6	16.8	7.0	3.7	44.1	1.9	Glass and Jenness (1971), Riedman and Ortiz (1979)
Weddell seal	15	9.8	14.7	12.4	1.8	39.0	2.3	Glass et al. (1967), Stull et al. (1967), Tedman (1980)
Otariids								
California sea lion	2	5.9	17.8	6.8	3.9	17.4	2.0	J. Bitman, D. L. Wood, S. J. Iverson, and O. T. Oftedal (unpublished)
Northern fur seal	3	6.3	19.3	9.9	2.0	31.6	2.7	Ashworth et al. (1966), Glass et al. (1967), Dosako et al. (1983)

[a]Column headings identify fatty acid chain length and number of double bonds.

TABLE X
Identification of Major Long-Chain Fatty Acids in Pinniped Milks That Exhibit Prolonged Retention Times on Gas-Liquid Chromatography Columns

Species	Percentage by weight[a]								References
	Total[b] >18:2	20:1	20:4	20:5	22:1	22:5	22:6	Other majors[c]	
					Phocids				
Grey seal	28.3	2.7	1.8	8.8	0.5	3.5	8.1	—	Ackman and Burgher (1963)
Harp seal	41.6	13.9	1.1	9.0	5.2	3.4	6.5	—	Jangaard and Ke (1968)
	28.6	11.7	0.9?	4.3	4.4	1.5	2.5	—	Cook and Baker (1969)
	31.8	8.3	0.8?	0.5	9.5	6.2	—	24:1	Van Horn and Baker (1971)
Hooded seal	36.7	15.0	1.6	6.8	5.0	1.7	3.9	—	Jangaard and Ke (1968)
Northern elephant seal	24.7	12.6	3.4	2.6	—	0.9	4.3	—	Riedman and Ortiz (1979)
Weddell seal	20.2	6.9	0.2	4.7	1.5	1.0	4.0	—	Stull et al. (1967)
					Otariids				
California sea lion	41.7	4.4	2.4	7.7	3.0	3.0	15.9	—	J. Bitman, D. L. Wood, S. J. Iverson, and O. T. Oftedal (unpublished)
Northern fur seal	32.1	4.1	—	—	5.4	—	—	18:3, 22:2	Ashworth et al. (1966)
	22.1	6.8	2.4	4.8	1.0	1.5	3.7	—	Dosako et al. (1983)

[a]Column headings identify fatty acid chain length and number of double bonds.
[b]Combined weight percentage of all fatty acids with retention times greater than that of 18:2.
[c]Includes other long-chain, unsaturated fatty acids present at levels greater than 2.0% by weight.

many pinniped milks are either devoid of lactose or contain insignificant amounts (Pilson and Kelly, 1962; Kerry and Messer, 1968; Johnson, 1975). Indirect calculation of milk carbohydrate content by subtracting fat, protein, and ash from dry matter level is apt to overestimate true carbohydrate content (R. E. A. Stewart et al., 1983) and is responsible for some erroneous values (for example, Amoroso et al., 1951; Lavigne et al., 1982). Myoinositol, a carbohydrate with vitaminlike activity in a few mammalian species (Cody, 1984), has been found at high concentration (0.12%) in northern fur seal milk (Dosako et al., 1983) and is probably present in southern elephant seal milk (Peaker and Goode, 1978).

3.2.4. Minerals

In the milks of terrestrial mammals calcium and phosphorus are usually present in a weight ratio of about 1.6:1 (Jenness, 1979), similar to estimated requirements of 1:1 to 2:1 in most species. By contrast, inverse Ca/P ratios of 0.5:1 to 0.9:1 have been reported for California sea lions, northern fur seals, harp seals, and northern elephant seals (Table XI). In California sea lions, phosphorus levels are apparently low in the immediate postpartum period, such that the Ca/P ratio may be as high as 2.8:1 (Table XI), but this ratio inverts by the second month after birth. The adaptive significance of an inverse Ca/P ratio is not clear. Jenness (1979) noted that pinnipeds had much less calcium and phosphorus in milk than would be predicted based on casein content.

Sodium (Na), potassium (K), and chloride (Cl) are predominantly present in ionic form in the aqueous phase of milk. In most species the isosmotic condition of milk relative to blood is maintained by the interplay of lactose and mineral ions in controlling the flow of water into mammary secretory cells and alveolar lumina (Peaker, 1977). The virtual absence of lactose from pinniped milks suggests that secretory mechanisms may operate differently in controlling secretion of the aqueous phase. The ratio of potassium to sodium should be similar to that of intracellular fluids, about 3:1 on a molar basis or 5:1 on a weight basis, and this is in fact observed in many species (Peaker, 1977). Data on pinniped milks indicate that the weight ratio of K/Na is much lower, about 1.0–2.4 (Table XI). It is difficult to collect milk by mammary expression in pinnipeds without potential contamination from dried salts (especially sodium) on the teat, skin, and pelage. This may explain the particularly high sodium levels (800–1000 mg/kg) and low K/Na ratios (0.7–1.0) reported by Riedman and Ortiz (1979) for northern elephant seal milk (Table XI). In California sea lions, prewashing of the

TABLE XI
Major Minerals in Pinniped Milks[a]

Species[b]	Lactation stage (days PP)	n	mg/kg Ca	mg/kg P	Ca/P	mg/kg K	mg/kg Na	K/Na	References
				Phocids					
Harp seal	2–6	4	723	1200	0.60	1084	618	1.8	Van Horn and Baker (1971)
Northern elephant seal	?	7	511	939	0.54	—	—	—	R. Jenness (personal communication)
	0–7	5	—	—	—	1060	1020	1.0	Riedman and Ortiz (1979)
	8–14	9	—	—	—	820	990	0.8	Riedman and Ortiz (1979)
	15–21	10	—	—	—	720	770	0.9	Riedman and Ortiz (1979)
	22–28	12	—	—	—	560	860	0.7	Riedman and Ortiz (1979)
	21–28	4	—	—	—	1000	670	1.5	LeBoeuf and Ortiz (1977)
Southern elephant seal	5–17	9	—	—	—	1360	990	1.6	Peaker and Goode (1978)
Weddell seal	15–43	5	—	—	—	570	530	1.1	Tedman (1980)
				Otariids					
California sea lion	<7	3	520	187	2.8	2400	1210	2.0	M. E. Allen, O. T. Oftedal, and D. J. Boness (unpublished)
	30–60	4	885	1003	0.88	2030	1060	1.9	M. E. Allen, O. T. Oftedal, and D. J. Boness (unpublished)
	90–120	3	683	1060	0.64	1620	670	2.4	M. E. Allen, O. T. Oftedal, and D. J. Boness (unpublished)
Northern fur seal	30–60	1	567	1193	0.48	838	521	1.6	Dosako et al. (1983)
Galapagos fur seal	1–30	17	630	—	—	1670	—	—	Trillmich and Lechner (1986)

[a]Column headings refer to calcium (Ca), phosphorus (P), potassium (K), and sodium (Na).
[b]All available sources included, except Cook and Baker (1969), which contains analyses in disagreement with subsequent reports from the same laboratory, and is apparently erroneous.

teat area with distilled water and collection of large volumes may have reduced contamination, as evidenced by the higher K/Na ratio (Table XI). Alternatively, interspecies differences in K/Na ratios may occur.

Peaker and Goode (1978) reported that southern elephant seal milk was somewhat hyperosmotic to blood, and Tedman (1980) found rather high osmolality in four of five samples of Weddell seal milk. Possible effects of freezing of samples and salt contamination cannot be ruled out, however. The osmolality of the aqueous phase of a sample of California sea lion milk that was measured shortly after collection and had not been frozen was 286 mosm/kg (M. E. Allen, O. T. Oftedal, and D. J. Boness, unpublished results), well within the expected range for fluids isoosmotic to blood. Further study of the osmolality of pinniped milks is needed.

Because chloride is found almost entirely in the aqueous phase, chloride data can be extrapolated to a whole-milk basis by dividing by milk water content. Chloride concentrations in milks of the northern fur seal, southern elephant seal, and Weddell seal average about 1000–1300 mg/kg (Ashworth et al., 1966; Peaker and Goode, 1978; Tedman, 1980; Dosako et al., 1983). Milk magnesium levels average about 120–170 mg/kg in California sea lions, northern fur seals, and harp seals (Van Horn and Baker, 1971; Dosako et al., 1983; M. E. Allen, O. T. Oftedal, and D. J. Boness, unpublished results). Iron, copper, zinc, and selenium were measured at levels averaging 5.4, 2.6, 7.7, and 0.45 mg/kg in California sea lion milk (D. Allen, O. T. Oftedal, and D. J. Boness, unpublished results, $n = 12$); roughly similar levels have been reported for northern fur seals (Dosako et al., 1983). The extent to which minor contamination may have affected trace mineral levels cannot be ascertained.

FIGURE 5. Mammary anatomy and ultrastructure in the lactating Weddell seal. (a) Adult female, showing location of semiprotruded nipples (N) near umbilicus (U); (bar = 20 cm). (b) Resin cast of milk duct system or mammary gland (N = nipple; bar = 3 cm). (c) Mammary tissue showing a lobule (L) filled with alveoli (hematoxylin and eosin; bar = 400 μm). (d) Mammary tissue showing an alveolus with fat droplets (F) (toluidine blue stain; bar = 20 μm). (e) Electron micrograph of alveolar epithelium showing myoepithelial cells (M), epithelial cells (E), and fat droplets (F); (bar = 2.0 μm). (f) Electron micrograph of a secretory epithelial cell showing Golgi vesicles (GV) and protein micelles (P) in the lumen; (bar = 1.0 μm). (g) Electron micrograph of a secretory epithelial cell showing Golgi apparatus (G), rough endoplasmic reticulum (R), and lumen (L); (bar = 2.0 μm).

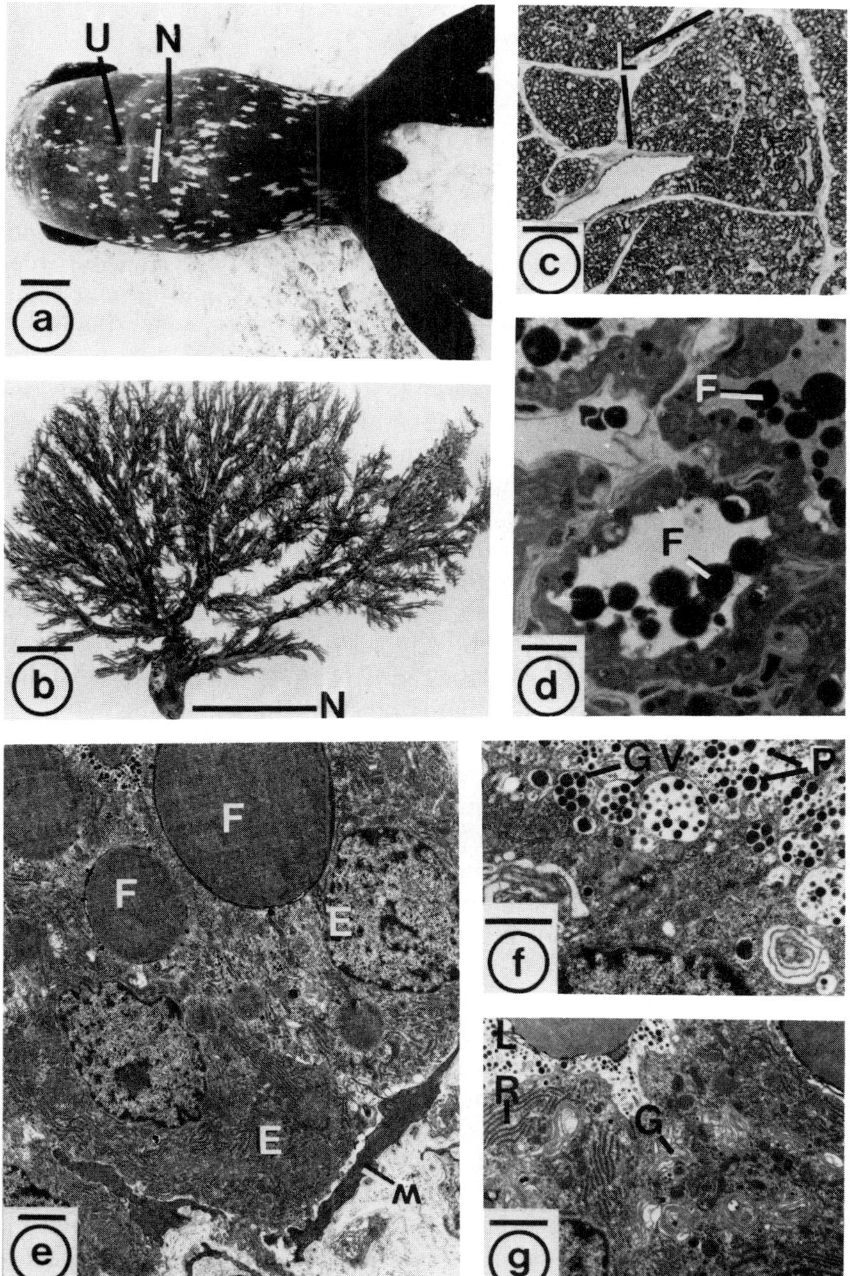
U
N
a
b
N
c
F
F
d
F
F
E
E
M
e
GV
P
f
L
R
G
g

4. MAMMARY ANATOMY AND FUNCTION

4.1. Gross Anatomy

The otariids and odobenids possess four mammary teats (nipples) located on the ventral abdominal wall anterior and posterior to the umbilicus (King, 1983). The bearded seal (Mansfield, 1967) and the monk seal (Kenyon and Rice, 1959; King, 1983) also possess four teats, whereas all other phocids have two abdominal teats adjacent to or posterior to the umbilicus (King, 1983) (Fig. 5a). Supernumerary teats have been reported in southern elephant seals (Laws, 1956), the northern fur seal (Scheffer, 1962), and the walrus (J. King, personal communication). The nipples are eversible under the influence of a sucking pup (Amoroso and Matthews, 1952), retracting below the general body level to be hidden by the pelage when sucking ends. In the Weddell seal the nipples protrude about 3.7 cm beyond the general body contour during nursing (Bertram, 1940). In some species, nipples may remain everted during off-teat intervals and continue to squirt milk (Fig. 1).

Information suggests that the mammary glands differ in shape and size between phocids and otariids. In phocids such as the harbor seal (Harrison, 1969), southern elephant seal (Laws, 1956; Bryden, 1968), and Weddell seal (Bertram, 1940; Tedman, 1977) the mammary glands form two flattened ellipsoid glands beneath the subcutaneous fat on the abdomen. In the northern fur seal the mammary glands spread like an extensive apron over the lower thorax, abdomen, and sides of the body (Scheffer, 1962). In all pinnipeds the mammary glands follow the general contour of the body and are not normally visible externally. However, in Weddell seals maternal fat loss becomes so marked near weaning that outlines of the mammary glands are often visible (Tedman, 1980). There have been no detailed descriptions of the mammary glands of otariids or odobenids. Most of our knowledge of pinniped mammary anatomy comes from observations on a few species of phocids.

The dimensions of lactating mammary glands have been measured following dissection in several species. The size of the gland depends on the size of the cow and on the time elapsed since the pup last sucked (Laws, 1956). In the Falkland Islands, for example, lactating mammary glands of a large southern elephant seal cow may be up to 82.5 cm long, 25 cm wide, and 11.2 cm thick, but only 57.5 by 20 by 7.5 cm in an average-sized cow ($n = 10$) (Laws, 1956). Laws estimated gland volume from these dimensions and concluded that volumes range from 2.0 to 5.3 liters. At Macquarie Island, estimated volumes for this species are similar, but somewhat smaller, 0.9–4.5 liters (Bryden, 1968). In Weddell seals lactating mammary glands may be up to 45 by 22.5 by 10 cm, with estimated volumes of 1.4–5.3 liters (Bertram, 1940; Tedman and Bryden, 1981a). Grey seal mammary glands are about 50 cm in length and 5 cm thick (Harrison, 1969), whereas hooded seal glands are about 78 cm long and 21 cm wide (W. D. Bowen, D. J. Boness, and O. T. Oftedal, unpublished, $n = 2$). Scheffer (1962) reports that mammary glands of the northern fur seal can be 60 cm long and 50 cm wide, with a depth of 1–2 cm. Given the small size of this otariid, the mammary surface area (up to 2190 cm^2; Scheffer, 1962) is remarkable. Scheffer (1962) suggests that mammary capacity may be 2–3 liters. Weight gains of lactating northern fur seals during foraging trips are of a similar magnitude (about 3–4 kg; Costa and Gentry, 1986) and may largely represent mammary filling at sea.

Among terrestrial mammals, interspecies differences in mammary tissue weight are related to body weight by an allometric equation (Hanwell and Peaker, 1977):

$$Y = 0.045(\text{body weight in kg})^{0.82}$$

Unfortunately, little is known of mammary weights in pinnipeds. Tedman (1985) reported a weight of 4 kg for an apparently normal mammary gland of a lactating Weddell seal. The combined weight of 8 kg is somewhat greater than the 6.1–6.7 kg predicted for a 400- to 450-kg female from the allometric equation. In hooded seals the discrepancy is even greater: a 166-kg female with a pup near weaning had a combined mammary weight of 5.7 kg, nearly twice the predicted value of 3.0 kg (W. D. Bowen, D. J. Boness, and O. T. Oftedal, unpublished). Further data are required to determine if pinnipeds typically have large mammary glands relative to body weight.

The milk duct system of the mammary glands of the Weddell seal and southern elephant seal resembles that found in other eutherians in functioning as a series of reservoirs for milk storage between and during suckling periods (Tedman and Bryden, 1981b). In adapting to the flattened nature of the mammary glands, there is a radiating arrangement of large canals running parallel to the skin surface and draining more or less straight to a gland cistern deep to the nipple (Tedman and Bryden, 1981b) (Fig. 5b). While the milk duct systems of nonlactating and lactating glands are similar, the increased thickness of lactating glands results from the development of alveoli, terminal ducts, and fine collecting ducts (Tedman and Bryden, 1981a). Laws (1956) reported a "small ampulla" beneath each nipple in the southern elephant seal and Murie (1874) found a "long capacious duct" deep to each of four teats in an adult male Steller sea lion.

4.2. Microscopic Anatomy

The histological appearance of nonlactating mammary glands of harp and hooded seals (Belov, 1971), southern elephant seals (Bryden and Tedman, 1974), and Weddell seals (Tedman and Bryden, 1981a) resembles that of glands at similar stages in other mammals (see review by Hollmann, 1974).

Mammary glands of newborn male and female harbor seals consist of ducts ending in tufts of short glands with no lumen and show no signs of being affected by the stimulus that causes marked change in the genital tract and associated glands at this time (Harrison, 1960). Amoroso et al. (1965) found a similar arrangement of mammary tissue in 2- to 13-day-old grey seals, a stillborn California sea lion, and a near-term harbor seal. In the adult Weddell seal the nonlactating mammary gland consists of branched and unbranched cords of epithelial cells and tubules compacted into widely spaced lobules (Tedman and Bryden, 1981a). Within lactating tissue alveoli are compacted together as lobules separated by thin connective tissue septa (Fig. 5c).

Alveoli possess a lumen bordered by a single layer of epithelial cells, which are surrounded by scattered myoepithelial cells (Tedman, 1980) (Fig. 5d). The mammary parenchyma shows a desynchronization of the secretory cycle, with alveoli varying from thin-walled units swollen with secretory products to thick-walled structures containing little secretion (Tedman, 1980).

According to Belov (1971), the mammary tissue from hooded seals displayed active liberation of secretion by the entire alveolar epithelium throughout lactation, with all alveoli distended and participating in a holocrine type of secretion. In this way functional cells are lost from the alveolar epithelium to be added to the secretory product. Belov claimed that harp seals also exhibit holocrine secretion, but that a desynchronization of secretory activity is indicated by the mixture of shrunken and distended alveoli observed in mammary tissue. The apparent synchronization of the cycle of secretory activity in mammary tissue of hooded seals led Belov to conclude that the mammary glands of this species function at a maximal level throughout the short nursing period. In the harp seal the secretory activity is less intense (desynchronization of secretory activity), but extended over a longer nursing period. Belov (1971) believed that the nursing periods for hooded and harp seals were 10–12 and 21 days, respectively, but the true lactation lengths are closer to 4 and 12 days, respectively (Table II).

In all terrestrial mammals so far examined (Wooding, 1977) plus various cetaceans (van Utrecht, 1968), the Weddell seal (Tedman, 1983), and the southern elephant seal (Bryden and Tedman, 1974) there is no evidence of a holocrine method of secretion in mammary tissue. The mode of secretion of major milk constituents such as fat and protein seems to be identical in all mammals. It is unlikely that hooded or harp seals would be different. It is important to examine mammary tissue from these species at the electron microscopic level to substantiate the mode of secretion.

4.3. Ultrastructure and Secretory Processes

The mammary parenchyma in nonlactating and lactating Weddell seals is almost identical in ultrastructure to that found in other mammals (Tedman, 1983) (Figs. 5e–5g). Consistent with a high level of secretory activity, epithelial cells characteristically contain well-developed mitochondria, Golgi apparatus, rough endoplasmic reticula (RER), microvilli, and a labyrinth of irregular cytoplasmic processes from the basal plasma membrane (Tedman, 1980).

In mammary secretory cells in the Weddell seal, as in other mammalian species, milk protein appears as electron-dense granules (micelles) within vacuoles (vesicles) that are associated with the Golgi apparatus and scattered throughout the cytoplasm, especially in the apical region (Tedman, 1980) (Figs. 5f and 5g). In other mammals these vesicles have been shown to contain milk proteins, citrate, inorganic ions, lactose, and water (Wooding, 1977). The contents of these vesicles are released into the alveolar lumen by the process of exocytosis when the vesicles fuse with the apical plasma membrane.

In the Weddell seal the relative volumes of most organelles in secretory epithelial cells of the mammary gland are similar to those in other eutherians, but the relative volumes of RER and Golgi vacuoles are reduced (Tedman, 1983). While the significance of the reduced volume of RER is not clear, the decrease in relative volume and absolute diameter of Golgi vacuoles may be related to reduced lactose and water content of Weddell seal milk in comparison to terrestrial mammals for which comparable data are available, such as the rat. The osmotic movement of water into the Golgi vacuoles during lactose formation and subsequent transport to the apical cell surface is believed to be the main mechanism by which bulk water movement into the milk occurs (Peaker, 1977). Consequently, the reduced volume of Golgi vacuoles in Weddell seal mammary tissue may reflect a relatively low rate of lactose synthesis and water transport compared to most terrestrial species.

However, Wooding (1977) concluded that there is no correlation between organelle content and gross milk composition in ten species of mammals, including a fur seal (species not identified). The values for relative volume of Golgi vesicles in the fur seal reported by Wooding (1977) were larger than those reported for the Weddell seal. Fur seal milk has even less lactose than Weddell seal milk. Mammary tissues from other pinniped species need to be examined to clarify this apparent anomaly.

In Weddell seal mammary tissue, fat droplets appear to be extruded from epithelial cells in the same manner as reported for numerous terrestrial mammals (Tedman, 1983). Fat droplets within mammary epithelial cells bulge into the apical plasma membrane and are ultimately extruded from the cell enveloped by a unit membrane derived from the apical plasma membrane and perhaps from Golgi vesicles associated with the lipid droplet (Mather and Keenan, 1983). The mode of secretion seems to fit the concept of an apocrine secretion in which there is loss of part of the secretory cell cytoplasm with the secretory product.

The fat droplets in milk samples, alveolar lumina, and epithelial cells have a larger mean diameter in Weddell seals than in other mammals that have been studied. Perhaps seals need to reduce the quantity of membrane required to package the large quantities of fat. The area of membrane per unit volume required to enclose fat droplets will be less for larger than for smaller droplets (Tedman, 1983). Golgi vacuoles fuse with apical plasma membranes during exocytosis and hence are believed to be responsible for replacement of apical plasma membrane lost during extrusion of fat droplets (Keenan et al., 1978). Reduced diameter and relative volume of Golgi vacuoles in Weddell seals will mean reduced membrane available for apical plasma membrane replacement.

However, according to Dylewski and Keenan (1983), there is evidence that most of the Golgi vesicular membrane material in bovine and rat epithelial cells may not be incorporated into apical plasma membrane. Rather, the Golgi vesicles fuse with each other, often forming vesicular chains of ten or more linked vesicles, of which only the terminal vesicle fuses with and becomes incorporated into the apical plasma membrane. The remaining linked vesicles disintegrate after the vesicular contents are passed into the alveolar lumen. Consequently, fundamental concepts of the role of Golgi vesicular membrane or plasma membrane turnover need to be examined in mammary tissue of not only the Weddell seal, but all marine mammals where large quantities of fat are being secreted into the milk.

It remains unclear just how the pinniped mammary gland can produce milk with such a high fat content. Holt (1983) viewed the concentration of lipid droplets in milk as depending on their secretion rate in relation to the volume of the aqueous phase. An increased rate of fat synthesis and secretion accompanied by a reduced volume of the aqueous phase could produce an elevated fat content. More detailed work on pinniped mammary glands may expand our understanding of the flexibility that exists within the functional capabilities of mammary tissue in general.

4.4. Immunologic Function

Seals, like other carnivores, possess an endotheliochorial placentation (Sinha and Erickson, 1974). Hence, as in the dog and cat, there is probably some form of antibody transfer prenatally across the placenta, and postnatally via mammary gland secretions—the colostrum and milk (Brambell, 1970). The mammary gland provides passive immune protection to the suckling infant and participates in immunologic defence of its own tissues against infection by microorganisms (Watson, 1980). While these functions vary in importance in different species, there are no data available on immunity transfer in pinnipeds and hence there are many areas that need to be investigated.

Plasma cells secrete particular classes of immunoglobulins, especially IgA (Pumphrey, 1977), which may have an immunoprotective role within the lumen of the gut of the suckling young, acting against a variety of enteric microorganisms (Watson, 1980). Pumphrey examined the distribution of plasma cells within mammary tissue from the grey seal, harbor seal, southern sea lion, South American fur seal, and Antarctic fur seal as part of a review involving 60 species of mammals. The centrilobular distribution of plasma cells in the pinnipeds examined by Pumphrey was the same as that reported by him for man, rhesus monkey, dog, cat, cape buffalo, and whale. Fewer plama cells were found in the mammary tissues of the fur seals than in harbor or grey seals, however. Pumphrey (1977) suggested that the unknown mechanism whereby fur seals avoid mammary involution during long feeding trips away from the young might prevent the plasma cells from congregating in the mammary tissue of these animals.

4.5. Maintenance of Lactation and Mammary Involution

The maintenance of milk secretion by the mammary gland is dependent upon and controlled by various hormones, the most important in nonruminants being prolactin (Tindal, 1978). Prolactin is released by the pituitary gland in response to sucking or milking as well as other tactile stimuli, but the pattern of release and the quantities involved vary in different species (Tindal, 1978). The role of prolactin and other hormones in the control of lactation in pinnipeds is not known.

Despite the variation in suckling habits and the diversity in gross structure of mammary glands in different species, there seems to be a common mechanism of milk ejection for all mammals (Cross, 1977). The evacuation of milk is assisted by the contraction of myoepithelial cells, which surround secretory alveoli in a basketlike arrangement, under the influence of oxytocin released from the maternal neurohypophysis during suckling (Cross, 1977). Stimulation of the nipple during suckling provides the main afferent impulse for initiation of milk ejection, but in many species the sight or sound of the young or other features of the nursing routine may trigger milk ejection (Lincoln and Paisley, 1982).

Compression of alveoli following myoepithelial cell contraction forces the milk into various milk ducts where it is available for removal from the gland by the sucking of the young. In species such as ruminants where much of the milk is stored in voluminous cisterns, the milk ejection reflex is not essential for the feeding of the young, although it assists in moving the milk from the alveoli to the milk ducts (Lincoln and Paisley, 1982). However, in many species the storage capacity of the milk ducts is not large and the milk ejection reflex forms an important part of the process of milk removal by the sucking young.

The anatomic arrangement of the duct system of the mammary glands of the pinnipeds that have been studied suggests that these species should be largely dependent on the milk ejection reflex. The existence of a milk ejection reflex in pinnipeds is indicated by the presence of myoepithelial cells in the mammary alveoli of the Weddell seal (Tedman, 1983) and the observation that exogenous oxytocin injections assist in milk collection.

In all mammals studied a regular suckling stimulus is vital for maintenance of mammary gland function, and without it regression or involution of the glandular tissue ensues (Cowie and Tindal, 1971). Involution follows a similar pattern in all species, but varies in duration. Removal of a sucking stimulus leads to milk accumulation, mammary gland distension, changes in milk composition, including a progressive increase in cellular content, and cellular degradation (Lascelles and Lee, 1978). Initial stages of cell degradation involving alterations to cell organelles occur within 1–3 days in a variety of mammals.

The mechanism whereby otariids avoid the onset of involution during the prolonged feeding trips at sea without exposure to suckling is unknown. It is possible that otariids reduce the effects of milk accumulation by combining a large storage volume with a relatively low rate of secretion of a high-nutrient-density milk. The extent to which mammary storage capacity in mammals is adapted to suckling interval is not known. Other mammals that suckle relatively infrequently produce concentrated milks (Oftedal, 1984a). The possibility that secretion rate varies between time on land and time at sea needs to be investigated. In mammals endogenous mammary factors such as prostaglandins and estradiol seem to play a significant role in the local control of the composition and rate of mammary secretion and may be particularly important (Maule Walker, 1984). The otariid mammary gland may prove to be a useful experimental model for understanding the physiological and biochemical regulation of milk secretion.

5. MILK PRODUCTION AND LACTATION ENERGETICS

5.1. Measurement of Milk Yield

Bryden (1968) made an early attempt to quantify milk production in the southern elephant seal on the basis of mammary volume, 12-h pup weight gains, and amounts of milk recovered from mammary glands and pup stomachs, but concluded that these estimates were not reliable. More accurate indirect estimates based on the energetics of maternal weight loss and pup weight gain have been published for the grey seal (Fedak and Anderson, 1982) and harp seal (R. E. A. Stewart and Lavigne, 1984). The applicability of isotope methods to pinnipeds under field conditions has led to a proliferation of milk production estimates in the past few years. Such data are now available for four phocids and two otariids (Tables XII and XIII).

The four phocids studied by isotope methods (Table XII) represent a wide range of lactation duration (4–45 days) and pup weight gain (about 2–6 kg/day). Interspecies comparison reveals a close correlation

TABLE XII
Milk Yields of Phocids Determined by Isotope Dilution Methods[a]

Species (lactation duration, days)	Period of study[b] (days PP)	n	Initial female weight (kg)	Pup gain[c] (kg/day)	Daily milk yield (kg/day)	Yield per pup gain (kg/kg)	Yield as percent female weight	Total yield[d] (kg)	References
Hooded seal (4)	0–4	7	179	5.71	8.52	1.55	4.8	27.6	O. T. Oftedal, D. J. Boness, and W. D. Bowen (unpublished observations)
Harp seal (12)	0–10	6	135	2.11	3.31	1.48	2.5	40.4	O. T. Oftedal, D. J. Boness, and W. D. Bowen (unpublished observations)
Northern elephant seal (27)	0–24	6	509	3.31	5.50	1.66	1.1	146	Costa et al. (1986); Ortiz et al. (1984)
Weddell seal (45)	8–38	9	447	1.98	3.54	1.96	1.0	159	Tedman and Green (1986)

[a]Values are means unless otherwise indicated.
[b]Range of pup ages over which isotope turnover measured; not all individuals were measured over the entire range.
[c]Mean growth rates for individual pups studied; species means may be somewhat different.
[d]Hooded seal data determined directly; data on other species calculated from mean daily yield and lactation duration.

TABLE XIII
Milk Yields of Otariids Determined by Isotope Dilution Methods[a]

Species (lactation duration, days)	Period of study (days PP)	n	Female weight (kg)	Pup gain (kg/day)	Daily milk yield (kg/day)	Yield per pup gain (kg/kg)	Yield as percent female weight	References
Northern fur seal	0–7	6	37.7	0.185	0.50	2.7	1.3	Costa and Gentry (1986)
(120)	10–60	13	37.7[b]	0.065[c]	0.450[d]	6.9	1.2	
California sea lion	0–30	22	88.1[e]	0.121	0.70	5.8	0.8	O. T. Oftedal,
(~300)	30–60	13	88.1	0.124	0.73	6.3	0.8	S. J. Iverson,
	90–120	5	88.1	0.085	1.01	11.9	1.1	and D. J. Boness (unpublished)

[a]Data are mean values.
[b]Average weight from perinatal period used, since subsequent weights not reported.
[c]Average weight gain of male and female pups, from Gentry et al. (1986).
[d]Calculated using 8 days as the average duration from the beginning of one suckling period to the beginning of the next.
[e]Average of all lactating females weighed 0–120 days postpartum (n = 8).

of estimated mean daily milk yield to rate of pup weight gain ($r = 0.995$, d.f. = 2, $P < 0.01$). The ratio of mean milk yield to mean pup gain ranges from 1.5 to 2.0 in the species studied (Table XII). Daily milk yields are equivalent to about 1.0–1.1% of initial maternal weight in the two larger species with longer lactations (northern elephant seal, Weddell seal), in contrast to 2.5% in the harp seal and 4.8% in the hooded seal with shorter lactation periods. Because milk production derives entirely from maternal nutrient stores, the rate of depletion of stores is apparently greater in those species with short lactation periods.

Milk production has been measured over the period of birth to weaning in the hooded seal; in the other species isotope dilution measurements cover a large part of lactation and can be extrapolated to the entire period (Table XII). Total yields increase with lactation duration, from 27.6 kg in the hooded seal to 159 kg in the Weddell seal. These yields are equivalent to 15, 30, 29, and 36% of initial female weight in hooded, harp, northern elephant, and Weddell seals, respectively. Because lactating phocids consume little or no food or water, female weight losses during lactation can be expected to exceed the above percentages by an amount dependent on metabolic processes and evaporative and other water losses. Serial weighings of females throughout lactation indicate weight losses of 39 and 59% in northern elephant and Weddell seals (Costa et al., 1986; Tedman and Green, 1986).

Milk intake estimates in otariids are restricted to the first 2 months of lactation in the northern fur seal and the first 4 months in the California sea lion (Table XIII). Average daily yields are a similar percentage of maternal weight (0.8–1.3%) as in northern elephant seals and Weddell seals. The average yield includes days the mother is at sea foraging, such that the delivery rate to the pup during periods of suckling will be much higher. In the northern fur seal there is little change in level of production from the initial period of continual maternal presence to the subsequent phase of feeding cycles. In the California sea lion, milk production appears to rise somewhat from the second to the fourth month of lactation (Table XIII).

The ratio of milk yield to pup weight gain may be higher in otariids than in phocids, but because available data cover only a part of the lactation period for just two otariids, a quantitative comparison cannot be made. For both California sea lions and northern fur seals, the greater growth efficiency in early lactation compared to later stages may be related to the continuous availability of the mother before she begins her foraging trips (Table XIII). In the northern fur seal, the ratio of milk yield to pup growth (2.7) in early lactation is comparable to that of phocids (Table XII). During the second month postpartum, both species ingest 6–7 kg milk per kg weight gain. The higher ratio (11.9) observed at 90–120 days in the California sea lion is probably age-related. In addition to the low growth rate in the five pups of this age used in the milk yield study, a similar decline in growth rate occurred in a larger sample of pups (D. J. Boness, O. T. Oftedal, and K. A. Ono, unpublished).

Milk delivery rates of 0.84, 2.82, and 6.48 kg/h nursing can be calculated for California sea lions, harp seals, and northern elephant seals, respectively, from data on daily yield (Tables XII and XIII) and cumulative time on teat (Table III). Absolute delivery rate appears to be a function of female size, but if delivery rate is expressed as a percentage of female weight, all three species deliver milk at a similar high rate (1.0, 2.1, and 1.3% of female weight per hour in California sea lions, harp seals, and northern elephant seals, respectively).

5.2. The Energetics of Lactation

From data on milk yield and milk energy content, it is possible to estimate both daily milk energy output and the total energy output for lactation (Table XIV). Among terrestrial mammals, peak milk energy output scales to metabolic size (weight$^{0.75}$), but the level of production is affected by litter size and mass (Oftedal, 1984b). Pinnipeds exhibit great variation in milk energy output per maternal metabolic mass, but this variation appears to be a function of lactation length (Table XIV). As lactation length increases, energy output per unit metabolic size falls from 991 kcal/kg$^{0.75}$ per day in the hooded seal to 181 kcal/kg$^{0.75}$ per day in the Weddell seal and to 85 kcal/kg$^{0.75}$ per day in the California sea lion. A Pearson's correlation analysis on these data, using log transformation, indicates a correlation between energy output and lactation length of 0.976 (d.f. $= 4$, $P < 0.001$). In this case phocids and otariids form a continuum.

Total output of milk energy follows the reverse trend—as lactation length increases, total energy output per maternal metabolic mass rises (Table XIV). Thus total milk energy output per kg$^{0.75}$ in Weddell seals (8.15 Mcal/kg$^{0.75}$) is about 2.5 times that of hooded seals (3.31 Mcal/kg$^{0.75}$). The data for the two otariids represent an extrapolation from the first 60 days to the entirety of lactation and may be substantially in error if either daily milk yield or the energy content of milk changes markedly in the latter part of lactation. In California sea lions both milk yield and milk energy content appear to increase in the fourth month (Tables VIII and XIV), such that the extrapolated value in Table XIV is probably an underestimate. Lactation length was assumed to be 300 days because

TABLE XIV
Estimates of Milk Energy Output in Pinnipeds[a]

Species (lactation duration, days)	Daily EO (Mcal)	Daily EO per MMBS (kcal/kg$^{0.75}$)	EO per pup gain (Mcal/kg)	Total energy output (Mcal)	Total EO per MMBS (Mcal/kg$^{0.75}$)	References
Hooded seal (4)	48.5	991	8.85	162	3.31	O. T. Oftedal, D. J. Boness, and W. D. Bowen (unpublished)
Harp seal (12)	16.1	408	7.66	193	4.88	O. T. Oftedal, D. J. Boness, and W. D. Bowen (unpublished)
Northern elephant seal (27)	23.0	215	6.95	610	5.69	Costa et al. (1986)
Weddell seal (45)	17.6	181	8.89	792	8.15	Tedman and Green (1986)
Northern fur seal[b] (120)	2.14	141	22.8	260?	17?	Costa and Gentry (1986)
California sea lion[c] (~300)	2.44	85	20.6	730?	25?	O. T. Oftedal, S. J. Iverson, and D. J. Boness (unpublished)

[a]Based on milk yields in Tables XII and XIII. EO, Energy output; MMBS, maternal metabolic size (kg$^{0.75}$).
[b]Daily energy output measured 0–60 days; extrapolated to 120 days to obtain rough estimate of total energy output (see text).
[c]Daily energy output measured 0–60 days; extrapolated to 300 days to obtain rough estimate of total energy output (see text).

some pups are beginning to feed at this time (D. J. Boness and O. T. Oftedal, unpublished). The estimated total energy yields for the northern fur seal (17 Mcal/kg$^{0.75}$) and California sea lion (25 Mcal/kg$^{0.75}$) are about twice and three times, respectively, that of the Weddell seal, suggesting that otariid lactation involves much greater energy outlays than does lactation in phocids.

By comparison, estimated total milk energy output in ungulates ranges from 7 to 13 Mcal/kg$^{0.75}$ (Oftedal, 1985). Thus, Weddell seals fall within the ungulate range, whereas the pack-ice species have much lower and the otariids much higher energy outputs per maternal metabolic mass.

Milk energy output can also be estimated indirectly from data on the storage and expenditure of energy by suckling pups (Fedak and Anderson, 1982; R. E. A. Stewart and Lavigne, 1984). In the absence of direct data on metabolic rates of suckling pups or on the efficiency of energy utilization for growth and metabolism, a number of assumptions and extrapolations have to be made. Nonetheless, estimates of daily milk energy output developed for both harp seals (563 kcal/kg$^{0.75}$ per day; R. E. A. Stewart and Lavigne, 1984) and grey seals (362 kcal/kg$^{0.75}$ per day; Fedak and Anderson, 1982) are similar in magnitude to values obtained from data on milk composition and yield (Table XIV). The correspondence of the two harp seal estimates (563 compared to 408 kcal/kg$^{0.75}$ per day) is improved if lactation length is taken as 12 days (Kovacs, 1986) rather than 9 days (R. E. A. Stewart and Lavigne, 1984); the recalculated value is then 422 kcal/kg$^{0.75}$ per day. R. E. A. Stewart and Lavigne (1984) estimate that total milk energy output over lactation is 194 Mcal in harp seals, an identical value to that estimated from milk yield (Table XIV). Grey seals lactate for 17 days (Table II), during which total milk energy output can be calculated as 289 Mcal (6.1 Mcal/kg$^{0.75}$; Fedak and Anderson, 1982). This value is only 10–20% higher than would be predicted from the estimates in Table XIV. The relatively good correspondence of estimates obtained by independent methods suggests that the errors of estimation are either small or similar for the two approaches.

6. SUMMARY OF THE LACTATIONAL STRATEGIES OF PINNIPEDS

6.1. Conflict between Foraging and Nursing

Reproductive strategies in pinnipeds are molded by the constraint that birth and early postnatal development must occur on a solid substrate of land or ice. With the exception of the walrus, nursing of pups always occurs on this solid substrate. Nursing and foraging at sea are thus mutually exclusive activities. Although terrestrial carnivores typically hunt prey and nurse their young in separate locations, feeding sites of pinnipeds may be far removed from suitable sites for pup rearing.

The three pinniped families have adopted three diverse strategies in coping with the potential conflict between foraging and nursing—female phocids typically abandon feeding during lactation, female otariids intersperse feeding periods with periods of nursing, whereas female odobenids take their young to sea at an early age and nurse them in the water. The major components of these three strategies will be briefly summarized. Variations among species in the behavioral, physiological, and anatomic aspects of these strategies have been reviewed in the body of this paper.

6.2. The Phocid Strategy: Fasting and Intensive Lactation

The phocid strategy of fasting during lactation requires that virtually all nutrients required for lactation be stored in maternal tissues prior to parturition. In the harp seal most energy accumulation occurs in the few months preceding parturition (W. D. Bowen, personal communication). Because energy and nutrient accumulation precedes lactation, and the lactational fast represents a loss of feeding opportunity, one would expect phocids to time parturition to coincide with the end of a period of seasonal food abundance. The short duration of the pupping season and the prevalence of migration in many phocids (Table II; Section 2.1) is consistent with the hypothesis that seasonality in food resources may be important.

Because the lactating phocid must support both milk production and maintenance metabolism by drawing on body reserves, any reduction in maintenance costs will increase the proportion of reserves that can be devoted to pup rearing. One way to reduce maintenance costs is to shorten the duration of lactation such that fewer days of maintenance metabolism must be supported. Relatively short lactation periods are characteristic of most phocids (Table II); at the extreme, the hooded seal has the shortest lactation of all mammals. Phocid pups are typically weaned abruptly upon female departure from the birth site (Section 2.5), and fast for 1–10 weeks before feeding on their own (Bowen et al., 1986b).

Abbreviation of the lactation period requires that the rate of energy transfer from mother to pup and the resultant rate of pup growth be accelerated, or the pup may not reach a weaning weight adequate for survival (Bowen et al., 1985; Kovacs and Lavigne, 1986). Indeed, both the relative rate of energy transfer (Section 5.2; Table XIV) and the relative rate of pup growth (Kovacs and Lavigne, 1986) are negatively correlated to duration of lactation in phocids. At the extreme, a hooded seal achieves a daily milk energy output equivalent to the combined peak production of milk energy by four to seven terrestrial carnivore mothers or seven to 15 ungulate mothers (Oftedal, 1984a, 1985). Females of species with longer lactation periods, such as northern elephant and Weddell seals, transfer energy to single pups at a rate equivalent to the peak output of terrestrial carnivores with large litters.

High rates of energy transfer to phocid pups are achieved by secretion of high-fat, high-energy milks (Table VIII) in substantial amounts (Table XII). Because milk lipids appear to derive from body stores with relatively little modification (Section 3.2), the energetic efficiency of fat secretion is undoubtedly high. The secretory mechanisms whereby milk of such high fat and low water content is produced are not understood, especially given the negligible amounts of sugars and unusual potassium-to-sodium ratios that are typical of both phocids and otariids (Tables VIII and XI). High rates of fat secretion may underlie the relatively large size of fat droplets in mammary secretory cells (Fig. 5), and the reduced relative volume of Golgi vacuoles in these cells may reflect low rates of aqueous phase secretion. In other regards, mammary tissue of phocids is comparable in gross anatomy and ultrastructure to other mammals (Sections 4.1–4.3). Limited data suggest that the relative size of phocid mammary glands exceeds the mammalian norm.

Given the rapid rate of energy transfer from mother to pup, one might expect phocid pups to spend more time in suckling activities than terrestrial mammals, but neither suckling frequencies nor suckling durations (Table III) are exceptional by comparison to other mammals. Thus, the high rates of transfer are a function of the high energy density of milk rather than modifications in suckling behavior. Female phocids remain in the vicinity of their pups throughout lactation, but in some species females enter the water and a minimal amount of maternal feeding may occur (Table IV). Water entry may be as much related to thermoregulation, sexual activity, or predator avoidance, however (Section 2.3.4).

6.3. The Otariid Strategy: Alternation of Foraging and Nursing

Nutrient stores of otariid females are apparently adequate to support lactation for about 1 week after birth; thereafter females depart the rookery on foraging trips of about 1–6 days duration (Table V). Foraging trips are alternated with time at the rookery when pups suckle. One-half to three-quarters of the lactation period of 4–12 months or longer (Table II) is spent at sea on foraging trips (Table V). Otariid pups devote more time to suckling when females are present than do phocid pups, which may help compensate for the briefness of maternal visits (Section 2.3.3).

The necessity of feeding has undoubtedly limited potential pupping sites to those within a reasonable swimming distance of foraging grounds; conversely, during lactation otariid females are unable to rely on pelagic resources at a distance from land. In contrast to phocids, otariids would be expected to time parturition such that seasonal food resources would become locally abundant during lactation, when successful foraging is essential to sustained milk production, pup growth, and pup survival.

The intermittent pattern of milk transfer results in a cycle of feeding and fasting for the otariid pup. Although milk intakes may be quite large when mother is in attendance, average intake calculations reflect the periods of fasting and are thus much lower than for phocid pups (Table XIII). Otariid pups require energy to support metabolic requirements on both fast days and days that milk is consumed. Maintenance requirements of pups are thus quite high and the efficiency of growth is consequently low (Oftedal et al., 1985). Suckling otariid pups ingest 6 kg or more milk/kg weight gain (Table XIII), as compared to 1.5–2.0 kg milk/kg weight gain in phocid pups (Table XII). It is not surprising that otariid pups exhibit slow growth rates and a prolonged period of dependence on maternal milk. Otariid pups are typically weaned at a weight somewhat greater than one-third of maternal weight (Gentry et al., 1986).

In most mammals a prolonged interval between suckling episodes leads to the onset of mammary involution with associated degradative changes in secretory epithelium. The mechanisms whereby female otariids avoid these changes during prolonged foraging bouts remain a mystery (Section 4.5). Incomplete data on mammary dimensions of the northern fur seal suggest that mammary gland volume may be quite large, but milk storage capacity of the mammary system has not been measured. The secretion of high-fat, energy-dense milk by otariids (Table IX) may be one means of minimizing mammary storage requirements during foraging trips, as well as facilitating rapid nutrient transfer to the pup during suckling periods ashore. It is not known whether secretion rates of milk or milk constituents change while the female is at sea or in response to suckling stimulii on land.

Although the average daily rate of milk energy transfer to pups is much lower in otariids than in phocids, the longer duration of lactation in otariids results in greater total energy outlays in milk over the course of lactation in this group (Table XIV). It appears that otariids have adopted a lactation strategy that is energetically expensive, but the costs are spread over time and can therefore be met by ongoing foraging activities.

6.4. The Odobenid Strategy: Aquatic Nursing

An obvious solution to the conflict between rearing pups on a solid substrate while foraging at sea is to evolve a pattern of aquatic rearing of the young. Yet no pinniped routinely gives birth at sea, and the sole extant odobenid, the walrus, is the only species that has evolved a pattern of aquatic suckling. Walrus pups abandon their birth sites on pack ice at an early age and accompany their mothers to sea, suckling both in the water and at haul-out sites (Fay, 1982; Miller and Boness, 1983). Thus, female walrus face neither the constraints on foraging activities and suckling patterns inherent in the otariid strategy of intermittent feeding nor the constraints on lactation duration arising from the phocid approach of fasting.

Unfortunately, relatively little is known about lactation in this unique species. Pups are apparently dependent on milk alone in the first 5 months after birth and then wean gradually to benthic invertebrates, being fully weaned by 2–3 years of age (Fay, 1982). This represents the longest lactation known for any pinniped. A few milk samples from the first month postpartum have been reported to contain moderate but variable amounts of fat (14–32%) and protein (7–11%), similar to early lactation samples in some otariids and phocids (Section 3.1.2). Some data are available on suckling duration on land and in water, but suckling frequency has not been studied (Section 2.3.3). Both mammary structure and function and lactation energetics of the walrus are unknown. Further study is needed to elucidate the consequences of aquatic suckling to the odobenid lactation strategy.

ACKNOWLEDGMENTS. We would like to thank our colleagues M. E. Allen, W. D. Bowen, M. M. Bryden, E. H. Miller, and M. H. Robinson for helpful comments on parts or all of this paper. M. M. Bryden kindly provided the photograph for Fig. 1, and A. Miller typed various sections of this paper on the word processor. Unpublished research results contained herein were obtained with grant support (to O. T. O. and D. J.

B.) from the Friends of the National Zoo and the Research Opportunities Fund of the Smithsonian Institution.

REFERENCES

Ackman, R. G., and Burgher, R. D., 1963, Component fatty acids of the milk of the grey (Atlantic) seal, *Can. J. Biochem. Physiol.* **41**:2501–2505.

Alcorn, D. J., 1984, The Hawaiian Monk Seal on Laysan Island: 1982, NOAA Technical Memorandum, National Marine Fisheries Service, Southwest Fisheries Center, No. 42, pp. 1–37.

Amoroso, E. C., and Matthews, L. H., 1952, Reproduction and lactation in the seal, in: *Report of the II International Congress of Physiology and Pathology of Animal Reproduction and of Artificial Insemination*, Volume 1, Copenhagen, pp. 193–201.

Amoroso, E. C., Goffin, A., Halley, G., Matthews, L. H., and Mathews, D. J., 1951, Lactation in the grey seal, *J. Physiol.* **113**:4–5.

Amoroso, E. C., Bourne, G. H., Harrison, R. L., Matthews, L. H., Rowlands, J. W., and Sloper, J. C., 1965, Reproductive and endocrine organs of foetal newborn and adult seals, *J. Zool. Lond.* **147**:430–486.

Anderson, S. S., and Fedak, M. A., 1985, Grey seal males: Energetic and behavioural links between size and sexual success, *Anim. Behav.* **33**:829–838.

Anderson, S. S., and Harwood, J., 1985, Time budgets and topography: How energy reserves and terrain determine the breeding behaviour of grey seals, *Anim. Behav.* **33**:1343–1348.

Anderson, S. S., Burton, R. W., and Summers, C. F., 1975, Behaviour of grey seals (*Halichoerus grypus*) during a breeding season on North Rona, *J. Zool. Lond.* **177**: 179–195.

Ashworth, U. S., Ramaiah, G. D., and Keyes, M. C., 1966, Species difference in the composition of milk with special reference to the northern fur seal, *J. Dairy Sci.* **49**: 1206–1211.

Bartholomew, G. A., 1959, Mother–young relations and the maturation of pup behavior in the Alaska fur seal, *Anim. Behav.* **7**:163–171.

Bartholomew, G. A., and Hoel, P. G., 1953, Reproductive behavior of the Alaska fur seal, *Callorhinus ursinus*, *J. Mammal.* **34**:417–437.

Belkin, A. N., Kosygin, G. M., and Panin, K. I., 1969, New materials on island seals (*Phoca insularis*), in: *Morskie Mlekopitayshchie* (V. A. Arsen'ev, B. A. Zenkovich, and K. K. Chapskii, eds.), Nauka, Moscow, pp. 157–175 [in Russian].

Belov, L., 1971, Histological organization of the mammary gland in the seals *Pagophilus groenlandicus* and *Cystophora cristata*, *J. Evol. Biochem. Physiol.* **7**:585–589.

Berger, J., 1979, Weaning conflict in desert and mountain bighorn sheep (*Ovis canadensis*): An ecological interpretation, *Z. Tierpsychol.* **50**:188–200.

Bertram, G. C. L., 1940, The biology of the Weddell and crabeater seals: With a study of the comparative behaviour of the Pinnipedia, in: *Scientific Reports British Graham Land Expedition, 1934–37*, Volume 1, pp. 1–139.

Best, H. A., 1974, A Preliminary Report on the Natural History of Hooker's Sea Lion at Enderby Island, New Zealand Ministry of Agriculture and Fisheries Report No. 132.

Bester, M. N., 1977, Habitat selection, seasonal population changes, and behaviour of the Amsterdam fur seal, *Arctocephalus tropicalis*, on Gough Island, Ph. D. thesis, University of Pretoria, Pretoria, South Africa.

Bester, M. N., 1981, Seasonal changes in the population composition of the fur seal, *Arctocephalus tropicalis*, at Gough Island, *S. Afr. J. Wildl. Res.* **11**:49–55.

Bigg, M. A., 1969, The Harbor Seal in British Columbia, *Bull. Fish. Res. Board Can.* **172**: 1–33.

Bigg, M. A., 1973, Census of the California sea lions on southern Vancouver Island, British Columbia, *J. Mammal.* **54**:285–287.

Bishop, S. C., 1967, Reproduction, age determination, and behavior of the harbor seal, *Phoca vitulina* L., in the Gulf of Alaska, M. Sc. Thesis, University of Alaska, Fairbanks.

Bock, H.-D., Wunsche, J. Hoffmann, B., Linke, K., and Fricke, G., 1968, Über die Aminosäuren-Zusammensetzung der Milch der Kegelrobbe, *Halichoerus grypus* Fab. und des Schweinswals, *Phocoena phocoena* L., *Zool. Gart. Hpz.* **35**:69–73.

Boness, D. J., 1979, The social system of the grey seal, *Halichoerus grypus* (Fab.), on Sable Island, Nova Scotia, Ph. D. thesis, Dalhousie University, Halifax, Nova Scotia.

Boness, D. J., and James, H., 1979, Reproductive behaviour of the grey seal (*Halichoerus grypus*) on Sable Island, Nova Scotia, *J. Zool. Lond.* **188**:477–500.

Boness, D. J., Anderson, S. S., and Cox, C. R., 1982, Functions of female aggression during the pupping and mating season of grey seals, *Halichoerus grypus* (Fabricius), *Can. J. Zool.* **60**:2270–2278.

Boness, D. J., Dabek, L, Ono, K. A., and Oftedal, O. T., 1985a, Female attendance behavior in California sea lions, in: *Abstracts Sixth Biennial Conference on the Biology of Marine Mammals*, Vancouver, British Columbia.

Boness, D. J., Oftedal, O. T., Ono, K. A., and Dabek, L., 1985b, Observations on yearling California sea lions at breeding beaches on San Nicolas Island, California, in: *Abstracts Sixth Biennial Conference on the Biology of Marine Mammals*, Vancouver, British Columbia.

Bonner, W. N., 1968, The Fur Seal of South Georgia, British Antarctic Survey Scientific Report No. 56, pp. 1–81.

Bonner, W. N., 1979a, Largha seal, *FAO Fish. Ser. No. 5* **1979**:63–66.

Bonner, W. N., 1979b, Harbour seal, *FAO Fish. Ser. No. 5* **1979**:58–63.

Bonner, W. N., 1984, Lactation strategies in pinnipeds: Problems for a marine mammalian group, *Symp. Zool. Soc. Lond.* **51**:253–272.

Bonner, W. N., and Witthames, S. R., 1974, Dispersal of common seals (*Phoca vitulina*), tagged in the Wash, East Anglia, *J. Zool. Lond.* **174**:528–531.

Boulva, J., 1979, Mediterranean monk seal, *FAO Fish. Ser. No. 5* **1979**:95–101.

Boulva, J., and McLaren, I. A., 1979, Biology of the Harbor Seal, *Phoca vitulina*, in Eastern Canada, *Bull. Fish. Res. Board Can.* **200**:1–24.

Bowen, W. D., and Sergeant, D. E., 1983, Mark-recapture estimates of harp seal pup (*Phoca groenlandica*) production in the northwest Atlantic, *Can. J. Fish. Aquat. Sci.* **40**:728–742.

Bowen, W. D., Oftedal, O. T., and Boness, D. J., 1985, Birth to weaning in four days: Remarkable growth in the hooded seal, *Cystophora cristata*, *Can. J. Zool.* **63**:2841–2846.

Bowen, W. D., Myers, R. A., and Hay, K., 1986a, Abundance of a dispersed, dynamic population: Hooded seals (*Cystophora cristata*) in the Northwest Atlantic, *Can. J. Fish. Aquat. Sci.*, in press.

Bowen, W. D., Boness, D. J. and Oftedal, O. T., 1986b, Mass transfer from mother to pup and subsequent mass loss by the weaned pup in the hooded seal, *Cystophora cristata*, *Can. J. Zool.*, in press.

Brambell, F. W. R., 1970, *The Transmission of Passive Immunity from Mother to Young*, North-Holland, Amsterdam.

Brodie, P., and Beck, B., 1983, Predation by sharks on the grey seal (*Halichoerus grypus*) in Eastern Canada, *Can. J. Fish. Aquat. Sci.* **40**:267–271.

Bryden, M. M., 1968, Lactation and suckling in relation to early growth of the southern elephant seal, *Aust. J. Zool.* **16**:739–747.

Bryden, M. M., and Stokes, G. B., 1969, Metabolism of fatty acids in the southern elephant seal, *Mirounga leonina* L., *Can. J. Biochem.* **47**:757–760.

Bryden, M. M., and Tedman, R. A., 1974, Studies of lactation in Weddell seals, *Antarct. J. U. S.* **9**:105–106.

Bunt, J. S., 1968, Microalgae of the Antarctic pack ice zone, in: *Symposium on Antarctic Oceanography*, Scott Polar Research Institute, Cambridge, England, pp. 198–218.

Burns, J. J., 1970, Remarks on the distribution and natural history of pagophilic pinnipeds in the Bering and Chukchi seas, *J. Mammal.* **51**:445–454.

Burns, J. J., 1981a, Bearded seal—*Erignathus barbatus*, in: *Handbook of Marine Mammals* (S. H. Ridgway and R. J. Harrison, eds.), Academic Press, New York, Volume 2, pp. 145–170.

Burns, J. J., 1981b, Ribbon seal—*Phoca fasciata*, in: *Handbook of Marine Mammals* (S. H. Ridgway and R. J. Harrison, eds.), Academic Press, New York, Volume 2, pp. 89–109.

Burns, J. J., and Frost, K. J., 1979, The Natural History and Ecology of the Bearded Seal, *Erignathus barbatus*, Final Report, Alaska Fish and Game, Contract #02-5-022-53, pp. 1–77.

Burns, J. J., Ray, G. C., Fay, F. H., and Shaughnessy, P., 1972, Adoption of a strange pup by the ice-inhabiting harbor seal, *Phoca vitulina largha*, *J. Mammal.* **53**:594–598.

Burton, R. W., Summers, C. F., and Anderson, S. S., 1975, Perinatal activities in the grey seal (*Halichoerus grupus*), *J. Zool. Lond.* **177**:197–201.

Campagna, C., 1985, The breeding cycle of the southern sea lion, *Otaria byronia*, *Mar. Mammal Sci.* **1**:210–219.

Carnio, J., 1982, Observations on the mother and young interactions in captive Cape fur seals (*Arctocephalus pusillus*), *Aquat. Mamm.* **9**:50–56.

Carrick, R., Csordas, S. E., Ingham, S. E., and Keith, K., 1962, Studies on the southern elephant seal, *Mirounga leonina*. 3. The annual cycle in relation to age and sex, *CSIRO Wildl. Res.* **7**:119–160.

Christenson, T. E., and LeBoeuf, B. J., 1978, Aggression in the female northern elephant seal, *Mirounga angustirostris*, *Behaviour* **64**:158–171.

Cody, M. M., 1984, Substances without vitamin status, in: *Handbook of Vitamins. Nutritional, Clinical, and Biochemical Aspects* (L. J. Machlin, ed.), Marcel Dekker, New York, pp. 571–585.

Condit, R., and LeBoeuf, B. J., 1984, Feeding habits and feeding grounds of the northern elephant seal, *J. Mammal.* **65**:281–291.

Condy, P. R., 1979, Annual cycle of the southern elephant seal, *Mirounga leonina* (Linn.) at Marion Island, *S. Afr. J. Zool.* **14**:95–102.

Condy, P. R., 1980, Postnatal development and growth in southern elephant seals, *Mirounga leonina*, at Marion Island, *S. Afr. J. Wildl. Res.* **8**:159–168.

Cook, H. W., and Baker, B. E., 1969, Seal milk. I. Harp seal (*Pagophilis groenlandicus*) milk: Composition and pesticide residue content, *Can. J. Zool.* **47**:1129–1132.

Costa, D. P., and Gentry, R. L., 1986, Free-ranging energetics of northern fur seals, in: *Fur Seals: Maternal Strategies on Land and at Sea* (R. L. Gentry and G. L. Kooyman, eds.), Princeton University Press, Princeton, pp. 79–101.

Costa, D. P., Thorson, P., Feldkamp, S., Gentry, R. L., DeLong, R., Antonelis, G., and

Croxall, J. P., 1985, At sea foraging energetics of three species of pinnipeds, *Fed. Proc.* **44**:1000.

Costa, D. P., LeBoeuf, B. J., Huntley, A. C., and Ortiz, C. L., 1986, The energetics of lactation in the northern elephant seal, *J. Zool. Lond.* **209**:21–33.

Coulson, J. C., and Hickling, G., 1964, The breeding biology of the grey seal, *Halichoerus grypus* (Fab.) on the Farne Islands, Northumberland, *J. Anim. Ecol.* **33**:483–512.

Cowie, A. T., and Tindal, J. S., 1971, *The Physiology of Lactation*, Edward Arnold, London.

Cross, B. A., 1977, Comparative physiology of milk removal, *Symp. Zool. Soc. Lond.* **41**: 193–210.

Croxall, J. P., Everson, I., Kooyman, G. L., Ricketts, C., and Davis, R. W., 1985, Fur seal diving behavior in relation to vertical distribution of krill, *J. Anim. Ecol.* **54**:1–8.

Curry-Lindahl, K., 1975, Ecology and conservation of the grey seal, *Halichoerus grypus*, common seal, *Phoca vitulina*, and ringed seal, *Phoca hispida* in the Baltic Sea, *Rapp. P. V. Reun. Cons. Int. Explor. Mer.* **169**:527–532.

David, J. H. M., and Rand, R. W., 1986, Attendance behavior of female South African fur seals, in: *Fur Seals: Maternal Strategies on Land and at Sea*, (R. L. Gentry and G. L. Kooyman, eds.), Princeton University Press, Princeton, pp. 126–141.

Davies, J. L., 1949, Observations of the grey seal (*Halichoerus grypus*) at Ramsey Island, Pembrokeshire, *Proc. Zool. Soc. Lond.* **119**:673–692.

Demaster, D. P., 1979, Weddell seal, *FAO Fish. Ser. No. 5* **1979**:130–134.

Doidge, D. W., McCann, T. S., and Croxall, J. P., 1986, Attendance behavior of Antarctic fur seals, in: *Fur Seals: Maternal Strategies on Land and at Sea* (R. L. Gentry and G. L. Kooyman, eds.), Princeton University Press, Princeton, pp. 102–114.

Dosako, S., Taneya, S., Kimura, T., Ohmori, T., Daikoku, H., Suzuki, N., Sawa, J., Kano, K., and Katayama, S., 1983, Milk of northern fur seal: Composition, especially carbohydrate and protein, *J. Dairy Sci.* **66**:2076–2083.

Drewett, R. F., 1983, Sucking, milk synthesis, and milk ejection in the Norway rat, in: *Parental Behaviour of Rodents* (R. W. Elwood, ed.), Wiley, London, pp. 181–203.

Dylewski, D. P., and Keenan, T. W., 1983, Compound exocytosis of casein micelles in mammary epithelial cells, *Eur. J. Cell. Biol.* **31**:114–124.

Edie, A. G., 1977, Distribution and movements of steller sea lion cows (*Eumetopias jubata*) on a pupping colony, M. Sc. Thesis, University of British Columbia, Vancouver, British Columbia.

FAO, 1982, *Mammals in the Seas, FAO Fisheries Ser. No. 5*, Volume 4.

Fay, F. H., 1982, Ecology and Biology of the Pacific Walrus, *Odobenus rosmarus divergens* Illiger, U. S. Fish and Wildlife Service, North American Fauna, No. 74, pp. 1–279.

Fedak, M. A., and Anderson, S. S., 1982, The energetics of lactation: Accurate measurements from a large wild mammal, *J. Zool. Lond.* **198**:473–479.

Fiscus, F. H., and Baines, G. A., 1966, Food and feeding behavior of Steller and California sea lions, *J. Mammal.* **47**:195–200.

Fogden, S. C. L., 1971, Mother–young behaviour at grey seal breeding beaches, *J. Zool. Lond.* **164**:61–92.

Freed, L. M., York, C. M., Hamosh, M., Sturman, J. T., Oftedal, O. T., and Hamosh, P., 1986, Bile salt stimulated lipase: The enzyme is present in nonprimate milk, in: *Human Lactation*, Volume 2: *Maternal-Environmental Factors* (M. Hamosh and A. S. Goldman, eds.), Plenum Press, New York, pp. 595–601.

Frost, K. J., and Lowery, L. F., 1980, Feeding of ribbon seals (*Phoca fasciata*) in the Bering Sea in spring, *Can. J. Zool.* **58**:1601–1607.

Gentry, R. L., 1970, Social behavior of Steller sea lions, Ph. D. thesis, University of California, Santa Cruz.

Gentry, R. L., and Holt, J. R., 1986, Attendance behavior of northern fur seals, in: *Fur Seals: Maternal Strategies on Land and at Sea* (R. L. Gentry and G. L. Kooyman, eds.), Princeton University Press, Princeton, pp. 41–60.

Gentry, R. L., Costa, D. P., Croxall, J. P., David, J. H. M., Davis, R. W., Kooyman, G. L., Majluf, P., McCann, T. S., and Trillmich, F., 1986, Synthesis and conclusions, in: *Fur Seals: Maternal Strategies on Land and at Sea* (R. L. Gentry and G. L. Kooyman, eds.), Princeton University Press, Princeton, pp. 220–278.

Glass, R. L., and Jenness, R., 1971, Comparative biochemical studies of milks—VI. Constituent fatty acids of milk fats of additional species, *Comp. Biochem. Physiol.* **38B:** 353–359.

Glass, R. L., Troolin, H. A., and Jenness, R., 1967, Comparative biochemical studies of milks—IV. Constituent fatty acids of milk fats, *Comp. Biochem. Physiol.* **22:**415–425.

Golt'sev, V. N., 1971, Feeding of the harbor seal, *Ekologiya* **2:**62–70.

Golt'sev, V. N., and Fedoseev, G. A., 1970, Dynamics of the age composition of rookeries and the replacement capacity of harbor seal populations, *TINRO* **71:**309–317 [Fisheries Research Board of Canada Translation Series No. 2401 (1974)].

Hall, W. G., 1983, Suckling isn't feeding, or is it? A search for developmental continuities, *Adv. Stud. Behav.* **13:**219–254.

Hamilton, J. E., 1934, The southern sea lion, *Otaria byronia, Discovery Rep.* **8:**269–318.

Hanwell, A., and Peaker, M., 1977, Physiological effects of lactation on the mother, *Symp. Zool. Soc. Lond.* **41:**297–312.

Harrison, R. J., 1960, Reproduction and reproductive organs in common seals (*Phoca vitulina*) in the Wash, East Anglia, *Mammalia* **24:**372–385.

Harrison, R. J., 1969, Reproduction and reproductive organs, in *Biology of Marine Mammals* (H. T. Anderson, ed.), Academic Press, New York, pp. 253–348.

Harrison, S. H., and Ridgway, R. J. (eds.), 1981, *Handbook of Marine Mammals,* Volumes 1 and 2, Academic Press, London.

Hewer, H. R., 1960, Behaviour of the grey seal (*Halichoerus grypus*) in the breeding season, *Mammalia* **24:**401–420.

Hewer, H. R., 1974, *British Seals,* Collins, London.

Higgins, L. V., 1984, Material behavior and attendance patterns of the Steller sea lion in California, M. Sc. thesis, University of California, Santa Cruz.

Hill, K. J., Noakes, D. E., and Lowe, R. E., 1970, Gastric digestive physiology of the calf and piglet, in: *Physiology of Digestion and Metabolism in the Ruminant* (A. T. Phillipson, ed.), Oriel Press, Newcastle, pp. 166–179.

Hollmann, K. H., 1974, Cytology and fine structure of the mammary gland, in: *Lactation: A Comprehensive Treatise,* Volume 1 (B. L. Larson and V. R. Smith, eds.), Academic Press, London, pp. 3–95.

Holt, C., 1983, Swelling of Golgi vesicles in mammary secretory cells and its relation to the yield and quantitative composition of milk, *J. Theor. Biol.* **101:**247–261.

Hook, O., and Johnels, A. G., 1972, The breeding and distribution of the grey seal (*Halichoerus grypus* Fab.) in the Baltic Sea, with observations on other seals of the area, *Proc. R. Soc. Ser. B* **182:**37–58.

Ivashin, M. V., Popov, L. A., and Tsapko, A. S., 1972, *Morskie Mlekopitayushchie,* Pishchevaia Promyshlennost, Moscow [Fisheries Research Board of Canada, Translation, Series No. 2783 (1973)].

Jangaard, P. M., and Ke, P. J., 1968, Principal fatty acids of depot fat and milk lipids from harp seal (*Pagophilis groenlandius*) and hooded seal (*Cystophora cristata*), J. Fish. Res. Board Can. **25**:2419–2426.

Jenness, R., 1979, Comparative aspects of milk proteins, *J. Dairy Res.* **46**:197–210.

Jenness, R., and Sloan, R. E., 1970, The composition of milks of various species: A review, *Dairy Sci. Abstr.* **32**:599–612.

Jenness, R., Erickson, A. W., and Craighead, J. J., 1972, Some comparative aspects of milk from four species of bears, *J. Mammal.* **53**:34–47.

Johnson, J. D., 1975, Comparative aspects of lactose digestion and synthesis in mammals—The California sea lion, *Mod. Probl. Paediatr.* **15**:194–200.

Johnson, J. D., Christiansen, R. O., and Kretchmer, N., 1972, Lactose synthetase in mammary gland of the California sea lion, *Biochem. Biophys. Res. Commun.* **47**:393–397.

Kaufman, G. W., Siniff, D. B., and Reichle, R., 1975, Colony behavior of Weddell seals, *Leptonychotes weddelli*, at Hutton Cliffs, Antarctica, *Rapp. P. V. Reun. Cons. Int. Explor. Mer.* **169**:228–246.

Keenan, T. W., Franke, W. W., Mather, I. H., and Moore, D. J., 1978, Endomembrane composition and function in milk formation, in: *Lactation: A Comprehensive Treatise*, Volume 4 (B. L. Larson, ed.), Academic Press, London, pp. 405–436.

Kenyon, K. W., and Rice, D. W., 1959, Life history of the Hawaiian monk seal, *Pac. Sci.* **13**:215–252.

Kerley, G. I. H., 1983, Comparison of seasonal haul-out patterns of fur seals, *Arctocephalus tropicalis* and *A. gazella*, on subantarctic Marion Island, *S. Afr. J. Wildl. Res.* **13**:71–77.

Kerry, K. R., and Messer, M., 1968, Intestinal glycosidases of three species of seals, *Comp. Biochem. Physiol.* **25**:437–446.

King, J. E., 1983, *Seals of the World*, 2nd ed., Oxford University Press, Oxford.

Knox, G. A., 1970, Antarctic marine ecosystems, in: *Antarctic Ecology*, Volume 1 (N. W. Holdgate, ed.), Academic Press, London, pp. 69–97.

Kooyman, G. L., and Drabek, C. M., 1968, Observations on the milk, blood, and urine constituents of the Weddell seal, *Physiol. Zool.* **41**:187–194.

Kormondy, E. J., 1969, *Concepts of Ecology*, Prentice-Hall, Englewood Cliffs, New Jersey.

Kovacs, K., 1986, Maternal investment and early behavioural development in harp seals (*Phoca groenlandica*) and grey seals (*Halichoerus grypus*), Ph. D. thesis, Guelph University, Guelph, Ontario.

Kovacs, K., and Lavigne, D. M., 1986, Maternal investment and neonatal growth in phocid seals, *J. Anim. Ecol.* **55**:(in press).

Lander, R. H., 1979, Alaskan or northern fur seal, *FAO Fish. Ser. No.* 5 **1979**:19–24.

Lander, R. H., and Kajimura, H., 1982, Status of northern fur seals, *FAO Fish. Ser. No.* 5 **1979**:319–347.

Lascelles, A. K., and Lee, C. S., 1978, Involution of the mammary gland, in: *Lactation: A Comprehensive Treatise*, Volume 4 (B. L. Larson, ed.), Academic Press, London, pp. 115–177.

Lauer, B. H., and Baker, B. E., 1977, Amino acid composition of casein isolated from the milk of different species, *Can. J. Zool.* **55**:231–236.

Lavigne, D. M., 1976, Counting harp seals with ultraviolet photography, *Polar Record* **18**:269–272.

Lavigne, D. M., Stewart, R. E. A., and Fletcher, F., 1982, Changes in composition and energy content of harp seal milk during lactation, *Physiol. Zool.* **55**:1–9.

Laws, R. M., 1956, The Elephant Seal (*Mirounga leonina* Linn). II General, Social, and Reproductive Behavior, Falkland Islands Dependency Survey, Scientific Report No. 13, pp. 1–88.

Laws, R. M., 1958, Growth rates and ages of crabeater seals, *Lobodon carcinophagus*, *Proc. Zool. Soc. Lond.* **130:**275–288.

Laws, R. M., 1959, Accelerated growth in seals, with special reference to the Phocidae, *Nor. Hvalfangst-Tid.* **48:**425–452.

LeBoeuf, B. J., 1974, Male–male competition and reproductive success in elephant seals, *Am. Zool.* **14:**163–176.

LeBoeuf, B. J., and Ortiz, C. L., 1977, Composition of elephant seal milk, *J. Mammal.* **58:** 683–685.

LeBoeuf, B. J., and Panken, K. J., 1977, Elephant seals breeding on the mainland in California, *Proc. Calif. Acad. Sci. Ser. 4* **41:**267–280.

LeBoeuf, B. J., Whiting, R. J., and Gantt, R. F., 1972, Perinatal behavior of northern elephant seal females and their young, *Behaviour* **43:**121–156.

Lincoln, D. W., and Paisley, A. C., 1982, Neuroendocrine control of milk ejection, *J. Reprod. Fertil.* **65:**571–586.

Lindsey, A. A., 1937, The Weddell seal in the Bay of Whales, Antarctica, *J. Mammal.* **18:**127–144.

Ling, J. K., and Walker, G. E., 1976, Seal studies in South Australia: Progress report for the year 1975, *S. Aust. Nat.* **50:**59–72.

Linzell, J. L., Peaker, M., and Taylor, J. C., 1975, The effects of prolactin and oxytocin on milk secretion and on the permeability of the mammary epithelium in the rabbit, *J. Physiol.* **253:**547–563.

Loughrey, A. G., 1959, Preliminary investigation of the Atlantic walrus, *Odobenus rosmarus rosmarus*, *Can. Wildl. Serv. Bull. Ser. No. 14* **1959:**1–23.

Macy, S. K., 1982, Mother–pup interactions in the northern fur seal, Ph. D. thesis, University of Washington, Seattle.

Mansfield, A. W., 1958, The biology of the Atlantic walrus, *Odobenus rosmarus rosmarus* (Linnaeus), in the Eastern Canadian Arctic, *Fish. Res. Board Can. Manuscr. Rep. Ser. No. 653* **1958:**1–146.

Mansfield, A. W., 1967, Seals of Arctic and Eastern Canada, 2nd ed., *Bull. Fish. Res. Board Can.* **137:**1–30.

Mansfield, A. W., and Beck, B., 1977, The grey seal in Eastern Canada, *Can. Fish. Mar. Serv. Tech. Rep.* **704:**1–81.

Marlow, B. J., 1975, The comparative behaviour of the Australian sea lions, *Neophoca cinerea* and *Phocartos hookeri*, *Mammalia* **39:**159–230.

Mate, B. R., and Gentry, R. L., 1979, Northern (Steller) sea lion, *FAO Fish. Ser. No. 5* **1979:**1–5.

Mather, I. H., and Keenan, T. W., 1983, Function of endomembranes and the cell surface in the secretion of organic milk constituents, in: *Biochemistry of Lactation* (T. B. Mepham, ed.), Elsevier, New York, pp. 231–283.

Maule Walker, F. M., 1984, Exocrine and endocrine secretions of the mammary gland: Local control in ruminants, *Symp. Zool. Soc. Lond.* **51:**171–191.

McCann, T. S., 1982, Aggressive and maternal activities of female southern elephant seals (*Mirounga leonina*), *Anim. Behav.* **30:**268–276.

McLaren, I. A., 1958, The Biology of the ringed seal (*Phoca groenlandica* Schreber) in the Eastern Canadian Arctic, *Bull. Fish. Res. Board Can.* **118:**1–97.

McLaren, I. A., 1960, On the origin of the Caspian and Baikal seals and the paleoclimatological implications, *Am. J. Sci.* **258:**35–50.

McNab, A. G., and Crawley, M. C., 1975, Mother and pup behaviour of the New Zealand fur seal, *Arctocephalus forsteri* (Lesson), *Mauri Ora* **3:**77–88.

Meara, M. L., 1952, The component fatty acids of the milk fat of a grey Atlantic seal, *Biochem. J.* **51**:190–193.

Miller, E. H., 1971, Social and thermoregulatory behaviour of the New Zealand fur seal (*Arctocephalus forsteri*), M. Sc. thesis, University of Cantebury, Christchurch.

Miller, E. H., 1975, Annual cycle of fur seals, *Arctocephalus forsteri* (Lesson), on the Open Bay Islands, New Zealand, *Pac. Sci.* **29**:139–152.

Miller, E. H., and Boness, D. J., 1983, Summer behaviour of the Atlantic walrus (*Odobenus rosmarus rosmarus*) on Coats Island, NWT, Canada, *Z. Säugetierkd.* **48**:298–313.

Murie, J., 1874, Researches upon the anatomy of the Pinnipedia, Part II, Descriptive anatomy of the sea-lion (*Otaria jubata*), *Trans. Zool. Soc. Lond.* **7**:527–596.

Naito, Y., and Nishiwaki, M., 1972, The growth of two species of the harbour seal in the adjacent waters of Hokkaido, *Whale Res. Inst. Sci. Rep.* **24**:127–144.

Oftedal, O. T., 1984a, Milk composition, milk yield, and energy output at peak lactation: A comparative review, *Symp. Zool. Soc. Lond.* **51**:33–85.

Oftedal, O. T., 1984b, Body size and reproductive strategy as correlates of milk energy output in lactating mammals, *Acta Zool. Fenn.* **171**:183–186.

Oftedal, O. T., 1985, Pregnancy and lactation, in: *The Bioenergetics of Wild Herbivores* (R. J. Hudson and R. G. White, eds.), CRC Press, Boca Raton, Florida, pp. 215–238.

Oftedal, O. T., Boness, D. J., and Iverson, S. J., 1983, The effect of sampling method and stage of lactation on the composition of California sea lion milk, in: *Abstracts Fifth Biennial Conference on the Biology of Marine Mammals*, Boston, Massachusetts, p. 73.

Oftedal, O. T., Iverson, S. J., and Boness, D. J., 1985, Energy intake in relation to growth rate of the California sea lion, in: *Abstracts Sixth Biennial Conference on the Biology of Marine Mammals*, Vancouver, British Columbia.

Ono, K. A., Boness, D. J., Oftedal, O. T., and Iverson, S. J., 1985, Effects of the 1983 El Niño on the mother-pup relationship in the California sea lion, *Abstracts Sixth Biennial Conference on the Biology of Marine Mammals*, Vancouver, British Columbia.

Ortiz, C. L., Costa, D. P., and LeBoeuf, B. J., 1978, Water and energy flux in elephant seals pups fasting under natural conditions, *Physiol. Zool.* **238**:166–178.

Ortiz, C. L., LeBoeuf, B. J., and Costa, D. P., 1984, Milk intake of elephant seal pups: An index of parental investment, *Am. Nat.* **124**:416–422.

Pastukhov, V. N., 1975a, Birth time and duration of pupping period of Baikal seals, *Morsk. Mlekopitayushchiye* **4**:41–43.

Pastukhov, V. N., 1975b, Number and distribution of the postparturent females of the Baikal seal, *Morsk. Mlekopitayushchiye* **4**:39–41.

Payne, M. R., 1979, Growth in the Antarctic fur seal, *Arctocephalus gazella, J. Zool. Lond.* **187**:1–20.

Payne, P. R., and Wheeler, E. F., 1968, Comparative nutrition in pregnancy and lactation, *Proc. Nutr. Soc.* **27**:129–138.

Peaker, M., 1977, The aqueous phase of milk: Ion and water transport, *Symp. Zool. Soc. Lond.* **41**:113–134.

Peaker, M., and Goode, J. A., 1978, The milk of the fur seal *Arctocephalus tropicalis gazella*, in particular the composition of the aqueous phase, *J. Zool. Lond.* **185**: 469–476.

Peterson, R. S., 1965, Behavior of the northern fur seal, D. Sc. thesis, Johns Hopkins University, Baltimore, Maryland.

Peterson, R. S., 1968, Social behavior in pinnipeds with particular reference to the northern fur seal, in: *Behavior and Physiology of Pinnipeds*, (R. J. Harrison, R. C. Hubbard,

R. C. Peterson, D. E. Rice, and R. J. Schusterman, eds.), Appleton-Century-Crofts, New York, pp. 1–53.

Peterson, R. S., and Bartholomew, G. A., 1967, *The Natural History and Behavior of the California Sea Lion*, American Society of Mammalogists.

Pierotti, R., and Pierotti, D., 1980, Effects of cold on the evolution of pinniped breeding systems, *Evolution* **34:**494–507.

Pierson, M. O., 1978, Breeding behavior of the Guadalupe fur seal, *Arctocephalus townsendi*, Ph. D. thesis, University of California, Santa Cruz.

Pilson, M. E. Q., and Kelly, A. L., 1962, Composition of the milk from *Zalophus californianus*, the California sea lion, *Science* **135:**104–105.

Pitcher, K. W., and Calkins, D. G., 1981, Reproductive biology of Steller sea lions in the Gulf of Alaska, *J. Mammal.* **62:**599–605.

Ponce de Leon, A., 1984, Lactancia y Composicion Cuantitativa de la Leche del Lobo Fino Sudamericano, *Arctocephalus australis* (Zimmerman, 1783), *Industria Lobera Pesquera Estado, Montevideo, Uruguay, Anales* **1984:**43–58.

Popov, L. A., 1979, Caspian seal, *FAO Fish. Ser. No.* 5 **1979:**63–66.

Popov, L. A., and Krylov, V. I., 1978, On the period of reproduction in the Weddell seal (*Leptonychotes weddelli*) on the King Georges Island (Waterloo, Antarctica), *Zool. Zh.* **57:**1250–1255.

Poulter, T. C., Pinney, T. C., Jennings, R., and Hubbard, R. C., 1965, The rearing of Steller sea lions, in: *Proceedings Second Annual Conference Biological Sonar and Diving Mammals* (C. E. Rice, ed.), Stanford Research Institute, Fremont, Calif., pp. 49–62.

Pumphrey, R. S. H., 1977, A comparative study of plasma cells in the mammary gland in pregnancy and lactation, *Symp. Zool. Soc. Lond.* **41:**261–276.

Rand, R. W., 1955, Reproduction in the Cape fur seal, *Arctocephalus pusillus* (Schreber), *Proc. Zool. Soc. Lond.* **124:**717–740.

Rand, R. W., 1959, The Cape fur seal (*Arctocephalus pusillus*). 2. Distribution, abundance and feeding habits off the southwestern coast of the Cape Province, S. *Afr. Dept. Comm. Ind. Div. Fish. Invt. Rep.* **34:**1–65.

Ray, C. E., 1976, Geography of phocid evolution, *Syst. Zool.* **25:**391–406.

Reinhardt, V., and Reinhardt, A., 1981, Natural sucking performance and age of weaning in Zebu cattle (*Bos indicus*), *J. Agric. Sci. Camb.* **96:**309–312.

Reiter, J., Stinson, N. L., and LeBoeuf, B. J., 1978, Northern elephant seal development: The transition from weaning to nutritional independence, *Behav. Ecol. Sociobiol.* **3:**337–367.

Renouf, D., 1984, The vocalization of the harbour seal pup (*Phoca vitulina*) and its role in the maintenance of contact with the mother, *J. Zool. Lond.* **202:**583–590.

Renouf, D., and Diemand, D., 1984, Behavioural interactions between harbour seal mothers and pups during weaning (Pinnipedia: Phocidae), *Mammalia* **48:**53–58.

Renouf, D., Gaborko, L., Galway, G., and Finlayson, R., 1981, The effects of disturbance on the daily movements of harbour seals and grey seals between the sea and their hauling grounds at Miquelon, *Appl. Anim. Ethol.* **7:**373–379.

Repenning, C. A., 1976, Adaptive evolution of sea lions and walruses, *Syst. Zool.* **25:** 375–391.

Repenning, C. A., Ray, C. E., and Grigorescu, D., 1979, Pinniped biogeography, in: *Historical Biogeography, Plate Tectonics and the Changing Environment* (J. Grey and A. J. Boucot, eds.), Oregon State University Press, Corvallis, pp. 357–369.

Riedman, M., and Ortiz, C. L., 1979, Changes in milk composition during lactation in the northern elephant seal, *Physiol. Zool.* **52:**240–249.

Roux, J. P., and Hess, A. D., 1984, The seasonal haul-out of the fur seal, *Arctocephalus tropicalis* (Gray, 1872) on Amsterdam Island, *Mammalia* **48:**377–389.

Salter, R., 1978, Normal Behavior and Disturbance of Walruses (*Odobenus rosmarus* L.) during Terrestrial Haul-out, Eastern Bathurst Island, N.W.T., Report to Polar Gas Environmental Program, LGL Ltd, Toronto, Ontario.

Sandegren, F. E., 1970, Breeding and maternal behavior of the Steller sea lion, M. Sc. thesis, University of Alaska, Fairbanks.

Scheffer, V. B., 1962, Pelage and Surface Topography of the Northern Fur Seal, U. S. Fish and Wildlife Service, North American Fauna, No. 64, pp. 1–206.

Scheffer, V. B., 1970, *The Year of the Seal*, Scribner's, New York.

Schmidt, D. V., Walker, L. E., and Ebner, K. E., 1971, Lactose synthetase activity in northern fur seal milk, *Biochim. Biophys. Acta* **252:**439–442.

Schroeder, C. R., and Wegeforth, H. M., 1935, The occurrence of gastric ulcers in sea mammals of the California coast, their etiology and pathology, *J. Am. Vet. Med. Assoc.* **87:**333–342.

Scolaro, J. A., 1976, Censo de Elefantes Marinos (*Mirounga leonina*) en el Territorio Continental Argentino, *Cent. Nac. Patagonica, Argentine Inf. Tec.* **1:**1–12.

Shaughnessy, P. D., 1974, An electrophoretic study of blood and milk proteins of the southern elephant seal, *Mirounga leonina, J. Mammal.* **55:**796–808.

Shepeleva, V. K., 1971, Adaptations of seals to life in the Arctic, in: *Morphology and Ecology of Marine Mammals* (K. K. Chapskii and V. E. Sokolov, eds.), Wiley, New York, pp. 1–58.

Shergin, A. L., Shergina, V. P., and Beloborofov, A. G., 1969, Composition of hooded seals' milk, *Dairy Sci. Abstr.* **31:**1086.

Sinha, A. A., and Erickson, A. W., 1974, Ultrastructure of the placenta of Antarctic seals during the first third of pregnancy, *Am. J. Anat.* **141:**263–280.

Siniff, D. B., Stirling, I., Bengston, J. L., and Reichle, R. A., 1979, Social and reproductive behavior of crabeater seals during the austral spring, *Can. J. Zool.* **57:**2243–2255.

Sivertsen, E., 1941, On the biology of the harp seal, *Phoca groenlandica*, Erxl. *Hvalraadets Skr.* **26:**1–164.

Smith, E. A., and Burton, R. W., 1970, Weddell seals of Signey Island, in: *Antarctic Ecology*, Volume 1 (N. W. Holdgate, ed.), Academic Press, London, pp. 415–428.

Smith, T. G., 1973, Population dynamics of the ringed seal, *Phoca hispida*, in the Canadian Eastern Arctic, *Bull. Fish. Res. Board Can.* **181:**1–55.

Smith, T. G., and Hammil, M. O., 1981, Ecology of the ringed seal, *Phoca hispida*, in its fast ice breeding habitat, *Can. J. Zool.* **59:**966–981.

Smith, T. G., and Stirling, I., 1978, Variation in the density of ringed seal (*Phoca hispida*) birth lairs in the Amundsen Gulf, NWT, *Can. J. Zool.* **56:**1066–1070.

Stewart, B. S., and Yochem, P., 1984, Seasonal abundance of pinnipeds at San Nicolas Island, California, 1980–1982, *Bull. S. Calif. Acad. Sci.* **83:**121–132.

Stewart, R. E. A., 1983, Behavioral and energetic aspects of reproductive effort in female harp seals, Ph. D. thesis, Guelph University, Guelph, Ontario.

Stewart, R. E. A., and Lavigne, D. M., 1984, Energy transfer and female condition in nursing harp seals *Phoca groenlandica, Holarct. Ecol.* **7:**182–194.

Stewart, R. E. A., and Murie, D. J., 1986, Food habits of lactating harp seals (*Phoca groenlandica*) in the Gulf of St. Lawrence in March, *J. Mammal.* **67:**186–188.

Stewart, R. E. A., Lightfoot, N., and Innes, S., 1981, Parturition in harp seals, *J. Mammal.* **62:**845–850.

Stewart, R. E. A., Webb, B. E., Lavigne, D. M., and Fletcher, F., 1983, Determining lactose content of harp seal milk, *Can. J. Zool.* **61:**1094–1100.

Stirling, I., 1970, Observations on the behaviour of the New Zealand fur seal (*Arctocephalus forsteri*), *J. Mammal.* **51:**766–778.

Stirling, I., 1971, Studies on the behaviour of the South Australian fur seal, *Arctocephalus forsteri* (Lesson). 2. Adult females and pups, *Aust. J. Zool.* **19:**267–273.

Stirling, I., 1975, Adoptive suckling in pinnipeds, *J. Aust. Mammal. Soc.* **1:**389–391.

Stirling, I., 1983, The evolution of mating systems in pinnipeds, in: *Advances in the Study of Mammalian Behavior* (J. F. Eisenberg and D. G. Kleiman, eds.), American Society of Mammalogists, pp. 489–527.

Stirling, I., and McEwan, E. H., 1975, Caloric value of whole ringed seals (*Phoca hispida*) in relation to polar bear (*Ursus maritimus*) ecology and hunting, *Can. J. Zool.* **8:** 1021–1027.

Stirling, I., and Warneke, R. M., 1971, Implications of a comparison of the airborne vocalizations and some aspects of the behaviour of two Australian fur seal species on the evolution and present taxonomy of the genus, *Aust. J. Zool.* **19:**227–241.

Stull, J. W., Brown, W. H., and Kooyman, G. L., 1967, Lipids of the Weddell seal, *Leptonychotes weddelli, J. Mammal.* **48:**642–645.

Tedman, R. A., 1977, The gross anatomy of the mammary glands of the Weddell seal, *Leptonychotes weddelli, J. Anat.* **124:**524.

Tedman, R. A., 1980, Lactation in the Weddell seal, *Leptonychotes weddelli*, Ph. D. thesis, University of Queensland, Australia.

Tedman, R. A., 1983, Ultrastructural morphology of the mammary gland with observations on the size distribution of fat droplets in milk of the Weddell seal, *Leptonychotes weddelli* (Pinnipedia), *J. Zool. Lond.* **200:**131–141.

Tedman, R. A., 1985, The Weddell seal, *Leptonychotes weddelli*, at McMurdo Sound, Antarctica: Milk production in relation to pup growth, in: *Studies of Sea Mammals in South Latitudes* (J. K. Ling and M. M. Bryden, eds.), South Australian Museum, Adelaide, pp. 41–52.

Tedman, R. A., and Bryden, M. M., 1979, Cow–pup behaviour of the Weddell seal, *Leptonychotes weddelli* (Pinnipedia), in McMurdo Sound, Antarctica, *Aust. Wildl. Res.* **6:**19–37.

Tedman, R. A., and Bryden, M. M., 1981a, The mammary gland of the Weddell seal, *Leptonychotes weddelli* (Pinnipedia), 1. Gross and microscopic anatomy, *Anat. Rec.* **199:**519–529.

Tedman, R. A., and Bryden, M. M., 1981b, Resin injection of milk duct and arterial systems of the mammary glands of Weddell and southern elephant seals, *Bull. Aust. Mammal. Soc.* **7:**61–62.

Tedman, R. A., and Green, B., 1986, Water and sodium fluxes and lactational energetics in suckling pups of Weddell seals (*Leptonychotes weddellii*), *J. Zool. Lond.*, in press.

Thomas, J. A., Pastukhov, V., Elsner, R., and Petrov, E., 1982, *Phoca siberica, Mammal. Species* **188:**1–6.

Tikhomirov, E. A., 1975, Biology of the ice forms of seals in the Pacific section of the Antarctic, *Rapp. P. V. Reun. Cons. Int. Explor. Mer* **169:**409–413.

Tindal, J. S., 1978, Neuroendocrine control of lactation, in: *Lactation/A Comprehensive Treatise*, Volume 4 (B. L. Larson, ed.), Academic Press, New York, pp. 67–114.

Trillmich, F., 1979, Sea lions and fur seals, *Not. Galapagos* **29:**8–14.

Trillmich, F., 1981, Mutual mother–pup recognition in Galapagos fur seals and sea lions: Cues used and functional significance, *Behaviour* **78:**21–42.

Trillmich, F., 1986a, Attendance behavior of Galapagos fur seals, in: *Fur Seals: Maternal Strategies on Land and at Sea* (R. L. Gentry and G. L. Kooyman, eds.), Princeton University Press, Princeton, pp. 168–185.

Trillmich, F., 1986b, Attendance behavior of Galapagos sea lions, in: *Fur Seals: Maternal Strategies on Land and at Sea* (R. L. Gentry and G. L. Kooyman, eds.), Princeton University Press, Princeton, pp. 196–207.

Trillmich, F., and Lechner, E., 1986, Milk of the Galapagos fur seal and sea lion, with a comparison of the milk of eared seals (Otariidae), *J. Zool. Lond.* **209:**271–277.

Trillmich, F., and Majluf, P., 1981, First observations on colony structure, behaviour, and vocal repertoire of the South American fur seal (*Arctocephalus australis* Zimmermann, 1783) in Peru, *Z. Säugetierkd.* **46:**310–322.

Trillmich, F., Majluf, P., Sanchez-Grinan, M., and Kooyman, G. L., 1986, Attendance and diving behavior of the South American fur seal during "El Nino" in 1983, in: *Fur Seals: Maternal Strategies on Land and at Sea* (R. L. Gentry and G. L. Kooyman, eds.), Princeton University Press, Princeton, pp. 153–167.

Van Utrecht, W. L., 1968, Notes on some aspects of the mammary glands in the fin whale, *Balaenoptera physalus* (L), with regard to the criterion 'lactating', *Nor. Hvalfangst-Tid.* **57:**1–13.

Van Horn, D. R., and Baker, B. E., 1971, Seal milk. II. Harp seal (*Pagophilis groenlandicus*) milk: Effects of stage of lactation on the composition of the milk, *Can. J. Zool.* **49:** 1085–1088.

Vaz Ferriera, R., 1956, Terrestrial Ethology of *Arctocephalus australis* in the Uruguayan Islands, Works Concerning Seals and Sea Lions, Oceanographics and Fisheries Service, No. 2, pp. 1–22.

Venables, U. M., and Venables, L. S. V., 1955, Observations on a breeding colony of the seal, *Phoca vitulina* in Shetland, *Proc. Zool. Soc. Lond.* **125:**521–532.

Watson, D. L., 1980, Immunological functions of the mammary gland and its secretion—Comparative review, *Aust. J. Biol. Sci.* **33:**403–422.

Whitworth, M. R., 1984, Maternal care and behavioural development in pikas, *Ochotona princeps, Anim. Behav.* **32:**743–752.

Wooding, F. B. P., 1977, Comparative mammary fine structure, *Symp. Zool. Soc. Lond.* **41:**1–41.

Worthy, G. A. J., and Lavigne, D. M., 1983a, Energetics of fasting and subsequent growth in harp seal pups, *Phoca groenlandica, Can. J. Zool.* **61:**447–456.

Worthy, G. A. J., and Lavigne, D. M., 1983b, Changes in energy stores during postnatal development of the harp seal, *Phoca groenlandica, J. Mammal.* **64:**89–96.

CHAPTER 7

THE SOCIAL STRUCTURE OF FREE-RANGING BOTTLENOSE DOLPHINS

RANDALL S. WELLS, MICHAEL D. SCOTT, and A. BLAIR IRVINE

1. INTRODUCTION

A description of the social structure of a species is a first step toward understanding its social organization and, ultimately, the evolutionary processes that shaped its social system. Since the mid-1960s the rapid accumulation of information from field studies of terrestrial mammals has made it possible to propose models to explain the evolution of mammalian social systems. These models have examined the species distribution of characteristics such as group size, group compositions, spatial patterns of individuals, and social interactions in relation to environmental variables (for example, Crook and Gartlan, 1966; Eisenberg et al., 1972; Clutton-Brock, 1974; Jarman, 1974; Emlen and Oring, 1977; Wrangham, 1980). Predictable patterns of organization have been found which provide insights into the adaptive significance of the social

RANDALL S. WELLS • Dolphin Biology Research Associates, Sarasota, Florida 34242, and Long Marine Laboratory, Institute of Marine Sciences, University of California, Santa Cruz, California 95060. MICHAEL D. SCOTT • Dolphin Biology Research Associates, Sarasota, Florida 34242, and InterAmerican Tropical Tuna Commission, Scripps Institute of Oceanography, La Jolla, California 92038. A. BLAIR IRVINE • Dolphin Biology Research Associates, Sarasota, Florida 34242, and Department of School and Community Health Education, University of Oregon, Eugene, Oregon 97401.

systems. Until recently, available information for cetaceans has been inadequate to allow construction of comparable models. A surge of systematic field studies of the behavior and ecology of cetaceans is beginning to provide the requisite information for examination of cetacean societies within a general mammalian context. To this end, this chapter presents the results of one study of the social structure of the bottlenose dolphin, *Tursiops truncatus.*

Tests of hypotheses about the evolution of mammalian societies in their natural environment can be difficult. Typically, experimental manipulations of social units or their environments are either undesirable or impractical. Thus, the field researcher must take advantage of natural experiments by carefully selecting research sites to control for the variables in question, and by remaining alert for serendipitous opportunities.

The approach of longitudinal studies, in which the researcher comes to recognize the individual members of a particular social unit, has proved extremely successful in studies of a variety of terrestrial mammals. In time, information on ages, sexes, familial relationships, and life history parameters is accumulated. Such data increase precision in the definition of the contexts of social interactions. All of these factors are important in shaping social systems. The data base thus compiled over a number of years can then be used as a baseline. Any deviations from this baseline that are observed can be treated as natural experiments.

There have been few longitudinal studies of cetaceans comparable to those of many terrestrial mammals, for several reasons. Studies of the social systems of delphinid cetaceans must contend with the additional complications of working with highly mobile animals in an aquatic environment. Individual identifications often require photographs of subtle natural markings. Most delphinids lack distinctive features identifying age or sex. Much of the available information on delphinid social structure is derived from fisheries reports, where sex and age compositions of schools have been determined from collected specimens (for example, pilot whales, *Globicephala macrorhynchus*, Kasuya and Marsh, 1984; striped dolphins, *Stenella coeruleoalba*, Miyazaki and Nishiwaki, 1978). Observations of social associations and interactions have been made for a few species, such as bottlenose dolphins (Würsig, 1978) and spinner dolphins (*Stenella longirostris,* Norris et al., 1985), but in most cases information on the ages, sexes, or relations of the participants has been limited. Typically, observation studies have not continued long enough to obtain correlative data on life histories and natural history.

Exceptions include several longitudinal studies of the killer whale

(*Orcinus orca*) and the bottlenose dolphin. These studies now provide information on social patterns of identifiable individuals, many of which are of known sex, age, and lineage. Photographic identification studies of killer whales in the waters of British Columbia, Canada, were pioneered by Bigg and colleagues in the early 1970s (Bigg et al., 1976; Bigg, 1982). Their efforts were soon joined by those of Balcomb and colleagues working with killer whale pods in the adjacent waters of Puget Sound, Washington (Balcomb et al., 1980). Both studies took advantage of the predictable presence of killer whale social units and the highly distinctive individual markings indicative of sex and age. Continuity of observations has made it possible to collect data on life histories and familial relationships, and has resulted in our present understanding of the important role of kinship in *Orcinus* pod structure.

Photographic identification techniques have been adapted for bottlenose dolphins as well (Würsig and Würsig, 1977). Unlike the killer whales, bottlenose dolphins offer few obvious clues to their sex, age, or lineage. However, short-term capture, sampling, and release of these dolphins can provide these data. A tagging study of bottlenose dolphins along the central west coast of Florida was initiated by Irvine and Wells in 1970. Opportunistic tagging and resighting efforts during the next 2 years (Irvine and Wells, 1972) and systematic tagging, radiotracking, and field observation during 1975–1976 resulted in the description of an apparently resident social unit. Individual home ranges and social association patterns were hypothesized to be largely dependent on the age and sex of the individual, resulting in a structured society that was stable over time (Wells, 1978; Wells et al., 1980; Irvine et al., 1981).

Continued research has made it possible to test these hypotheses. If the hypotheses were to be accepted, then we would expect sex and age segregation to remain the predominant pattern over time, in spite of changes in the overall membership of the resident social unit. We would also expect particular individuals to change their movement and association patterns in predictable ways as they made transitions from one age class to the next during the process of social maturation. This chapter reports on the results of these tests from observations conducted during 1980–1984. In addition, it elaborates upon new findings and develops testable hypotheses about the social structure of the bottlenose dolphins.

2. METHODS

The field work with bottlenose dolphins along the central west coast of Florida, which was initiated in 1970, is still in progress. The research program has two main components: (1) temporary captures for marking, measurements, and biological sampling, and (2) observations, including radiotracking, photographic identification censuses, and focal animal behavioral observations.

2.1. Study Area

The study area extends southward approximately 160 km from St. Petersburg to Ft. Myers Beach, Florida (Fig. 1). The area includes large bays, such as Tampa Bay, Sarasota Bay, Gasparilla Sound, Charlotte Harbor, and Pine Island Sound, along with their associated channels and shallow grassflats. These inshore waters are bounded on the west by a series of barrier islands, and communicate with the Gulf of Mexico through passes between the islands. Water depths vary from less than

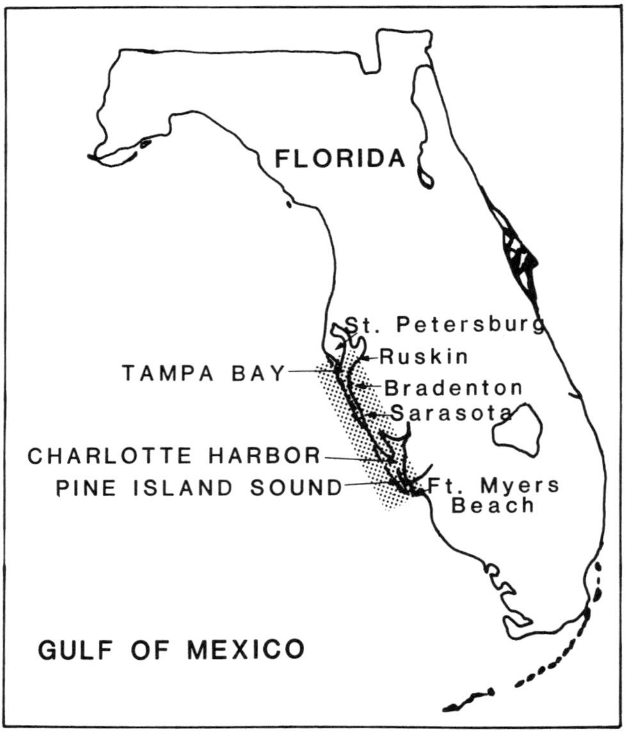

FIGURE 1. Study area along the central west coast of Florida. Shading encloses the entire area where research has been conducted, but most captures and observations have occurred off Bradenton and Sarasota.

1 m over the grassflats to 10 m or more in the passes and at the western extent of the study area, about 5 km offshore of the barrier islands. Tidal fluctuations are typically less than 1 m.

Most of the capture activities were concentrated in the 60-km-long area off Bradenton and Sarasota. But some capture operations have occurred from the southern edge of Tampa Bay, near Ruskin, through Pine Island Sound, near Ft. Myers Beach. Observations and photographic censuses have been conducted through the entire study area, but most of the effort has been concentrated from the southern portion of Tampa Bay southward to Siesta Key off Sarasota.

2.2. Capture Operations

Dolphins were captured in the study area during 1970–1971, 1975–1976, and 1984–1985. The seine net technique described by Asper (1975) was used in each case to encircle small schools of dolphins in shallow (<2 m deep) water. During 1970–1971, we accompanied a commercial dolphin collector, and measured, sexed, tagged, and released his "reject" animals. Twelve dolphins were tagged in the northern half of the study area and 18 in the southern half (Irvine and Wells, 1972).

All of the captures during January 1975–July 1976 were conducted in the northern half of the study area. Forty-seven individuals were measured, sexed, tagged, and released. Ten of these dolphins carried radiotransmitters, and these were tracked for up to 22 days (Wells, 1978; Wells et al., 1980; Irvine et al., 1981, 1982).

Seventy individuals were captured, measured, photographed, marked, and released during June 1984–July 1985. Sixty of these were handled during June/July capture operations in the northern half of the study area. Blood samples were obtained from all of these individuals. Estimated ages were determined from analyses of teeth from 32 individuals. Ten dolphins were tagged incidental to the efforts of a commercial dolphin collector in the adjacent waters of Pine Island Sound, Charlotte Harbor, and southern Tampa Bay during December 1984. Blood samples were obtained from nine of these dolphins, as well as four other dolphins that were collected at that time.

Blood samples were obtained for assessment of reproductive condition and for determination of genetic relationships. The samples were collected from vessels in the fluke. Serum from one aliquot was analyzed by radioimmunoassay for concentrations of testosterone, progesterone, and estradiol by V. L. Kirby at the San Diego Zoo Research Department. Another aliquot was analyzed by D. Duffield at Portland

State University. Electrophoresis and chromosome banding patterns were used to examine familial relationships and genetic patterns in social unit membership. Analyses of samples from June/July 1985 are still in progress at the time of this writing; therefore, only samples collected prior to June 1985 will be considered in this report.

2.3. Observations

Three kinds of systematic observations were conducted. Radiotracking provided information on daily movements, activities, and social associations. Identification censuses established the membership of the resident social unit and provided information on individual ranges and social associations. Focal animal behavioral observations provided activity pattern and social interaction data.

Observations of tagged dolphins during 1970–1971 were opportunistic, usually incidental to capture operations (Irvine and Wells, 1972). During 1975–1976, observations were made from a 7.3-m vessel. The boat was used for radiotracking and for systematic censuses over specified survey routes (Irvine et al., 1981). Photography was used extensively to confirm identifications of recognizable dolphins.

All of the new observation data reported here are the result of two kinds of efforts from small outboard-powered boats (4.3–5.2 m long) during April 1980–January 1984. First, photographic censuses used the survey routes established during 1975–1976 through Bradenton and Sarasota waters, and these routes were expanded to include adjacent waters. Date, time, location, total number of dolphins, number of calves, individual identifications, and behavioral and environmental data were recorded for each sighting. The dolphins' dorsal fins were photographed with Kodachrome 64 color slide film in 35 mm cameras equipped with 200- to 300-mm telephoto lenses, powerwinders, and databacks. The best photographs of each individual were copied and placed in an identification catalog for subsequent comparisons. Dolphins were identified from natural markings, scars from previous tags, or freezebrands on their dorsal fins (Fig. 2). These photographs were examined and classified three times before a sighting of a particular individual was entered into the computer data base.

Second, focal animal behavioral observations were made; this involved moving with dolphin groups containing identifiable members for prolonged periods (typically >15 min). The behaviors of a focal dolphin and its associates were narrated into a tape recorder, and in a number of cases simultaneous hydrophone recordings of the animals'

FIGURE 2. The notch and faint freezebrand on the dorsal fin in this 1983 photograph identify the dolphin as one that was marked as a calf in 1976. Her own calf, a yearling, surfaces alongside.

acoustic emissions were also made. Members of groups observed in this way were also photographed to confirm identifications.

2.4. Definitions and Data Analysis

2.4.1. Social Unit Categories

Variability in dolphin grouping patterns made classification of social units difficult. Three roughly hierarchical categories were used. The term "school" was used as an instantaneous sampling unit. "Group" referred to a kind of stable social unit. "Community" referred to a collection of groups.

Sighting data were categorized and tabulated using the procedures of Irvine et al. (1981). Dolphins sighted within an area of approximately

100 m radius were considered to be in a single school (= group, Irvine et al., 1981). Typically, these were the only dolphins in sight at any given time, and they were engaged in similar activities. This definition is in concordance with the usage of Norris et al. (1985, p. 33), who considered a school to be "any aggregation of aquatic animals that swim together as a unit [see Norris and Dohl 1980]. This means that any group of dolphins that we see traveling together for long enough to make us sure they represent a cohesive unit we call a school." The terms "school" and "sighting" are used interchangeably here in reference to sampling units in the field. These units are similar to small, changeable units of chimpanzees known as "parties" (Goodall, 1983). They are generally unlike the stable, permanent "pods" of killer whales (Bigg, 1982). Sightings of dolphins were retabulated if the same dolphins were resighted more than 1 h apart or if the school composition changed during observations.

The term "group" is used in this paper specifically in the context of a stable social unit. As described in Section 3.2.4, regularly associating dolphins, which shared congruent core areas, were considered to belong to particular groups. In this context, the term "group" will be accompanied by an identifier referring to the sex, age, and/or geographic range of the members. For example, on occasion all of the members of the Anna Maria female group were observed together. At other times the members were dispersed into several smaller units, which each met the criteria for separate schools. Thus, a group maintained compositional integrity over time, whereas schools were ephemeral.

The Sarasota community of dolphins, as defined by Wells (1986), was composed of all of the groups (and individuals that did not clearly belong to groups) that shared large portions of their ranges and interacted with each other to a much greater extent than with members of similar units in adjacent waters. This regional society could be considered to be a population, in the broadest sense (for example, Wilson, 1975). However, evidence indicates that this "population" was not a closed reproductive unit, and thus not a population in the strictest sense. The term community was used because it emphasized the geographic and social relationships of the individuals. The bottlenose dolphin community was similar to that of chimpanzees (Goodall, 1983) in that the ranges of the members were largely contained within a well-defined area, and most of the activities of the members occurred within the community range, but genetic exchange occurred between communities. Killer whale communities were apparently similar to that described here for bottlenose dolphins, except that different killer whale communities did not mix with each other (Bigg, 1982).

2.4.2. Dolphin Age Classes

Age classes were assigned on the basis of one or more of a variety of criteria, including known birth date or known season of birth, analysis of growth layer groups in tooth sections (Hohn, 1980), blood serum concentrations of testosterone, progesterone, or estradiol (measured during the height of the breeding season; Irvine et al., 1981), body length relative to a maturity curve based on Sarasota dolphins, and independence from or association with a presumed mother or calf.

Calves were small dolphins regularly accompanying larger dolphins presumed to be their mothers. Calves were divided into four age classes: 0, calves from birth through their first year; 1, from 1 to 2 years of age; 2, from 2 to 3 years of age; and 3, greater than 3 years old.

Two classes of immature males were identified. Subadult males were those dolphins independent of a presumed mother, who generally were less than 245 cm in length, and who were less than 8 years old. Males meeting these criteria typically had testosterone concentrations of less than 1.0 ng/ml when measured during the 1984 breeding season. Because sexual maturation is a gradual process, a transitional stage was identified for the males. Maturing males were those dolphins who were independent, 245–249 cm long, and approximately 8–9 years old. During the 1984 breeding season, testosterone concentrations for dolphins meeting these criteria were typically 1.0–10.0 ng/ml.

Males were considered adults if they were more than 249 cm in length and 10 years old or older. Males meeting these criteria typically had testosterone concentrations greater than 10 ng/ml when measured during the breeding season in 1984. This concentration is considered to be characteristic of adult males (Ridgway and Benirschke, 1977). Harrison and Ridgway (1971) and Harrison et al. (1972) suggested that males do not reach sexual maturity until 10 years of age. Sergeant et al. (1973) reported that males from northeast Florida mature at about 12 years of age. Perrin and Reilly (1984) considered the average age at sexual maturity for males to be 11 years. Hohn (1985) cautions, however, that comparison of life history parameters produced by different studies may not be valid, due to different methods of age determination and analytical techniques. Sexually mature adults were not necessarily socially mature.

Subadult females were independent young animals that had not yet been observed with a calf and did not produce a calf within 1 year of the end of the observations reported here. Subadult females were typically less than 230 cm long and less than 7 years old.

Females were classified as adult on the basis of having been seen with calves believed to be their own. Depending upon the author, sexual maturity for females is believed to occur between the ages of 5 and 12 years (Harrison and Ridgway, 1971; Harrison et al., 1972; Sergeant et al., 1973; Perrin and Reilly, 1984). Data from Sarasota suggest that some females matured at about 6–7 years of age. Progesterone concentrations for a 6.4-year-old Sarasota female (#21) were indicative of sexual maturity. Female #02, which was tagged as a calf in 1976, was approximately 7.5 years old in 1982 (from the body length versus age curve) when she gave birth. This suggests that #02 was mature by 6.5 years of age.

Adult females were categorized by reproductive condition as (1) receptive, (2) pregnant, (3) with a calf, or (4) single, during the period when they were not accompanied by a calf. Dolphins were categorized as receptive or pregnant a posteriori, following the birth of a calf, based on a 12-month gestation period (Ridgway and Benirschke, 1977). A female was considered to be pregnant for the 12-month period preceding birth. She was considered receptive during the season in which she became pregnant.

2.4.3. Social Association Analyses

Social associations between pairs of individuals were quantified using a coefficient of association (Schaller, 1972),

$$\text{coefficient of association} = 2J/(a + b)$$

where J is the number of joint sightings, scored once for each sighting that included both dolphin A and dolphin B, a is the total number of sightings of dolphin A, and b is the total number of sightings of dolphin B. Coefficients were calculated for all possible pairs of identifiable dolphins of known sex and age. Coefficients ranged from 0.00 for two dolphins that were never seen together to 1.00 for pairs that were always seen together.

3. RESULTS

3.1. Data Base

Most of the analyses reported here deal with data collected during April 1980–January 1984. However, since comparisons are made to previously published data from the same population (Wells, 1978; Wells et al., 1980; Irvine et al., 1981), and because previously unpublished sightings during 1976–1978 are included, the entire observational data base will be described. Observations from vessels were conducted during radiotracking (1975–1976) and censuses (1975–1984). Observations occurred on 423 census days during 9 years, distributed as follows: 1975 (109 days), 1976 (107 days), 1977 (6 days), 1978 (1 day), 1980 (22 days), 1981 (21 days), 1982 (77 days), 1983 (72 days), and 1984 (8 days).

During 1975–1978, 695 dolphin sightings representing a total of approximately 3413 dolphins were recorded. Of these, 914 were identifications of 47 tagged and 12 naturally marked dolphins. During April 1980–January 1984, 1074 sightings were recorded, representing approximately 7806 dolphins. Four hundred sixty-six individuals were identified, including 421 naturally marked dolphins and 45 previously tagged dolphins and their offspring. Of these, 116 were seen five or more times (up to 96 times); these accounted for 49% (3842 dolphins) of all dolphins sighted. Sightings of 77 identifiable dolphins of known sex and estimated age (Tables I and II) form the data base for most of the analyses in this report. Repeated sightings of these 77 individuals accounted for 35% (3883 dolphins) of the total of 11,219 dolphins recorded during 1975–1984.

3.2. Characteristics of the Sarasota Bottlenose Dolphin Society

Inshore bottlenose dolphins near Sarasota, Florida, constitute a resident community of approximately 100 individuals. These animals inhabited a 40-km section of coastal and associated inshore waters. This community shall be referred to as the "Sarasota community" throughout the remainder of this report. The home range of the Sarasota community has apparently remained relatively unchanged and the community size has remained constant from at least 1976 through 1983 (Wells, 1986).

The Sarasota community has remained relatively discrete from adjacent communities in the Gulf of Mexico and Tampa Bay in observations made to date. Mixing with identifiable residents of these adjacent communities occurred in 17% of sightings during 1980–1984, but annual rates of immigration and emigration appeared to be less than 2–3% (Wells, 1986). Thus, the vast majority of social interactions of Sarasota community members involved other long-term members of the same community.

The community is used as the framework within which the social structure is considered in this report because of its relatively permanent residency and its discreteness from other, apparently similar units. The structure of societies is strongly influenced by environmental and social

TABLE I
Identifiable Female Bottlenose Dolphins of Known or Estimated Age

Identification no.	Age class[a]				Basis of age estimate[b]	Number of sightings							Relations[c]
	70–71	75–78	80–81	82–84		75–78	80	81	82	83	84	Sum	
124		C0			LBA	7						7	M8
18				C0, 1	LBA				34	32	4	70	M2
61				C0	LBHA					17	3	20	M32
20				C0, 1	LBHA				24	39	4	67	M4
2		C	S, A	A	LBHAIC	10	7	9	40	32	4	102	M22, C18
51				C0	LBA					33	1	34	M28
25				C1–3	LTHA				23	34	2	59	M23
15				C0–2	LBTHA				30	38	7	75	M1
13			C0, 1	C1–3	LBTHA		3	5	23	42	4	77	M8
62		S	S		LI	20	8	9				37	
14		S	A	A	LTHI	21	9	5	22	34	6	97	
9		S	A	A	LTIC	1	3	7	38	33	1	83	C19
125		S			LI	34						34	
21			S	S	LTHI		12	9	18	24	3	66	
8		A	A	A	LTCI	20	7	8	23	42	4	104	C124, 13
45		A	A	A	LC	2	7	4	13	12	1	39	C127, 46
4		S	A	A	LTHIC	38	7	8	34	39	4	130	C20
23				A	LTHC				26	36	2	64	C25
5	A	A	A	A	LHC	16	11	11	23	41	6	108	C128, 6
35		A	A	A	LHC	1	3	10	6	24	1	45	C30

68			A	A	LTC		4	1	4	6		15	C121
16		A	A	A	LTC	24	11	8	21	30	4	98	C117
53[d]		A	A	A	LC	1	3	6	23			33	C122
42		A	A	A	LC	1	3	5	10	22		41	C66, 67
22		A	A	A	LHC	10	5	6	20	31	3	75	C2
1		A	A	A	LTHC	83	11	10	30	38	7	179	C15
28		S	A	A	LTIC	2	5	4	7	35	1	54	C123, 51
32		A	A	A	LHC	1	3	8	13	20	3	48	C61
126		A			LC	33						33	C129
48	A	A	A	A	LC	1	4	8	8	16	1	38	C86
7	A	A	A	A	LTHC	37	9	8	24	40	5	123	C11, 12
10		A	A	A	LTC	20	4	8	25	38	5	100	C3, 17
26		A	A	A	LC	16	6	12	14	26	1	75	C33
52			A	A	LHC		1	4	4	24	1	34	
24	A	A	A	A	LC	24	7	12	22	20	2	87	
47			A	A	C	9	6	5	11	16		47	C54
50			A	A	C			4	19	12	1	36	C57
72[e]			A	A	C		1	6	1	5		13	C115
75[e]			A	A	C			4	6	2		12	C116
91				A	C					6	1	7	C120
79			A	A	C		1	1	3	5	1	11	C119
93[e]			A	A	C			3	1	4		8	C118

[a]Age classes: C0, first-year calf; C1, second-year calf, etc., S, subadult, M, maturing, A, adult.
[b]Age determinations: L, from body length; B, known date or season of birth; T, from tooth sections; H, from hormone measurements; A, regularly accompanied by presumed mother; I, independent of presumed mother; C, regularly accompanied by a calf.
[c]Relations: M identifies the dolphin's mother, C identifies the dolphin's calf.
[d]Became diseased and disappeared in 1982.
[e]Considered to belong to the Passage Key Inlet–Tampa Bay community.

TABLE II
Identifiable Male Bottlenose Dolphins of Known or Estimated Age

Identification no.	Age class[a]				Basis of age estimate[b]	Number of sightings							Relations[c]
	70–71	75–78	80–81	82–84		75–78	80	81	82	83	84	Sum	
127		C			LA	1						1	M45
128		C0			LBA	6						6	M5
30		C	C3	C3	LHA	1	5	9	7	28	2	52	M35
86		C	C3		LA	1	5	2				8	M48
129		C			LA	30						30	M126
27	C1	S	A	A	LTHAI	45	7	5	11	27	3	98	
11		C0, 1	S	S, M	LTHAI	24	12	5	26	35	2	104	M7
130[d]		C			L	2						2	
3		C2	S	S, M	LTHAI	19	14	9	34	34	2	112	M10
6			C0, 1	C1–3	LBHA		10	9	23	41	6	89	M5
12			C0	C1–3	LBTHA			8	24	40	5	77	M7
17			C0	C0–2	LBTA			4	25	38	5	72	M10
49		S	S, M	M, A	LHI	76	6	5	9	15	1	112	
131		S			LI	15						15	
56		S	S, M	M, A	LI	53	9	6	9	4		81	
132		S			LTHI	1						1	
38[e]		S	S, M	M, A	LTI	31	10	9	3	18	1	72	
133[d]		S			LI	21						21	

134[d]		M			LTI	31						31
44			S	S	LI		1	4	18	16		39
135		M			LI	3						3
36			S	S	LHI			2	16	23	1	42
136	A	A			LI	13						13
74			A	A	LHI		1		7	3	1	12
78				A	LTHI				5	6		11
137[f]		A			LTI	52						52
70[g]				A	LI				4	11		15
73	A	A	A	A	LHI	19			5	6	1	31
58		A	A	A	LHI	13	6	5	1	14		39
138		A			LI	6						6
112		A	A	A	LI	10				1		11
31				S, M	LTHI				17	32	2	51
34			A	A	LTI		6	3	14	21	2	46
139	A	A			LI	1						1
41			A	A	LI	1	6	2	9	22	2	42

[a]Age classes: C0, first-year calf; C1, second-year calf, etc.; S, subadult; M, maturing; A, adult.
[b]Age determinations: L, from body length; B, known date or season of birth; T, from tooth sections; H, from hormone measurements; A, regularly accompanied by presumed mother; I, independent of presumed mother.
[c]Relations: M, identifies the dolphin's mother.
[d]Died in 1976.
[e]Died in 1984.
[f]Died in 1977.
[g]Considered to belong to the Passage Key Inlet–Tampa Bay community.

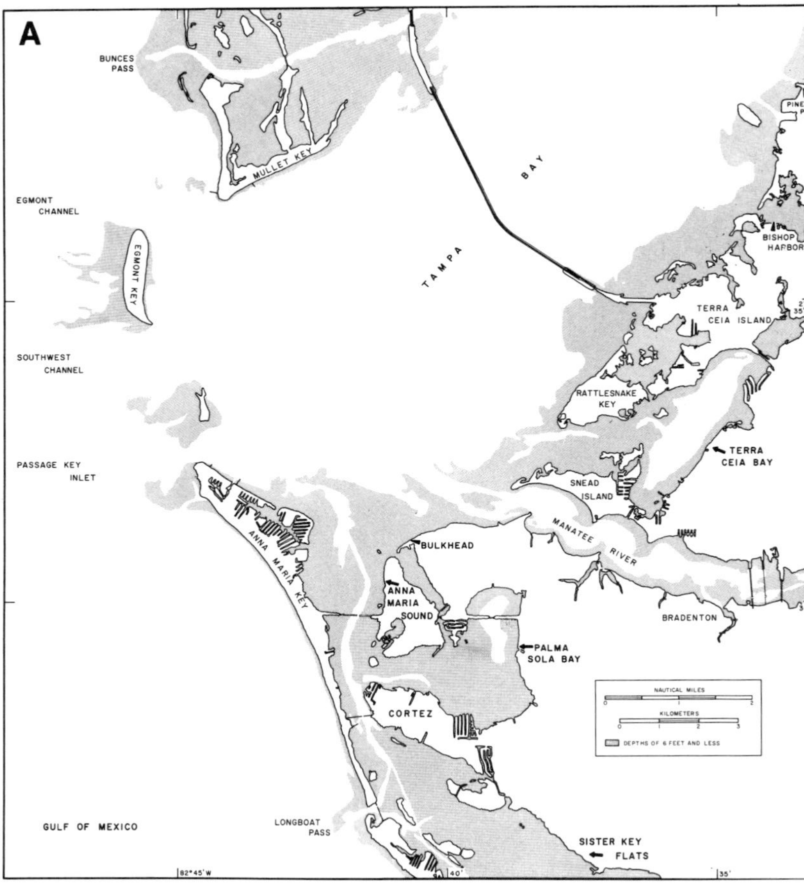

FIGURE 3. Details of the home range of the Sarasota population and surrounding waters. (A) The northern half of the primary study area; (B) the southern half.

pressures. The following summary of home range and social unit features will place the structure in its proper perspective.

3.2.1. Home Range Characteristics

The Sarasota community home range extended southward from the southern edge of Tampa Bay to Siesta Key, off Sarasota, Florida (Fig. 3). This range included all of the shallow (typically less than 4 m deep) inshore waters to the east of the barrier island chain, and Gulf of Mexico coastal waters within approximately 1 km of shore. The inshore waters

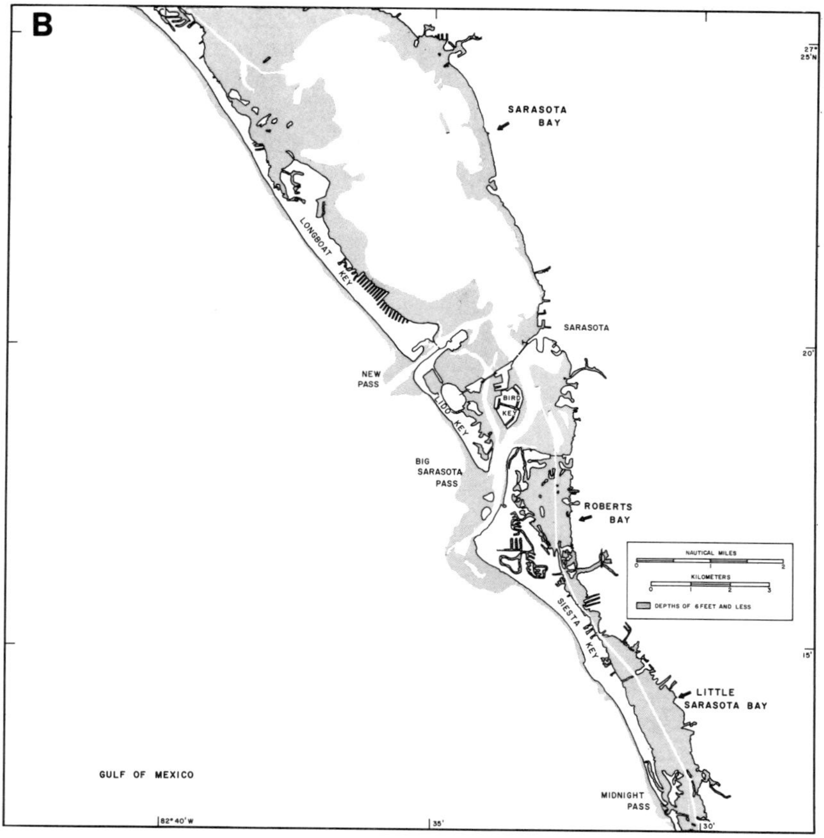

FIGURE 3. (*continued*)

were typified by shallow, protected bays of 3–4 m depth (Terra Ceia Bay, Palma Sola Bay, Sarasota Bay) and included the mouth of the Manatee River. Extensive areas of highly productive grassflats of 1–2 m depth (Anna Maria Sound, Sister Key Flats), traversed by narrow channels of 3 m depth, exist primarily in the northern half of the home range. Narrow passes of up to 10 m depth connect the inshore and Gulf coastal waters. In the Gulf, a gently sloping sandy bottom, with occasional low (less than 1 m high) rocky reefs, gradually reaches depths of about 8–10 m at the western extent of the Sarasota community's typical movements. The northern border of the home range was defined by the deep (4–10 m depth) waters of Tampa Bay.

The present Sarasota community home range, as defined by Wells

(1986), differs from that described by Wells (1978), Wells et al. (1980), and Irvine et al. (1981) only in the addition of Terra Ceia Bay at the northern extent of the inshore portion of the range. Members of the Sarasota community have been seen regularly in Terra Ceia Bay since 1980. However, because surveys through the Bay were infrequent prior to 1980, it cannot be determined if this was an actual range extension or a sampling artifact.

Long-term residency of the Sarasota community was indicated from regular resightings of a number of individuals (Wells, 1986). Seven of 12 (58%) of the Sarasota dolphins tagged during 1970–1971 were identified in the area during 1985. Of the 48 dolphins tagged during 1970–1976 (less the known mortalities), 66% were identified during 1980–1985, in spite of the fact that the tags had been removed in 1976. This value was probably low, since marked changes in dorsal fin morphology during early growth may have made some of the young dolphins unrecognizable between 1976 and 1980.

The Sarasota community members were year-round residents of the home range described above. Seasonal shifts in habitat use during 1975–1976 were described by Irvine et al. (1981). During the summer months, the residents emphasized the use of the shallow grassflats during their daily activities. During the winter, the passes, channels, and shallow Gulf waters were emphasized. This pattern was maintained during 1980–1984. Irvine et al. (1981) and Wells et al. (1980) speculated that the shifts were due at least in part to seasonal movements of the primary prey fish, mullet (*Mugil cephalus*), and seasonal changes in abundance of various shark species, especially the bull shark (*Carcharhinus leucas*), which preys on dolphins.

3.2.2. Community Demography

The Sarasota community size seems to have been relatively stable over many years, as indicated by mark–recapture analyses. Irvine et al. (1981) estimated that the community contained approximately 102 dolphins (95% *CL* = 90–117) in 1976. In 1983, the community size was estimated to be 98 dolphins (95% *CL* = 89–108; Wells, 1986).

The study community consisted of all ages and sexes. Based on similarities in ranges and association patterns, 73 of the 77 dolphins in the data base (Tables I and II) were considered to be members of the Sarasota community (Wells, 1986). Table III summarizes the sex and age structure of the community during 1975–1976 and 1983–1984, based on that portion of the community that has been captured for age estimation and on the observed presence of calves with females.

In both samples, the adult portion of the community appeared to be skewed toward females, whereas the subadult or maturing segment of the community was skewed toward males. Several factors may be contributing to these differences, assuming the relatively closed community proposed by Wells (1986). Females may tend to live longer than males. Of the individuals identified during 1975–1976, 88% of the females were reidentified during 1980–1983, whereas only 46% of the males were reidentified. Preliminary analyses of growth layer groups in teeth collected from 35 dolphins through 1985 found the oldest female to be approximately 44 years old, with four females over 30 years old and four females between 20 and 30 years of age (mean = 17.5 years, n = 19). The oldest male measured to date in the community was approximately 28 years old, and only two males were older than 20 years of age (mean = 12.3 years, n = 16).

Differential mortality was indicated from the few deaths of identifiable dolphins recorded to date. Carcasses of five males (two adults and three subadults) have been examined. Through 1985, no carcasses of identifiable females have been recovered, though one adult female (#53) developed a widespread skin disorder in 1982 and has not been seen since. However, the present samples of both teeth and carcasses are too small to be conclusive on this point. Additional tooth sample collections are planned for 1986–1987.

Differences in ages at sexual maturity probably accounted for some of the sex ratio differences. Females appeared to be recruited into the breeding community at an earlier age than males. If females mature

TABLE III
Sex and Age Composition of the Sarasota Community during 1975–1976 and 1983–1984

Class	1975–1976 Males:Females	1975–1976 Total	1983–1984 Males:Females	1983–1984 Total
Adult	7:16 (30%:70%)	23	11:26 (30%:70%)	37
Subadult or maturing	9:5 (64%:36%)	14	5:1 (83%:17%)	6
Calf	8:2 (80%:20%)	10	4:7 (36%:64%)	11

earlier than males, then there should be fewer subadult females than males in the community at any given time. Assuming females have equal or longer average life spans than males, then this would result in more adult females than males at any given time.

3.2.3. Social Unit Characteristics

The members of the Sarasota community were never all seen together in a single cohesive school. Rather, they were typically dispersed into a variety of schools distributed through the approximately 85-km^2 community home range, as described by Wells et al. (1980) and Irvine et al. (1981). Associations between individuals were relatively fluid, within limits. Typically, small units of persistent or frequently recurrent associates swam through the range, often joining other, similar schools for periods of minutes or hours. These schools occasionally mixed with members of adjacent communities to the north and west (17% of sightings; Wells, 1986).

This paper departs from earlier descriptions (Wells, 1978; Wells et al., 1980; Irvine et al., 1981) by examining the determinants of group structure in the detail that could only be allowed by continued long-term observations. The composition and size of social units appeared to be related to a number of factors, including sex, age, familial relationships, and reproductive condition.

3.2.3a. Composition of Schools. Sexual segregation was a marked feature of the dolphin schools observed both during 1970–1976 (Wells et al., 1980) and 1980–1984. Schools were composed typically of independent (noncalf) members of a single sex. Subgroups within schools also tended to be composed of a single sex. Social symmetry was thus allowed.

Two different samples were examined for patterns of composition. The first sample consisted of 177 sightings of Sarasota community members only, in which all of the noncalves were of known sex and estimated age. The second sample consisted of 536 sightings in which only Sarasota community members were identified, and in which age and sex data were available for at least some of the group members. Sixty-six percent of the 2745 noncalves in this sample were of known sex and estimated age.

Of the 177 sightings in the first sample, only 20% were of mixed sex. Sixteen percent of the schools were all male, and 64% were all female. The results from this sample may be somewhat biased because the schools from which complete sex/age information was most frequently available tended to be small (mean = 4.06 dolphins, SD = 3.585, n = 177, range = 1–22).

Sexual segregation was also suggested from analysis of the second sample. Mixed sexes were observed in only 31% of the sightings. Males only were identified in 23% of the sightings; females only were identified in 46% of the sightings.

School composition was also related to individual age and reproductive condition. The first sample was analyzed for association patterns between the eight noncalf sex/age classes. Coefficients of association were calculated for each class based on the frequencies of joint occurrences within the 177 groups (Table IV). Chi-square analyses were conducted on the frequencies of joint occurrences of classes, testing the null hypothesis that associations were evenly distributed across all classes. For χ^2 analysis, classes were combined on the basis of reproductive condition and age to provide sufficiently large expected frequencies in all cells to be able to proceed with the analysis (Zar, 1974). The combinations were: (1) immatures, including subadult males and females and maturing males, (2) adult males, (3) adult females without calves (singles) and receptive females, and (4) pregnant females and females with calves. The results of these analyses are summarized in Table IV. The χ^2 analyses were subsequently subdivided for analysis of the relative contributions of particular classes to overall significant differences.

While associations between immatures and adults did occur, age segregation appeared to be the general rule, as suggested by Wells et al. (1980). Class associations were not evenly distributed across all classes (Table IV). Immature dolphins tended to be seen with other immatures more frequently than they were seen with any adult classes.

In the first sample, immature males were seen with an adult male and receptive females on only one occasion, on 13 July 1983. These animals interacted briefly. The sighting consisted of a young adult male trailing a single (possibly receptive) female, two receptive females, one male calf, and two maturing males. One of the receptive females (#7), with a 2-year-old male calf (#12), was approached closely by the maturing male pair, which included her 8.5-year-old son (#11). In an apparent display of aggression toward her elder son, #7 tailslapped repeatedly as he approached. Soon, #7 and #12 left the area. Thus, consideration of the associations within this anomalous school on the strict basis of sex, age, and reproductive condition was confounded by familial relationships and possible dispersive interactions within the school.

TABLE IV
Dolphin Group Composition Patterns As Measured by Coefficients of Association for Classes Based on Age, Sex, and Reproductive Condition[a]

Class	Subadult female	Subadult male	Maturing male	Adult male	Single female	Receptive female	Pregnant female	Female + calf
Subadult female	0.00	0.18	0.31	0.10	0.08	0.06	0.05	0.05
Subadult male	0.18	0.05	0.47	0.00	0.17	0.04	0.04	0.11
Maturing male	0.31	0.47	0.29	0.04	0.10	0.04	0.00	0.06
Adult male	0.10	0.00	0.04	0.00	0.23	0.29	0.10	0.16
Single female	0.08	0.17	0.10	0.23	0.45	0.51	0.23	0.35
Receptive female	0.06	0.04	0.04	0.29	0.51	0.36	0.11	0.26
Pregnant female	0.05	0.04	0.00	0.10	0.23	0.11	0.25	0.25
Female + calf	0.05	0.11	0.06	0.16	0.35	0.26	0.25	0.42
χ^2	10.63	19.84	55.59	16.30	15.56	21.57	8.29	14.45
d.f.	3	3	3	3	3	3	3	3
Significance	$P < 0.025$	$P < 0.001$	$P < 0.001$	$P < 0.001$	$P < 0.005$	$P < 0.001$	$P < 0.05$	$P < 0.005$

[a]Coefficients were calculated from 177 groups of completely known composition.

The adult male class shared its highest coefficient of association with receptive adult females (Table IV). Adult male affinities with female classes were roughly correlated with the probability of receptive females existing within the class. Affinities decreased from receptive females through single females, females with calves, and pregnant females and subadult females. Some single adult females may have been receptive, but this was not detected, because they were not impregnated or because their calves were lost. Similarly, some females with older calves may have been receptive, but they did not give birth to a viable calf 1 year later. By definition, pregnant or subadult females could not be considered receptive. The frequency of adult male associations with receptive and single females was much higher than that expected from the null hypothesis.

Adult male affinities for male classes were low (Table IV). While this general trend was also clear from observations outside of the first sample considered here, the extremely low values (0.00) for affinities with subadult males and other adults in Table IV appeared to be artifacts of the sample. Adult males were seen with subadult males in cases not included in the first sample. As discussed below, persistent adult male pairings resulted in some of the highest individual affinities recorded in this study. Thus, although class associations between adult males and other males may have been infrequent, they were not nonexistent.

The various classes of adult females showed relatively high affinities for each other and low affinities for adult males and immatures of both sexes (Table IV). Among the adult female classes, affinities appeared to be closely related to the presence, absence, or imminence of a calf. Single adult females were most closely affiliated with receptive females, and vice versa. Pregnant females were closely affiliated with all adult female classes except receptive females. Females with calves were most closely associated with other females with calves, but they were frequently seen with the other adult female classes as well.

In summary, composition patterns for schools within the Sarasota community were related to dolphin age, sex, and reproductive condition. Immature males and females tended to be seen with other immatures more frequently than with adults of either sex. Adult males associated with receptive adult females to a greater degree than with any other class. Adult female classes tended to associate with adult females more than with any other classes, with the closest associations being between females of similar reproductive condition.

3.2.3b. School Size. The number of dolphins per school was quite variable within the Sarasota community. During 1975 and 1976, the mean value was 4.8 dolphins per sighting (Irvine et al., 1981). The mean number of dolphins per sighting during 1980–1984, for those schools in which only Sarasota community members were identified, was 7.04 ($SD = 6.008$, $n = 536$, range = 1–39).

Numbers of dolphins per school varied significantly relative to time of day, physiography, and composition, based on earlier analyses (Wells et al., 1980). More detailed analyses of the relationships between size and composition were possible from the 1980–1984 data base than from the earlier data set. This resulted from improved information on ages and reproductive conditions provided from long-term observations and a capture/release program.

School size varied significantly with dolphin sex, age, and reproductive condition during 1980–1984. Frequency distributions for school size by age, sex, and reproductive class (Figs. 4 and 5) were computed from the data set of all sightings that included at least one known member of the Sarasota community (Tables I and II). Differences in the numbers of dolphins per sighting were found for older age classes. There were no significant differences between male and female calves, either overall or when subdivided by calf age class. The average number of dolphins per sighting was not significantly different for male and female subadults, but sightings containing subadults of either sex were significantly larger than those containing maturing males (Kruskal–Wallis test with Dunn's multiple comparison, $P < 0.05$). The number of dolphins per sighting for all four classes of adult females combined was significantly larger than that of adult males (Mann–Whitney $U = 65938$, $n = 199, 756$, $P < 0.01$).

Male dolphins tended to be found in progressively smaller schools from birth until they approached sexual maturity. Mean school size varied from a high of 17.80 dolphins per sighting for schools containing females with calves less than 1 year old, to a low value of 8.39 dolphins per sighting for maturing males (Fig. 4). The two youngest male calf classes were found in significantly larger schools than were any of the noncalf male classes. The two oldest male calf classes were found in significantly larger schools than were the maturing males (Kruskal–Wallis test with Dunn's multiple comparison, $P < 0.05$). The difference in school size between the younger calves and the adult males may be due in part to the fact that females with young calves were generally nonreceptive; thus, the adult males and young calves were not together often. In contrast, mothers of older calves were more likely to be receptive, which may in part account for the similarities in school sizes for adult males and the oldest male calves. Within the male calf classes, there was a significant linear relationship between increasing calf age and decreasing number of dolphins per sighting (Fig. 4; Spear-

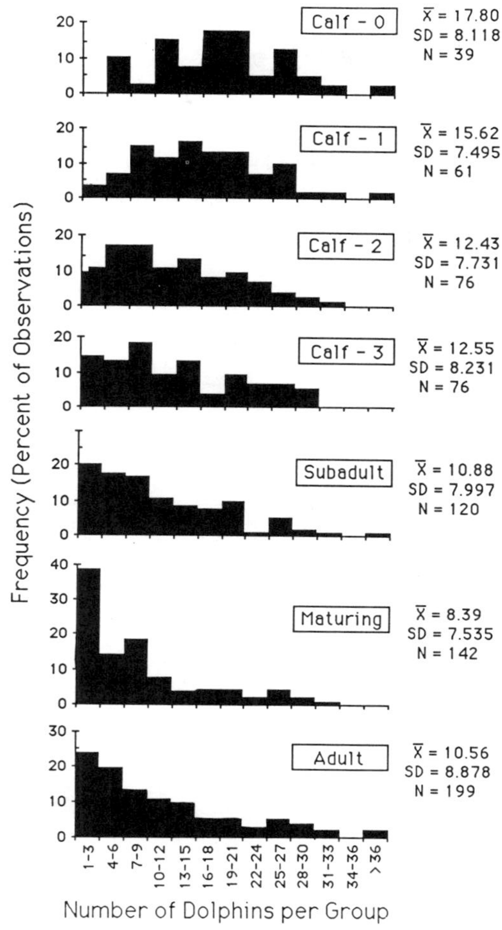

FIGURE 4. Frequency distributions of the numbers of dolphins per sighting for male dolphins of different age classes. The sample size is the number of sightings that contained each age class.

man rank correlation coefficient $r = -0.23$, $n = 250$, $P < 0.001$). Maturing males tended to be found in the smallest schools of any classes, with associations of three or fewer individuals being most common.

The average number of dolphins per sighting varied with age for females in a manner similar to that noted for males. Mean school size ranged from a high of 14.49 dolphins per sighting for sightings containing females with calves less than 1 year old, to an average value of 10.24 dolphins per sighting for females with calves of all ages (Fig. 5). The two classes of the youngest female calves were found with significantly more dolphins per sighting than were adult females with calves of all ages (Kruskal–Wallis test with Dunn's multiple comparison,

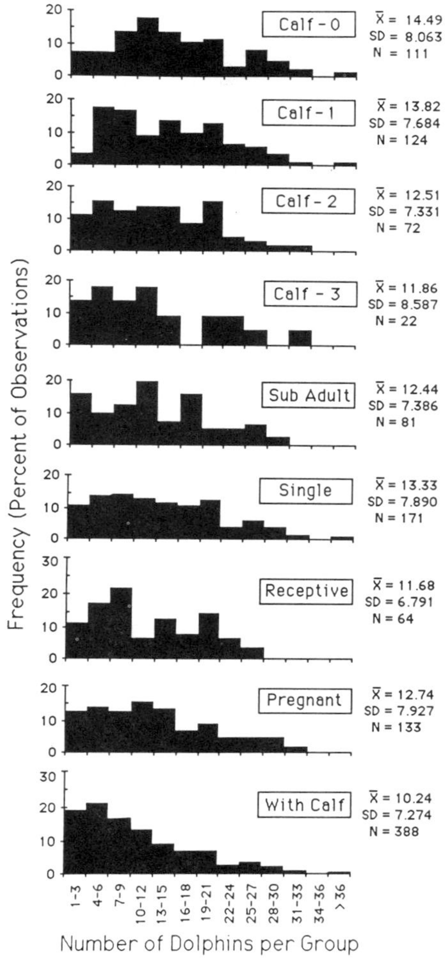

FIGURE 5. Frequency distributions of the numbers of dolphins per sighting for female dolphins of different classes of age or reproductive condition. The sample size is the number of observed groups that contained each class.

$P < 0.05$). As in the case of the male calves, a significant inverse linear relationship was found between female calf age and average school size (Fig. 5, Spearman rank correlation coefficient $r = -0.11$, $n = 329$, $P < 0.05$). The female-with-calf class includes a number of females with older calves, born prior to 1980, for which no other age or sex information is available. This lends additional support to the notion of decreasing school size with calf age. Schools containing single females and/or pregnant females were also significantly larger than those of females and their calves (Kruskal–Wallis test with Dunn's multiple comparison, $P < 0.05$). The lack of significant differences between the sizes of schools containing older calves, receptive females, and females with calves of all ages may be related in part to the inclusion of undetected receptive females in the classes involving females and calves.

In summary, bottlenose dolphin school size tended to decrease with dolphin age until the individuals approached sexual maturity. Young calves tended to be found in the largest schools. Adult males were found in somewhat larger schools than subadult or maturing males. For females, school size was relatively constant across subadult and adult classes. However, females with calves of all ages tended to be found in somewhat smaller schools than were other, noncalf females.

3.2.4. Patterns of Individual Associations

Individual dolphins were seen with a variety of associates. During 1980–1984, each of 38 Sarasota community noncalves was seen with an average of 60.55 different identifiable noncalves (SD = 13.866, range = 25–91). This total may be low, since sightings may have included some individuals that were not identified either because they lacked distinctive markings or because they did not appear in photographs. On average, unidentified noncalves comprised 18.5% of the dolphins in the 536 sightings in which only Sarasota community members were identified.

The Sarasota community members also showed much variation in the frequency of association with particular individuals as well as in the number of associates. As with many of the other features of the schools, the frequency of association, as described by coefficients of association, varied with sex and age.

3.2.4a. Females with Calves. Females and their young calves formed the most consistent associations within the Sarasota community. Mothers and calves were seen together in nearly every sighting of the mother during the first 3–4 years of the calf's life (Fig. 6). The variation in the affinities shown for young calves (Fig. 6) may be explained at least in part by two factors. First, newborn calves surfacing on the far side of their mothers can be difficult to see and thus may not have been recorded. Second, young calves typically lacked individually distinctive markings, and were usually identified on the basis of the female with which they were most closely associated. When young calves were swimming together in "playpens" (groups of interacting calves of similar age surrounded by their mothers), or when one female was apparently "babysitting" for another's calf, it was difficult to identify each individual calf.

Separation of the mother and calf typically occurred after 3 years

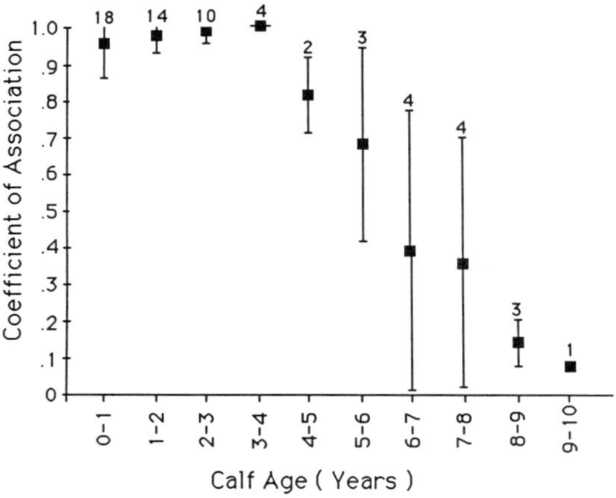

FIGURE 6. Comparison of coefficients of association between mothers and their calves over time. Mean, standard deviation, and number of mother–calf pairs are presented for each calf age.

or more. In Fig. 6, the coefficients of association between mothers and calves gradually decreased after the calf reached 3–4 years of age, but the calves were still seen with their mothers on occasion up to at least 9–10 years of age. The precise ages at separation have been documented for four calves, but these separations occurred after the January 1984 cutoff for data included in association analyses. Thus, coefficients of association with their mothers following separation were not available. The mean age at separation for these four calves was 3.5 years (SD = 0.41).

Three years seemed to be the minimum age of separation. Only four other calves were observed to cease swimming with their mothers at less than 4 years of age. Two of these disappeared within the first year of life, presumably prior to weaning, and therefore they probably died. The fates of the other two remain unknown, since the separation occurred between 1976 and 1980, when few observations were made. At this writing, six other calves were still with their mothers (with coefficients of association >0.96) more than 3.5 years after birth.

The impetus for separation of mothers and calves remains unclear. Pregnancy does not appear to be the underlying cause of separation. In two cases, mothers gave birth to a new calf within 1–5 months following separation. New calves were not immediately forthcoming in four other cases. In two other cases older siblings remained with their mothers following the birth of new calves.

Parent–offspring conflict was suggested by extensive toothrakes on at least two calves. Immediately prior to separation, one 3.5-year-old female calf exhibited fresh rakes over much of her dorsum anterior to her dorsal fin. The spacing between rakes (8–9 mm) was similar to that

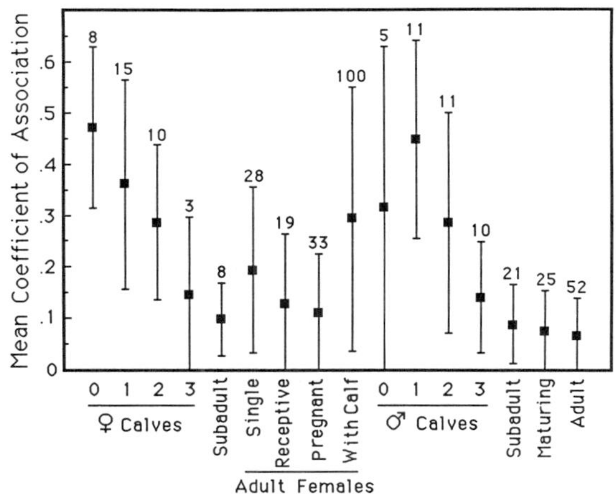

FIGURE 7. Mean coefficients of association of first-year female calves with all other sex, age, and reproductive condition classes. Mean, standard deviation, and number of pairwise associations are presented for interactions between first-year females and each other class.

between her mother's teeth. A 4.5-year-old male calf that was still swimming alongside his mother 17 months after the birth of a sister showed extensive toothrakes over much of his body. However, in neither case could the mothers be definitely implicated as the sources of the rakes.

Calf age appears to be one factor involved in the ordering of schools containing females and calves. Female calves less than 1 year old showed a higher degree of association with other calves less than 1 year old than with older calves or subadults (Fig. 7). A highly significant inverse correlation was found between coefficient of association and dolphin age from newborn through subadult (Spearman rank correlation coefficient $r = 0.60$, $t = -7.56$, $n = 102$, $P < 0.001$).

The creation of a well-defined cohort of newborn calves, such as that produced by a seasonal concentration of births, would facilitate the formation of nursery schools. Calving appeared to be highly seasonal in the Sarasota community (Fig. 8). Though calves may have been born throughout the year, most births occurred during the late spring through early autumn, as also shown by Irvine et al. (1981).

Reproductive seasonality appeared to be closely tied to environmental factors such as water temperature. The number of births was significantly correlated with water temperature (Spearman rank correlation coefficient $r = 0.93$, $z = 3.08$, $P < 0.001$). Births tended to occur when water temperatures exceeded 27°C, and at least 38% of the

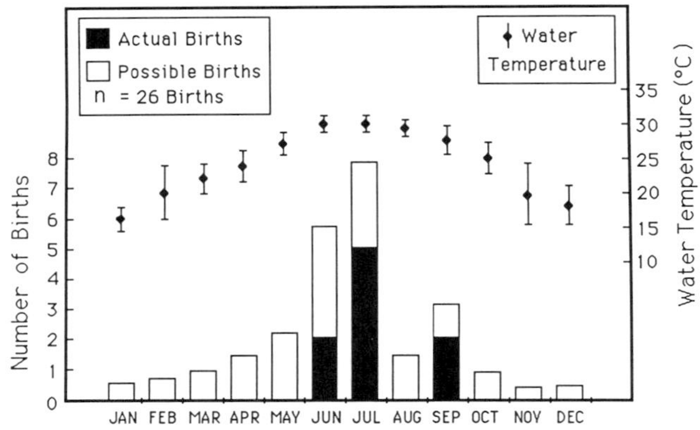

FIGURE 8. Seasonality of bottlenose dolphin births near Sarasota, Florida during 1976–1985 relative to month and water temperature. "Actual births" refer to those documented within a given month. "Possible births" were scored when the interval between the last sighting of a mother without a calf and her first sighting with a new calf spanned more than 1 month. A possible birth was considered to have an equal probability of occurrence at any given time during the interval. Mean water temperatures were calculated from records from Palma Sola Bay during 1975–1982.

births occurred when the temperatures were in excess of 30°C. Annual water temperature cycles in the Sarasota area were highly predictable, typically ranging from lows of 13–14°C to highs of 30–31°C. The coincidence of births with the warm extreme of the thermal cycle suggests that thermoregulation requirements or associated energetic limitations may be important driving forces in determining the seasonality of reproduction. A difference between body and ambient temperatures of 7–10°C in the summer compared to 23–24°C in the winter means much less energy must be expended by the mother during the summer to maintain both her own and, indirectly, her newborn's body temperature.

The attraction between mothers with new calves may be related to a similar degree of dependence by their calves. Though precocial in terms of swimming ability at birth, newborn calves share similar limitations in their abilities, and each calf places similar energetic demands on its mother. Also, during the summer months mullet (one of the primary prey species) are abundant over the shallow grassflats frequented by the dolphins. These shallow grassflats should provide protection from large predatory sharks during the first few months of life (Wells et al., 1980).

3.2.4b. Female Associations with Females. One of the most striking features of female–female interactions was that nearly every female

Females Observed During 1975-78

Dolphins	24	26	125	62	08	14	22	16	10	01	126	05	07	04
24														
26	.80													
125	.77	.61												
62	.65	.51	.72											
08	.33	.17	.22	.32										
14	.21	.13	.17	.24	.47									
22	.13		.11	.14	.21	.17								
16	.09	.06	.15	.15	.05	.05	.27							
10					.11	.18	.07	.15						
01	.23	.09	.31	.19	.13	.08	.19	.23	.05					
126	.11		.17	.20	.12	.20	.48	.23	.07	.28				
05	.11		.09	.07	.25	.29	.19	.26	.26	.24	.19			
07			.09			.18	.11	.03	.14	.14	.15	.17		
04			.04	.06		.11			.06	.06	.19	.04	.11	

Females Observed During 1975-78

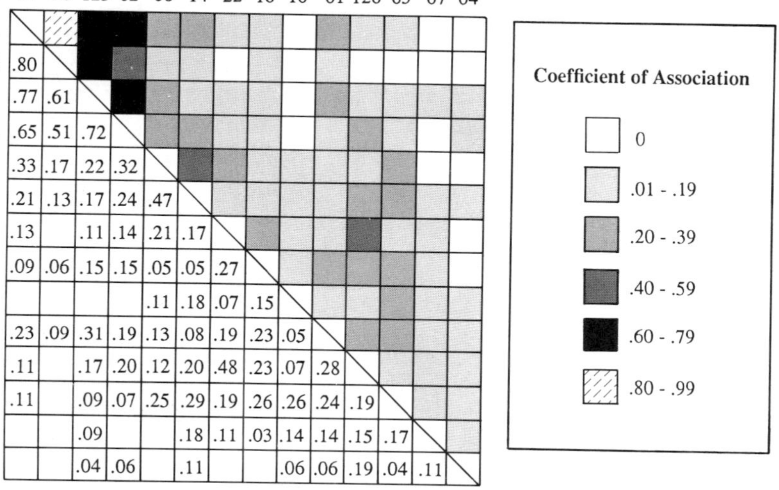

FIGURE 9. Coefficients of association for pairs of female dolphins observed during 1975–1978. The coefficients given in the bottom half of the matrix are presented graphically with corresponding degrees of shading in the upper half.

was seen with each other female member of the Sarasota community at some time during the study. During 1975–1978, 82.4% of all possible pairwise female–female associations were recorded (Fig. 9). During 1980–1984, 95.8% of these possible associations were observed (Fig. 10). The lower value from 1975–1978 was probably an artifact of the observation effort. Half of the females in Fig. 9 were identified for the first time during the last 4 months of the primary 1975–1976 field effort, whereas all of the females in Fig. 10 were observed repeatedly over multiple years.

The females were quite variable in their frequencies of association with other females. Clusters of females with higher coefficients of association with each other than with the other females are evident in Figs. 9 and 10. The most obvious cluster from 1975–1976 included two adults, #24 and #26, and two subadults, #125 and #62. The mean coefficient of association for this cluster was 0.68 (SD = 0.103, n = 12). During 1980–1984, three of the four dolphins in this cluster (#24,

Females Observed During 1980-84 (columns); **Females Observed During 1980-84** (rows)

Dolphins	68	52	35	42	47	48	24	26	62	21	28	08	14	22	16	23	10	01	05	53	07	09	04	02	50	32	45
68																											
52	.04																										
35	.10	.05																									
42	.11	.22	.24																								
47	.23	.08	.15	.26																							
48	.15	.03	.20	.18	.16																						
24	.10	.08	.24	.23	.28	.62																					
26	.14	.09	.27	.20	.33	.63	.72																				
62	.18	.10	.20	.33	.29	.44	.47	.49																			
21	.15	.10	.18	.13	.23	.19	.23	.27	.38																		
28	.09	.09	.19	.13	.18	.18	.21	.20	.08	.12																	
08	.02	.10	.08	.10	.08	.10	.11	.10	.13	.12	.19																
14	.04	.07	.10	.14	.14	.14	.23	.19	.33	.11	.21	.26															
22		.02	.04	.08	.10	.08	.16	.18	.21	.06	.09	.16	.36														
16	.05	.06	.09	.14	.14	.18	.22	.23	.34	.16	.16	.25	.43	.58													
23	.05	.06	.06	.10	.15	.07	.22	.15		.06	.15	.29	.37	.37	.32												
10	.02	.05	.03	.12	.07	.15	.21	.22	.21	.11	.09	.24	.41	.33	.44	.36											
01	.02	.03	.11	.12	.09	.12	.16	.14	.16	.11	.16	.32	.44	.35	.41	.45	.71										
05	.08	.03	.13	.08	.11	.11	.16	.16	.26	.17	.21	.31	.36	.28	.43	.42	.51	.57									
53			.04	.16	.15	.08	.11	.13	.08	.11	.04	.26	.38	.35	.33	.43	.38	.48	.37								
07	.04	.05	.06	.06	.08	.11	.16	.15	.13	.05	.19	.37	.40	.31	.36	.29	.48	.57	.47	.30							
09	.02	.12	.02	.13	.15	.10	.14	.10	.15	.11	.13	.34	.33	.33	.27	.41	.27	.33	.26	.46	.21						
04	.04	.11	.10	.18	.17	.12	.17	.16	.26	.15	.24	.32	.37	.22	.25	.33	.33	.37	.35	.30	.43	.43					
02	.06	.11	.04	.09	.15	.08	.16	.13	.36	.18	.13	.32	.26	.31	.16	.30	.20	.25	.24	.37	.24	.54	.40				
50		.12	.03	.12	.18	.09	.09	.09	.15	.11	.10	.18	.18	.17	.18	.23	.13	.15	.14	.39	.05	.46	.23	.31			
32		.15	.07	.12	.09	.05	.04	.06	.07	.12	.14	.14	.24	.21	.22	.18	.08	.11	.12	.21	.08	.20	.16	.14	.23		
45	.04	.06								.02		.10	.04	.02		.05	.05	.05	.02		.02	.08	.05	.08	.19	.07	

FIGURE 10. Coefficients of association for pairs of female dolphins observed during 1980–1984. The coefficients given in the bottom half of the matrix are presented graphically with corresponding degrees of shading in the upper half.

#26, #62) were identified, and they still associated closely with each other, with a mean coefficient of association of 0.56 (SD = 0.124, n = 6). In Fig. 10, these three animals appear to form the "core" associations within a larger cluster of dolphins with similar association patterns. A second, even larger cluster is also evident in Fig. 10.

The members of the clusters were examined for shared characteristics in addition to association patterns. Familial relationships were identified for six of the members of the larger of the two clusters in Fig. 10. A unique marker chromosome was found in four adult females and one of their calves (#08, #14, #23, #05, #13), indicating a high degree of relatedness (Duffield et al., 1985). The mean coefficient of association between the three related females identified during 1975–1978 was 0.34 (SD = 0.105, n = 6). During 1980–1984, the mean coefficient of association for all four of these related adults was again 0.34 (SD = 0.056, n = 12). Two other females (#22 and #02) in the larger cluster were identified as a mother–calf pair in 1976. Both have given birth to new calves since 1982, and have remained in the cluster. The coefficient of association between these two females was 1.00 in 1976 and 0.31 in 1980–1984. The age difference between the youngest calf and the oldest adult female with the chromosome marker (30 years), the long-term presence of the dolphins in the area (since 1970 for #05), and the similarities in the coefficients of association of the adults over at least 9 years suggest that familial relationships may be important determinants in the formation of these clusters.

The females in the clusters of similar associations shown in Figs. 9 and 10 typically shared congruent home ranges (Wells, 1986). The home ranges of the members of different clusters were overlapping, but different core areas were emphasized in the animals' day-to-day activities. Wells (1986) identified particular female groups with nonoverlapping core areas based on these clusters. In general, a female was considered to belong to a particular group if (1) she shared her home range with the other group members, (2) she had a coefficient of association of at least 0.31 (from the value for related females) with one or more group members, and (3) her mean coefficient of association with all of the group members was within one standard deviation of the overall mean association coefficient for all of the group members with each other. Some stable groups existed over many years. However, these groups were not discrete, permanent units of constant composition, and not all of the community members belonged to distinguishable groups. The formulation of a hard and fast definition for a less than absolute behavioral tendency is fraught with difficulty. Thus, the working definition for a group should be applied cautiously.

Two female groups were identified from the clusters in Fig. 10. These two groups accounted for 78% of the females with sufficient numbers of sighting records (≥15) to be included in Fig. 10. These groups were named on the basis of the unique regions in which the activities of each group were concentrated (Fig. 11). The Anna Maria female group consisted of seven identifiable noncalves: #42, #47, #48, #24, #26, #62, and #21 (mean coefficient of association = 0.34, SD = 0.165). The Palma Sola female group consisted of 14 identifiable noncalves: #08, #14, #22, #16, #23, #10, #01, #05, #53, #07, #09, #04, #02, and #50 (mean coefficient of association = 0.33, SD = 0.109). Analysis using the Kruskal–Wallis test with Dunn's multiple

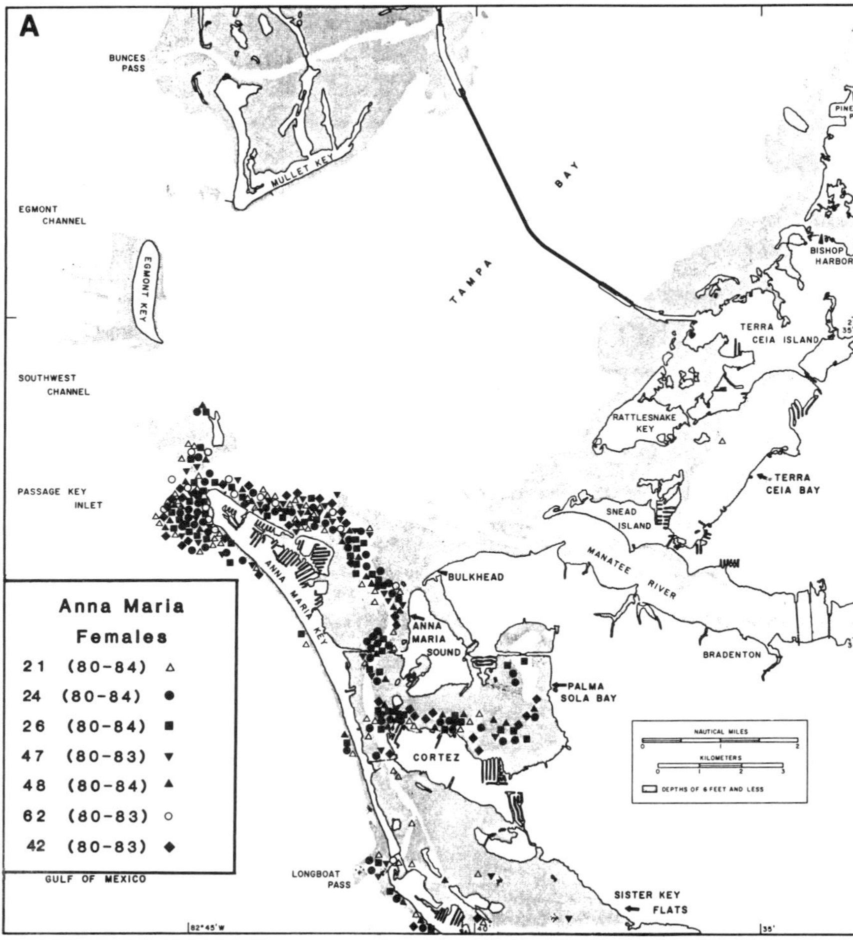

FIGURE 11. Sightings of representative dolphins of the Anna Maria (A) and Palma Sola (B) female groups in the northern half of the community home range. Note the differential use of Palma Sola Bay and the waters around the northern half of Anna Maria Key.

comparison showed significant differences ($P < 0.05$) for the coefficients of association of the Anna Maria group members for each other compared to their coefficients of association for Palma Sola group members, and vice versa. The within-group coefficients of association were similar for both female groups.

A third female group within the Sarasota community was provisionally identified. The Manatee River female group consisted of two

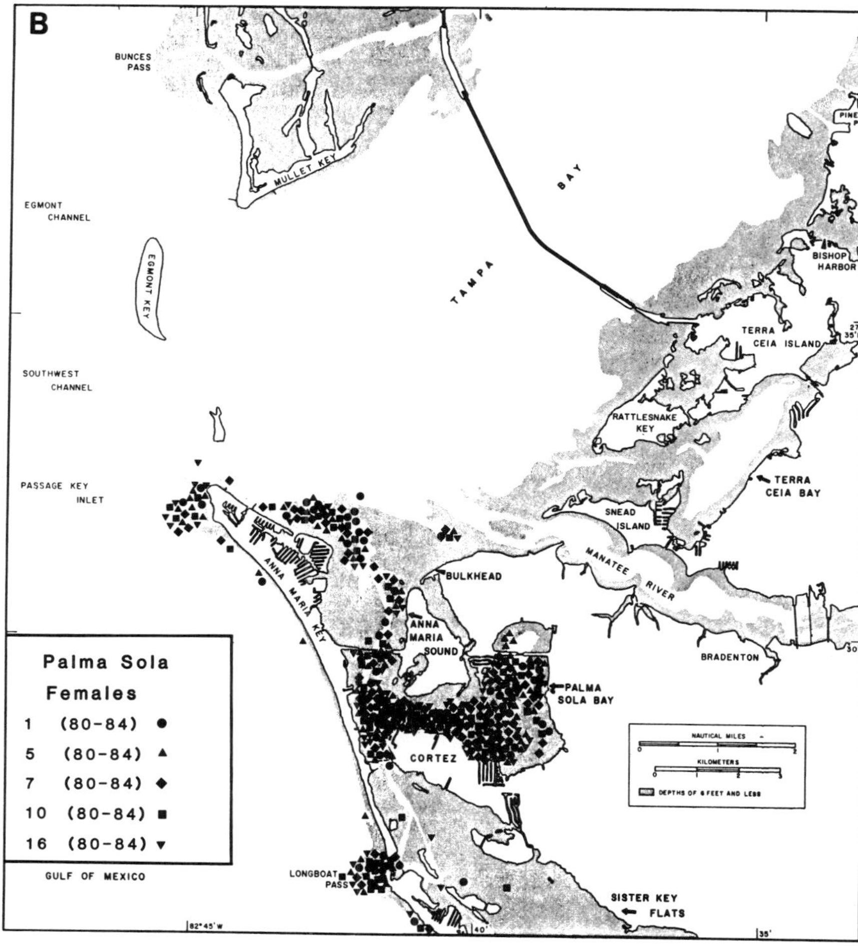

FIGURE 11. (*continued*)

identifiable adult females (#79 and #91) with a coefficient of association of 0.92. These dolphins were not included in Fig. 10 because of low numbers of sighting records (11 and seven, respectively). These two females used many of the same areas used by other groups, but they were seen most frequently to the northeast of the other groups, in the Manatee River and Terra Ceia Bay.

Five other adult females did not fit clearly into the female groups described above. Four of these (#28, #35, #52, and #68) swam often with the groups, and their home ranges were within the community range (Wells, 1986), but their associations with any particular female or group were not sufficiently frequent to constitute group membership. Female #45 spent most of her time in the southern portion of the community range, but she swam with other females when they moved through the southern range or when she traveled to the northern range.

Members of particular female groups were rarely all present in a single school, at least in the case of a large group such as the Palma Sola females. Rather, group members often were observed in several different schools distributed through the group's range. Within these female groups, associations tended to follow patterns relating to reproductive condition. As described above, the presence and age of calves were important factors, with calfless females swimming together and females with young calves tending to swim together. Changes in reproductive status tended to change female association patterns. For example, the association between mother (#22) and her subadult daughter (#02) decreased from 0.77 to 0.17 after #02 gave birth to a calf. During the first year of her calf's life, the coefficients of association of #02 with two other Palma Sola female group members with young of the year (#04 and #09) increased from 0.29 and 0.39 prior to birth to 0.57 and 0.68, respectively. Thus, although familial relationships may play an important role in determining female group membership, other factors, such as reproductive condition, appear to play the predominant role in determining associates within a group.

3.2.4c. Male Associations with Males. Male–male interactions were much more limited than female–female interactions. During 1975–1978, only 41.0% of the possible pairwise male–male associations were recorded (Fig. 12). For 1980–1984, this figure increased to 76.2% (Fig. 13). It seems likely that this increase was largely a sampling artifact, resulting from differences in duration of observation periods. Over half of the males in Fig. 12 were initially identified during the final 6 months of the primary 1975–1976 field effort, whereas all of the males in Fig. 13 were observed repeatedly over multiple years.

As indicated above by the analyses of class association patterns in general, the association patterns of individual males appeared to be strongly correlated with the ages of the individuals. Most male–male associations were between males of similar age. Males tended to become more restricted in their associations with other males of the same age class after the onset of sexual maturity. In both the 1975–1978 and 1980–1984 samples, adult males associated with few other adult males (40.0 and 61.9% of all possible pairwise associations, respectively). In contrast, younger males were observed in 85.7% (Fig. 12) and 92.9%

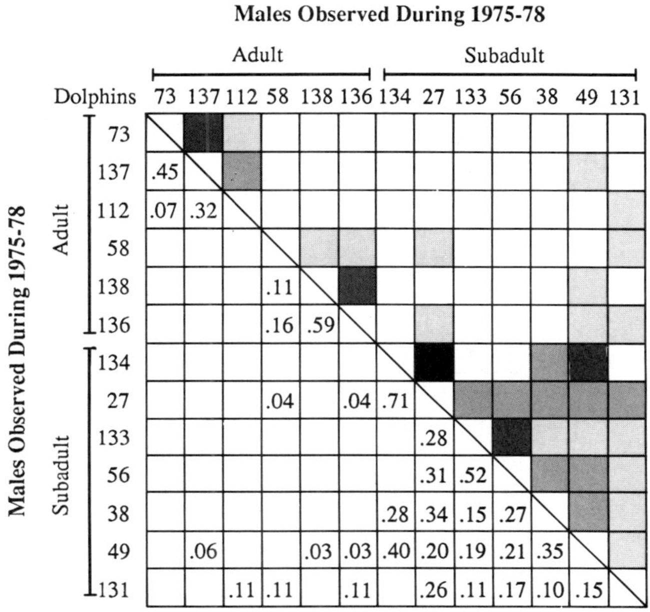

FIGURE 12. Coefficients of association for pairs of male dolphins observed during 1975–1978. The coefficients given in the bottom half of the matrix are presented graphically with corresponding degrees of shading in the upper half.

(Fig. 13) of all possible pairwise associations with other young males. The association coefficients for individuals of each male class over the entire period 1980–1984 were examined with a Kruskal–Wallis test and Dunn's multiple comparison. The coefficients of association between adult males (mean = 0.08, $SD = 0.162$, $n = 40$) were significantly ($P < 0.05$) less than those between males that matured during this period (mean = 0.29, $SD = 0.137$, $n = 6$) or those between immature males (mean = 0.20, $SD = 0.223$, $n = 20$). The within-age-class coefficients for the two young male categories were not significantly different from each other. From Figs. 12 and 13 it is evident that the most frequent interactions between males were between small clusters of individuals of the same age classes. Younger males demonstrated a clear preference for associating with a number of other males of similar age; individual adult males showed a consistently low frequency of association with any males except a select few other adults (Figs. 12 and 13).

Associations between individual males of different age classes occurred, but they were infrequent. During 1975–1978, adult and subadult

Males Observed During 1980-84

Males Observed During 1980-84	Dolphins	73	74	41	58	34	78	27	56	38	49	03	11	31	36	44
		Adult							Subadult →Maturing →Adult			Subadult →Maturing				Subadult
Adult	73															
	74	.75														
	41	.04														
	58			.12												
	34	.03	.07													
	78			.05	.08	.13										
	27		.03	.02	.08	.08	.12									
Subadult →Maturing →Adult	56				.04	.08		.15								
	38		.04		.03	.14		.21	.46							
	49		.04	.03	.03	.10		.16	.16	.26						
Subadult →Maturing	03	.02	.02	.05	.05	.14		.12	.22	.36	.23					
	11	.02	.02	.02	.06	.16	.03	.17	.15	.35	.22	.80				
	31			.02		.11	.03	.09	.06	.11	.08	.17	.16			
	36	.04		.03		.10	.04	.10	.03	.08	.17	.13	.13	.24		
Subadult	44	.04	.04	.10	.06	.07	.04	.07			.03	.03	.03	.02	.30	

FIGURE 13. Coefficients of association for pairs of male dolphins observed during 1980–1984. The coefficients are overall values, based on the entire 4-year period. Thus, some of the calculations are for males that made transitions through more than one age class. The coefficients given in the bottom half of the matrix are presented graphically with corresponding degrees of shading in the upper half.

males were observed together in only 19.0% of all possible pairwise interactions. Adult males were observed during 1980–1984 in 73.2% of all possible pairwise associations with younger males. The mean association coefficient for the adult male interactions with other adult males was not significantly different from those for associations between adult males and either of the younger male categories. However, the within-age-class coefficients for the younger males were significantly ($P < 0.05$) greater than their coefficients of association with adult males.

Groups of regular associates were evident for many males. The characteristics of these male groups were similar to those of the female groups. Group members moved through similar ranges (Wells, 1986). Whereas females showed a high degree of site fidelity to a relatively limited area, with occasional excursions to other areas, males visited the extremes of the community range more frequently than did females. Maturing or adult males occasionally disappeared from the normal range of the community for months at a time. Wells et al. (1980) reported a pattern of differential use of the waters of the community home range by adult males compared to subadult males. This pattern was not clearly evident during 1980–1984.

Adult male groups tended to be considerably smaller than female groups. Individual adult males were typically observed either as the only adult male in a school or as a member of a persistent pair (or, in some cases, trio) of dolphins. For example, in 90.2% of his sightings adult male #41 was the only adult male identified, and in 76.9% of his sightings #58 was the only adult male identified. In contrast, relatively high coefficients of association (0.45–0.75) were measured for the pairings of #73 and #137, #136 and #138 during 1975–1976, and #73 and #74 during 1980–1984 (Figs. 12 and 13).

Pairs of closely associated adult males may have been more common than is indicated in Figs. 12 and 13. Other pairs of close associates were believed to have been composed of two adult males, but the sexes and ages of some of the individuals were not confirmed. Dolphin #34, one of the largest and oldest males handled during the study, shared a coefficient of association of 0.94 with #39. Dolphin #39 was believed to be a male based on the facts that it appeared even larger than #34, it was heavily scarred, and it was not observed with a calf presumed to be its own at any time during a period of more than 6 years. Such large size and heavy scarring have been considered characteristic of sexually mature males (Ridgway and Benirschke, 1977). Similarly, two other large, heavily scarred dolphins (#40 and #43) observed together consistently since 1975 and which were never seen with calves believed to be their own shared a coefficient of association of 0.96. During 1986 it was determined that both #40 and #43 were indeed adult males. These two cases represented the highest coefficients of association for any pairs of dolphins in the Sarasota community, with the exception of mothers with their young calves. In another case, male #27 shared a coefficient of association of 0.55 with another large, heavily scarred dolphin that had been identified regularly since 1970, but was never with a calf of its own.

Some adult male pairs formed originally when the animals were subadults. Males were observed as members of small, persistent groups (= primary groups; Wells et al., 1980) well before they reached sexual maturity (Figs. 12 and 13). In each case, the highest affinities were with other animals of nearly the same age. Subadult males were in schools in which only other subadult or maturing males were identified in 29.4% of the schools in which subadult males were observed. Long-lasting, closely associated pairs (with their coefficients of association) included #27 and #134 (0.71) and #56 and #133 (0.52) during 1975–1978, and #03 and #11 (0.80) during 1980–1984. The #03–#11 pair already existed in 1980, when the dolphins were 6.5 and 5.5 years old, respectively. The mothers of these dolphins were rarely seen together during the brief time that both were identifiable in 1976, but they showed a high association coefficient (0.48) from 1980–1984. A persistent trio, composed of males #56 and #38 and a dolphin of similar size but unknown sex, #37, had a mean coefficient of association of 0.58 (SD = 0.134) during 1980–1984. Unlike the adult males, all of the younger males were typically observed with a number of other males of the same age. Those males that were not part of particular groups were generally sighted with a variety of males of similar age. Groups of close associates frequently interacted with the other groups of similar-age males. Extensive socializing was observed in these young male schools. The socializing bouts were manifested by much contact, simultaneous erections, chases, aerial behaviors, and splashing.

The importance of close associations to some of the individual males was suggested by two facts. First, the durations of the bonds tended to be prolonged. Dolphins #136 and #138 were observed together during 1970–1971 and during 1975–1976 (Irvine and Wells, 1972; Wells et al., 1980). Dolphins #73 and #74 were together at least from 1982 through 1985. Dolphins #56, #38, and #37 were closely associated from 1980 through 1982. The pair #38 and #37 remained closely associated until the death of #38 in 1984. Dolphins #03 and #11 have remained closely associated at least from 1980 through 1985.

Second, when one member was lost from a pair, the remaining member often formed a new bond of similar strength with another male. Adult male #73 formed a new pair with #74 following the death of his previous partner, #137. Similarly, #56 formed the trio with #38 and #37 following the death of his previous partner, #133. Likewise, #27 formed a new pair with #60 following the death of his previous partner, #134.

Speculations on the adaptive value of these male pair bonds are possible based on anecdotal field observations, but available data are insufficient to be conclusive. If some male pairs are composed of related individuals (a possibility suggested by the fact that the mothers of at least one male pair swim in the same band), then the maintenance of the pair may serve to increase the fitness of both individuals, as in the case of lions (Schaller, 1972).

On several occasions, a male pair has been observed apparently separating an individual female from a school. One particularly intriguing incident occurred on 4 July 1985. Six dolphins, including two pairs of young adult males 10–13 years old and two females 7–10 years old, moved slowly northward past our observation boat. They moved over shallow grassflats, feeding on mullet by knocking them from the water with their flukes. Two large, older adult males (#73 and #74) approached rapidly from the south. When the dolphins were resighted several minutes later, only mature female #21 from the original school was in sight, with #73 and #74. Following a brief bout of socializing culminated by tailslaps against one of the males by the female, the two males apparently chased the female at moderate speed for 2.8 km, flanking her on each side and slightly behind her or moving up alongside her. The three dolphins swam toward a feeding school of females and calves. The nursery school left the area, and the trio began feeding.

In summary, patterns of male interactions with other males apparently took either of two forms. Some adult males remained relatively isolated from other adult males, whereas other adult males formed close associations with a few individuals of similar age and sex, similar to those typical of younger males.

3.2.4d. Associations between Females and Males. Interactions between the sexes involved most of the members of the Sarasota community (Figs. 14 and 15). During 1980–1984, 83.7% of all possible pairwise associations between males and females were recorded. There were no significant differences in the numbers of different females with which males of different ages associated.

Interactions between the sexes were less frequent than interactions within sex and age classes (Figs. 14 and 15). The frequencies of male interactions with females varied significantly with age of the males (Kruskal–Wallis test with Dunn's multiple comparison, $P < 0.05$). Significantly higher mean coefficients of association with females were measured for adult males (0.08, $SD = 0.067$, $n = 189$) and immature males (0.10, $SD = 0.062$, $n = 135$) than for transitional males, the males that matured during 1980–1984 (0.03, $SD = 0.033$, $n = 81$).

Interactions between subadult dolphins and members of the opposite gender of any age were relatively common. Of all groups in which subadult males were identified, 55.5% also included adult females. Of these mixed groups, 34.8% also included adult males. The cooccurrence of subadult males and adult females was independent of female reproductive seasonality (contingency $\chi^2 = 0.02$, $P < 0.75$).

Adult females were recorded from 44.4% sightings of groups containing maturing males. In 28.6% of these mixed sightings, adult males were also present. Frequency of sightings of maturing males with adult

Males Observed During 1975-78

Rows: Females Observed During 1975-78

	Adult						Subadult						
Dolphin	73	137	112	58	138	136	134	27	133	56	38	49	131
24		.26	.06										
26	.59	.26	.08										.08
125		.25	.06									.04	.07
62	.56	.20						.07		.06	.07		.08
08	.32	.05	.07										.36
14	.25			.07		.07		.07			.04	.08	.29
22	.14	.07	.20										.10
16	.15	.09	.27	.06				.03	.05	.03			.07
10		.05	.07	.08		.08		.14					.18
01	.29	.18	.19			.09		.05				.09	.06
126	.13	.08	.26					.06	.06	.06	.03	.11	.30
05	.13							.03				.03	.16
07		.06		.05		.05		.06	.07	.03	.04	.07	.19
04				.05		.05	.51	.30	.24	.16	.46		.36

FIGURE 14. Coefficients of association for male–female pairs observed during 1975–1978. The high coefficients obscured by shading are as follows: 24–73 = 0.62, 125–73 = 0.61, and 04–49 = 0.66.

females was independent of female reproductive seasonality (contingency $\chi^2 = 1.99$, $P < 0.25$).

Individual subadult females were also seen with subadult and maturing males. During 1975–1978, the coefficients of association for subadult female #04 with subadult males were equal to or greater than the coefficients for the subadult males with each other (Fig. 14). Similarly, during 1980–1984 the coefficients of association for subadult female #21 with the immature males were within the range of those of the immature males for each other.

Adult males were associated with female dolphins more frequently than were younger males, and their association frequencies were correlated with female reproductive seasonality. Adult males did not remain with any particular school of females for an extended period. Rather, they tended to move from female school to female school. Adult females were identified in 65.3% of the groups in which adult males were identified. Adult males and females were together during the May–October breeding season significantly more frequently than would

Males Observed During 1980-84

	Adult							Subadult →Maturing →Adult			Subadult →Maturing				Subadult
Dolphins	73	74	41	58	34	78	27	56	38	49	03	11	31	36	44
68			.03	.05	.10	.10	.03			.04	.07	.06	.03	.08	.11
52			.05	.07			.09			.06	.06	.07	.03	.08	.25
35			.14	.06	.13		.02				.07	.08	.07	.19	.19
42	.04	.04	.10	.03	.02		.04				.05	.02		.15	.38
47	.08	.04	.05	.19	.07	.11	.09	.03	.03	.14	.08	.05		.19	.13
48	.08	.08	.15	.06	.12	.06	.07			.06	.05	.03	.05	.03	.16
24	.08	.08	.17	.07	.07	.07	.05			.06	.06	.04	.04	.06	.18
26	.09	.09	.10	.09	.10	.04	.04			.04	.07	.03	.04	.06	.14
62			.24	.50			.14	.06	.06		.05				.29
21	.08	.05	.19	.15	.09		.12	.02	.04	.10	.23	.19	.06	.17	.19
28	.06	.03	.19	.08	.12	.04	.13		.02	.14	.07	.06	.09	.11	.11
08	.02	.02	.10	.11	.02	.08	.16	.04	.03	.05	.10	.12	.07	.07	.10
14	.05	.02	.22	.18	.08	.11	.09	.04	.03	.11	.13	.10	.09	.13	.09
22	.03		.21	.11	.04	.09	.10			.02	.17	.12	.08	.08	.04
16	.05	.02	.33	.20	.03	.06	.14	.06	.05	.07	.20	.14	.08	.04	.07
23	.05	.05	.17	.20	.08	.08	.10			.05	.13	.11	.02	.10	.10
10	.07	.04	.18	.08	.05	.10	.09	.02	.02	.03	.13	.08	.08	.02	.07
01		.02	.20	.08	.07	.09	.11	.07	.06	.05	.18	.14	.06	.05	.09
05	.04	.02	.18	.15	.07	.06	.11	.07	.06	.06	.18	.12	.07	.03	.05
53			.08	.14	.11	.22	.11	.07	.07	.04	.14	.08	.15	.13	.22
07	.04	.02	.22	.09	.06	.08	.13	.05	.05	.08	.17	.13	.10	.05	.08
09	.04	.04	.08	.04	.06	.07	.12	.02	.02	.07	.16	.14	.08	.08	.13
04	.10	.06	.18	.14	.04	.05	.14	.03	.02	.06	.16	.16	.11	.09	.08
02	.02	.02	.15	.10	.03	.09	.12		.06	.06	.14	.14	.09	.12	.12
50	.04	.04	.09	.04	.05	.09	.12			.03	.10	.10		.08	.11
32	.03		.09	.11			.08			.05	.10	.09	.07	.14	.02
45			.03				.07	.03						.06	

Females Observed During 1980-84

FIGURE 15. Coefficients of association for male–female pairs observed during 1980–1984.

be expected if the associations were distributed evenly throughout the year (contingency $\chi^2 = 6.04$, $P < 0.025$).

At least two very different patterns of association with females were apparent for different adult males. The resident male pattern, exemplified most clearly by #41 and #58, was characterized by associations between a single adult male with one or more females at any given time. The single male tended to remain in the limited area within which the female groups concentrated their activities. The roving male pattern, as exemplified by #34, #73, and #74, was typified by males that ranged over a large area, beyond the regions in which the largest female groups concentrated their activities. The roving males were frequently observed traveling through the channels of the community home range, spending brief periods of time with females they encountered. Evidence suggests that the tight multimale units described above were characteristic of the roving male pattern, but confirmation of this will require confirmation of the sex of members of the pairs that follow this pattern, #39, #40, and #43. During 1986 it was determined that #40 and #43 were adult males.

The differences between these two patterns can be described best by comparing the two extreme cases for which the most sighting data were available, dolphins #41 and #34. Ninety-three percent of resident male #41's sightings were within the region used most frequently by the Anna Maria and Palma Sola female groups. Dolphin #34 ranged to the north and south of this area significantly more frequently than did #41; only 67% of his sightings were within the core areas of the main female groups (contingency $\chi^2 = 8.29$, $P < 0.025$). Both were large males: #41 was 266 cm long, and #34 was 273 cm long and approximately 26 years old in 1985. In 87.5% of his sightings with adult females during the breeding season, #41 was the only adult male identified. In contrast, #34 shared a coefficient of association of 0.94 with #39, a presumed (but unconfirmed) adult male. Other adult males were identified in 38.9% of the schools in which #34 was with adult females during the breeding season.

The males #41 and #34 showed marked differences in their associations with females. Resident male #41 had the greatest mean coefficient of association (0.15, $SD = 0.072$, $n = 27$) for an adult male with females during 1980–1984. One of the lowest mean values belonged to roving male #34 (0.06, $SD = 0.039$, $n = 27$), and the difference between the two males was significant (Kruskal–Wallis test with Dunn's multiple comparison, $P < 0.05$). Resident males #41 and #58 were observed with receptive females on 26.8% and 23.1% of their sightings, respectively. In contrast, roving males #34, #73, and #74 were not observed with known receptive females during 1980–1984. The difference in the patterns for resident male #41 and roving male #34 was highly significant (contingency $\chi^2 = 15.54$, $P << 0.001$).

During the breeding season, resident male #41 was observed with significantly more adult females per sighting (mean = 4.82, $SD = 3.316$, $n = 34$) than was #34 (mean = 2.17, $SD = 3.621$, $n = 36$; Kruskal–Wallis test with Dunn's multiple comparison, $P < 0.05$). In those cases when #41 was not the only adult male with a school of females, the number of females in the school was much larger than his overall mean (mean = 8.00, $SD = 2.580$, $n = 4$). This was also true for those cases when #34 and #39 were associated with adult females and other adult males (mean = 5.00, $SD = 3.916$, $n = 7$). These observations were suggestive of a pattern in which each single or multiadult male unit may have typically been accompanying an optimal number of females.

In summary, associations occurred between members of all age classes of males and females. Frequencies of associations were independent of female reproductive seasonality for subadult and maturing males, but associations between adult males and females were most frequent during the breeding season. Male associations with females became less frequent as subadult males matured, but they became more frequent during adulthood. Adult males apparently engaged in at least two different patterns in their associations with adult females. Single, resident males remained in areas frequented by large numbers of females and associated with many of them. Roving multimale groups tended to range over large areas and associated with small numbers of adult females at any given time. Only resident males have been observed with receptive females.

4. DISCUSSION

The data from 1980–1984 support the hypothesis of a stable sex- and age-based social structure for the Sarasota bottlenose dolphin community. In addition, improved means of recognizing familial relationships and reproductive condition have facilitated refinement of hypotheses about the structural basis of the dolphin society. The features of this society can be summarized as follows.

4.1. Social Structure Summary

The Sarasota community represents a relatively closed society (Wells, 1986). The community home range appears to be well-defined and permanent. Dispersal of young out of the community has not yet been documented. Several offspring of community members have remained with the natal community through sexual maturity. Permanent immigration and emigration of members of other age classes appears to be infrequent. However, short-term absences (days to months) from the region have been recorded for maturing and adult males.

In spite of the relatively closed nature of the community, genetic heterogeneity within the community is high (Duffield and Wells, 1986). Wells (1986) suggested that such high heterogeneity might be maintained if male movements outside of normal community ranges were a common feature of coastal bottlenose dolphin societies and if fertile matings occurred during the excursions. Thus, Sarasota females might be expected to be impregnated by males from other communities on occasion.

This community of about 100 animals was composed of smaller assemblages of socially interacting individuals. These groups were organized on the basis of age, sex, familial relationships, and reproductive condition. Groups of regular, long-term associates were generally of the same age and sex. Members of particular groups shared congruent ranges.

Female groups tended to be larger than male groups. One female group, at least, was composed of family members. Three generations of related females have been recorded from this group. Typically, all of the members of a particular female group did not swim together at any given time. Rather, several schools of females of similar reproductive condition were dispersed over the group's range. Different female groups had overlapping ranges, and females from different groups often swam with each other in these areas of overlap. However, different groups had different core areas in which most of their activities were concentrated.

Young dolphins remained with their natal groups for at least the first 3–4 years of their lives. During this time they typically swam alongside their mothers, with "babysitters," or in "playpens" with other calves, with their mothers nearby. Mothers and their calves were observed with progressively smaller numbers of animals per sighting with increasing calf age. Mothers with calves less than 1 year old tended to be found together more than with any other age or sex class. Male offspring were sighted with their mothers on occasion 10 or more years after birth.

Upon leaving their natal groups, calves joined subadult groups. Both males and females comprised these groups, but males greatly outnumbered females. These small groups frequently swam with similar groups, interspersing bouts of traveling and feeding with intensive socializing. Subadults used much the same ranges as the adult females, but they visited the extremes of the ranges more frequently.

Upon reaching sexual maturity, some, if not all, females rejoined their natal groups. A number of close male–male associations begun as subadults or earlier were maintained through the maturing phase and into adulthood. Maturing males were found in small groups of pairs or trios. Pairs of males formed the closest affinities of any unit except for mothers with their young calves.

Adult male groups concentrated most of their activities within the range of a single community. However, their ranges included those of a number of female groups, perhaps from more than one community. Associations between adult males and younger dolphins were infrequent. Associations between adult males and adult females were most frequent during the breeding season, and then they were most closely associated with receptive females. Adult males apparently engaged in at least two patterns of association with females. In the first, large, single males resided in the areas used most extensively by the major female groups. All of the associations with presumed receptive females involved these males. The second pattern involved pairs of adult males roving through large areas, associating briefly with one or a small number of females at a time.

4.2. Mating System

The mating system of wild dolphins has defied accurate description to date. Sexual behavior, including copulation, is frequently observed under a variety of circumstances in captivity and in the wild. Both heterosexual and homosexual behavior is evident during all seasons (McBride and Hebb, 1948; Brown and Norris, 1956; Saayman and Tayler, 1977; Ostman, 1985). Copulations occur between dolphins of all age classes (Tavolga, 1966). Two-day-old male bottlenose dolphins have exhibited erections, and copulations involving dolphins within their first few months of life have been recorded (McBride and Kritzler, 1951; M. C. Caldwell and Caldwell, 1972). The early development of sexual behavior, many years before sexual maturity, suggests that it is quite important in the lives of the animals. However, the frequency of occurrence of sexual behavior between adults outside of the breeding season, between individuals of different species, between individuals of the same sex, and involving immature individuals indicates that sexual behavior can occur in both reproductive and nonreproductive contexts.

Descriptions of mating systems of dolphins have suffered from an inability to factor out sexual interactions occurring within strictly social contexts from those involved in reproductive contexts. Seasonal variations in reproductive condition are well documented. However, the animals exhibit few visual indications of reproductive condition. In captivity, it has been possible to monitor reproductive condition precisely through regular measurements of steroid hormone concentrations in blood (*Tursiops:* Kirby and Ridgway, 1984; *Stenella:* Wells, 1984). Thus, to define mating systems, definition of partners during coinciding periods of estrus and maximum testicular activity would be of interest. The determination of paternities through analysis of blood samples from calves resulting from these combinations would be of particular interest. Unfortunately, all of the captive experiments to date that have involved simultaneous systematic behavioral observations and hormonal measurements have occurred in situations where only a single adult male was present.

In nature, four generalized mating systems are possible: monogamy, polyandry, polygyny, and promiscuity. Ideally, identification of a mating system is based on measurements of copulation frequencies and reproductive success (LeBoeuf, 1978). These kinds of data are not yet available for free-ranging dolphins. But the mating systems may be inferred because the four mating systems differ in the patterns of relative parental investment of each sex. Because parental investment is manifested in many animals by social bonding, inferences about dolphin mating systems may be drawn from the association patterns of adults.

The Sarasota dolphins do not exhibit any patterns of long-term pair bonding between adult males and females that would be diagnostic of a monogamous system. Individual males or small groups of males typically joined larger groups of females for only brief periods. Different adult males associated with a given adult, and presumably receptive, female from one day to the next.

A number of facts argue for a polygamous breeding system. The association of adult males and females mostly during the breeding season is indicative of low levels of male parental investment. Kenagy and Trombulak (1986) predicted that dolphins would exhibit extreme multimale breeding because of their unusually large testes. They calculated relative testes size for three delphinid species (not including *Tursiops*), and found the values to be 7.18–13.10 times the predicted size based on the general allometric relationship of mammalian testes mass/body mass. Their sample size was small, but the trend was consistent across five species of small odontocetes.

Members of the Genus *Tursiops* appeared to follow the general delphinid pattern for relative testes size. Data on testes mass and body mass for two adult male *Tursiops aduncus* (Ross, 1977) were used to calculate their position on a scale of relative testes size. The resulting values were 5.98 and 7.74 times the predicted values. In general, greater relative testes size is associated in mammals with multimale breeding systems. This is apparently a result of the selective pressures of multiple inseminations, sperm competition within the female reproductive tract, spontaneous ovulations, and seasonal reproduction (Harcourt et al., 1981; Harvey and Harcourt, 1984; Kenagy and Trombulak, 1986).

The bottlenose dolphin also shows extremely high sperm concentrations in the ejaculate. Schroeder et al. (1983) reported sperm counts of up to 6.298×10^{10} sperm per cm^3 of ejaculate for captive male bottlenose dolphins. This concentration is among the highest recorded for any mammal, and is approximately 300 times the mean concentration for humans and 100 times that of chimpanzees (Smith, 1984).

The fluid association patterns, in conjunction with the unusually high sperm-producing capabilities of the bottlenose dolphins, support a polygamous mating system hypothesis for these animals. Skewed adult sex ratios and differential mortality rates relative to sex, such as those observed for the Sarasota dolphins, are also characteristic of polygamous mating systems (Trivers, 1972, 1985; Ralls et al., 1980).

Polygamy is correlated with sexual dimorphism in many mammals. As Ralls et al. (1980, p. 239) state, "In species having promiscuous or polygynous mating systems, the number of matings a male may attain is potentially large and there is an increased selective premium on morphological, behavioral, and physiological characteristics that function in epigamic and intrasexual interactions directly related to mating." Some sexual size dimorphism exists in bottlenose dolphins, but it is muted in comparison with that of some other odontocetes, such as pilot whales, killer whales, and sperm whales. Adult male bottlenose dolphins tend to attain greater lengths than females (D. K. Caldwell and Caldwell, 1972). That the dimorphism may be a result of intrasexual selection is indicated by the heightened male aggression that has been commonly reported for *Tursiops* reaching sexual maturity (McBride and Hebb, 1948; McBride and Kritzler, 1951; Tavolga, 1966; Norris, 1967; M. C. Caldwell and Caldwell, 1977; Wood, 1977). However, data for polygynous pinnipeds suggest that size dimorphism may not be as important in sexual selection in water as on land (Stirling, 1983).

The variability of associations between adult males and receptive females does not facilitate distinguishing between the possibilities of one male mating with multiple females or one female mating with multiple males. But there is a strong tendency toward polygynous mating systems in the Mammalia (Eisenberg, 1981). In the extreme case, polygyny implies that a single male will control access to all of the receptive females in a group throughout the period of receptivity. Unfortunately, there are no data for the precise duration of the period of female bottlenose dolphin receptivity, though Sawyer-Steffan et al. (1983) and Kirby and Ridgway (1984) reported multiple ovulations by captive female *Tursiops* during a single breeding season.

One cannot rule out the possibility of a given male controlling access to a female during a relatively brief period of receptivity. In fact, captive adult males have been reported to maintain much longer associations with adult females during the breeding season than during the rest of the year (days or weeks versus minutes or hours; Tavolga and Essapian, 1957). However, given the brevity of the associations of adult males and females in the wild, and the fact that bottlenose dolphins appear to be spontaneous rather than induced ovulators (Kirby and Ridgway, 1984), it seems unlikely that a male could be assured of being present during the entire period of receptivity of a particular female. Perhaps such factors as the unrestricted movements of the females, the presence of other adult males, and/or simultaneous existence of receptive females in different schools inhibit the expression in the wild of the longer pair bonds reported from captivity.

Promiscuous and other polygamous mating systems can be distinguished on the basis of the formation of at least temporary pair bonds in the latter (Wilson, 1975, p. 327). The day-to-day changes in associations between adult male and female dolphins during the breeding season indicate that neither strict polygyny nor polyandry is the rule. Thus, we are left with promiscuity as the most likely mating system. The frequent movements of adult males from one female school to another during the breeding season suggest that the males are searching for mating opportunities. Because reproduction is roughly seasonal, and multiple ovulations per female may occur, it is likely that more than one female in the community may be receptive at any given time. Rather than investing their efforts in monopolizing access to any given receptive female, the Sarasota males appeared to be maximizing their opportunities for contact with a number of receptive females.

Promiscuous matings are not necessarily random matings. In fact, they are usually highly selective (Wilson, 1975). In theory, mating attempts by a number of males with a given female dolphin may also provide the female with opportunities to choose from a selection of males. Tayler and Saayman (1972) reported competition between adult females for attention from the adult males during the breeding season.

Observations from captivity support the hypothesis of a promiscuous mating system. Tayler and Saayman (1972, p. 21) observed a pair of captive males with females during the breeding season, and described a "rotating, alternating kind [of mating system] lacking in permanent sexual bonds" for *Tursiops aduncus*. McBride (1940) reported that two adult males captured together courted the same female, and engaged in practically no fighting. In both of these accounts, dominance of one adult male over the other was well established, but both males had mating opportunities. These anecdotes are of particular interest because of the existence in the Sarasota community of closely bonded male pairs that persist even during the breeding season.

Tests of the hypothesis of a promiscuous mating system require measurements of the relative reproductive success of males. Such measurements are possible through examination of genetic patterns from blood samples. Comparisons of results of chromosome banding pattern and electrophoretic analyses for calves, their mothers, and possible fathers can provide an indication of patterns of paternity. A demonstration of the impregnation of the receptive females in the community by a variety of males during a given breeding season would support the hypothesis of promiscuity. Additional support would be obtained if the number of calves sired by a particular male was independent of the amount of time the male associated with receptive females.

The promiscuity hypothesis would have to be reexamined relative to polygyny if, for example, all of the calves produced within a female group in a given year were sired by a single male. Bottlenose dolphins exhibit a high degree of behavioral plasticity (Shane et al., 1986). It is possible that this plasticity could be translated into the existence of more than one mating system at any given time, as demonstrated for chimpanzees (Tutin, 1979) and other animals (Emlen and Oring, 1977). Tests of the promiscuity hypothesis are in progress at this writing. Future research will emphasize focal animal behavioral observations of males over long periods. This will allow more precise definition of periods of association with females, and examination of male–male interactions for indications that direct or indirect control of access to females may be exerted.

4.3. Rearing System

The rearing of the young appeared to be entirely the responsibility of females, and it included both maternal and allomaternal care. The parental investment of the males apparently ends with conception. Such differential investment would be expected in a promiscuous mating system, where the nature of the mating system precluded paternal identification of their own offspring. The mother–calf bond is strong, with close associations lasting several years beyond weaning.

The nursery school appears to provide a protected environment within which early postnatal development and learning occur. McBride and Kritzler (1951) observed that the parent and other adult female captive dolphins showed great solicitude for the protection of the newborn young. Leatherwood (1977) noted that subgroups containing small bottlenose dolphin calves were located away from the perimeter of free-ranging schools, and mothers shielded the young from aversive stimuli. The protective function of dolphin schools has been discussed at length by Norris and Dohl (1980) and Wells et al. (1980). As summarized by Norris and Dohl (p. 248, 1980), "The variation in school size is exactly what would be predicted if predation were the basis for the degree of schooling tendency (Williams, 1964; Hamilton, 1971)."

Nursery schools were prominent features of the Sarasota community's social structure. Schools containing the youngest calves were, on average, the largest schools observed off Sarasota. School size decreased regularly with increasing calf age (Fig. 5). It may be hypothesized that this pattern might be expected if there existed an optimal school size for females without dependent calves. This optimal school size of approximately 10–12 dolphins might be based at least in part on the need for protection from predators and mediated by resource availability. As the calves' dependence decreased and they began to contribute to the integrated functions of the schools, the overall school size decreased, but the total number of contributing members remained constant.

One measure of the effectiveness of nursery schools in protection from predation is the incidence of shark bite scarring. Of 86 dolphins examined during captures from 1975–1985, 22 were calves, and none of these calves exhibited obvious shark bite scars. Of the remaining noncalves, 21.9% wore obvious shark bite scars, and others were scarred in a manner suggestive of shark bites. An alternative explanation for the low incidence of scarring on calves is that these small animals were more likely to be killed than were larger dolphins when they were attacked. However, the disappearance of only two dependent calves has been documented during 1980–1984. In contrast, 16 calves were known to survive beyond 3 years of age during the same period. Thus, it seems that the nursery schools provided an effective defense against predation.

The lengthy period of close association between mother and calf beyond the 1.5-year nursing period (McBride and Kritzler, 1951) indicates that the bond is based on more than simply the nutritional needs of the calf. Norris and Dohl (1980) discussed the importance of learning in the developmental process of schooling cetaceans. Many, if not most, of the activities of everyday dolphin life appear to have a learned component. The 3-year or longer period of the mother–calf bond is likely when the calves learn or refine their knowledge of such things as echolocation, other individual and group foraging and feeding techniques (Norris and Prescott, 1961), recognition of the 100 or so other community members, home range features and limits, recognition of adjacent communities, resource distribution, predator avoidance, patterns of social interaction, and sensory integration with other school members.

4.4. Comparisons with Other Mammals

Beyond certain basic differences imposed by living in different media, large mammals in both aquatic and terrestrial environments face many of the same environmental pressures related to sociality. The pervasiveness of patterns of social responses to similar ecological factors both within and across taxonomic groups of terrestrial mammals (for example, Eisenberg et al., 1972; Crook et al., 1976) suggests that analogous social solutions should be exhibited by some aquatic mammals as well. Thus, it is perhaps not surprising to find that coastal bottlenose dolphins share many features of their social structure in common with a large, terrestrial carnivore such as the Serengeti lion, *Panthera leo*. Both species are top-level predators in relatively open habitats. Both species engage in cooperative hunting, but for both species, prey capture is done typically on an individual basis. Increased hunting efficiency through cooperation is apparently one of the factors that has led to group formation in lions (Kleiman and Eisenberg, 1973); it is likely that the same is true for dolphins (Norris and Dohl, 1980).

Schaller (1972, p. 33) described "prides" of lions as the resident lionesses and their cubs, which share the pride area and interact peacefully. Schaller also included attending males in this definition, but his subsequent description of male behavior indicated that males could be considered separately, as in the case of the Sarasota dolphins. Much like the female groups of dolphins, different prides had ranges that overlapped, but each emphasized different core areas. Other pride features include: (1) all pride lionesses are directly related and consist of daughters, mothers, grandmothers, and perhaps another generation, (2) no matter how widely females are scattered or how frequently they meet each of the other members, they still constitute a closed social unit that strange lionesses are not permitted to join, (3) the most potent factor for drawing lionesses together is the presence of small cubs, (4) although each pride member readily associates with every other member, some, especially members of the same age and sex, are together more often than others, (5) males may be alone or together, with females or without them, (6) males often form long-lasting companionships with brothers or male pride-mates of similar age, (7) male bonds with females are not as close as with other males; male contacts with females remain casual except during courtship, (8) both nomad and resident lions inhabited the Serengeti; estrus females accept either resident or nomad males, and (9) males remain with females during the period of estrus.

Some of the pride features listed above are directly comparable to dolphin group features; other comparisons require additional data. The description of lion prides, where female group composition varied from day to day within the more permanent structure of an extended family appears to parallel dolphin group structure much more closely than do the more stable female-centered family units within kinship groups of elephants (*Loxodonta africana*, Douglas-Hamilton, 1973; Douglas-Hamilton and Douglas-Hamilton, 1975). More information on dolphin familial relationships and interactions are required to accurately place the dolphin groups in perspective relative to other mammalian social units.

In spite of the many similarities in their social structures, lions and dolphins apparently differ in their mating systems. Lions are polygynous, whereas bottlenose dolphins appear to be promiscuous. The marked differences may be related to several factors. First, defense of multiparous females may be a better investment for male lions than defense of the uniparous females would be in the dolphins. Second, the open habitat of the Serengeti may facilitate visual detection of competitors by males for the control of access to an estrus female lion. Murky coastal waters and brief copulations might render such detection and defense much more difficult for bottlenose dolphins.

The apparent convergence of the social structures of dolphins and lions is intriguing. However, at our present level of knowledge about the behavioral and ecological relationships of free-ranging dolphins these comparisons are necessarily simplistic. More rigorous comparisons with lions and other mammals can lead to well-founded hypotheses of the evolutionary basis of the bottlenose dolphin social system. Field studies to develop and test these hypotheses will provide a more complete understanding of the lives of these animals within a general mammalian context.

ACKNOWLEDGMENTS. Field work during 1980–1986 was supported in part by Earthwatch/Center for Field Research, the Southeast and Southwest Fisheries Centers of the National Marine Fisheries Service, the University of California at Santa Cruz, the Sirenia Research Group of the U. S. Fish and Wildlife Service, the Inter-American Tropical Tuna Commission, and private contributions of funds, equipment, and time to Dolphin Biology Research Associates.

The efforts of many people have seen this research program successfully through its first 16 years. The technical expertise and seminal advice of J. Chamberlin-Lea, D. Duffield, A. Hohn, V. Kirby, and J. Sweeney have greatly expanded both the data base of dolphin ages, relationships, and reproductive conditions for the Sarasota community, and, consequently, our understanding of the dolphins' social system. Major contributions of field support were provided by R. Arden, D. Black, K. Fischer, L. Fulford, S. Kruse, L. Mayall, P. Page, E. Patterson, A. Read, R. Spaulding, P. Tyack, F. Wells, J. Wells, F. Worl, J. Zaias, and 93 Earthwatch volunteers. Figure preparation was done in part by K. Miller and E. Mathews. Review of the manuscript by K. Norris, B. LeBoeuf, W. Perrin, and A. Zihlman is greatly appreciated.

REFERENCES

Asper, E. D., 1975, Techniques of live capture of smaller Cetacea, *J. Fish. Res. Board Can.* **32**:1191–1196.

Balcomb, K. C., Boran, J. R., Osborne, R. W., and Haenel, N. J., 1980, Observations of Killer Whales (*Orcinus orca*) in Greater Puget Sound, State of Washington, National Technical Information Service PB-224728, U.S. Department of Commerce, Springfield, Virginia, pp. 1–42.

Bigg, M. A., 1982, An assessment of killer whale (*Orcinus orca*) stocks off Vancouver Island, British Columbia, *Rep. Int. Whaling Comm.* **32**:655–666.

Bigg, M. A., McAskie, I. B., and Ellis, G., 1976, Abundance and Movements of Killer Whales off Eastern and Southern Vancouver Island with Comments on Management, Preliminary Report, Arctic Biological Station, Ste. Anne de Bellevue, Quebec, Canada.

Brown, D. H., and Norris, K. S., 1956, Observations of captive and wild cetaceans, *J. Mammal.* **37**:311–326.

Caldwell, D. K., and Caldwell, M. C., 1972, *The World of the Bottlenosed Dolphin*, Lippincott, New York.

Caldwell, M. C., and Caldwell, D. K., 1972, Behavior of marine mammals, in: *Mammals of the Sea: Biology and Medicine* (S. H. Ridgway, ed.), C. C. Thomas, Springfield, Illinois, pp. 419–465.

Caldwell, M. C., and Caldwell, D. K., 1977, Social interactions and reproduction in the Atlantic bottlenosed dolphin, in: *Breeding Dolphins: Present Status, Suggestions for the Future* (S. H. Ridgway and K. W. Benirschke, eds.), National Technical Information Service PB-273673, U. S. Department of Commerce, Springfield, Virginia, pp. 133–142.

Clutton-Brock, T. H., 1974, Primate social organization and ecology, *Nature* **250**:539–542.

Crook, J. H., and Gartlan, J. S., 1966, Evolution of primate societies, *Nature* **210**:1200–1203.

Crook, J. H., Ellis, J. E., and Goss-Custard, J. D., 1976, Mammalian social systems: Structure and function, *Anim. Behav.* **24**:261–274.

Douglas-Hamilton, I., 1973, On the ecology and behaviour of the Lake Manyara elephants, *East Afr. Wildl. J.* **11**:401–403.

Douglas-Hamilton, I., and Douglas-Hamilton, O., 1975, *Among the Elephants*, Penguin Books, Harmondsworth, England.

Duffield, D. A. and Wells, R. S., 1986, Population Structure of Bottlenose Dolphins: (2) Genetic Studies of Bottlenose Dolphins Along the Central West Coast of Florida, Final Report to National Marine Fisheries Service, Southeast Fisheries Center, Contract No. 45-WCNF-5-00366, pp. 1–10.

Duffield, D. A., Chamberlin-Lea, J., Wells, R. S., and Scott, M. D., 1985, Inheritance of an extra chromosome in a resident female band of bottlenose dolphins in Sarasota, Florida (Abstract), in: *Proceedings 6th Biennial Conference on the Biology of Marine Mammals*, Vancouver, British Columbia.

Eisenberg, J. F., 1981, *The Mammalian Radiations*, University of Chicago Press, Chicago.

Eisenberg, J. F., Muckenhirn, N. A., and Rudran, R., 1972, The relation between ecology and social structure in primates, *Science* **1976**:863–874.

Emlen, S. T., and Oring, L. W., 1977, Ecology, sexual selection, and the evolution of mating systems, *Science* **197**:215–223.

Goodall, J., 1983, Population dynamics during a 15 year period in one community of free-living chimpanzees in the Gombe National Park, Tanzania, *Z. Tierpsychol.* **61**:1–60.

Hamilton, W. D., 1971, Geometry for the selfish herd, *J. Theor. Biol.* **31**:295–311.

Harcourt, A. H., Harvey, P. H., Larson, S. G., and Short, R. V., 1981, Testis weight, body weight and breeding system in primates, *Nature* **293**:55–57.

Harrison, R. J., and Ridgway, S. H., 1971, Gonadal activity in some bottlenose dolphins (*Tursiops truncatus*), *J. Zool. Lond.* **165**:355–366.

Harrison, R. J., Brownell, R. L., Jr., and Boice, R. C., 1972, Reproduction and gonadal appearances in some odontocetes, in: *Functional Anatomy of Marine Mammals*, Volume 1 (R. J. Harrison, ed.), Academic Press, London, pp. 362–429.

Harvey, P. H., and Harcourt, A. H., 1984, Sperm competition, testes size, and breeding systems in primates, in: *Sperm Competition and the Evolution of Animal Mating Systems* (R. L. Smith, ed.), Academic Press, Orlando, pp. 589–600.

Hohn, A. A., 1980, Age determination and age related factors in the teeth of western North Atlantic bottlenose dolphins, *Sci. Rep. Whales Res. Inst.* **32**:39–66.

Hohn, A. A., 1985, Comparison of methods to estimate the average age at sexual maturation in dolphins (Abstract), in: *Proceedings 6th Biennial Conference on the Biology of Marine Mammals*, Vancouver, British Columbia.

Irvine, A. B., and Wells, R. S., 1972, Results of attempts to tag Atlantic bottlenosed dolphins, *Tursiops truncatus*, *Cetology* **13**:1–5.

Irvine, A. B., Scott, M. D., Wells, R. S., and Kaufmann, J. H., 1981, Movements and activities of the Atlantic bottlenose dolphin, *Tursiops truncatus*, near Sarasota, Florida, *Fish. Bull. U. S.* **79**:671–688.

Irvine, A. B., Wells, R. S., and Scott, M. D., 1982, An evaluation of techniques for tagging small odontocete cetaceans, *Fish. Bull. U. S.* **80**:135–143.

Jarman, P. J., 1974, The social organisation of antelope in relation to their ecology, *Behaviour* **58:**215–267.

Kasuya, T., and Marsh, H., 1984, Life history and reproductive biology of the short-finned pilot whale, *Globicephala macrorhynchus*, off the Pacific coast of Japan, in: *Reproduction in Whales, Dolphins and Porpoises* (W. F. Perrin, R. L. Brownell, Jr., and D. P. DeMaster, eds.), Reports of the International Whaling Commission, Special Issue 6, Cambridge, England, pp. 259–310.

Kenagy, G. J., and Trombulak, S. C., 1986, Size and function of mammalian testes in relation to body size, *J. Mammal.* **67:**1–22.

Kirby, V. L., and Ridgway, S. H., 1984, Hormonal evidence of spontaneous ovulation in captive dolphins, *Tursiops truncatus* and *Delphinus delphis*, in: *Reproduction in Whales, Dolphins and Porpoises*, (W. F. Perrin, R. L. Brownell, Jr., and D. P. DeMaster, eds.), Reports of the International Whaling Commission, Special Issue 6, Cambridge, England, pp. 459–464.

Kleiman, D. G., and Eisenberg, J. F., 1973, Comparisons of canid and felid social systems from an evolutionary perspective, *Anim. Behav.* **21:**637–659.

Leatherwood, J. S., 1977, Some preliminary impressions on the numbers and social behavior of free-swimming, bottlenosed dolphin calves (*Tursiops truncatus*) in the northern Gulf of Mexico, in: *Breeding Dolphins: Present Status, Suggestions for the Future* (S. H. Ridgway and K. W. Benirschke, eds.), National Technical Information Service PB-273673, U.S. Department of Commerce, Springfield, Virginia, pp. 143–167.

LeBoeuf, B. J., 1978, Sex and evolution, in: *Sex and Behavior* (T. E. McGill, D. A. Dewsbury, and B. D. Sachs, eds.), Plenum Press, New York, pp. 3–33.

McBride, A. F., 1940, Meet Mr. Porpoise, *Nat. Hist.* **45:**16–29.

McBride, A. F., and Hebb, D. O., 1948, Behavior of the captive bottlenose dolphin, *Tursiops truncatus*, *J. Comp. Physiol. Psychol.* **41:**111–123.

McBride, A. F., and Kritzler, H., 1951, Observations on pregnancy, parturition, and postnatal behavior in the bottlenose dolphin, *J. Mammal.* **32:**251–266.

Miyazaki, N., and Nishiwaki, M., 1978, School structure of the striped dolphin off the Pacific coast of Japan, *Sci. Rep. Whales Res. Inst.* **30:**65–115.

Norris, K. S., 1967, Aggressive behavior in Cetacea, in: *Aggression and Defense* (C. D. Clemente and D. B. Lindsley, eds.), University of California Press, Berkeley, pp. 225–241.

Norris, K. S., and Dohl, T. P., 1980, The structure and functions of cetacean schools, in: *Cetacean Behavior: Mechanisms and Functions* (L. M. Herman, ed.) Wiley, New York, pp. 211–261.

Norris, K. S., and Prescott, J. H., 1961, Observations on Pacific cetaceans of Californian and Mexican waters, *Univ. Calif. Publ. Zool.* **63:**291–402.

Norris, K. S., Würsig, B., Wells, R. S., Würsig, M., Brownlee, S. M., Johnson, C., and Solow, J., 1985, The Behavior of the Hawaiian Spinner Dolphin, *Stenella longirostris*, National Marine Fisheries Service, Southwest Fisheries Center Administrative Report LJ-85-06C.

Ostman, J., 1985, An ethogram for dolphin social behavior, and observations on changes in aggressive and homosexual behavior among two subadult male bottlenose dolphins (*Tursiops truncatus*) in a captive colony, M. A. thesis, San Francisco State University.

Perrin, W. F., and Reilly, S. B., 1984, Reproductive parameters of dolphins and small whales of the family Delphinidae, in: *Reproduction in Whales, Dolphins and Por-*

poises (W. F. Perrin, R. L. Brownell, and D. P. DeMaster, eds.), Reports of the International Whaling Commission, Special Issue 6, Cambridge, England, pp. 97–133.

Ralls, K., Brownell, R. L., Jr., and Ballou, J., 1980, Differential mortality by sex and age in mammals, with specific reference to the sperm whale, in: *Reports of the International Whaling Commission, Special Issue 2*, pp. 233–243.

Ridgway, S. H., and Benirschke, K. (eds.), 1977, *Breeding Dolphins: Present Status, Suggestions for the Future*, National Technical Information Service PB-273673, U.S. Department of Commerce, Springfield, Virginia.

Ross, G. J. B., 1977, The taxonomy of bottlenosed dolphins *Tursiops* sp. in South African waters, with notes on their biology, *Ann. Cape Prov. Mus. Nat. Hist.* **11:**135–194.

Saayman, G. S., and Tayler, C. K., 1977, Observations on the sexual behavior of Indian Ocean bottlenosed dolphins (*Tursiops aduncus*), in: *Breeding Dolphins: Present Status, Suggestions for the Future* (S. H. Ridgway and K. W. Benirschke, eds.), National Technical Information Service PB-273673, U.S. Department of Commerce, Springfield, Virginia, pp. 113–129.

Sawyer-Steffan, J. E., Kirby, V. L., and Gilmartin, W. G., 1983, Progesterone and estrogens in the pregnant and non-pregnant dolphin, *Tursiops truncatus*, and the effect of induced ovulation, *Biol. Reprod.* **28:**897–901.

Schaller, G. B., 1972, *The Serengeti Lion: A Study of Predator–Prey Relations*, University of Chicago Press, Chicago.

Schroeder, J. P., Keller, K., and Kirby, V. L., 1983, Testosterone levels and sperm production in *Tursiops truncatus* (Abstract), in: *Proceedings 5th Biennial Conference on the Biology of Marine Mammals*, Boston, p. 91.

Sergeant, D. E., Caldwell, D. K., and Caldwell, M. C., 1973, Age, growth, and maturity of bottlenosed dolphins (*Tursiops truncatus*) from northeast Florida, *J. Fish. Res. Board Can.* **30:**1009–1011.

Shane, S. H., Wells, R. S., and Würsig, B., 1986, Ecology, behavior, and social organization of the bottlenose dolphin: A review, *Mar. Mammal Sci.* **2:**34–63.

Smith, R. L., 1984, Human sperm competition, in: *Sperm Competition and the Evolution of Animal Mating Systems* (R. L. Smith, ed.), Academic Press, Orlando, pp. 601–659.

Stirling, I., 1983, The evolution of mating systems in pinnipeds, in: *Advances in the Study of Mammalian Behavior* (J. F. Eisenberg and D. G. Kleiman, eds.), *American Society of Mammalogists*, pp. 489–527.

Tavolga, M. C., 1966, Behavior of the bottlenose dolphin (*Tursiops truncatus*): Social interactions in a captive colony, in: *Whales, Dolphins, and Porpoises* (K. S. Norris, ed.), University of California Press, Berkeley, pp. 718–730.

Tavolga, M. C., and Essapian, F. S., 1957, The behavior of the bottlenosed dolphin (*Tursiops truncatus*): Mating, pregnancy, parturition and mother–infant behavior, *Zoologica* **42:**11–31.

Tayler, C. K., and Saayman, G. S., 1972, The social organization of dolphins, (*Tursiops aduncus*) and baboons (*Papio ursinus*): Some comparisons and assessments, *Ann. Cape Prov. Mus. Nat. Hist.* **9:**11–49.

Trivers, R. L., 1972, Parental investment and sexual selection, in: *Sexual Selection and the Descent of Man 1871–1971* (B. Campbell, ed.), Aldine, Chicago, pp. 136–179.

Trivers, R. L., 1985, *Social Evolution*, Benjamin/Cummings, Menlo Park, California.

Tutin, C. E. G., 1979, Mating patterns and reproductive strategies in a community of wild chimpanzees (*Pan troglodytes schweinfurthii*), *Behav. Ecol. Sociobiol.* **6:**29–38.

Wells, R. S., 1978, Home range characteristics and group composition of Atlantic bottlenose dolphins, *Tursiops truncatus*, on the west coast of Florida, M. Sc. thesis, University of Florida, Gainesville.

Wells, R. S., 1984, Reproductive behavior and hormonal correlates in Hawaiian spinner dolphins, *Stenella longirostris*, in: *Reproduction in Whales, Dolphins and Porpoises* (W. F. Perrin, R. L. Brownell, and D. P. DeMaster, eds.), Reports of the International Whaling Commission, Special Issue 6, Cambridge, England, pp. 465–472.

Wells, R. S., 1986, Population structure of bottlenose dolphins: (1) Behavioral studies along the central west coast of Florida, Final Report to the National Marine Fisheries Service, Southeast Fisheries Center, Contract No. 45-WCNF-5-00366, pp. 1–70.

Wells, R. S., Irvine, A. B., and Scott, M. D., 1980, The social ecology of inshore odontocetes, in: *Cetacean Behavior: Mechanisms and Functions* (L. M. Herman, ed.), Wiley, New York, pp. 263–317.

Williams, G. C., 1964, Measurement of consociation among fishes and comments on the evolution of schooling, *Mich. State Mus. Publ. Biol. Sci.* **2**:351–383.

Wilson, E. O., 1975, *Sociobiology: The New Synthesis*, Harvard University Press, Cambridge, Massachusetts.

Wood, F. G., 1977, Births of dolphins at Marineland of Florida, 1939–1969, and comments on problems involved in breeding of small Cetacea, in: *Breeding Dolphins: Present Status, Suggestions for the Future* (S. H. Ridgway and K. W. Benirschke, eds.), National Technical Information Service PB-273673, U.S. Department of Commerce, Springfield, Virginia, pp. 47–60.

Wrangham, R. W., 1980, An ecological model of female-bonded primate groups, *Behaviour* **75**:262–299.

Würsig, B., 1978, Occurrence and group organization of Atlantic bottlenose porpoises (*Tursiops truncatus*) in an Argentine Bay, *Biol. Bull.* **154**:348–359.

Würsig, B., and Würsig, M., 1977, The photographic determination of group size, composition, and stability of coastal porpoises (*Tursiops truncatus*), *Science* **198**:755–756.

Zar, J. H., 1974, *Biostatistical Analysis*, Prentice-Hall, Englewood Cliffs, New Jersey.

CHAPTER 8

ECOLOGICAL, MORPHOLOGICAL, AND BEHAVIORAL CONVERGENCE IN ROCK-DWELLING MAMMALS

MICHAEL A. MARES
and THOMAS E. LACHER, JR.

1. INTRODUCTION

Much recent research has attempted to document and clarify the process of convergent evolution at various levels of biological organization. Studies have ranged from detailed examples of convergence at the species level (for example, Pianka and Pianka, 1970), through examples of similar evolutionary pathways having been taken by extensive subsets of communities developing within similar macrohabitats (for example, Mares, 1976, 1980, 1983), to an assessment of generally similar patterns of development of entire disjunct ecosystems sharing numerous abiotic parameters (for example, Orians and Solbrig, 1977). There has been some controversy regarding the level of ecological similarity that has developed within unrelated faunas (for example, Schall and Pianka, 1978), and whether or not convergence actually occurs among a broad array of species evolving within apparently similar micro- and macrohabitats in response to what would presumably be broadly similar selective regimes.

MICHAEL A. MARES • Stovall Museum and Department of Zoology, University of Oklahoma, Norman, Oklahoma 73019. THOMAS E. LACHER, JR. • Huxley College of Environmental Studies, Western Washington University, Bellingham, Washington 98225.

One uncommon habitat that occurs with regularity throughout the world is that of isolated accumulations of rocks and boulders. These may appear as massive boulder piles rising above an otherwise level surface—for example, outcrops in the scrubland of northeastern Brazil (Fig. 1) or the kopjes of the scrub plains of East Africa (Fig. 2) (Hoeck, 1975)—or as extensive rock faces with interspersed boulder piles that characterize such habitats as the Andean ridges of central South America (for example, Fig. 3). There are also other rock habitats frequented by mammals, including the talus slopes formed from extruding rock that are common in the western mountains of North America (Fig. 4).

Because these localized rocky microhabitats often exhibit a microclimate distinct from the surrounding region, they frequently support a characteristic flora and fauna. For example, in the Caatinga of northeastern Brazil, a semiarid region renowned for its unpredictable extreme droughts (Markham, 1972), the rock outcrops offer a microclimate that is more mesic than the adjacent thorn scrub, and generally supports green vegetation throughout the year (for example, Lacher, 1981; Streilein, 1982a, 1982b). These isolated boulder piles, which encompass only a minuscule portion of the areal extent of the Caatinga biome, in fact make an unusually important contribution to the faunal diversity of this region. For example, a number of mammal species tend to be more common in the rockpiles than in other areas (Mares et al., 1981a), while during the driest times of the year these mesic microhabitats are the only areas where some species are able to persist (Streilein, 1982a).

Certain species of animals occur only in the rockpiles of the Caatinga, for example, the buthid scorpions *Rhopalurus rochai* and *R. laticaudata* (L. Vitt, personal communication), the iguanid lizard *Platynotus semitaeniatus* (Vanzolini et al., 1980; Vitt, 1981), the molossid bat *Molossops mattogrossensis* (Willig, 1982, 1983), and the caviid rodent *Kerodon rupestris*, which is the only endemic mammal occurring in the Caatinga (Lacher, 1981). Thus, rockpiles within the Caatinga offer a highly specialized environment that is distinctive in structure, climate, vegetation, and faunal composition from the surrounding area. As such, we might expect natural selection to favor particular traits that facilitate survival within such a distinctive habitat. Moreover, we might expect that other rocky habitats scattered throughout the world would evince similarly distinctive microclimates offering favorable conditions to those organisms able to adapt to this specialized environment.

In this paper, we examine the fauna of largely rodentiform medium-sized mammals inhabiting rocky areas like those described above. Because this microhabitat is so widely dispersed, the fauna we examine is taxonomically diverse, including four orders, 11 families, 17 genera,

FIGURE 1. An isolated rockpile, or lajeiro, in the Caatinga of northeastern Brazil. Male caviid rodents, *Kerodon rupestris*, defend boulder piles within the larger rock agglomeration. The relatively lusher vegetation of the boulder pile is evident when compared to the surrounding thorn scrub. A second caviid species, *Galea spixii*, inhabits the scrub areas around and between the boulders. (Photograph by Thomas E. Lacher, Jr.)

FIGURE 2. Inselbergs, or kopjes, in East Africa. These habitats rise abruptly from the intervening grassland (or, in some areas, thornscrub) and support a distinctive flora. Hyraxes (*Procavia* and *Heterohyrax*) defend isolated boulder piles against conspecifics. (Photograph by Ron Tilson.)

FIGURE 3. The montane rocky habitat of *Lagidium* in southern Argentina (Mendoza Province), with the extensive boulder faces visible along the ridge line of the mountains in the center of the photograph. (Photograph by Michael A. Mares.)

FIGURE 4. Talus slopes on the Cascade Mountains of Washington in the northwestern United States. The isolated extrusions of talus rock are defended by *Marmota caligata* and are also inhabited by pikas (*Ochotona princeps*). The rockpiles primarily offer protection from adverse weather and from predation for both of these taxa. (Photograph by John Miles.)

and 29 species from all continents except Antarctica. Nevertheless, all species examined share the problems inherent in adapting to the specialized rockpile habitat. Our aim is one of assessing the degree of convergent evolution that might have occurred among the members of this disparate assemblage and we will examine morphological, ecological, reproductive, and behavioral aspects of their biology.

2. METHODS

2.1. Introduction

The major sources of data utilized in this analysis are the many papers published on the biology of the various species (citations below). In addition, we have first-hand experience with several species, especially *Kerodon rupestris* (for example, Lacher, 1979, 1980, 1981; Lacher et al., 1982), *Galea spixii* (Lacher, 1981; Mares et al., 1981a), and *Lagidium viscacia* (one of us, M. A. M., has observed *Lagidium* in the field in Argentina).

We were especially concerned with examining as many important ecological, morphological, reproductive, and behavioral characteristics as possible for the organisms included in this study. We here detail the characteristics we examined and the evolutionary importance of these traits. We have attempted to piece together disparate observations on the various species to ascertain whether or not a clear pattern of adaptation emerges. We collected the same information (ecology, morphology, and behavior) on a second group of non-rock-dwelling mammals. This group, which contained four orders, eight families, and eight genera, consisted of species chosen because each was phylogenetically closely related to a rock specialist and/or occurred in the same general area as the rock-dwelling species. We hypothesized that, despite the phylogenetic similarity, this group would share fewer ecological and behavioral traits with the rock-dwellers than the rock-dwellers would share among themselves. These results will obtain if the level of convergence is sufficient to override phylogeny. We have restricted our analyses to a consideration of animals ranging from guinea-pig size to small wallabies because smaller species may be small enough to perceive the rock pile habitat as being relatively less patchy in distribution. A major part of our hypothesis is predicated on the patchiness of the rock habitat influencing the evolution of certain traits.

Sixteen taxa of rock-dwelling mammals and eight taxa of non-rock-dwellers were compared using multivariate methods. Nineteen qualitative characters reflecting ecological, morphological, and behavioral adaptations were chosen. Data were recorded either as multistate variables or as presence, absence, or no comparison (NC) for each character of every taxon and were obtained from available literature sources. Simple matching coefficients were produced as measures of resemblance. These coefficients are pair functions that measure the agreement between pairs of OTUs over an array of two- or multistate characters (Sneath and Sokal, 1973). The unweighted pair-group method using arithmetic averages (UPGMA) was used as the clustering algorithm and a cophenetic correlation coefficient was generated to indicate how accurately the resulting phenogram reflected the original resemblance matrix.

2.2. Characters Examined

We describe characters that appear to have a certain a priori importance in the adaptation of medium-sized mammals to a saxicolous existence. Alternate states of each character are also described.

2.2.1. Ecological Characteristics

2.2.1a. Habitat. The habitat type within which the rocks are located might affect the degree to which organisms become specialized for life in the rocks. Rocky habitats will likely be an important mesic enclave in habitats that tend to be dry (that is, rocks may not offer such enclaves in mesic savannas, but may in dry grasslands). Rocks may not offer many advantages in mature forests, other than as a temporary shelter. We examined species for occurrence in one of five broad habitat classes: montane, forest, shrubland, grassland, and desert.

2.2.1b. Rainfall. Because the importance of rocks as an enclave may be associated with rainfall, we classified the habitat of each species by the amount of rainfall: <400 mm, 400–1000 mm, >1000 mm.

2.2.1c. Diet. We examined diet only at the broadest levels, being concerned with whether a species was herbivorous, insectivorous, or omnivorous. Rockpiles often offer mesic microclimates where green vegetation can flourish in an otherwise xeric environment and where insects might be more abundant than in the surrounding habitat. However, all species selected for our analyses were herbivores, and thus this trait was not utilized in analyses.

2.2.1d. Patchy Habitat—Defensible Resource. Emlen and Oring (1977) have hypothesized that in microhabitats that are especially patchy, the habitat itself (or portions thereof) can be an important limiting resource. If this resource occurs in patches that are sufficiently small to be defended by a single male, it is theoretically possible for a male to control access to the resource and thereby accrue a group of females for breeding purposes. Emlen and Oring termed this process resource defense polygny.

2.2.1e. Lookout Posts. Wilson (1975) noted that lookout posts often characterize open country animals. These are apparently used primarily to detect potential predatory attacks. An important aspect of rock outcrops is protection from predation and the behavior of using rocks as lookout posts may be prevalent in rock-dwellers.

2.2.1f. Arboreality. Species that live in rocky areas often have little available forage. They either need to range out of the rocky areas, thereby losing the protection of the rocks, or they may forage in the trees that are frequently encountered growing in the cracks and fissures of kopjes and outcrops. We hypothesize that rock-dwelling mammals will show pronounced climbing ability.

2.2.1g. Seasonality in Precipitation. Because rock outcrops are frequently mesic refugia, they might be a valuable habitat resource in areas that have seasonal rainfall patterns. We expect that rock specialists will be most common in regions having marked wet and dry seasons.

2.2.1h. Seasonality in Temperature. Areas that have year-round high temperatures have a higher evaporation potential and therefore effectively lower annual precipitation. This would enhance the value of rock outcrops as refugia, and rock specialists would be more likely to evolve under this condition.

2.2.2. Reproductive and Morphological Characteristics

2.2.2a. Litter Size. The number of young that are produced at birth is an important ecological characteristic of a species. Basic adaptational strategies such as *r*- or *K*-selected life history parameters (MacArthur, 1972) can be reflected in the litter size. We chose two categories: generally one offspring; greater than one offspring. Rock-dwellers, utilizing a more stable resource base, might be expected to have smaller litter sizes.

2.2.2b. Precocial–Altricial Young. Eisenberg (1981) has an extensive discussion on the evolution of precocial young. Altricial young might be expected in species having extended parental care (Eisenberg, 1981). Precocial young might be expected in more open, unprotected habitats.

2.2.2c. Reduced Tail. The tail of mammals can be modified to fit many needs, from that of a prehensile appendage to one offering a site for fat storage. In many arboreal small mammals the tail is an important balancing organ and may be exceptionally elongated, whereas in other species, tail length may be reduced, or the tail may be eliminated altogether. Long tails would seem to be a disadvantage to animals that live in rock fissures.

2.2.2d. Padded Feet. In rock-dwelling mammals, the feet may be exceptionally padded. Whether or not such padding functions to absorb shocks, to increase traction, or both is unclear. Nevertheless, mammals inhabiting rocky areas often develop a jumping behavior that is used in springing from one rock to another, perhaps necessitating some type of foot protection and enhanced traction.

2.2.2e. Clawless–Reduced Claws–Grooming Claw. Rock-inhabiting mammals often evidence a reduction in the size and length of claws. It might be expected that long claws would be of limited use to an animal inhabiting a substrate that is as hard and abrasive as a rock, particularly when a major locomotor pattern may involve extended jumps from one rock to another. Nevertheless, since mammals use the feet and claws for most of their grooming, we may expect a compromise between nail loss and reduction, such as by the development of a single grooming claw or similar structure.

2.2.3. Behavioral Characteristics

2.2.3a. Sociality and Mating System. Included within these general categories are such traits as a solitary existence, coloniality, family groups, or multiple females. The association between an open country existence and coloniality has already been noted. Also included are various mating systems, such as monogamy, promiscuity, and polygyny.

2.2.3b. Alarm Whistle. Alarm whistles to signal impending danger are utilized by animals sitting at lookout posts in open habitats having rocky promontories or other elevated sites rising above the general level of the land. Such alarm calls often result in all animals seeking cover or freezing in a position of alertness (Wilson, 1975).

2.2.3c. Communal Defecation. In at least some species of rock-dwelling mammals, there appears to be a tendency for fecal material to be deposited in communal piles. Such piles may act as olfactory cues associated with territorial marking, especially since rock surfaces offer a poor substrate for the more traditional scent marking via glandular secretion.

3. SPECIES ACCOUNTS

All species considered in this report are listed in Table I. Given that this is a disparate assemblage of poorly known species, we here review aspects of their biology.

3.1. Order Marsupialia

3.1.1. Family Petauridae

Pseudocheirus dahli. The nocturnal, rock-haunting, ring-tailed possum is a large-bodied (325–450 mm), long-tailed (225–275 mm) marsupial. The body is gray with a rufous rump. There are five bluntly clawed digits on the fore and hind feet. The face is strongly reminiscent of the face of the hyrax. Newborn are altricial, and litter size is unknown (Walker, 1975). They are found in rock outcrops in savannas of northern and northwestern Australia (Corbet and Hill, 1980). When pursued, they flee into narrow crevices, where they lodge themselves. They are arboreal, and forage on leaves and fruit. It has been suggested that they form pairs or family groups, but data are scanty with regard to their social system (Troughton, 1943; Walker, 1975).

3.1.2. Family Macropodidae

Peradorcas concinna. Head–body length is 310–400 mm; tail length is 315–345 mm. *Peradorcas* has padded hind feet, and the hind toes have reduced claws. Normally one altricial offspring is born per litter (Walker, 1975). *Peradorcas* occur in rocky outcrops and on boulder-strewn hillsides in northern Australia (Walker, 1975; Corbet and Hill, 1980). Little is known of the habits of the little rock wallaby. They are apparently herbivorous, and emerge from the rocks at night to feed. They will thump their feet as a warning signal when disturbed (Troughton, 1943; Walker, 1975).

Petrogale spp. There are seven species of *Petrogale: P. brachyotis, P. burbidgei, P. godmani, P. penicillata, P. purpureicollis, P. rothschildi,* and *P. xanthopus*. Head–body length varies from 500 to 800 mm, and tail length ranges from 400 to 700 mm. The hind feet are heavily padded, with a granulated sole. The hind claws are reduced. There is usually one young per litter (Walker, 1975). Species of *Petrogale* inhabit rocky ridges and boulder piles throughout Australia, and are also found on a number of coastal islands. Ring-tailed rock wallabies

TABLE I
Species of Midsize Rock-Dwelling and Non-Rock-Dwelling Mammals Considered in This Report[a]

Species	Rock-dwelling	Species	Rock-dwelling
Marsupialia		Rodentia (*continued*)	
Petauridae		Caviidae	
Pseudocheirus dahli	+ *	*Kerodon rupestris*	+ *
Macropodidae		*Galea spixii*	− *
Peradorcas concinna	+ *	Chinchillidae	
Petrogale brachyotis	+	*Chinchilla brevicaudata*	+
Petrogale burbidgei	+	*Chinchilla lanigera*	+ *
Petrogale godmani	+	*Lagidium peruanum*	+ *
Petrogale penicillata	+ *	*Lagostomus maximus*	− *
Petrogale purpureicollis	+	Petromyidae	
Petrogale rothschildi	+	*Petromus typicus*	+ *
Petrogale xanthopus	+	Ctenodactylidae	
Onychogalea unguifer	− *	*Ctenodactylus gundi*	+ *
Hyracoidea		*Ctenodactylus vali*	+
Procaviidae		*Felovia vae*	+
Heterohyrax brucei	+ *	*Massoutiera mzabi*	+
Procavia johnstoni	+ *	*Pectinator spekei*	+
Dendrohyrax dorsalis	− *	Lagomorpha	
Rodentia		Leporidae	
Sciuridae		*Lepus capensis*	− *
Marmota monax	− *	*Pronolagus crassicaudatus*	+
Marmota caligata	+ *	*Pronolagus randensis*	+
Marmota olympus	+ *	*Pronolagus rupestris*	+ *
Cricetidae		Ochotonidae	
Neotoma albigula	− *	*Ochotona princeps*	+ *
Neotoma cinerea	+ *	*Ochotona pusilla*	− *

[a] +, Affirmative; −, negative. An asterisk indicates species used in multivariate analyses.

are nocturnal, and only emerge during the day to bask. They feed on grasses and the roots and bark of trees and shrubs (Troughton, 1943; Tyndale-Biscoe, 1973; Walker, 1975; Corbet and Hill, 1980).

Onychogalea unguifer. The northern nail-tailed wallaby, or karrabul, has about the same size range (head–body length 450–670 mm; tail length 330–665 mm) as *Petrogale.* The tail is tipped with a horny, nail-like structure. They have long, well-developed nails. They give birth to a single young (Ride, 1970; Nowak and Paradiso, 1983). *Onychogalea unguifer* is found in woodland and shrub savannas in northern Australia (Ride, 1970). Little is known of their ecology and behavior. They are mostly nocturnal and forage on grasses and herbaceous vegetation. When disturbed, they emit a brief warning cry and flee into thick vegetation. They are said to be solitary (Troughton, 1943; Ride, 1970; Nowak and Paradiso, 1983).

3.2. Order Hyracoidea

3.2.1. Family Procaviidae

Heterohyrax brucei. Head–body length is 305–380 mm and there is no external tail. The feet possess leathery pads and have reduced claws. There are usually one to three young per litter, with an average of 1.7 (Hoeck, 1982). Rock hyraxes are found primarily in rocky outcrops called kopjes, which are common in the savannas of East Africa. Their range extends from southeastern Egypt to the Transvaal of South Africa (Corbet and Hill, 1980; Hoeck, 1982). Hyraxes are highly specialized for the rocky kopje habitat. They forage on the leaves of trees in the kopjes as well as on vegetation adjacent to the rocks. Their social organization has been described in detail by Hoeck (1975; Hoeck et al., 1982). Briefly, solitary males defend a clump of rocks and maintain a small harem of females (three to seven). Juveniles of both sexes also form part of the social group. The males accumulate multiple females, ostensibly through defense of a critical resource.

Procavia johnstoni. This species is much like *Heterohyrax,* only larger (305–550 mm head–body length). *Procavia* occupies kopjes and other rocky outcrops throughout much of Africa, the Sinai, and neighboring regions as far north as Syria. They are often sympatric with *Heterohyrax* (Hoeck, 1975; Corbet and Hill, 1980). Hoeck (1975, 1982; Hoeck et al., 1982) states that the basic ecology and social organization of *Procavia* are essentially the same as observed in *Heterohyrax.*

Dendrohyrax dorsalis. The tree hyrax is much the same size as the rock hyrax (400–600 mm head–body length) and has a small vestigial tail (10–30 mm). As with other hyraxes, the feet are covered with leathery pads, and reduced claws are present. Litter size is smaller than in other hyraxes, with one, or occasionally two, precocial young born per litter (Jones, 1978; Nowak and Paradiso, 1983). Tree hyraxes are not rock-dwellers, but are arboreal and occur in a variety of tropical forest formations in western equatorial Africa (Kingdon, 1971; Jones, 1978). Their social organization seems markedly different from the polygynous rock hyraxes. According to Jones (1978) and Hoeck et al. (1982), tree hyraxes appear to be solitary. Each individual has a home tree contained within a small home range. *Dendrohyrax* share some behavioral attributes with the rock hyraxes, including an alarm whistle. Communal defecation sites do not occur, but individuals defecate in restricted spots.

3.3. Order Rodentia

3.3.1. Family Sciuridae

Marmota spp. Marmots are large (300–600 mm head–body length) rodents with a relatively short tail (100–250 mm). The four species discussed in this review (*M. monax*, *M. flaviventris*, *M. caligata*, *M. olympus*) are very similar in external morphology; *M. monax* is larger than the other species. The soles of the feet are padded; adults have well-developed claws. Litter size ranges from two to nine. Newborn are altricial (Walker, 1975). All but *M. monax* occur in mountainous areas in the western United States and Canada. Three of the species (excluding *M. monax*) are commonly found in high-altitude meadows with talus slopes (Walker, 1975; Cowan and Guiguet, 1978; Frase and Hoffman, 1980). The montane species have been studied extensively (Downhower and Armitage, 1971; Barash, 1973a,b, 1974a,b; Andersen et al., 1976). They utilize talus areas in open meadows as a refuge from predators and as an overwintering hibernation site. Since these talus sites are small in areal extent, one adult male can successfully defend an area large enough to accommodate more than one female and her most recent offspring, and thus a harem accrues. The size of the harem seem to be positively correlated with the length of the growing season; they are largest for *M. flaviventris* and smallest for *M. olympus*. *Marmota monax* inhabits the lowland eastern deciduous forest of North America and the coniferous forests of southern Canada and Alaska (Lee and Funderburg, 1982). It does not defend a particular resource and does not accrue a harem (Bronson, 1964).

3.3.2. Family Cricetidae

Neotoma cinerea. This species of woodrat (180–250 mm head–body length) differs from all other woodrat species in that it possesses a bushy tail (130–190 mm). Large adults weigh about 300 g (females) to over 400 g (males). The soles of the feet are furred. Litter size varies from two to six altricial young (Burt and Grossenheider, 1976; Cowan and Guiguet, 1978; Escherich, 1981). It inhabits cliffs, ravines, talus, and mountainous areas from the southwestern United States to the Canadian Yukon. Escherich (1981) showed that bushy-tailed woodrats live in rock clumps, which are defended by a single adult male. Like the montane marmots, *Neotoma* males maintain harems of up to three females. Within these harems, clumps of feces and urine accumulate over the years, forming congealed masses. *Neotoma* also scentmark extensively. Woodrats are primarily herbivorous.

Neotoma albigula. The white-throated woodrat is considerably smaller than *N. cinerea* (head–body length 190–216 mm). The tail (140–186 mm) is not bushy. The soles of the feet are naked to the heel (Bailey, 1971). Litter size is two, with occasional births of three (Richardson, 1943). Young are altricial. *Neotoma albigula* inhabits desert scrublands where prickly-pear cacti are present (Brown et al., 1972). These woodrats are occasionally found in rocky areas as well (Bailey, 1905). The species is restricted in distribution to the arid and semiarid regions of the American Southwest (Nowak and Paradiso, 1983). *Neotoma albigula* is solitary and nocturnal, and is aggressive even in captivity (Howe, 1978). It is herbivorous.

3.3.3. Family Caviidae

Kerodon rupestris. The mocó resembles other guinea-pigs in general body form, but certain obvious differences make it easily distinguishable. *Kerodon* is thinner, with longer legs and an elongated rostrum. It is 350 mm in head–body length and its mass is up to 1000 g. *Kerodon* is tailless. The feet are clawless, and the soles are heavily padded. The innermost of the three hind digits has a well-developed grooming claw. The highly precocial young are born in litters of one or two (Lacher, 1980, 1981). *Kerodon* inhabits boulder piles and rocky outcrops in the Caatinga of northeastern Brazil (Lacher, 1980, 1981). Lacher (1981) found that *Kerodon* are rock specialists and seldom leave the rocks. They are primarily diurnal, but remain active throughout the day. They are herbivorous, feeding on the leaves and bark of trees and vines in the rock outcrops and are excellent climbers. Available information suggests that adult males defend boulder piles and accumulate harems (Lacher, 1981). Adult sex ratios in the field are skewed, with one adult male per five females. When disturbed, *Kerodon* move to high vantage points in the rocks and give a sharp alarm whistle. Mocós have communal defecation sites.

Galea spixii. The preá of northeastern Brazil resembles the domestic guinea-pig, although it is somewhat smaller (150–250 mm head–body length). It has no tail. The fore and hind feet are strongly clawed. The young are precocial, and litter size ranges from one to seven, with a mean of three (Lacher, 1980, 1981; Nowak and Paradiso, 1983). *Galea spixii* inhabits the semiarid thornscrub of northeastern Brazil. Although it occurs synectically with *Kerodon,* it uses the rock piles as temporary refugia only. *Galea* do not climb trees, and run across the rocks only with difficulty. They are diurnal and herbivorous, feeding primarily on grasses. Their mating system is promiscuous. Alarm whistles and communal defecation sites are absent (Lacher, 1980, 1981).

3.3.4. Family Chinchillidae

Chinchilla spp. Head–body length is 225–380 mm, and there is a medium-length (75–150 mm), bushy tail. Adult mass varies from 500 to 1000 g. There are currently two recognized species, *C. brevicaudata* and *C. lanigera*. Chinchillas have five front toes and four hind toes; one is rudimentary on both the fore and hind feet. The soles of the feet are well padded and the claws are weak. Litter size ranges from one to six. The young are precocial (Osgood, 1943; Walker, 1975; Honacki et al., 1982). Chinchillas live in rocky areas in the Andes of Chile, Argentina, Bolivia, and Peru, from the coastal mountains of Chile up to approximately 4000 m elevation (Osgood, 1943; Cabrera and Yepes, 1960; Fischer, 1978). Little detailed information is available on wild chinchillas. Fischer (1978) reports that the coastal chinchilla (*C. lanigera*) browses primarily on *Prosopis chilensis,* whereas *C. brevicaudata* grazes on grasses (*Festuca* and *Distichia*) and shrubs. Chinchillas live in and among rock crevices and rocky outcrops. They have been observed in associations of as many as 100 individuals and are reported to be monogamous (Walker, 1975).

Lagidium peruanum. The mountain vizcacha is a large (300–400 mm head–body length) rodent resembling a long-tailed (230–320 mm) rabbit. There are four toes on the fore feet and three on the hind feet. The soles are well padded. Mountain vizcachas give birth to a single precocial young (Pearson, 1948; Walker, 1975). *Lagidium* species range from 900 m elevation in Argentina to over 5000 m. *Lagidium peruanum* is a high-altitude species found from 4000 to 5000 m in the mountains of the central Andes of Peru. Vizcachas are found in islands of rock interspersed in the high-altitude plains of the central Andes (Pearson, 1948; Walker, 1975). The only extensive field study of *L. peruanum* was done by Pearson (1948). Mountain vizcachas forage on grasses and shrubs in close proximity to the home rocks. They select a wide variety of green vegetation. Their social organization has not been adequately described. Pearson (1948) states that vizcachas live in large colonies of up to 75 individuals that seem to be composed of associations of smaller (two to five individuals) family groups. There is only one adult male and one adult female per group. They are not territorial, but show communal defecation. Vizcachas give an alarm whistle when disturbed, which has been described as a high, clear twitter (Eisenberg, 1974).

Lagostomus maximus. Plains vizcachas are large (470–660 mm head–body length; 150–200 mm tail length), heavy-bodied rodents. The tail is proportionately shorter than it is in *Chinchilla* and *Lagidium* and the feet are much less padded. The claws are strong, since *Lagostomus* is a good burrower. Litter size ranges from one to four, with a mean of two. The young are precocial (Nowak and Paradiso, 1983). Plains vizcachas occupy thorny scrublands from southern Paraguay to central Argentina, occurring primarily in the Chaco (Mares et al., 1981b). *Lagostomus* is a crepuscular herbivore. It appears to be colonial, but the mating system has not been described. It grunts when alarmed (Mares et al., 1981b), and has several other vocalizations (Eisenberg, 1974).

3.3.5. Family Petromyidae

Petromus typicus. The dassie rat is the only species in the family. Adults are gray, midsized (140–250 mm head–body length) rodents with a hairy tail (130–180 mm). *Petromus* has a flattened skull and flexible ribs that allow the animal to squeeze itself into very narrow crevices in the rocks. The teats are located near the shoulder blades, so that the young can suckle while the female is in the crevice. The feet are narrow, with four digits on the front feet and five on the hind feet. Claws are reduced, and the hind feet possess some bristlelike hairs. One to two precocial young are born once a year (Walker, 1975). Their habitat includes rocky areas in Namibia and southern Angola (Corbet and Hill, 1980). Dassie rats are diurnal, emerging from crevices in the rocks to sun themselves and feed. They forage on grasses, leaves, seeds, and fruits, at times climbing up into bushes to gather food. They are quite agile on the rocks, and when jumping from boulder to boulder, they will spread their flattened bodies and glide. When disturbed, dassie rats utter a sharp alarm whistle (Walker, 1975).

3.3.6. Family Ctenodactylidae

Ctenodactylus, Felovia, Massoutiera, Pectinator spp. There are five known species of gundis, the genus *Ctenodactylus* having two species (Honacki et al., 1982). There are a number of basic characteristics shared by the genera. Head–body length is 160–240 mm, and the short tail is 10–50 mm long. The skull is flattened. They resemble guinea-pigs in external appearance. The feet are highly modified, with the soles being well padded, and the upper parts of the feet having a number of stiff, white bristles. On the hind feet, these bristles form a comblike arrangement that the gundis use to groom their fur. Litter size is small (one to three) and the newborn are very precocial (Rosevear, 1969; George, 1974; Walker, 1975). They inhabit rock outcrops, escarpments, and boulder piles throughout northern Africa from Morocco to Ethiopia (Rosevear, 1969; George, 1974). Virtually nothing is known of the behavior of ctenodactylids. George (1974) reported they are diurnal, being especially active in the early morning. They are strictly herbivorous, foraging on the leaves, flowers, seeds, and stalks of plants that grow in and around the rock outcrops. They give a sharp alarm whistle when disturbed and seek shelter in narrow crevices in the rocks. They have communal defecation sites.

3.4. Order Lagomorpha

3.4.1. Family Ochotonidae

Ochotona spp. Pikas are medium-sized (125–300 mm head–body length), tailless, rodentiform lagomorphs that resemble large-eared guineapigs. The feet have five digits, and the soles are heavily padded with fur. The tips of the digits have soft, leathery pads, and the claws are weak. Litter size varies from two to six and breeding is highly seasonal. The young are altricial (Walker, 1975). Pikas are found from northern and central Asia to western North America. They are primarily montane in distribution and occur in two different kinds of high-altitude habitat: talus slopes and open plains and steppes (Corbet and Hill, 1980; Smith 1981a). There have been many papers published on pikas (Kawamichi, 1970, 1971, 1976; Sharp, 1973; Tapper, 1973; Smith, 1974a,b, 1978, 1979, 1980, 1981a,b; Svendsen, 1979). The ecology and behavior of the talus-dwelling forms seem fairly consistent. Rock pikas are solitary and territorial. They show little female–female overlap in territories; male territories overlap with both sexes. *Ochotona princeps* is facultatively monogamous (Smith, 1981b; Smith and Ivins, 1984). The Asian rock pikas live in widely spaced family groups, making them distinctly more sociable than the North American pikas. The steppe-dwelling pikas, like *O. pusilla*, live in large colonies comprised of individual family groups that live separately from other such groups. The families are closely spaced, and extremely high densities are attained. Steppe pikas have larger and more frequent litters than the rock pikas. Only in steppe pikas do the young of the year breed. Pikas do not hibernate. All species spend a great deal of time and effort amassing the characteristic hay piles that provide food during the winter months.

3.4.2. Family Leporidae

Pronolagus spp. Rock hares, also called red hares, are 350–500 mm long with a short tail 50–100 mm. There are three species: *P. crassicaudatus*, *P. randensis*, and *P. rupestris*. None has been studied in detail, but some information is available for *P. rupestris* (Kingdon, 1974; Walker, 1975; Corbet and Hill, 1980). They have relatively short ears for rabbits, and short feet with blunt claws. The soles of the feet are padded with dense, greasy hair. Their padding improves traction on the rocks. The body color is grizzled brown, and russet on the limbs and tail. Young are altricial. Litter size is unknown. *Pronolagus* are found only in rocky areas in southern and eastern African savannas (Kingdon, 1974; Walker, 1975). Rock hares are nocturnal, and like many other rock-dwelling mammals, utilize the rocks for shelter. They emerge from the rocks in early evening to forage on grasses and twigs. They are alert and wary while foraging, and if disturbed, they retreat to high points in the rock pile, which they use as observation posts. When pursued, they utter a loud series of screams. They always defecate in one of a number of communal defecation spots. Nothing is known of their social organization, population dynamics, or mating system (Kingdon, 1974; Walker, 1975).

Lepus capensis. The brown hare (head–body length 400–600 mm; tail length 70–135 mm) is a fairly generalized *Lepus*. The claws are not reduced as in *Pronolagus*. The young are precocial and the litter size is usually one or two, but may be as large as seven (Kingdon, 1974). The brown hare inhabits open and arid habitats throughout Africa (Kingdon, 1974). They are crepuscular and diurnal, herbivorous, and solitary. Mating is promiscuous (Kingdon, 1974).

4. RESULTS

Various ecological parameters of the 24 diverse taxa are compared in Table II. Clearly, certain traits characterize the assemblage of rock-dwelling species. Approximately half of the taxa are arboreal, frequenting the trees and shrubs that are commonly associated with isolated rock piles, while all but one non-rock-dweller are terrestrial. All species of rock-dwellers are associated with a habitat that is distinctly patchy in its distribution, while most species for which information is available utilize lookout posts. The great majority of rock-dwelling taxa studied thus far are able to utilize the rockpiles as defensible resources. Non-rock-dwellers generally differ from the rock-dwelling species in these latter three traits.

Reproductive and morphological traits of the 24 taxa are compared in Table III. Rock-dwellers may have relatively smaller litters. Precocial young are no more common to rock-dwellers than to non-rock-dwellers. Tail reduction is equally common to both groups. Finally, the feet of rock-dwellers show a number of specializations. Fully 94% of the taxa have padded feet, and 75% exhibit nail reduction. Almost one-third (31%) of the genera possess grooming claws or some structural analogue. Among non-rock-dwellers, padded feet, reduced claws, and a specialized grooming structure are less common.

Behavioral traits of 24 taxa are listed in Table IV, although data are lacking for many species. Most rock-dwellers (70% of species for which data are available) exhibit the uncommon social pattern of harem formation, whereas only 10% are solitary taxa. Polygyny is common (78%). All species utilize warning vocalizations and most (74%) defecate in communal sites.

When all characteristics of all species (Tables II–IV) were analyzed in a multivariate manner using cluster analysis, two major assemblages were formed (Fig. 5). With only one exception, species that are specialized for existence in rockpiles were grouped together, whereas those species that are not rock-dwellers formed a second major cluster. Thus, even though the non-rock-dwelling species were phylogenetically closely related to the rock-specialized species, the similarities of adaptations to the rock habitat overrode similar phylogenetic backgrounds. The single exception is *Ochotona pusilla*, a non-rock-dwelling pika of the Old World that was only very loosely associated with the rock-dwellers. This is not too surprising, given the fact that numerous important categories of information were unavailable for this species (Tables II–IV). For example, we lacked information on the mating system, the use of lookout posts, the use of alarm vocalization, and the use of communal defecation sites for this species.

Several smaller clusters were formed within the rock specialist–non-rock specialist dichotomy, and these frequently followed taxonomic affinity. Thus, the three marsupial species formed one cluster, whereas the chinchillids formed a second. The African *Petromus* and *Ctenodactylus* were clustered, and *Ochotona pusilla* was placed in the subcluster formed by the African, chinchillid, and marsupial species. Although this subcluster was clearly composed of rock specialists, it is likely that these species are associated as much by what we lack in the way of information on their biology as what they share in the way of adaptations. In particular, we lack information for most species in this group on the use of lookout posts, the defensibility of their resources, social structure, mating system, and the use of communal defecation sites.

Two other subclusters are evident within the rock-specialist assemblage. The hyraxes and the caviid *Kerodon*, all tailless species, form a very tight cluster. Clearly, these very distantly related species have evolved a host of similar morphological, ecological, and behavioral adaptations in response to the rock-dwelling niche each fills on a different continent.

In the final rock-specialist subcluster, the montane marmots are grouped together closely and are associated with *Neotoma cinerea*. These species are then associated with a pairing composed of the rock-specialized lagomorphs *Pronolagus* and *Ochotona princeps*. All species in this cluster have tails, although the tail of pikas is rudimentary.

The second major cluster, the non-rock specialists, contains the marsupial *Onychogalea*, which is loosely associated with the chinchillid *Lagostomus*. Both species are grouped with a cluster composed of *Marmota monax*, *Neotoma albigula*, *Galea spixii*, and *Lepus capensis*. Finally, the tree hyrax, *Dendrohyrax*, is loosely clustered with the remainder of the non-rock-dwellers. There is a pronounced separation between the rock-specialist and non-rock-specialist clusters and the cophenetic correlation coefficient of 0.778 indicates little distortion in the two-dimensional representation of the multivariate relationship. This is especially true considering the qualitative nature of the traits.

5. DISCUSSION

5.1. Ecological Traits

A number of interesting points emerge from our analyses. Although we examined two infraclasses (Metatheria and Eutheria), four orders, and 11 families, there was a great deal of similarity evident in numerous

TABLE II
Ecological Characteristics of Rock-Dwelling Mammals and Several of Their Close Relatives

Species	Characteristics[a]							
	Habitat	Rainfall	Arboreal	Patchiness	Lookout posts	Defensible resource	Temperature seasonality	Rainfall seasonality
Rock-dwellers								
Pseudocheirus dahli	3	2	2	2	0	0	1	2
Peradorcas concinna	3	2	1	2	0	0	1	2
Petrogale penicillata	3	2	2	2	0	0	1	2
Heterohyrax brucei	2	2	2	2	2	2	1	2
Procavia johnstoni	2	2	2	2	2	2	1	2
Marmota flaviventris	5	2	1	2	2	2	2	2
Marmota caligata	5	1	1	2	2	2	2	2
Marmota olympus	5	1	1	2	2	2	2	1
Neotoma cinerea	5	2	2	2	2	2	2	2
Kerodon rupestris	3	2	2	2	2	2	1	2
Chinchilla lanigera	5	3	1	2	0	0	2	2

Lagidium peruanum	5	3	1	2	2	0	2	2
Petromus typicus	1	3	2	2	0	0	1	1
Ctenodactylus gundi	1	3	1	2	2	0	1	1
Pronolagus rupestris	3	2	1	2	2	0	2	2
Ochotona princeps	5	2	1	2	2	2	2	2
				Non-rock-dwellers				
Onychogalea unguifer	3	2	1	1	1	0	1	2
Dendrohyrax dorsalis	4	1	2	1	1	1	1	2
Marmota monax	4	1	1	1	1	1	2	1
Neotoma albigula	1	3	1	1	1	1	2	2
Galea spixii	3	2	1	1	1	1	1	2
Lagostomus maximus	3	2	1	1	1	1	2	2
Lepus capensis	2	2	1	1	1	1	2	2
Ochotona pusilla	2	2	1	1	0	1	2	1

[a]See text for explanation of characteristics. Ecological characteristic codes are as follows: Habitat: montane (5), forest (4), scrubland (3), grassland (2), desert (1). Rainfall: <400 mm/year (3), 400–1000 mm/year (2), >1000 mm/year (1). Arboreality: present (2), absent (1). Habitat patchiness: present (2), absent (1). Lookout posts: present (2), absent (1), no data (0). Defensible resource: present (2), absent (1), no data (0). Seasonality in temperature: present (2), absent (1). Seasonality in rainfall: present (2), absent (1).

TABLE III
Reproductive and Morphological Characteristics of Rock-Dwelling Mammals and Several of Their Close Relatives

Species	Characteristics[a]					
	Litter size	Young at birth	Reduced tail	Padded feet	Reduced claws	Grooming structure
			Rock-dwellers			
Pseudocheirus dahli	0	1	1	2	2	1
Peradorcas concinna	1	1	1	2	2	1
Petrogale penicillata	1	1	1	2	2	1
Heterohyrax brucei	2	2	2	2	2	2
Procavia johnstoni	2	2	2	2	2	2
Marmota flaviventris	2	1	2	2	1	1
Marmota caligata	2	1	2	2	1	1
Marmota olympus	2	1	2	2	1	1
Neotoma cinerea	2	1	1	2[b]	1	1
Kerodon rupestris	1	2	2	2	2	2
Chinchilla lanigera	2	2	1	2	2	1
Lagidium peruanum	1	2	1	2	2	1

Petromus typicus	1	2	1	0	2	2[c]
Ctenodactylus gundi	2	2	2	2	2	2[c]
Pronolagus rupestris	0	1	2	2[b]	2	1
Ochotona princeps	2	1	2	2[b]	2	1
			Non-rock-dwellers			
Onychogalea unguifer	1	1	1	1	1	1
Dendrohyrax dorsalis	2	2	1[d]	2	2	2
Marmota monax	2	1	2	2	1	1
Neotoma albigula	2	1	1	1	1	1
Galea spixii	2	2	2	1[e]	1	1
Lagostomus maximus	2	2	1	1	1	1
Lepus capensis	2	2	2	2	1	1
Ochotona pusilla	2	1	2	2[b]	2	1

[a]See text for explanation of characteristics. Reproductive and morphological characteristic codes are as follows: Litter size: >1 (2), = 1 (1), no data (0). Young at birth: precocial (2), altricial (1). The remaining traits are: present (2), absent (1), no data (0).
[b]Feet well-furred.
[c]Hardened bristles on feet for grooming, rather than specialized grooming claw.
[d]Although *Dendrohyrax* has a vestigial tail, the other hyraxes lack a tail entirely.
[e]*Galea spixii* has the hind feet much less padded than *Kerodon*, its confamilial.

TABLE IV
Behavioral Characteristics of Rock-Dwelling Mammals and Several of their Close Relatives

Species	Characteristics[a]			
	Sociality	Mating system	Alarm vocalizations	Communal defecation
Rock-dwellers				
Pseudocheirus dahli	0	0	0	0
Peradorcas concinna	0	0	2[b]	0
Petrogale penicillata	0	0	2[b]	0
Heterohyrax brucei	1	2	2	2
Procavia johnstoni	1	2	2	2
Marmota flaviventris	1	2	2	1
Marmota caligata	1	2	2	1
Marmota olympus	1	2	2	1
Neotoma cinerea	1	2	0	0
Kerodon rupestris	1	2	2	2
Chinchilla lanigera	2	3	0	0
Lagidium peruanum	2	0	2	2
Petromus typicus	0	0	2	2
Ctenodactylus gundi	0	0	2	2
Pronolagus rupestris	0	0	2	2
Ochotona princeps	4	1	2	2
Non-rock-dwellers				
Onychogalea unguifer	4	0	2[c]	1
Dendrohyrax dorsalis	4	1	2	1
Marmota monax	4	1	1	1
Neotoma albigula	4	1	1	1
Galea spixii	4	1	1	1
Lagostomus maximus	3	0	2	1
Lepus capensis	4	1	1	1
Ochotona pusilla	2	0	0	0

[a]See text for explanation of characteristics. Behavioral characteristic codes are as follows: Sociality: solitary (4), colonial (3), family group (2), multiple females per male (1), no data (0). Mating system: monogamous (3), polygynous (2), promiscuous (1), no data (0). Alarm vocalization: present (2), absent (1), no data (0). Communal defecation: present (2), absent (1), no data (0).
[b]Warning sound generated by thumping hind feet on rock surface (Troughton, 1943).
[c]Emits vocalization when frightened that may function as alarm call (Nowak and Paradiso, 1983).

characteristics of the rock-dwelling members of these diverse taxa. Considering the great phylogenetic differences between the taxa, it is perhaps surprising that so many traits should be shared. The rock-dwelling taxa were selected only because they were found in a similar specific habitat type, isolated rock areas—this was the only shared trait that characterized the species a priori.

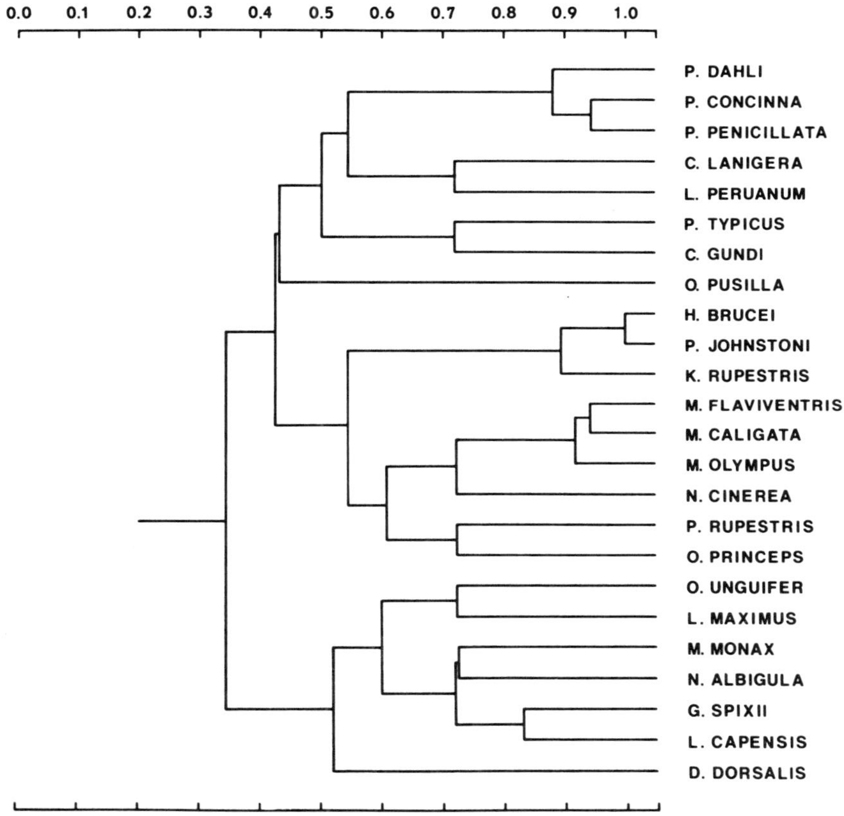

FIGURE 5. A phenogram showing the degree of similarity among diverse taxa of rock-dwelling mammals. Clusters were based on an analysis of the traits listed in Tables I–III. Two major clusters are evident, with species showing the most pronounced specialization to the rock habitat comprising the first cluster.

Rockpiles offer a habitat that provides a mesic microclimate where numerous plants can exist in regions that otherwise support more sparse vegetation. In Brazil, for example, or on the plains of East Africa isolated rockpiles support a forestlike vegetation within either a thorn scrub or a grassland macrohabitat (for example, Lacher, 1981; Hoeck, 1982). In some areas, where the difference between the vegetation growing in association with rockpiles and that growing in the surrounding habitat is less extreme, rockpiles still appear to support a more luxuriant plant cover. It is this regularity of plant resources that has been exploited by almost all rock-dwellers. We suppose that the rockpiles also offer a ready microhabitat for insects, but, by virtue of the delimited nature of the habitat, the insect resource base is probably too low to support any animals but those of the smallest body size. All species we examined were herbivorous, although their non-rock-dwelling relatives were also herbivores.

Almost one-half of the rock-inhabiting taxa could be considered significantly arboreal; however, this number is misleading. Several taxa occur in habitats that lack trees entirely. *Lagidium* and *Chinchilla*, for example, inhabit the high, treeless Andes of southern South America. *Marmota* and *Ochotona* are found in the treeless alpine meadows of the mountains of western North America, whereas ctenodactylids occur in areas of the North African desert that are largely devoid of trees. Thus, the development of arboreality seems to occur with great frequency where there are trees associated with the rockpiles. If taxa that occur in treeless areas are excluded from consideration, seven of nine species are arboreal.

Most taxa examined appear to utilize the rockpiles as lookout posts, although information on behavioral characteristics is quite sketchy in some cases. Furthermore, most of these taxa also utilize alarm vocalizations to warn of impending danger. The various rock habitats considered in this paper usually rise above the surrounding vegetation. The kopjes of Africa, for example, offer elevated lookout posts that are much higher than the grassy plains, whereas the boulder piles frequented by *Kerodon* in Brazil have their highest points above the surrounding thorn scrub (Lacher, 1981). Alarm calls allow the animals to return to the safety of the rockpiles, where few predators can enter.

5.2. Morphological Traits

Various morphological traits seem to have evolved in response to the saxicolous habit, although in some cases the similarities are more apparent than real. In many taxa, for example, the tail is reduced in length or absent entirely. It is difficult to tell whether or not a reduced tail is an adaptation to life in rocky areas. In some cases (for example, hyraxes and caviids), tail reduction appears to be related to phylogenetic history; hyraxes and caviids other than those associated with the rock habitat also show tail reduction. Moreover, it is difficult to see why tail reduction might be associated with arboreality; quite the converse would be expected. Most likely, tail reduction seems to have some importance merely because several taxa that frequent rocky areas hap-

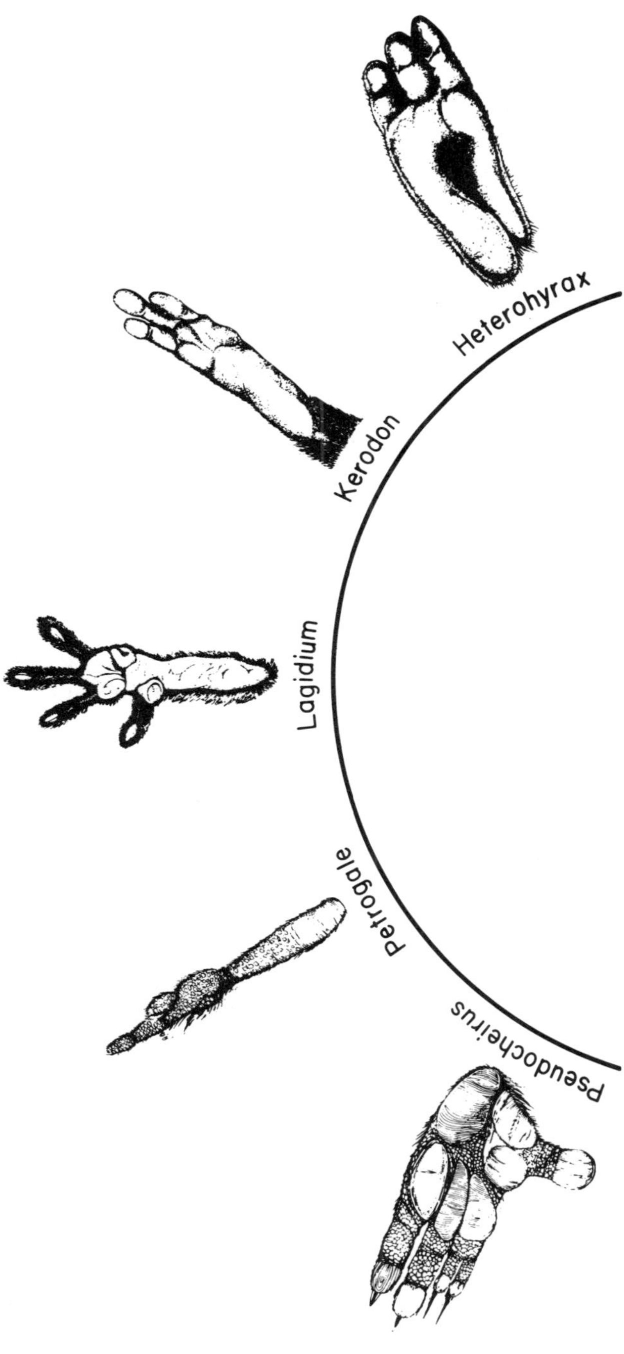

FIGURE 6. An example of adaptations of the hind feet of rock-dwelling mammals. The marsupial *Pseudocheirus* possesses numerous ridges on the foot that presumably assist in traction on rock surfaces, while pads are also present. The marsupial *Petrogale* and the rodent *Kerodon* both shown pronounced toe reduction and leatherlike soles, with the former having numerous protuberances, which appear to be for increased traction. The South American chinchilla, *Lagidium*, shows toe reduction, padded feet, and stiff fringe hairs on the hind foot, while the rock dassie, *Heterohyrax*, has hind feet with reduced toe number, leatherlike pads, and a central suction apparatus that aids in locomotion on rock surfaces.

pen to be members of phylogenetic assemblages that evidence tail reduction.

There are several morphological traits that do have a clear association with existence in rockpiles, however. Most taxa considered have feet that are well padded. In some cases, the feet possess unusual specializations for existence on a rocky substrate (Fig. 6). Marsupials of the genera *Petrogale* and *Peradorcas*, for example, not only have heavily padded feet, but have granulated soles. Rock hyraxes have padded feet and a suction cup arrangement on the feet that allows the animals to cling to bare rock. *Kerodon* also have very thickened, leatherlike soles that allow for traction on rock surfaces while cushioning shocks when jumps are made from one boulder to another. *Pronolagus* and *Ochotona*, two rock-dwelling lagomorphs included in our analyses, have feet that are heavily padded with hair, as does *Neotoma cinerea*.

Most of the taxa that have developed specialized padding on the feet have also undergone a reduction of the claws; 73% of the animals that have padded feet also have small and/or blunt claws. Only five of the taxa considered, however (or 42% of those exhibiting nail reduction), have developed specialized grooming claws or analogous structures. The hyraxes and the caviid possess a pectinate grooming claw on the hind feet, whereas *Ctenodactylus* and *Petromus* have developed stiffened bristles on the hind feet that are used to groom the fur.

5.3. Mating System Evolution

For many of the taxa examined here, the rockpiles form a habitat component that is, in principle, defensible by a single male. Clearly, many of the animals we considered have evolved numerous specializations to life in these boulder piles. For at least eight taxa (Table II), the localized nature of the rockpiles, and their apparently critically important role in the biology of the species, constitutes a situation where a vital and limited resource can be controlled by a single individual. Thus, the rockpiles fall into the category of having a high environmental potential for polygamy (Emlen and Oring, 1977). In an area where a critical resource has a patchy distribution and can be monopolized by a single individual, the potential is high for a dominant male to utilize differential access to the resource to accrue a harem. In Jamaican fruit bats (*Artibeus jamaicensis*), for example, a single male can control access to roosting sites in a hollow tree and this results in each male having a harem (Morrison, 1979). It is significant that in areas where roosting occurs in the open forest, rather than in hollow trees, this same species does not maintain harems (Morrison, 1979). This lability in mating systems within a species in different localities supports the contention that the mating system observed for a species has a strong environmental component. Such adaptive variability in social organization is shown by several species (for example, acorn woodpeckers, *Melanerpes formicivorus*, Stacey and Bock, 1978; savannah sparrows, *Passerculus sandwichensis*, Weatherhead, 1979).

Table IV shows a number of traits that are associated with sociality. In light of the Emlen and Oring (1977) hypothesis, we might expect rock-dwelling mammals to show a disproportionate occurrence of harem-based mating systems. This indeed appears to be the case. Although many of the species considered have not been studied from the standpoint of modern behavioral theory, seven species show a polygynous mating system. Thus, 70% of the species that have been studied may maintain harems. In view of the fact that some lability in the mating system is expected, depending on the defensibility of the particular rockpile, this high percentage of such a generally uncommon behavioral attribute is startling and warrants further discussion.

Kerodon, the caviid rock specialist of the Brazilian Caatinga, has been shown by Lacher (1981) to maintain harems. A second caviid (*Galea spixii*) lives in the same habitat as *Kerodon*, but inhabits the surrounding thorn scrub rather than the rockpiles. *Galea* lacks the various morphological specializations to life in a rocky habitat, such as padded feet, grooming claws, and reduced nails. In addition, there appear to be no patchily distributed resources in *Galea*'s habitat that are defensible by a single animal and that are critical to *Galea*'s reproductive success. Rather than the harem polygyny of *Kerodon*, *Galea* shows hierarchical promiscuity (Wittenberger, 1979). We emphasize that *Galea* and *Kerodon* are closely related caviids living sympatrically in the Caatinga. The major environmental variable differing between the genera is the rockpile habitat, and the response of the two taxa to the presence or absence of such a critical and defensible resource is evident in their mating system.

In East Africa, the thorn scrub region is quite similar physiognomically and climatologically to the Brazilian Caatinga. The rockpiles there are inhabited by two species of hyraxes, *Procavia johnstoni* and *Heterohyrax brucei*. Both genera use the rock fissures for predator avoidance and for shelter. Like *Kerodon*, both genera use the rocks for lookout posts and give alarm calls. Both show morphological features that are strikingly like those of *Kerodon*, including the various specializations of the feet and general body form. Hoeck (1975) has shown that both species of hyraxes defend the rockpiles from conspecifics and differentially admit females to use this vital resource. In concordance with the expectations of the Emlen–Oring model, both hyraxes have a harem-based polygynous mating system. These rock specialists provide strong support for Emlen and Oring's model for the evolution of polygyny, and we believe their rationale may be convincingly extended to other rock-dwelling mammals.

Barash (1973a,b; 1974a,b) summarized the information on social organization in four species of North American marmots (*Marmota*) and concluded that it was possible to explain between-species differences in social organization in terms of the length of the growing season. In the productive, low-elevational deciduous woodlands of eastern North America, young ground hogs (*M. monax*) are able to mature and disperse during the long summer season. The species is thus aggressive and solitary with a promiscuous mating system based on a dominance hierarchy. In contrast, young *M. olympus*, which experience an abbreviated growing season at high elevations, cannot acquire sufficient energy in the summer to accumulate the necessary fat reserves that ensure survival during hibernation and, at the same time, supply the energy needs for growth, maturity, and dispersal. Barash reported that this species shows delayed maturity, reduced aggression, and coloniality based on extended family groups. Barash (1974a) notes cautiously that "the adaptive values may appear too obvious." We suspect this observation to be correct and can incorporate the four marmots into the more general scheme based on the Emlen–Oring model. *Marmota monax* differs from the other marmots in being essentially a forest species, whereas the three montane species (*M. flaviventris, M. caligata,* and *M. olympus*) occupy meadow habitats, which may be broken into patches by intervening montane forest. However, montane marmots require not only meadows, but meadows that contain talus slopes with extruding rock piles (Downhower and Armitage, 1971; Andersen et al., 1976). Because the soil is so shallow at these high altitudes, the rocks provide the only reliable refuge from both predators and the deep winter cold and also serve as lookout posts for predator detection. These rockpiles constitute an important resource, one, moreover, that is patchily distributed and capable of monopolization by individual males. Females need the resource for hibernation and for raising young and so, by the Emlen and Oring arguments of resource defense polygyny, we would expect that each of the highland marmots would show some version of harem polygyny. Indeed, *M. flaviventris* is known to maintain harems and the "colonies" of *M. olympus* and *M. caligata* appear, in fact, to be harems of varying size (Downhower and Armitage, 1971). In the woodland habitat of *M. monax* there appear to be no critical resources that are potentially monopolizable by males and, as noted, the species does not maintain harems, but has a promiscuous system organized by male and female dominance hierarchies (for example, Bronson, 1964).

The marmot data alone are not sufficient to evaluate the respective merits of the "season-length" and "resource defensibility" theories of sociality and harem formation, because the ecological variables are confounded with season length and altitude. Such is not the case with the comparison made between *Kerodon* and *Galea*, however. The major climatic and seasonal variables are held constant (the animals live only a few meters apart) and phylogenetic differences are small; resource defensibility is strongly implicated as a determinant of the mating system. We also note here that no other caviid maintains harems or inhabits isolated rockpiles (Rood, 1970, 1972; Dubost and Genest, 1974; Wilson and Kleiman, 1974). In a similar vein, we note that the tree-dwelling hyrax, *Dendrohyrax arboreus*, of central and southern Africa appears to be essentially solitary in its forested habitat, even in the highland forest on the slopes of Mt. Kenya and the Ruwenzori Mountains (Coe, 1962; Hoffmeister, 1967). We doubt that its food (fruits, leaves, berries, insects) or its places of refuge (tree hollows or trees with dense foliage) constitute a monopolizable resource at any altitude, and the solitary existence and lack of a harem are consistent with the Emlen–Oring model. Of course the fact that both of the rock-dwelling hyraxes *do* defend rockpiles and *do* maintain harems is also significant.

We are aware of some rock specialists that do not appear to show harem polygyny and suggest that, for these species, the rock habitat does not constitute a defensible resource. We further suggest that the mating system that characterizes each species is consistent with that predicted by the Emlen–Oring model.

Pikas (*Ochotona princeps*) inhabit montane talus slopes very similar to those supporting *M. caligata* and *M. olympus* (Sharp, 1973; Tapper, 1973; Smith, 1974a,b, 1978, 1979, 1980, 1981b). Pikas show a very consistent pattern of social organization throughout their distribution. Both sexes are solitary and maintain territories of equal size. There is little female–female overlap in territories; however, male territories show overlap with both sexes, especially during the breeding season. Smith and Ivins (1984) have shown that pikas are facultatively monogamous. Such a mating system is quite different from the harem-based mating system present in the talus-dwelling marmots. There are, however, a number of complicating factors.

Pikas do not hibernate (no lagomorph hibernates) and are peculiar in accumulating large piles of hay during the short growing season upon which they feed during the winter. This collecting of hay is very time-consuming (Kawamichi, 1976) and likely places severe constraints on the daily time budgets of both sexes. In effect, food is quite scarce for pikas during the growing season, so that much time and effort must be spent in gathering the food supply for the rest of the year, which is spent under the snow-covered rocks. A male pika would have difficulty defending an area with sufficient resources for both himself and a harem. Optimal reproductive success, given the environmental and physiological constraints, is apparently obtained via the formation of monogamous pair bonds.

Finally, mountain vizcachas (Chinchillidae: *Lagidium*) in the altiplano of Peru maintain a colonial social system based on monogamous family groups (Pearson, 1948). These rodents are obligate rock-dwellers and show several morphological adaptations to this habit. We suggest that the enormous extent of the boulder areas in these regions may preclude their defense by males, because burrows are placed under boulders, which may be part of a continuous ridge extending for many miles along mountain slopes. Although they are colonial, *Lagidium* do not maintain harems. Some populations of *Lagidium* occupy areas with more isolated rockpiles (for example, in Tafí del Valle, Tucumán, Argentina) and we may predict that populations in such habitats will show harem polygyny. Such variability within a species is to be expected as critical ecological variables change. In fact, a situation similar to the one discussed here (the control of mating systems by a single environmental factor) has recently been reported in *Dascyllus marginatus*, a coral reef fish (Fricke, 1980). This species lives in pairs, harems, or multimale groups, with group size and the number of sexually active males in a group being positively correlated with the size of the coral reef. The situation in rock-dwelling mammals, although complicated by additional morphological, behavioral, or physiological constraints, shows a similar tendency.

Our analysis provides strong support for the explanatory framework proposed by Emlen and Oring. The success of their model derives from its having avoided being based on major habitat correlates of breeding systems such as forest or open country, which may or may not be critical variables for the organisms themselves with regard to eventual reproductive success (which is the final arbiter of successful adaptation). Instead, the model attempts to focus the general question, "How may an individual maximize, relative to other members of the same sex, access to resources that are critical to reproductive success?" The nature of the resource will differ between (and within) species, even between those occupying superficially very similar habitats. These points are well illustrated by the rock-dwelling mammals considered above, and their fit with the model promotes the prediction that populations of the unstudied rock-inhabiting ctenodactylid rodents (gundis) of North Africa, for which rockpiles seem to constitute a defensible resource (Kock and Schomber, 1960), will also show harem polygyny.

5.4. Multivariate Analysis

The phenogram shown in Fig. 5 is quite interesting. There is almost a perfect division of species into two groups—those that inhabit rockpiles and those that do not. The only exception, *Ochotona pusilla*, a non-rock specialist, was very loosely clustered with the rock-dwelling species. Significantly, we lacked critical behavioral data on this species of pika that might be expected to have affected its placement in the phenogram, especially such traits as type of mating system, alarm calls, and other characteristics that appear to correlate with the rock habitat.

Within the rock-specialist assemblage, close phylogenetic relationship is evident in some subclusters, such as the grouping of marsupials, chinchillids, or marmots. The fact that *Kerodon*, a caviid rodent, was closely clustered with the hyraxes, however, shows how similarity of adaptations to a particular habitat and mode of existence can override phylogenetic similarity. Indeed, the entire phenogram reflects how selection for a suite of characteristics associated with life in isolated rockpiles can mold similar adaptive types from diverse stocks of organisms. The multivariate analysis is an example of this phenomenon for a broad array of taxa and characteristics.

5.5. Convergent Evolution

We selected the taxa included in these analyses on the basis of a single shared characteristic—occurrence in isolated rockpiles. Thus, one parameter of the physical environment was the only a priori unifying link in this disparate assemblage of mammals. The multivariate analysis and Tables II–IV show that there was a great deal of similarity among taxa. Perhaps the most pronounced level of convergent evolution is shown in the fourth subcluster, where members of different orders of mammals (hyraxes and rodents) are very closely clustered. The similarities between rock hyraxes and *Kerodon* have been commented on briefly by Lacher (1981), but these unrelated mammals share a very large number of traits, many of which were not included in our analyses (Fig. 7). Besides occupying almost identical habitats, their general body form is remarkably similar. Both groups defecate in piles on boulders, are herbivorous, use alarm whistles, have small litters of precocial young, have rather long snouts, which they poke into vegetation while foraging,

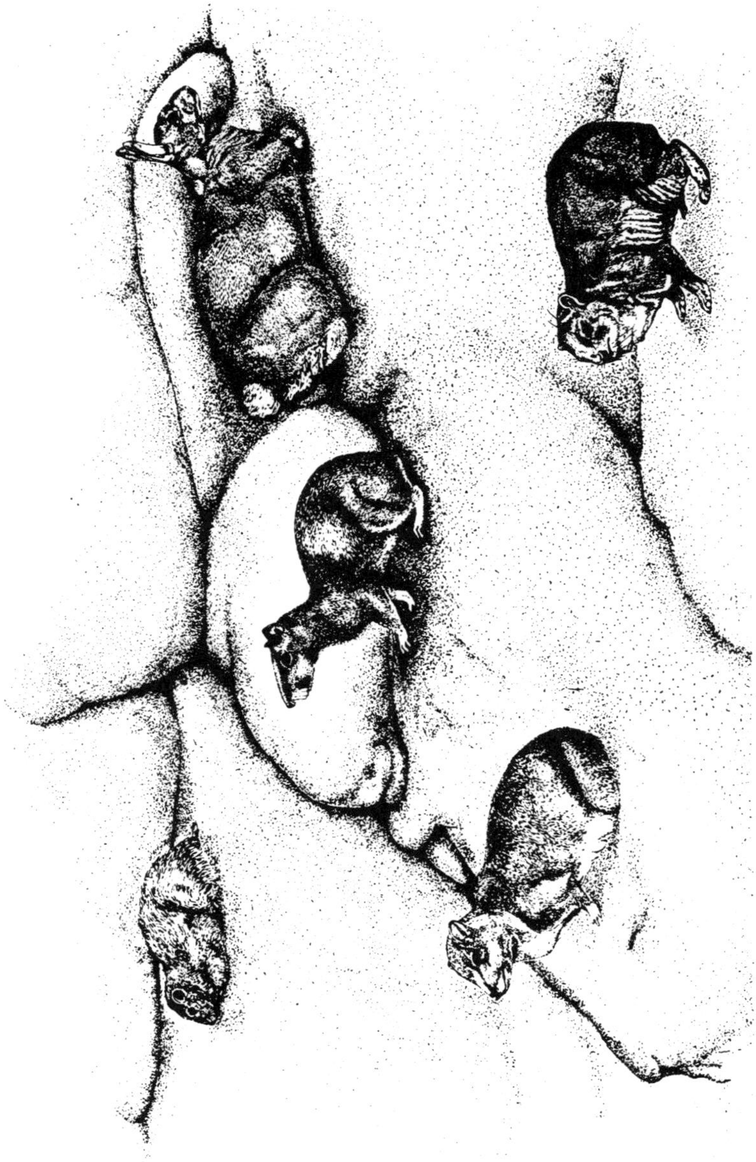

FIGURE 7. Several species of rock-adapted mammals that show a pronounced degree of convergent evolution. Clockwise from upper left are the ctenodactylid rodent *Ctenodactylus*, the lagomorph *Pronolagus*, the hyraxes *Procavia* and *Heterohyrax*, and, in the center, the caviid rodent *Kerodon*. The animals are approximately drawn to scale, except the hyraxes, which are about half-size.

have leatherlike pads on the feet, lack a tail, defend individual boulder piles, accrue groups of females by allowing differential access to the rockpiles to males and females, have harem polygyny, and share many other characteristics. Many of these same traits seem to be shared with taxa that have been studied only superficially, such as *Ctenodactylus*, *Petromus*, and *Pronolagus*.

Significantly, all other caviid rodents inhabit nonrocky areas and show few of *Kerodon*'s adaptations that appear to be associated with the rocky habitat. Even *Galea spixii*, a caviid that literally lives next to *Kerodon*'s boulder piles, shows none of the latter's specializations. Similarly, the rock hyraxes differ significantly in aspects of morphology, ecology, and behavior from other hyraxes of the genus *Dendrohyrax*. In fact, in many characteristics, *Kerodon* and the hyraxes are more similar to each other than they are to closely related taxa. Indeed, many of the characteristics shared among the taxa in the first large cluster of Fig. 5 are traits that are not shared with taxa to which they are closely related. For example, *Neotoma cinerea* is a harem polygynyst, whereas other *Neotoma* are not (for example, Escherich, 1981); the three *Marmota* species considered are quite different behaviorally from the groundhog, *M. monax* (for example, Barash, 1974b). While overall morphology has not always converged to the degree evident among members of the hyrax–*Kerodon* subcluster, various attributes of the taxa we examined are shared among all members of the rock-specialist assemblage.

We suggest that, given the very distant phylogenetic relationships of the taxa herein considered, these similarities are the result of convergent evolution. The selective factors favoring existence in isolated rockpiles were very likely quite similar for the diverse taxa that were adapting to this unusual microhabitat, and the results of this selection pressure were also quite similar in many cases. Apparently, the ways in which a mammal can evolve toward a specialized existence in isolated rockpiles are rather limited, and very similar adaptations must appear if a species is to be a successful colonizer of these habitats. It is interesting that the phenotypic plasticity of such distantly related groups of mammals is sufficiently high that surprisingly similar phenotypes can be formed from genetic complements that might be expected to have been canalized in different directions.

6. CONCLUSIONS

A broad comparison was made of mammals that have specialized for existence on isolated rock and boulder piles in diverse locations throughout the world and for close relatives of rock-inhabiting species that did not show specializations for a rock existence. Comparisons were made for selected ecological, morphological, and behavioral characteristics. Rock-dwelling mammals are shown to be remarkably similar, considering the phylogenetic dissimilarity of the species examined (marsupials and placentals representing two subclasses, four orders, 12 families, and 17 genera were included). Various analyses indicated that the rock specialists were generally more similar to each other in the traits examined than they were to their near relatives, if the latter had not evolved for life in isolated rockpiles. Basically, rock specialists are herbivorous, feeding on the green vegetation usually associated with the rocky microhabitat. A tendency to climb trees has also developed along with the herbivorous habit. Morphological similarities of these species are pronounced, especially in structures associated with the feet. Generally the feet are padded, with nail reduction also common. In several groups, a specialized grooming claw, or analogous structure, is present. Behaviorally, the taxa examined are quite similar. Most utilize the rocks as lookout posts and give alarm whistles when disturbed. A surprisingly large percentage of the species examined have developed a social organization based on harem polygyny. This unusual mating system has possibly developed because the rocks represent a vital resource that is controllable by a single male. Such a result is in accord with the hypothesis proffered by Emlen and Oring (1977) concerning resource defense polygyny. Our results show that overall convergent evolution in these rock-dwelling mammals is quite pronounced, with adaptations to the saxicolous habit occurring in all major categories of traits that were examined. The results demonstrate how a single environmental factor (isolated rockpiles) can have an enormous influence on a great many characteristics of species. Furthermore, there is a high degree of repeatability in the traits that have evolved among different mammals faced with strongly similar selective pressures.

ACKNOWLEDGMENTS. We thank Paul Handford, M. R. Willig, and Janet Braun, who read and commented on various drafts of the manuscript. Janet Braun aided in the multivariate analysis of the data. Zenith Marsh made the two drawings. Ron Tilson and John Miles loaned us photographs of rock mammal habitats. Brazilian research was supported by a grant from the Brazilian National Academy of Sciences, project number 85 ("Ecology, evolution and zoogeography of mammals") as part of the larger program, "Ecological studies of the semi-arid region of northeastern Brazil." Dr. Aristides A. Leão, President of the Academy, and Dr. Paulo Vanzolini were most helpful during our stay in Brazil. Dr.

Craig Black and the Carnegie Museum of Natural History helped to support the research. Dr. Andrew Smith made several unpublished manuscripts available to us.

REFERENCES

Andersen, D. C., Armitage, K. B., and Hoffman, R. S., 1976, Socioecology of marmots: Female reproductive strategies, *Ecology* **57**:552–560.

Bailey, V., 1905, Biological survey of Texas, *N. Am. Fauna* **25**:1–222.

Bailey, V., 1971, *Mammals of the Southwestern United States (with Special Reference to New Mexico)*, Dover, New York [reprint of 1931 ed.].

Barash, D. P., 1973a, The social biology of the Olympic marmot (*Marmota olympus*), *Anim. Behav. Monogr.* **6**:172–245.

Barash, D. P., 1973b, Social variety in the yellow-bellied marmot (*Marmota flaviventris*), *Anim. Behav.* **21**:579–584.

Barash, D. P., 1974a, The social behavior of the hoary marmot (*Marmota caligata*), *Anim. Behav.* **22**:256–261.

Barash, D. P., 1974b, The evolution of marmot societies: A general theory, *Science* **185**:415–420.

Bronson, F. H., 1964, Agonistic behavior in woodchucks, *Anim. Behav.* **12**:470–478.

Brown, J. H., Lieberman, G. A., and Dengler, W. F., 1972, Woodrats and cholla: Dependence of small mammal populations on the density of cacti, *Ecology* **53**:310–313.

Burt, W. H., and Grossenheider, R. P., 1976, *A Field Guide to the Mammals*, Houghton Mifflin, Boston.

Cabrera, A., and Yepes, J., 1960, *Mamíferos Sud Americanos*, Ediar, Buenos Aires.

Coe, M. J., 1962, Notes on the habits of the Mount Kenya hyrax (*Procavia johnstoni mackinderi:* Thomas), *Proc. Zool. Soc. Lond.* **138**:639–644.

Corbet, G. B., and Hill, J. E., 1980, *A World List of Mammalian Species*, Comstock, London.

Cowan, I. M., and Guiguet, C. J., 1978, *The Mammals of British Columbia*, British Columbia Provincial Museum, Victoria.

Downhower, J. F., and Armitage, K. B., 1971, The yellow-bellied marmot and the evolution of polygamy, *Am. Nat.* **105**:355–370.

Dubost, G., and Genest, H., 1974, Le comportement social d'une colonie de Maras *Dolichotis patagonum* L. dans le Parc de Branfère, *Z. Tierpsychol.* **35**:225–302.

Eisenberg, J. F., 1974, The function and motivational basis of hystricomorph vocalizations, in: *The Biology of Hystricomorph Rodents* (I. W. Rowlands and B. J. Weir, eds.), Academic Press, London, pp. 211–244.

Eisenberg, J. F., 1981, *The Mammalian Radiations*, University of Chicago Press, Chicago.

Emlen, S. T., and Oring, L. W., 1977, Ecology, sexual selection, and the evolution of mating systems, *Science* **197**:215–223.

Escherich, P. C., 1981, Social biology of the bushy-tailed woodrat, *Neotoma cinerea*, *Univ. Calif. Publ. Zool.* **100**:1–132.

Fischer, G. M., 1978, *Los Pequeños Mamíferos de Chile*, Gayana, Concepción.

Frase, B. A., and Hoffmann, R. S., 1980, Marmota flaviventris, *Mammal. Species* **135**:1–8.

Fricke, H. W., 1980, Control of different mating systems in coral reef fish by one environmental factor, *Anim. Behav.* **28**:561–569.

George, W., 1974, Notes on the ecology of gundis (F. Ctenodactylidae), in: *The Biology of Hystricomorph Rodents* (I. W. Rowlands and B. J. Weir, eds.), Academic Press, London, pp. 143–160.

Hoeck, H., 1975, Differential feeding behavior of the sympatric hyrax *Procavia johnstoni* and *Heterohyrax brucei*, *Oecologia* **22**:15–47.

Hoeck, H., 1982, Population dynamics, dispersal, and genetic isolation in two species of hyrax (*Heterohyrax brucei* and *Procavia johnstoni*) on habitat islands in the Serengeti, *Z. Tierpsychol.* **59**:177–210.

Hoeck, H. N., Klein, H., and Hoeck, P., 1982, Flexible social organization in hyrax, *Z. Tierpsychol.* **59**:265–298.

Hoffmeister, D. F., 1967, Tubulidentates, proboscideans, and hyracoideans, in: *Recent Mammals of the World* (S. Anderson and J. K. Jones, Jr., eds.), Ronald Press, New York, pp. 355–365.

Honacki, J. H., Kinman, K. E., and Koepple, J. W., 1982, *Mammal Species of the World: A Taxonomic and Geographic Reference*, Allen Press, Lawrence.

Howe, R. J., 1978, Agonistic behavior of three sympatric species of woodrats (*Neotoma mexicana, N. albigula,* and *N. stephensi*), *J. Mammal.* **59**:780–786.

Jones, C., 1978, *Dendrohyrax dorsalis, Mammal. Species* **113**:1–4.

Kawamichi, T., 1970, Social pattern of the Japanese pika, *Ochotona hyperborea yesoensis,* preliminary report, *J. Fac. Sci. Hokkaido Univ. Ser. VI Zool.* **17**:462–473.

Kawamichi, T., 1971, Daily activities and social pattern of two Himalayan pikas, *Ochotona macrotis* and *O. roylei,* observed at Mt. Everest, *J. Fac. Sci. Hokkaido Univ. Ser. VI Zool.* **17**:587–609.

Kawamichi, T., 1976, Hay territory and dominance rank of pikas (*Ochotona princeps*), *J. Mammal.* **57**:133–148.

Kingdon, J., 1971, *East African mammals,* I, Academic Press, London.

Kingdon, J., 1974, *East African mammals,* IIB, Academic Press, London.

Kock, D., and Schomber, H. W., 1960, Our days among the gundis (*Ctenodactylus gundi*), *Afr. Wild Life* **14**:199–203.

Lacher, T. E., Jr., 1979, Rates of growth in *Kerodon rupestris* and an assessment of its potential as a domesticated food source, *Pap. Avulsos Zool.* (São Paulo) **33**:67–76.

Lacher, T. E., Jr., 1980, The comparative social behavior of *Kerodon rupestris* and *Galea spixii* in the xeric caatinga of northeastern Brazil, Ph. D. thesis, University of Pittsburgh.

Lacher, T. E., Jr., 1981, The comparative social behavior of *Kerodon rupestris* and *Galea spixii* and the evolution of behavior in the caviidae, *Bull. Carnegie Mus. Nat. Hist.* **17**:1–71.

Lacher, T. E., Jr., 1981, The comparative social behavior of *Kerodon rupestris* and *Galea spixii* and the evolution of behavior in the Caviidae, *Bull. Carnegie Mus. Nat. Hist.* **17**:1–71.

Lee, D. S., and Funderburg, J. B., 1982, Marmots, in: *Mammals of North America* (J. A. Chapman and G. A. Feldhamer, eds.), Johns Hopkins University Press, Baltimore, pp. 176–191.

MacArthur, R., 1972, *Geographical Ecology,* Harper and Row, New York.

Mares, M. A., 1976, Convergent evolution of desert rodents: Multivariate analysis and zoogeographic implications, *Paleobiology* **2**:39–63.

Mares, M. A., 1980, Convergent evolution among desert rodents: A global perspective, *Bull. Carnegie Mus. Nat. Hist.* **16**:1–51.

Mares, M. A., 1983, Desert rodent adaptation and community structure, *Great Basin Nat. Mem.* **7**:30–43.

Mares, M. A., Willig, M. R., Streilein, K. E., and Lacher, T. E., Jr., 1981a, The mammals of northeastern Brazil: A preliminary assessment, *Ann. Carnegie Mus.* **50**:80–137.

Mares, M. A., Ojeda, R., and Kosco, M., 1981b, Observations on the distribution and ecology of the mammals of Salta Province, Argentina, *Ann. Carnegie Mus.* **50:**151–206.

Markham, C. G., 1972, *Aspectos Climatológicos da Seca no Brasil-Nordeste,* Sudene Assessouria Técnica, Divisão de Documentacão, Recife.

Morrison, D. W., 1979, Apparent male defense of tree hollows in the fruit bat, *Artibeus jamaicensis, J. Mammal.* **60:**11–15.

Nowak, R. M., and Paradiso, J. L., 1983, *Walker's Mammals of the World,* Johns Hopkins University Press, Baltimore.

Orians, G. H., and Solbrig, O. T., 1977, *Convergent Evolution in Warm Deserts,* Dowden, Hutchison, and Ross, Stroudsburg.

Osgood, W. H., 1943, The mammals of Chile, *Zool. Ser. Field Mus. Nat. Hist.* **30:**1–268.

Pearson, O. P., 1948, Life history of mountain viscachas in Peru, *J. Mammal.* **29:**345–373.

Pianka, E. R., and Pianka, H. D., 1970, The ecology of *Moloch horridus* (Lacertilia: Agamidae) in Western Australia, *Copeia* **1970:**90–103.

Richardson, W. B., 1943, Woodrats (*Neotoma albigula*): Their growth and development, *J. Mammal.* **24:**130–143.

Ride, W. D. L., 1970, *A Guide to the Native Mammals of Australia,* Oxford University Press, Melbourne.

Rood, J. P., 1970, Ecology and social behavior of the desert cavy (*Microcavia australis*), *Am. Midl. Nat.* **83:**415–454.

Rood, J. P., 1972, Ecological and behavioral comparisons of three genera of Argentine cavies, *Anim. Behav. Monogr.* **5:**1–83.

Rosevear, D. R., 1969, *The Rodents of West Africa,* British Museum (Natural History), London.

Schall, J. J., and Pianka, E. R., 1978, Geographical trends in numbers of species, *Science* **201:**679–686.

Sharp, P. L., 1973, Behavior of the pika, (*Ochotona princeps*), in the Kananaskis region of Alberta, M. Sc. thesis, University of Alberta, Edmonton.

Smith, A. T., 1974a, The distribution and dispersal of pikas: Consequences of insular population structure, *Ecology* **55:**1112–1119.

Smith, A. T., 1974b, The distribution and dispersal of pikas: Influences of behavior and climate, *Ecology* **55:**1368–1376.

Smith, A. T., 1978, Comparative demography of pikas (*Ochotona*): Effect of spatial and temporal age-specific mortality, *Ecology* **59:**133–139.

Smith, A. T., 1979, Equilibrial population structure of pikas, *Murrelet* **60:**30–31.

Smith, A. T., 1980, Temporal changes in insular populations of the pika (*Ochotona princeps*), *Ecology* **61:**8–13.

Smith, A. T., 1981a, Population dynamics of pikas (genus *Ochotona*), in: *Proceedings of the World Lagomorph Conference* (K. Myers and C. D. MacInnes, eds.), Guelph University Press, Guelph, Ontario, pp. 572–586.

Smith, A. T., 1981b, Territoriality and social behavior of *Ochotona princeps,* in: *Proceedings of the World Lagomorph Conference* (K. Myers and C. D. MacInnes, eds.), Guelph University Press, Guelph, Ontario, pp. 310–323.

Smith, A. T., and Ivins, B., 1984, Spatial relationships and social organization in adult pikas: A facultatively monogamous mammal, *Z. Tierpsychol.* **66:**289–308.

Sneath, P. H. A., and Sokal, R. R., 1973, *Numerical Taxonomy,* Freeman, San Francisco.

Stacey, P. B., and Bock, C. E., 1978, Social plasticity in the acorn woodpecker, *Science* **202:**1298–1300.

Streilein, K. E., 1982a, Ecology of small mammals in the semiarid Brazilian Caatinga. I. Climate and faunal composition, *Ann. Carnegie Mus.* **51:**79–107.

Streilein, K. E., 1982b, The ecology of small mammals in the semiarid Brazilian Caatinga. IV. Habitat selection, *Ann. Carnegie Mus.* **51**:331–343.

Svendsen, G. E., 1979, Territoriality and behavior in a population of pikas (*Ochotona princeps*), *J. Mammal.* **60**:324–330.

Tapper, S. C., 1973, The spatial organization of pikas (*Ochotona*), and its effect on population recruitment, Ph. D. thesis, University of Alberta, Edmonton.

Troughton, E., 1943, *Furred Mammals of Australia*, Angus and Robertson, Sydney.

Tyndale-Biscoe, H., 1973, *Life of Marsupials*, Elsevier, New York.

Vanzolini, P. E., Ramos-Costa, A. M. M., and Vitt, L. J., 1980, *Repteis das Caatingas*, Academia Brasileira de Ciências, Rio de Janeiro.

Vitt, L. J., 1981, Lizard reproduction: Habitat specificity and constraints on relative clutch mass, *Am. Nat.* **117**:506–514.

Walker, E., 1975, *Mammals of the World*, Johns Hopkins University Press, Baltimore.

Weatherhead, P. J., 1979, Ecological correlates of monogamy in tundra-breeding savannah sparrows, *Auk* **96**:391–401.

Willig, M. R., 1982, A comparative ecological study of Caatingas and Cerrado Chiropteran communities: Composition, structure, morphometrics and reproduction, Ph. D. thesis, University of Pittsburgh.

Willig, M. R., 1983, Composition, microgeographic variation, and sexual dimorphism in Caatingas and Cerrado bat communities from Northeast Brazil, *Bull. Carnegie Mus. Nat. Hist.* **23**:1–131.

Wilson, E. O., 1975, *Sociobiology*, Harvard University Press, Cambridge, Massachusetts.

Wilson, S. C., and Kleiman, D. G., 1974, Eliciting play: A comparative study, *Am. Zool.* **14**:341–370.

Wittenberger, J. F., 1979, The evolution of mating systems in birds and mammals, in: *Handbook of Behavioral Neurobiology*, Volume 3 (P. Marler and J. Vandenbergh, eds.), Plenum Press, New York. pp. 271–349.

CHAPTER 9

ANTS AND TERMITES AS FOOD

Patterns of Mammalian Myrmecophagy

KENT H. REDFORD

1. INTRODUCTION

Many mammals, ranging from antelope to elephant shrews, eat ants and termites. The specialized mammalian myrmecophages (ant- and termite-eaters) such as aardvarks and pangolins have long drawn attention, but little work has been done on general predator–prey relationships between mammals and ants and termites.

Mammals probably rank second behind ants as the most important predators of both termites and other ants (Deligne et al., 1981; Wilson, 1971). The extent to which mammals prey on ants and termites is in large part due to the abundance of these social insects (Erwin, 1983; Fittkau and Klinge, 1973; Levings, 1983). An overall view of mammalian myrmecophagy has never been presented and even basic questions, such as the extent to which "ant-eating" is the same as "termite-eating," have never been addressed.

The purpose of this paper is (1) to document the extent of mam-

KENT H. REDFORD • Museum of Comparative Zoology, Harvard University, Cambridge, Massachusetts 02138, and Department of Zoological Research, National Zoological Park, Smithsonian Institution, Washington, D. C. 20008. *Present address:* Center for Latin American Studies, University of Florida, Gainesville, Florida 32611.

malian predation on ants and termites, and (2) to examine the ways in which social insect defenses affect this predation.

2. METHODS

Using records obtained from a literature search, four categories of predation were defined based on the percent of ants and/or termites in the diet of a predator (1–10, 11–50, 51–90, and 91–100%) (Table I). If no indication was given in the original study as to whether percent volume or percent occurrence had been measured in the samples examined, I used "occurrence." The type (stomachs, feces, or observations) and number of samples on which the dietary categorization was based are also indicated. When data were unavailable to place a species in one of the categories of myrmecophagy, it was scored simply as consuming ants and/or termites. Finally, if ant or termite immatures or eggs were recorded in the diet of a species, or if it was seen to open nests, the species was scored as "breaking into" ant and/or termite nests.

The generic names of the ants and termites consumed, when reported, are available from the author.

The species of ants and termites eaten by mammals were classified according to their soldier-based defense systems; the patterns of predation on species possessing different defense systems was analyzed.

The diet of myrmecophages varies among individuals, among populations, and seasonally, but this variation is not accounted for in the collected data. This review is an overall view of the phenomenon of mammalian myrmecophagy and should serve to pinpoint areas for further, more specific research.

3. RESULTS

3.1. Mammalian Predators

Table I presents data for 216 mammalian species in 43 families and 13 orders (taxonomy follows Honacki *et al.*, 1982). This list is not exhaustive, since there are probably numerous unrecorded species of small marsupials, rodents, and insectivores that also eat ants and termites. However, the list probably includes most mammalian genera with species whose diet includes more than 50% ants and termites and all species whose known diet includes more than 90% ants and termites.

For all species, 36.4% were recorded as eating ants and no termites,

16.8% as eating termites and no ants, and 46.7% as eating both ants and termites. Obligate ant-eating species fall into two major categories: (1) mainly arboreal or (2) from the temperate zone. The majority of the 37 species recorded as exclusive termite-eaters fall into two major categories: (1) small terrestrial carnivores or (2) species in which the consumption of ants was not mentioned or for which studies dealt specifically with termite consumption (for example, Mathews, 1977). This second factor indicates that several of the species classified as exclusively termite-eaters probably eat ants as well. Additional research based on larger sample sizes is likely to reveal that many species not recorded here as eating termites do in fact include them in their diet. Overall, the data suggest that a mammal that eats termites probably also eats ants, but rarely vice versa.

From data on the percent occurrence of ants and termites in the diet, it is possible to classify 93 species of mammals. A similar analysis using percent volume allows classification of 53 species. In both classifications the number of species consuming ants and termites (64.4 and 43.4%, respectively) is much greater than the number consuming just ants (20 and 32.1%, respectively) or just termites (16.6 and 24.5%, respectively). In both systems (occurrence and volume) the two categories containing the most mammalian species are "0–10% ants only" (10 and 10, respectively) and "91–100% ants and termites" (16 and 13, respectively). The category of species eating only termites is the smallest. Very few species consume 91–100% ants or 91–100% termites (four in both classification systems). The small number of mammal species in whose diet ants and termites are of minor importance is probably a function of fewer studies on mammals that eat small proportions of social insects.

It is surprising that analysis of percent volume and percent occurrence produce such similar results. Differences are only apparent in the number of species listed as eating 91–100% ants and termites.

Data on predation of ant and termite nests show that most mammals (50%) open only ant nests. Just one of the 44 species for which there were data opened only termite nests (2.3%), whereas 47.7% opened both ant and termite nests.

3.2. Ants As Prey

Species from 47 genera of ants from eight of the nine subfamilies were recorded in the diets of 53 different species of mammals. No preference for members of particular subfamilies is evident, based on percentage of all genera in the subfamilies as an expected value and

TABLE I
Mammals That Eat Ants and Termites[a]

Mammal species	Source of classification			Composition of diet[b]				Nest opened	References
				Percent volume		Percent occurrence			
	Stomachs (*n*)	Feces (*n*)	Observation	Ants	Termites	Ants	Termites		
Monotremata									
Tachyglossidae									
Tachyglossus aculeatus		x	x	4	4	4	4	A	Griffiths (1968, 1978), Griffiths and Simpson (1966), Jones (1923 in Van Deusen and George, 1969)
Marsupialia									
Didelphidae									
Caluromys philander	37					2 >	2		Charles-Dominique et al. (1981)
Didelphis albiventris	4				1		x		Mathews (1977)
Didelphis marsupialis	1				1		x		Mathews (1977)
Marmosa murina	x					x			Charles-Dominique et al. (1981)
Marmosa sp.	1						x		Mathews (1977)
Marmosa sp.	1						x		Mathews (1977)
Metachirus nudicaudatus	8				3		x		Mathews (1977)
Monodelphis domestica	1						x		Mathews (1977)
Philander opossum	x					x	x		Charles-Dominique et al. (1981)

Dasyuridae										
Antechinus stuartii	309 (both)					1			A	Hall (1980)
Antechinus swainsoni	451 (both)					1				Hall (1980)
Myrmecobiidae										
Myrmecobius fasciatus		252	x	2	3	2	≪	3		Calaby (1960), Fleay (1942), Friend (1982)
Permalidae										
Isoodon obesulus	11+					x			A	Heinsohn (1966), Stoddart and Braithewaite (1979)
Thalacomyidae										
Macrotis lagotis		14+	x	3	3	3	<	3		Newsome (1962), Smyth and Philpott (1968)
Edentata										
Myrmecophagidae										
Cyclopes didactylus			x	4		4		2	A	Snethlage (pers. comm. in Bequaert, 1922), Husson (1978), Montgomery (1979), Montgomery (1983), Montgomery (1985), R. Best and A. Harada (unpublished)
		4		4		4				
Myrmecophaga tridactyla	3+	x	x	4	4	4		4	A,T	Carvalho (1966), Carvalho and Kloss (1951), Coles

(Continued)

TABLE I (*Continued*)

Mammal species	Source of classification			Composition of diet[b]				Nest opened	References
				Percent volume		Percent occurrence			
	Stomachs (*n*)	Feces (*n*)	Observation	Ants	Termites	Ants	Termites		
Myrmecophagidae									
Myrmecophaga tridactyla (*continued*)									(1980), Heyde (1949 in Husson, 1978), Kappler (1887), Montgomery (1979), Montgomery and Lubin (1977), Redford (1985a), Rengger (1830), Schaller (1983), Shaw et al. (1985), Silveira (1969), Weber (1972), Zietz (1872)
Tamandua mexicana	x		x	4	4	4	4	A,T	Lubin and Montgomery (1981), Montgomery (1979), Montgomery and Lubin (1977), Silveira (1968)

Tamandua tetradactyla	x		x	4	4	4	4	A,T	Enke (1978), Lubin and Montgomery (1981), Mathews (1977), Montgomery (1979), Montgomery and Lubin (1977), Redford (1983), Weber (1972)
Dasypodidae									
Cabassous centralis				4	4	4	4	T	Merrit (1985), Redford (1985b)
Cabassous unicinctus	x		x	4	4	4	4	A	Mathews (1977), Redford (1985b)
Chaetophractus vellerosus	684+		x	1		x		A	Greegor (1974), Redford (1985b)
Chlamyphorus truncatus	x		x	3		x			Moeller (1975), Redford (1985b), Walker (1975)
Dasypus novemcinctus	250+		x	2+	2+	x	x	A	Baker (1943), Fitch et al. (1952), Kalmbach (1944), Mathews (1977), Newman (1913), Redford (1985b)
Dasypus sabanicola	x			3	3	x	x		M. Baretto et al. (unpublished), Pacheco and Naranjo (1978)
Euphractus sexcinctus	12+		x	2	2	x	x	A	Redford (1985b), Schaller (1983)

(*Continued*)

TABLE I (*Continued*)

Mammal species	Source of classification			Composition of diet[b]				Nest opened	References
				Percent volume		Percent occurrence			
	Stomachs (*n*)	Feces (*n*)	Observation	Ants	Termites	Ants	Termites		
Dasypodidae (*continued*)									
Priodontes maximus	x		x	4	4	4	4	A,T	Beebe (1922), Moeller (1975), Redford (1985b), M. Baretto et al. (unpublished)
Tolypeutes matacus	9+		x	4	4	4	4		Redford (1985b), Schaller (1983)
Zaedyus pichiy	x		x	2	2	x	x	A	Redford (1985b)
Insectivora									
Tenrecidae									
Tenrec ecaudatus	29			2	2	2 >	2	A,T	Nicoll (1982)
Erinaceidae									
Echinosorex gymnurus	7					x	x		Davis (1962)
Erinaceus frontalis	4+						x		Hegh (1922), Smithers (1971)
Hylomys suillus	1					x			Beng (1965)
Soricidae									
Suncus ?griffithi	2					x	x		Roonwal (1949)
Talpidae									
Parascalops breweri	100					3			Eadie (1939)
Scalopus aquaticus	100					2			Scheffer (1910 in Bequaert, 1922)

Scandentia							
Tupaiidae							
Dendrogale melanura	3				x		Lim and Heyneman (1968)
Ptilocerus lowii	9				x		Lim (1967)
Tupaia glis	36	193	1+	1+	x	x	Davis (1962), Langham (1982), Thomas and Wroughton (1909)
	24		1	2			Langham (1983)
Tupaia minor	1				x		Lim (1965)
Tupaia montana	22				x		Lim and Heyneman (1968)
Tupaia tana	11				x	x	Davis (1962), Lim (1965)
Chiroptera							
Emballonuridae							
Emballonura nigrescens	44				3	1	Vestgens and Hall (1977)
Megadermatidae							
Macroderma gigas	10				1	1	Vestgens and Hall (1977)
Vespertillionidae							
Chalinolobus gouldii	148				1		Vestgens and Hall (1977)
Chalinolobus nigrogriseus	17				1		Vestgens and Hall (1977)
Glischropus tylopus	29				1		Beng (1965)
Nicticeius balstoni	25				2	1	Vestgens and Hall (1977)
Nicticeius sp.	12				2	1	Vestgens and Hall (1977)

(Continued)

TABLE I (*Continued*)

Mammal species	Source of classification			Composition of diet[b]				Nest opened	References
				Percent volume		Percent occurrence			
	Stomachs (n)	Feces (n)	Observation	Ants	Termites	Ants	Termites		
Vespertillionidae (*continued*)									
Nyctophilus geoffroyi	36					1			Vestgens and Hall (1977)
Molossidae									
Mormopterus loriae	2					2			Vestgens and Hall (1977)
Mormopterus planiceps	16					2			Vestgens and Hall (1977)
Primates									
Lorisidae									
Perodicticus potto	42			3		x			Charles-Dominique (1971), Jewell and Oates (1969)
Galagidae									
Galago alleni	13			1		x	x		Charles-Dominique (1971), Jewell and Oates (1969)
Galago senegalensis			x				x		Kingdon (1971)
Otolemur (=Galago) crassicaudatus			x				x		Wilson and Clark (1977)
Callithricidae									
Cebuella pygmaea			x			x			Ramirez et al. (1977)
Saguinus fuscicollis	11					x			Izawa (1975)

Cebidae								
Aotus trivirgatus	2		1	x				Izawa and Yoneda (1981)
Callicebus moloch	4		1	x				Izawa and Yoneda (1981), Schaller (1983)
Callicebus torquatus	5			x			A	Izawa (1975)
Cebus albifrons	5			x			A	Izawa (1975), Izawa and Yoneda (1981)
		x		x	$\gg$	x		Terborgh (1983)
Cebus apella	16			x		x	A,T	Izawa (1979), Izawa and Yoneda (1981)
		x		x	$\gg$	x		Terborgh (1983)
Cebus capucinus		x		x			A	Freese (1976)
Cebus nigrivittatus		x		x		x		Robinson (1986)
Chiropotes satanus	4			x				Ayres and Nessimian (1982), R. Mittermeier and W. Konstant (pers. comm.)
Lagothrix lagotricha	6			x				Izawa (1975)
Pithecia monachus	9			x				Izawa (1975)
Pithecia hirsuta	2			x				Izawa and Yoneda (1981)
Saimiri sciureus	13			x			A	Izawa (1975)
Cercopithecidae								
Cercocebus albigena		x		x			A	Waser (1977), D. McKey (pers. comm.)
Cercopithecus aethiops	x	x				x		Smithers (1971), Struhsaker (1967)
Cercopithecus albigena	15			4		1	A,T	Gautier-Hion (1978)
Cercopithecus ascanius	100			1				Haddow (1952)

(*Continued*)

TABLE I (*Continued*)

Mammal species	Source of classification			Composition of diet[b]				Nest opened	References
				Percent volume		Percent occurrence			
	Stomachs (*n*)	Feces (*n*)	Observation	Ants	Termites	Ants	Termites		
Cercopithecidae (*continued*)									
Cercopithecus cephus	62					2			Gautier-Hion (1978, 1980)
Cercopithecus mitis			x				x		Kingdon (1971)
Cercopithecus nictitans	100					3	1	A	Gautier-Hion (1978, 1980)
Cercopithecus pogonias	52					2	1		Gautier-Hion (1978, 1980)
Miopithecus talapoin	9					2			Gautier-Hion (1978)
Erythrocebus patas			x			x			Hall (1965)
Macaca sinica			x			x			Hladik and Hladik (1972)
Papio hamadryas (=*ursinus*)			x				x		Hamilton et al. (1978)
Papio hamadryas (=*cynocephalus*)			?x				x		Uehara (1982)
Hylobates pileatus			x				x		S. Srikosamatara (unpublished)
Hylobates sp.			x			x	x		Gittins and Raemekers (1980)
Pongidae									
Pan trogolodytes		x	x			1 >	1	A,T	Lang (pers. comm. in Bequaert, 1922), Goodall (1963), Hladik (1973),

								McBeath and McGrew (1982), McGrew (1974), McGrew et al. (1979), Nishida (1973), Nishida and Hiraiwa (1982), Nishida and Uehara (1980), Suzuki (1966), Uehara (1982)
Pongo pygmaeus		x	x	x		x	A,T	Galdikas-Brindamour and Brindamour (1975), MacKinnon (1974), Rijksen (1978)
Hominidae								
Homo sapiens			x	x		x	A,T	Bodenheimer (1951), Hegh (1922), K. H. Redford (unpublished), Ruddle (1973)
Carnivora								
Canidae								
Canis adustus	49			1	<	1		Rahm and Christiansen (1963), Smithers (1971)
Canis latrans			x	x				Parks (1929 in Chew, 1979)
Canis mesomelas	76	87		1	<	1		Bothma (1959), Hall-Martin and Bothma (1980), Smithers (1971), Stuart (1976)

(*Continued*)

TABLE I (*Continued*)

Mammal species	Source of classification: Stomachs (*n*)	Feces (*n*)	Observation	Composition of diet[b]: Percent volume: Ants	Termites	Percent occurrence: Ants	Termites	Nest opened	References
Canidae (*continued*)									
Cerdocyon thous		x	x				x		J. Dietz (pers. comm.)
Otocyon megalotis	63	267+	x		3	2	3		Bothma (1959, 1971), Kingdon (1977), Koop and Velimirov (1982), Nel (1978), Smithers (1971), Waser (1980)
Vulpes chama	89			1	1	1	1		Bothma (1971), Smithers (1971)
Vulpes vulpes	409		x	1	1				McIntosh (1963), Prakash (1959)
Ursidae									
Helarctos malaynus			x			x	x	A,T	Fetherstonhaugh (1940), Lekagul and McNeely (1977)
Melursus ursinus		231	x			2	< 2	A,T	Eisenberg and Lockhart (1972), Laurie and Seidensticker (1977), Prater (1965), Schaller (1967)

Ursus americanus		x	x	1		x		A	Beeman and Pelton (1980), Muir (1916 in Bequaert, 1922)
Ursus arctos			x			x		A	Elgmork (1981), Novikov (1962)
Ursus thibetanus			x			x	x		Novikov (1962), Prater (1965)
Procyonidae									
Nasua nasua	2					x	x		Schaller (1983)
Nasua narica			x			x	x	A,T	Kaufmann (1962)
Procyon lotor						x			Llewellyn and Uhler (1952)
	330 (both)								
Mustelidae									
Meles meles	x					1		A	Neal (1977)
Mellivora capensis		x	x			x	x	A,T	Kingdon (1977)
Mustela frenata	x					x			Errington (1936 in Svendsen, 1982)
Spilogale ?interrupta	x					x			Crabb (1941)
Taxidea taxus						x		A	Chew (1979), Errington (1937)
Viverridae									
Civettictis civetta	17				1		x		Bothma (1965), Bothma (1971)
Genetta genetta	78	x					2		Smithers (1971), du Toit (1980)
Genetta tigrina	51						2		Rowe-Rowe (1978), Smithers (1971)
Hemigalus derbyanus	12					x			Davis (1962)
Viverra tangalunga	6						1		Davis (1962)
Herpestidae									
Bdeogale crassicauda	1+					3	3		W. F. H. Ansell (1965), Kingdon (1977)

(Continued)

TABLE I (Continued)

Mammal species	Source of classification			Composition of diet[b]				Nest opened	References
				Percent volume		Percent occurrence			
	Stomachs (n)	Feces (n)	Observation	Ants	Termites	Ants	Termites		
Herpestidae (*continued*)									
Bdeogale nigripes	9	40				2 >	1		Lang (pers. comm. in Bequaert, 1922), Kingdon (1977)
Cynictis penicillata	133	x	x			1 ≪	2		Earle (1981), Herzig-Straschil (1977), Rowe-Rowe (1978), Smithers (1971), du Toit (1980)
	156					2	3		Lynch (1980)
Helogale parvula	21						2		Smithers (1971)
Herpestes edwardsi	3						2		Prakash (1959)
Herpestes pulverulentus		x					1		du Toit (1980)
Herpestes sanguineus	33						2		Herzig-Straschil (1977), Rowe-Rowe (1978), Smithers (1971)
Ichneumia albicauda	8+	12				2 <	2		Rowe-Rowe (1978), Smithers (1971), Waser (1980)
Mungos mungo		120				3	2		Neal (1970), Rood (1975)
Paracynictis selousi	34						1		Smithers (1971)
Rhynchogale melleri	2				4		4		Ansell (1960)
Suricata suricatta	18	x					2		Herzig-Straschil (1977), Smithers

										(1971), du Toit (1980)
	98					2		2		Lynch (1980)
Protelidae										
Proteles cristatus	54	102	x	1	4	2	≪	4		Ansell (1964), Azzaroli and Simonetta (1966), Bothma (1965, 1971), Dean (1978), Ketelhodt (1966), Kingdon (1977), Kruuk and Sands (1972), Rowe-Rowe (1978), Smithers (1971), P. Richardson (unpublished)
Hyaenidae										
Hyaena brunnea		383				1		2		Mills and Mills (1978)
Felidae										
Felis serval			x					x		Kingdon (1977)
Felis silvestris	80+							1		Prakash (1959), Smithers (1971)
Tubulidentata										
Orycteropidae										
Orycteropus afer	11	x	x	4	4	4		4	A,T	Lang (pers. comm. in Bequaert, 1922), Bourliere and Verschuren (1972), Fossati (1937), Kingdon (1971), Melton (1976a,

(Continued)

TABLE I (*Continued*)

Mammal species	Source of classification			Composition of diet[b]				Nest opened	References
				Percent volume		Percent occurrence			
	Stomachs (*n*)	Feces (*n*)	Observation	Ants	Termites	Ants	Termites		
Orycteropidae									
Orycteropus afer (*continued*)									1976b), Pages (1970), Smithers (1971), Urbain (1954), Verheyen (1951)
		15+				3	2	T	Melton (1976a)
		9+				2	3	T	Melton (1976a)
Artiodactyla									
Tragulidae									
Hyemoschus aquaticus	3		x			x			Dubost (1963, 1965)
Bovidae									
Cephalophus monticola	x			2		x			Grimm (1970)
Pholiodota									
Manidae									
Manis crassicaudata	2+		x	4	4	4	4	A,T	Hutton (1949), Roberts (1977), Roberts and Vielliard (1971)
Manis gigantea	2+		x	4	4	4	4	A,T	Forel 1909 (in Bequaert, 1922), Buttikofer (1890), Kingdon (1971), Pages (1970)

Manis javanica	x		x	4	4	4		4	A,T	Beebe (1914 in Bequaert, 1922), Kreyenberg (1907), Schultze (1914)
Manis pentadactyla						4		4	A,T	Allen (1938), Sowerby (1925)
Manis temminckii	5		x	4	4	4		4	A,T	Kingdon (1971), Smithers (1971), Sweeney (1956), Vossler (1907)
Manis tetradactyla		x	x			4			A	Pages (1965, 1970)
Manis tricuspis			x			x		x	A,T	Lang (personal communication in Bequaert, 1922), Kingdon (1971), Menzies (1962, 1966), Pages (1965, 1970, 1975)
Rodentia										
Sciuridae										
Callosciurus nigrovittatus	2					x				Lim and Heyneman (1968)
Callosciurus notatus	21					x	>	x		Lim (1965), Lim and Heyneman (1968)
Callosciurus prevosti	5					1		1		Davis (1962)
Epixerus ebii	4			1		x				Emmons (1975, 1980)
Exilosciurus exilis	4					3				Davis (1962)
Exilosciurus whiteheadi	3					x				Lim and Heyneman (1968)
Funambulus pennanti			x					x		Prakash (1959), Bartlett (1899 in Roberts, 1977)
Funisciurus anerythrus	12					4		3		Emmons (1975, 1980)

(*Continued*)

TABLE I (*Continued*)

Mammal species	Source of classification: Stomachs (*n*)	Feces (*n*)	Observation	Composition of diet[b]: Percent volume: Ants	Percent volume: Termites	Percent occurrence: Ants	Percent occurrence: Termites	Nest opened	References
Sciuridae (*continued*)									
Funisciurus isabella	14					3 >	3		Emmons (1975, 1980)
Funisciurus lemniscatus	15					3	4		Emmons (1975, 1980)
Funisciurus pyrrhopus	12					4 <	4	T	Emmons (1975, 1980), Kingdon (1974b), Rahm and Christiaensen (1963)
Heliosciurus gambianus							x		Delany (1964)
Heliosciurus rufobranchium	14					2 >	2		Emmons (1980)
Iomys horsfieldi	6					x			Lim and Heyneman (1968)
Myosciurus pumilio	3					2	3		Emmons (1980)
Paraxerus boehmi	30					x			Kingdon (1974b)
Paraxerus cepapi						x	x		Viljoen (1977 in de-Graaf, 1981)
Paraxerus lucifer						x	x		Kingdon (1974b)
Paraxerus poensis	8					3	2		Emmons (1980)
Protoxerus stangeri	14					2 >	2		Emmons (1980)
Rhinosciurus laticaudatus			x			x	x		Davis (1962), Harrison (1950), Lekagul and McNeely (1977)
Rhinosciurus ?tupaioides						x			Thomas and Wroughton (1909)

Taxon									Reference
Sciurus vulgaris						x			Wichman (1954 in Landry, 1970)
Spermophilus spp.						x			Bailey (1893 in Landry, 1970)
Tamias minimus						x			Schmidt (1931 in Landry, 1970)
Cricetidae									
Cricetomys gambianus	32		x			x		x	W. F. H. Ansell (1964), Iwuala et al. (1980)
Dendromus melanotis								x	de Graaf (1981)
Deomys ferrugineus	5+					x		x	Delany (1964), Genest-Villard (1980), Hatt (1940), Kingdon (1974b)
Juscelinomys candango	x					x			Moojen (1965)
Onochomys torridus		x				1			Horner et al. (1965)
Oxymycterus roberti	28				2	x		x	Borchert and Hansen (1983)
Oxymycterus rufus	12					4			Barlow (1969)
Oxymycterus spp.	4					x		x	Mathews (1977)
Peromyscus leucopus	272			1		x			Whitaker (1966), Calhoun (1941 in Landry, 1970)
Peromyscus maniculatus	444			1		x			Whitaker (1966)
Prionomys batesi	x			4		4			Genest-Villard (1980)
Saccostomus campestris						x		x	Delany (1964), de Graaf (1981)
Scapteromys tumidus	11					2		1	Barlow (1969)
Tatera indica	154							x	Prakash (1959)
Tatera valida	1					x			Delany (1964)
Taterillus emini	1					x			Hatt (1940)
Taterillus gracilis	x					x	<	x	Hubert et al. (1981)

(Continued)

TABLE I (*Continued*)

Mammal species	Source of classification			Composition of diet[b]				Nest opened	References
	Stomachs (*n*)	Feces (*n*)	Observation	Percent volume		Percent occurrence			
				Ants	Termites	Ants	Termites		
Muridae									
Acomys cahirinus	44					x			Hanney (1965)
Acomys spinosissimus						x	x		Vesey-Fitzgerald (1966 in de Graaf, 1981)
Arvicanthis niloticus	x					x <	x		Hubert et al. (1981)
Hybomys trivirgatus	45					x	x		Cole (1975)
Hybomys univittatus	x					x	x		Genest-Villard (1980)
Hylomyscus stella	x						2		Genest-Villard (1980)
Lemniscomys striatus	x					2 <	2		Field (1975), Genest-Villard (1980)
Leopoldamys (=*Rattus*) *sabanus*	15			2	2	x	x		Harrison (1954)
Lophuromys flavopunctatus	121					1 >	1		Delany (1964), Hanney (1964)
Lophuromys sikapusi	40+			3		3 ≫	2		Cole (1975), Delany (1964), Genest-Villard (1980), Kingdon (1974b), Verheyen and Verschuren (1966)
Maxomys (=*Rattus*) *alticola*	1					x			Lim and Heyneman (1968)
Maxomys whiteheadi	8					x			Lim (1965)
Mus minutoides	55					x	x		Hanney (1965), Iwuala et al. (1980)

Mus musculus	478	1		x				Whitaker (1966)
Mus setulosus	x			1		1		Genest-Villard (1980)
Mus triton	36			1			A	Hanney (1965)
Praomys erythroleucus	x			4	<	4		Hubert et al. (1981)
Praomys natalensis	55+	1	3	1		3		Delany (1964), Field (1975), Iwuala et al. (1980)
Praomys stella	20			x		x		Cole (1975)
Praomys tulbergi	154			x		x		Cole (1975), Genest-Villard (1980)
Rattus argentiventer	1			x				Lim and Heyneman (1968)
Rattus (=Stenomys) mulleri	14			x		x		Harrison (1954), Lim (1965)
Rattus rattus argentiventer	25		2			x		Harrison (1954)
Rattus rattus jalorensis	43		3			x		Harrison (1954)
Rattus rattus rattus	24			x				Iwuala et al. (1980)
Rattus tiomanicus	2					x		Lim and Heyneman (1968)
Stochomys defua	23			x		x		Cole (1975)
Uranomys ruddi	2					x		Hanney (1965)
Gliridae								
Dryomys nitedula				x			A	Jeitteles (1967 in Landry, 1970)
Graphiurus murinus						x		de Graaf (1981)
Dipodidae								
Allactaga sibiricus				x				Ognev (1940 in Landry, 1970)
Bathyergidae								
Cryptomys sp.						x		Rosevaer (1969 in Landry, 1970)

(Continued)

TABLE I (*Continued*)

Mammal species	Source of classification			Composition of diet[b]					References
				Percent volume		Percent occurrence			
	Stomachs (*n*)	Feces (*n*)	Observation	Ants	Termites	Ants	Termites	Nest opened	
Macroscelidea									
Macroscelidae									
Elephantulus intufi						x			Smithers (1971)
Elephantulus rufescens	18		x		3	2	x		Rathbun (1979)
Petrodromus tetradactylus	34		x			x	x		A. D. H. Ansell and Ansell (1969), Kingdon (1974a), Rathbun (1979), Sheppe (1973)
Rhynchocyon chrysopygus	21		x			x	x		Rathbun (1979)
Rhynchocyon cirnei						x	x		Kingdon (1974a)

[a]Taxonomy after Honacki et al. (1982).
[b]Key: 1, 0–10%. 2, 11–50%. 3, 51–90%. 4, 91–100%. x, Unknown. $>$, more ants than termites. $\gg$, many more ants than termites. $<$, more termites than ants. $\ll$, many more termites than ants.

the number of genera containing recorded prey species eaten as an observed value. There is a tendency for the subfamilies with a small number of genera to be overrepresented in mammalian diets; this is probably a function of the types of mammalian predators that have been studied. Excluding subfamilies with less than seven genera, the results show that formicines and myrmicines are slightly underrepresented in mammalian diets, dolichoderines are equally represented, and ponerines are overrepresented.

Comparing the numbers of mammalian species preying on species within each ant genus within each subfamily yields a different picture. Formicines and dolichoderines are substantially more preferred than would be expected from the number of genera consumed, whereas ponerines and myrmicines are less preferred. This measure of preference can be used to compare the effectiveness of various forms of ant defense in deterring mammalian predation, because the individual defense of ants is fairly consistent within formicine subfamilies. Dolichoderines and formicines lack a functional sting, and their strictly chemical defenses appear to be relatively less effective in deterring mammalian predation than defense by stinging, but no defense completely prevents predation.

3.3. Termites As Prey

Thirty-eight genera representing three of the six families of termites were recorded as mammalian prey; the family Termitidae is analyzed by subfamily with four of the subfamilies represented. Hodotermitids, Rhinotermitids, and species in the genera in the subfamily Amitermitinae are preyed upon as would be expected from their proportion in the order; the Termitinae are greatly underrepresented, whereas the Macrotermitinae and the Nasutitermitinae are overrepresented. This may of course represent a bias based on the species of predators studied.

However, when considering the numbers of mammalian species that prey on each termite genus, the relative preference changes. Using this index, one finds that the hodotermitids and the macrotermitids are the only two groups that have a higher than expected value based on the number of genera consumed.

The subfamilies must be divided to analyze the comparative effectiveness of the different termite defenses. This is because soldier-based defensive systems in termites vary among genera within a subfamily. The resulting analysis shows that the most frequently consumed species are those with a mixed mandibulate–chemical defense. However, when analyzing the total number of times mammalian predators were reported to eat species in termite genera within a given type of defense, it is apparent that no major differences exist between types of defenses.

4. DISCUSSION

4.1. Eating Ants

4.1.1. Mammals As Predators

Of the mammals covered in this survey, 180 species (83.3%) eat ants. Undoubtedly the number of all mammalian species that eat ants is much greater. Overall, ants are taken by predators foraging both arboreally and terrestrially, but in tropical areas they are primarily taken by arboreal myrmecophages. Erwin (1983) and Adis et al. (1984) found that ants constituted the greatest number of individual insects and the greatest biomass in the canopy of neotropical forests. Considering this ubiquity, it is not surprising that primates, arboreal squirrels, and bats are largely ant-eaters.

Prior to Vestgens and Hall's (1977) study on Australasian bats, the extent to which ants were an important source of chiropteran food was underestimated. Nine species of bats were recorded as consuming worker ants and in two species ants were quite common in the stomachs examined.

Ubiquity also may account for the frequency with which primates consume ants. In addition, primates, unlike bats, have the capacity to open ant nests and consume ant larvae and eggs as well as adults. This habit is common in monkeys of the genera *Cebus* and *Cercopithecus*. Twig-nesting ants such as *Crematogaster*, *Pseudomyrmex*, and *Cephalotes* are particularly susceptible to having their nests raided.

Two species of mammals known to consume 91–100% ants by volume, the silky anteater *Cyclopes didactylus* and the mouse *Prionomys batesi*, are arboreal feeders. A third species of arboreal feeder, the pangolin *Manis tetradactyla*, probably belongs in this group, although data are lacking. A recent study (Smith, 1982) on the striped possum (*Dactylopsila trivirgata*) indicates that this species may also join the guild of arboreal anteaters (96% by occurrence of the stomach contents of two individuals was ants; this species not included in Table I). Ten of the 14 species for which ants occupy more than 51% of the diet feed arboreally.

Significant consumption of ants is not confined to arboreal species; the rodents *Funisciurus anerythrus*, *Lophuromys sikapusi*, and *Oxy-*

mycterus rufus and the mongoose *Mungos* are all terrestrial, or largely so, and ants are very commonly found in their diets. The fossorial mole *Parascalops* also finds large numbers of ants underground.

Most myrmecophages eat ants opportunistically. This group of species, in whose diet ants are found less than half the time, is undoubtedly underrepresented in this study. Species such as the elephant shrew (*Elephantulus rufescens*) or the deer mouse (*Peromyscus leucopus*) take a broad range of arthropods, as well as other types of food, and probably consume ants as they encounter them. A striking example of this is the duiker (*Cephalophus*), which was found to consume many individuals of a solitary foraging ponerine. These opportunistic ant-eaters would be expected to take a large range of ant species. It is probable that some of the species for which ants appear to contribute a major part of the diet, such as the rodent *Paraxerus poensis,* might simply be responding opportunistically to very high densities of ants in the places where they forage.

Other predators whose diet does not include many ants seek out and break into ant nests. This is particularly common in temperate bears and mustelids. The frequency of temperate zone species in the category of "ants only" is undoubtedly a function of the scarcity of termites in these colder areas.

For predators that consume large quantities of ants, the insects are clearly of important nutritional value. This is probably also the case for species that consume smaller quantities, particularly those that are otherwise predominantly plant-feeders. Thus, for the duiker (ants = 11% dry weight of diet; Grimm, 1970) and chimpanzees (ants = about 4% dry weight of diet; Hladik, 1973), ants probably provide vital nutritional elements.

4.1.2. Ants As Prey

Colony defense in ants includes stinging, biting, and spraying chemicals. Large-colony species can rapidly mobilize, cover an attacker in seconds, or run rapidly carrying their eggs and pupae. Ants can nest in the branch tips of tall trees, deep in tree trunks, or behind the thick clay walls of termite mounds. A great deal of work has been done on the effectiveness of such behaviors in deterring predation by other arthropods (Hermann and Blum, 1981; Wilson, 1971), but their effectiveness against mammals is poorly understood.

This review shows that no single type of defense, either by the individual ant or by the colony, prevents all mammalian predation. Subfamilies can be grouped according to the types of defense exhibited by their members. An analysis of the patterns of mammalian predation on representatives of these groups gives the same results as an analysis of predation on species in the separate families.

4.1.2a. "Army Ant" Defense. Army ants, both Old and New World (Ecitoninae and Dorylinae), with their huge swarms of highly aggressive biting and stinging workers, are commonly found in mammalian diets. In fact, some mammals, such as the mongoose *Bdeogale nigripes,* seem to specialize on the driver ant, *Dorylus* (*Anomma*) *nigricans* (Lang, personal communication in Bequaert, 1922; Kingdon, 1977). The fact that these mammals prefer this type of ant is attested to by their fondness for *Myrmicaria* (Myrmicinae), which is similar in aggressiveness to *D. nigricans* (Arnold, 1915).

Chimpanzees exploit the aggressiveness of *Dorylus* by "fishing" for army ants using a long stick. Even though not sufficient to deter predation, the defenses of *Dorylus* are effective in limiting it, because chimpanzees are very wary when fishing for army ants and are clearly affected by their bites (McGrew, 1974). Most of the species that prey on *Dorylus* and *Eciton* consume large quantities of ants—this is not a type of ant taken by opportunistic anteaters.

4.1.2b. "Ponerine" Defense. This group of ants includes the subfamilies Myrmiciinae and Ponerinae, generally solitary foragers living in relatively small colonies (Arnold, 1915; Bolton, 1973; Wilson, 1971). Species from this group are taken by all types of ant-eaters, from the specialized ant-eating echidna (*Tachyglossus*) to opportunistic predators like the jackal (*Canis mesomelas*) and the tenrec (*Tenrec*). The large size, tendency to forage and nest in many different types of locations, and cosmopolitan distribution of such ants make them available to arboreal and terrestrial predators. Not enough data are available to assess how frequently ponerine and myrmiciinae nests are raided, but the majority of the recorded predation was probably on foragers. Solitary foraging by most ponerine ants probably limits the extent of predation and gives these ants a role much more like that of other solitary arthropods in mammalian diets.

4.1.2c. Stinging and Biting Defense. The third group of ants, most of which are small-bodied and capable of biting and stinging simultaneously, contains the subfamilies Pseudomyrmecinae and Myrmecinae. This category contains the greatest number of ant genera. Of the three ant genera whose species are consumed by the most species of mammals, two, *Crematogaster* and *Pheidole,* fall in this category of defense.

Virtually all of the species of Pseudomyrmecines are arboreal nesters, as are *Cephalotes, Cataulacus,* and *Crematogaster*. Many of these arboreal species are vulnerable to nest predation, particularly genera, which nest in twigs and rotting wood. Frequently, the brood is preferred, as Nishida and Hiraiwa (1982) report for chimpanzees that open *Crematogaster* nests, brush away the adults, and eat the eggs and larvae.

Crematogaster, very common in West African rain forests, is reported by Pages (1970) to be the only food taken by the pangolin *Manis tetradactyla*. It would be interesting to know whether or not this extreme specialization holds true in other localities. It is possible that *M. tetradactyla* was responding to a very high abundance of *Crematogaster* in the areas where Pages worked. Another extreme ant specialist, the mouse *Prionomys,* in captivity will eat only *Tetramorium* and *Macromischoides,* both genera of this group (Harrison, 1950).

Genera such as *Solenopsis* have very aggressive workers that pour out of a broken nest and deliver painful stings. Despite this, they are commonly eaten by sloth bears (*Melursus*) and undoubtedly by many other species as well. Schaller (1983) has suggested that the aggressiveness of *Solenopsis,* living commensally in termite mounds, causes giant anteaters (*Myrmecophaga*) to stop feeding on termites.

Fungus-growing ants, represented by *Trachymyrmex, Acromyrmex,* and *Atta,* cannot sting, but have heavily armored exoskeletons. There has been controversy in the literature as to whether or not these species are eaten by mammals (Redford, 1983), but as shown here, they are eaten by several different species, including giant anteaters and giant armadillos.

4.1.2d. Biting and Chemical Defense. Ants in this group, belonging to the subfamilies Dolichoderinae and Formicinae, have no functional sting and rely instead on biting and/or a chemical defense. Ants of the genus *Camponotus* spray formic acid through the acidopore (Hermann and Blum, 1981), but this chemical defense appears to be ineffective in preventing mammalian predation and species in this genus are the most frequently preyed upon among all ant genera. This analysis, conducted at the generic level, underestimates the frequency with which species of *Camponotus* are eaten. Chimpanzees from a single site have been recorded consuming four different species of this genus (Nishida and Hiraiwa, 1982). Species of the genus *Camponotus* are probably the most common ants in the world, despite their vulnerability to mammalian predation.

Camponotus individuals tend to be very aggressive. Chimpanzees, particularly young individuals, occasionally stop preying on a nest because of attacks from them (Nishida and Hiraiwa, 1982). Chimpanzees and orangutans can use this aggressiveness to their advantage by disturbing a nest and preying on the soldiers that emerge (Nishida, 1973; Rijksen, 1978).

Weaver ants (*Oecophylla* and some species of *Polyrhachis*) are also aggressive formicines preyed upon by mammals. The soldiers of *Oecophylla* are potentially effective against primates and three species use different techniques to avoid the soldiers. Macaques (*Macaca sinica*) break the nest from a tree, drop it to the ground, and consume the brood after the adults have dispersed (Hladik, 1977b). Chimpanzees approach an *Oecophylla* nest rapidly, crush the nest, and pick out the brood and dead adults (Nishida and Hiraiwa, 1982). Orangutans have been seen to use a third technique, stripping the leaves down the branch holding an *Oecophylla* nest and sealing it within a protective layer of leaves. The nest is then held tightly, allowing only a few adults at a time to leave the nest, which are then eaten (Rijksen, 1978). Like other primates, humans in many parts of the world also eat ants, particularly *Oecophylla* and *Camponotus,* which are used in curries and as a "spicy snack" (Bodenheimer, 1951).

Species in the dolichoderine genera *Iridomyrmex, Dolichoderus,* and *Azteca,* like the myrmecine *Solenopsis,* are known for their aggressive nest defense (Griffiths and Simpson, 1966; Lubin and Montgomery, 1981). Yet, despite this defense, accompanied by a strong bite, *Dolichoderus* and *Azteca* are important prey for *Tamandua* anteaters. These mammals are visibly bothered by the attacks of such ants, and spend much time grooming them off, yet consume them in large numbers (Lubin and Montgomery, 1981).

Species in three ant genera, *Plagiolepis, Anoplolepis,* and *Myrmecocystus,* have a system of storing nectar in workers with expanded gasters (repletes). These "honey-pot" ants are well known in the southwestern United States, where badgers and coyotes (Chew, 1979) and humans (Bodenheimer, 1951) dig up the repletes for food. A similar phenomenon apparently takes place in Africa, where the highly myrmecophagous pangolins (*Manis gigantea* and *M. temmincki*) and aardvark (*Orycteropus*) are predators of *Plagiolepis* and *Anoplolepis* (Forel 1909 in Bequaert, 1922; Smithers, 1971).

Mammals from a large number of families all over the world eat ants, some in large numbers. The foraging behavior of the predator seems largely to determine the taxonomic identity and therefore the defense type of its ant prey. There are probably few ant species not consumed by some mammalian predator. Rather than preventing predation, the defenses of ants seem to serve to limit the predation. Ants can largely control the extent and impact that predation may have on their colony by rapidly responding to a predatory attempt. This response can take the form of attacking the predator as in *Dorylus* and *Azteca* or fleeing from it as in *Cephalotes*. In the case of the swarming defense, getting eaten may very well be part of the defense of the colony.

4.2. Eating Termites

4.2.1. Mammals As Predators

Sixty-four percent of the 216 mammalian species listed eat termites, as compared with 83.3% that eat ants. Of these 138, only 37 were recorded as eating termites alone. Many of these species probably also eat ants, though data to check this are not available. The largest group recorded as exclusively termite-eating (37.8%) is composed of small carnivores; 11 of the 14 species are East African and eight of these are mongooses. Several mongooses, such as *Mungos* and *Ichneumia*, do eat ants, and further work will probably show that most of the small carnivores listed here eat ants as well as termites. The exception to this may be the two cats, *Felis serval* and *F. silvestris*. It seems likely that most of the mammals recorded as primarily or exclusively termite-eaters fail to take ants because the ants available to them are often fast-fleeing. Termites, once located, are easier to eat than ants, a conclusion substantiated by Rathbun's (1979) study on elephant shrews (*Elephantulus* and *Rhynchocyon*).

The association between arboreality and ant-eating is complemented by an association between terrestriality and termite-eating. At least 34 of the 46 mammal species that eat termites primarily or exclusively are mainly or entirely terrestrial. No monkeys are recorded eating more than 0–10% termites (occurrence or volume). This association can be seen in the nine species of West African squirrels studied by Emmons (1975, 1980), in which the two species with the highest percent occurrence of termites were primarily terrestrial. However, terrestriality does not dictate a diet of termites rather than ants, as shown by the terrestrial *Epixerus*, which eats a few ants and no termites.

Five of the seven species recorded as eating over 50% termites also eat ants, though in lesser abundance. The two exceptions, the mongoose *Rhyncogale* ($n = 2$ stomachs) and the oppossum *Metachirus* ($n = 8$ stomachs), fall into the category mentioned earlier; that is, species that probably will be shown on further study to eat ants. Data on the aardwolf (*Proteles*), the numbat (*Myrmecobius*), and the bat-eared fox (*Otocyon*) are sufficient to show that these species consistently consume many more termites than ants.

Based on an examination of 252 numbat scats, Calaby (1960) showed that 85% of *Myrmecobius'* diet is termites (ten genera) and 15% ants (14 genera); he concluded that the ants were taken while the numbat was feeding on termites. It is possible that this may not always be true and ants may sometimes be sought after and eaten (J. A. Friend, personal communication).

Bat-eared foxes show a preference for harvester termites (*Hodotermes*), although they have been shown to eat *Macrotermes* as well. Most investigators who have examined *Otocyon* food habits report this species to eat many more termites than ants (Bothma, 1959; Smithers, 1971; Kingdon, 1977), but Nel (1978) reports ants to be a common food in summer months.

The third species, the aardwolf, is probably the most specialized of the termite-eating mammals. In many areas of East Africa the single species *Trinervitermes bettonianus* makes up more than 90% of the diet of *Proteles* (Bothma and Nel, 1980; Bothma et al., 1984; Kruuk and Sands, 1972). In other areas, *T. bettonianus* is replaced by other species of *Trinervitermes* (Bothma, 1965, 1971; Smithers, 1971), or by the very similar *Fulleritermes* (Dean, 1978). The only other species recorded in the aardwolf diet are harvester termites (*Hodotermes*) or three species of macrotermitids (Dean, 1978; Kruuk and Sands, 1972; Smithers, 1971). Ants were recorded only rarely, and in 79 fecal samples surveyed by Kruuk and Sands ants were never recorded at more than 14% occurrence.

The mammalian species whose diets consist of less than 50% termites are typified only by their distribution in warmer areas where termite biomass is high. There are five rodents whose diet consists of 11–50% termites; probably many more will be found to utilize these abundant social insects as food. Bats are usually not considered to be termite-eaters, but Vestgens and Hall (1977) found that three species ate termites regularly, including one *Macroderma* in whose stomach they found more than 50 *Nasutitermes* soldiers.

4.2.2. Termites As Food

Termites display many forms of individual and colony defense similar to those of ants. Some species nest in tall trees and others build extremely hard mounds; some forage under protective soil galleries or in the safety of wood. Some termite soldiers have strong biting mandibles, which can be accompanied by noxious chemical secretions. But perhaps the most advanced form of social insect defense is found in the species with nasute soldiers that can fire a potentially noxious or repellent chemical without contacting the potential predator (Deligne et al., 1981; Prestwich, 1984). As with ants, none of the general forms of termite soldier defense serve to prevent mammalian predation.

4.2.2a. No Soldiers. Three of the termite genera with species consumed by mammals have no soldiers. At least two of these genera are largely geophagous (Mill, 1982a), resulting in extremely high ash contents (Redford and Dorea, 1984). They have been recorded in only minor numbers in the diets of the giant anteater (Redford, 1985a), whereas the much more opportunistic burrowing mouse (*Oxymycterus*) eats them more commonly and with a higher preference value (Mathews, 1977; Redford, 1984b). It is probable that more specialized myrmecophages, which eat only ants and termites, are strongly influenced by the poor nutritional quality of some social insect species (Redford and Dorea, 1984).

4.2.2b. Strictly Mandibulate Defense. Eight genera of termites with species eaten by mammals have soldiers with a defense based solely on biting, snapping, or slashing mandibles (Deligne et al., 1981). The actual configuration of these mandibles varies greatly, with some species having long, slender mandibles and others short, biting mandibles. Nineteen of the 28 cases of mammalian predation on termites with a mandibular defense were on hodotermitids. As shown earlier, this is much higher than would be expected for the family. *Microcerotermes* was the only genus in this defense group with species that were consumed by arboreal mammals.

Anacanthotermes on the Indian subcontinent is similar in form and habit to *Hodotermes* and is preyed on by a set of mammals similar to the African hodotermitid predators. Fifteen of the 19 predators of hodotermitids were terrestrial carnivores, two were primates, which preyed on these termites on the ground, and two were rodents.

Two lines of evidence point to the ineffectiveness of mandibulate soldiers against mammals. First, the proportion of soldiers to workers is much lower in these genera than in others (Coles de Negret and Redford, 1982), and second, mandibulate soldiers often flee from mammalian predators. As an example of the ineffectiveness of such soldiers, *Hodotermes* soldiers rarely or never venture above ground, and thus the workers forage "unprotected" (Wilson and Clark, 1977). When attacked, workers rapidly flee down their exit holes.

4.2.2c. Mixed Mandibulate and Chemical Defense. The most frequently preyed upon group of termites have soldiers that combine biting mandibles with a poorly developed chemical defense. For these soldiers, physical contact with the predator is necessary for their chemicals to be effective. Six of the 15 genera with soldiers in this category belong to the subfamily Macrotermitinae, with *Macrotermes* the most frequently preyed on genus (15 cases), followed by *Odontotermes* (13 cases). The macrotermitids are dominant termites in Africa and Asia and are consumed by a variety of terrestrial predators, from hedgehogs to bears.

One hundred percent of the incidences of mammalian predation on the mandibulate–chemical termites were by terrestrial mammals. The South American termite ecological equivalents of the macrotermitids are the genera *Syntermes*, *Cornitermes*, and related genera. The research to date on South American mammals is far from complete and undoubtedly further work will show these termites to be frequently eaten by mammals (Redford, 1983).

Soldiers with a mandibulate–chemical defense are among the largest termite soldiers and can deliver a nasty bite. In his work on the stomach contents of small rodents and marsupials, Mathews (1977) found *Syntermes* and *Cornitermes* soldiers scarce or absent, although workers were present. His conclusion was that the soldiers were effective in limiting the extent of mammalian predation. I found similar results in laboratory tests with the burrowing mouse (*Oxymycterus*) (Redford, 1984b). This effectiveness is clearly related to the size of the predator, though, because *Syntermes* soldiers were frequently consumed by giant anteaters (Redford, 1985a). The response of such soldiers depends on the size of the disturbance—they flee from a large disturbance and recruit to a small one.

As a result of their aggressiveness, *Macrotermes* and *Pseudoacanthotermes* soldiers are frequently consumed by chimpanzees and humans (Bodenheimer, 1951; Uehara, 1982). Both predators insert sticks or other "fishing" tools into termite foraging holes or mounds, where the tools are attacked by soldiers reacting to the disturbance. The predators then remove the tool and eat the attached soldiers.

The frequency with which termites with mixed defenses are eaten can partially be explained by their foraging habits. Like hodotermitids, these termites tend to forage on the surface, although sometimes covered by soil sheeting (Kruuk and Sands, 1972). This soil sheeting may reduce predation from some predators, such as the aardwolf. However, it provides no protection from other predators, such as the elephant shrew (Rathbun, 1979). Large colony size, surface foraging, ubiquity, and large individual size all make these termites good mammalian food.

Macrotermitids taken from the nest are a source of food for some predators. The extent and importance of such nest predation are unclear. The hard mounds of some *Macrotermes* and *Cornitermes* are sufficient to stop all but the most powerful diggers. Many authors report giant pangolins and aardvarks (and some armadillos for *Cornitermes;* Redford, 1984a) as important nest predators.

4.2.2d. Strictly Chemical Defense. Termites with a strictly chemical defense have a higher proportion of soldiers to workers and more aggressive soldiers than termites of other defense types (Coles de Negret and Redford, 1982). These soldiers recruit to a disturbance and discharge chemicals, conducting a form of synergistic chemical warfare. There has been a great deal of research on termites with a chemical defense (Mill, 1982a; Prestwich, 1984) and the conclusion has been that such defenses are very efficient at deterring predation by other arthropods. Lubin and Montgomery (1981) and Kruuk and Sands (1972) have shown or suggested that such a defense is also highly effective against mammals. But as in all the other ant and termite defenses discussed, a chemical defense does not prevent predation, but rather limits it. As shown in this survey, nine of these termite species are eaten, with several of them serving as major sources of food for specialized myrmecophages.

The more specialized myrmecophages eat a large number of termites with chemical defenses—19 of the 28 cases of predation are by mammals whose diet consists of 91–100% termites. Again, most of the cases of predation recorded are by terrestrial predators; the major exception is predation on *Nasutitermes*. Species of this cosmopolitan genus nest and forage at all levels and are consumed at most of them. In virtually all areas they are probably the most "apparent," if not dominant, arboreal termites (Mill, 1982b).

Coles (1980), Lubin and Montgomery (1981), and K. H. Redford (unpublished results) have all found that nasute soldiers and/or their chemical secretions depress mammalian feeding. However, Redford and Dorea (1984) have shown that nasute soldiers and workers both have high nutritional value. The way in which a mammal may feed on termites that have a high food value but also have soldiers with an effective chemical defense has been suggested by Kruuk and Sands (1972) for aardwolves preying on *Trinervitermes*, by Lubin and Montgomery (1981) for *Tamandua* anteaters preying on *Nasutitermes*, and by Redford (1985) for giant anteaters preying on *Velocitermes*. In all cases the predators appear to feed in short bouts, a method that minimizes their encountering and/or eating soldiers. Such feeding behavior would minimize their exposure to the chemicals of the soldiers and increase the number of workers per soldier ingested. Short feeding bouts may also reflect an animal's limited capacity to digest toxic chemicals, as suggested by Freeland and Janzen's (1974) study of herbivorous mammals, which noted that herbivores can detoxify chemicals, but only in small quantities, and thus can ingest only small amounts of a given toxin-bearing food. This could apply to mammals ingesting nasute soldiers.

In response to the behavior of their prey, myrmecophagous mammals could be expected to feed for short periods of time, to ingest more workers per soldier than the average colony proportion, to seek areas with a "favorable" worker-to-soldier ratio (such as termite feeding locations), and to leave a feeding site when potential prey are still available. Such behaviors have been observed for aardwolves, giant anteaters, numbats, and two species of *Tamandua* anteaters, all feeding on nasute termites.

In summary, none of the four major types of termite defenses prevent mammalian predation. Instead, like the defenses of ants, they limit predation by decreasing the food value of the colony on which the mammals are feeding. This can be done either by running away (*Hodotermes*) and removing the source of food or by recruiting soldiers and having workers flee (*Trinervitermes*) and thus decreasing the quality of the food. Both behaviors would explain the short bouts characteristic of mammals feeding on termites and ants. As an example, the average feeding time of bat-eared foxes on *Hodotermes* was 7.4 sec ($n = 16$, range = 1–58; Nel, 1978) and the average time for aardwolf feeding on *Trinervitermes* was 24 sec ($n = 79$, range = 3–7; Kruuk and Sands, 1972).

4.3. Eating Ants and Termites: Flexibility and Variation

Most mammals that eat large quantities of ants also eat large quantities of termites. They are specialized not on prey of a particular taxonomy, but on prey of a certain type—small, fairly homogeneous, social arthropods. There are often disagreements in the literature about which type of social insect is of greater importance in the diets of myrmecophages with little appreciation of the dietary variation within a species. As Redford (1985a) has shown, the giant anteater in some areas feeds largely on termites and in others largely on ants. Griffiths (1968) showed a similar finding for the echidna (*Tachyglossus*): in some areas of Australia, ants composed 98% of the fecal samples, while in others they occupied only 18%. Newsome (1962), working on rabbit-eared bandicoots, and Griffiths (1968), working on echidnas, have suggested that termites may be eaten by these animals in dry areas because of their higher water content.

Intraspecific variation in diet can occur in several ways. Nel (1978) has shown that in different areas the proportion of termites to ants in the diet of bat-eared foxes varies. Melton (1976b) states that aardvarks eat termites in the wet season and ants in the dry season. Seasonal differences have also been documented for aardwolves (Kruuk and Sands, 1972) and mongoose (Lynch, 1980). In a very interesting example, Mills and Mills (1978) have shown that hyena cubs eat more termites (hodotermitids) than do adult hyenas because cubs are more active during the day in winter, thus overlapping with the foraging period of the termites. Bothma and Nel (1980) have shown that the relative proportions of *Hodotermes* and *Trinervitermes* in aardwolf diets vary between years.

A final type of variation can exist between individuals and has been shown in bat-eared foxes (Nel, 1978), pangolins (Lang, personal communication in Bequaert, 1922), and *Tamandua* anteaters (Lubin and Montgomery, 1981). This last instance of interindividual variation is of particular interest; within one population one individual *Tamandua tetradactyla* preferred termites (90% of feeding bouts on termites), while another preferred ants (76% of feeding bouts on ants) (Lubin and Montgomery, 1981). *Tamandua* anteaters also varied individually in degree of arboreality, a difference that correlated with eating different proportions of the same species of ants (Montgomery and Lubin, 1977).

In many cases myrmecophages choose ants or termites not on an absolute scale of preference, but relative to what is available to them. For example, Kruuk and Sands (1972) have shown that aardwolves have a much more varied diet in the wet season, when more termites forage unprotected above ground. This is concomitant with a decrease in the above-ground activity of its major prey species, *Trinervitermes*. Both factors result in a seasonal change in diet. Uehara (1982), working on chimpanzees, and Field (1975), working on mice (*Lemniscomys*), have both shown that termites are eaten by these animals when they are most available. This is probably the circumstance under which most mammals eat termites and ants and reflects the opportunistic nature of much of this predation. The relative availability of ants versus termites may explain why some predators, such as most arboreal monkeys, eat mostly ants and others, such as savanna-dwelling carnivores, eat termites.

Opportunistic predation is not confined to those species that eat small quantities of ants, but is practiced by specialized myrmecophages as well. Numbats will feed on termites exposed by echidnas (Friend, 1982), aardwolves on termites exposed by aardvark diggings (Bothma and Nel, 1980), and giant anteaters on termites exposed by the diggings of armadillos (K. H. Redford, unpublished observations). These observations again reflect the relative nature of myrmecophage preference for certain species of prey and how preference is a function of "availability."

Individual differences in feeding may be nonresource-related. Many of the highly myrmecophagous mammals care for their young in a way that would enable "cultural" transmission of food preferences and account for such nonresource-related individual differences (McGrew, 1983). This form of care, which involves the mother carrying the infant with her as she moves about her home range, is found in three genera of anteaters, probably all seven species of pangolins, sloth bears, and to a lesser extent echidnas. Accompanying the mother while she feeds and eventually feeding with her, the young may learn the same preferences for food type or foraging location exhibited by its mother.

The flexibility of preference exhibited by highly myrmecophagous mammals and the ubiquity of ants and termites in many parts of the world allow the broad distribution of the guild of ant- and termite-eating mammals. These two factors probably also account for the enormous range of such species as the giant anteater, the giant armadillo, and the aardvark, all three of which are found from dry thorn scrub to wet tropical forest. In each habitat they are probably selecting from among the available species of ants and termites.

4.4. Other Types of Food

Predation on the winged reproductives of ants and termites (alates) outside the nest is different than predation on the worker and soldier caste. Alates are much more nutritious than workers and soldiers (Redford and Dorea, 1984) and when located outside the nest are virtually defenseless. The enormous numbers of alates that emerge seasonally are preyed upon heavily by birds, arthropods, spiders, reptiles, and mammals.

Alates within a nest can change the food value of the ant or termite colony. Griffiths and Simpson (1966) have shown this for echidnas and Lubin and Montgomery (1981) for *Tamandua* anteaters. In both cases the presence of alates or prealates in an ant or termite nest was strongly correlated with longer feeding bouts on that nest. In fact, *Tamandua* anteaters usually attacked healthy *Nasutitermes* nests only when alates were present. Echidnas apparently never attack *Iridomyrmex* nests except when alates are present.

Highly specialized myrmecophages do consume prey other than ants and termites. Beetle larvae and adults are perhaps the most common additional type of food, having been found in the diets of echidnas, anteaters, pangolins, aardwolves, bat-eared foxes, and numbats. Occasional mass emergences of moths provide food for echidnas and aardwolves. In addition, a few observations have been made of highly myrmecophagous mammals consuming prey very different from ants and termites: aardvarks and naked-tailed armadillos have been found eating mice (Heck 1940 in Redford, 1985b; Smithers, 1971); aardwolves have been found to eat carrion (Bothma, 1965); bat-eared foxes eat many small vertebrates and fruit (Bothma and Nel, 1980); captive giant anteaters appear to relish mice (K. H. Redford, unpublished observations); and captive *Tamandua* anteaters kill and consume lizards (P. Parker, personal communication). These incidents have always been regarded as anomalous, but in fact such feeding is probably more common than suspected. Many animals live and nest in and under termite mounds and myrmecophagous mammals digging into the mound might consume them if they could capture them.

Wasps, bees, and earthworms are also frequently eaten by ant- and termite-eating mammals. Wasps and bees are common prey for temperate carnivores and tropical arboreal foragers, such as monkeys (Redford, 1983; Terborgh, 1983). Earthworms are often very abundant (Edwards and Lofty, 1977), as well as slow-moving, defenseless, and easily encountered. A high incidence of earthworms in the diet of the viverrid *Eupleres* (Albignac, 1974), the fox *Vulpes* (Macdonald, 1980), and the opossum *Philander* (Charles-Dominique et al., 1981) probably reflects a high biomass and ease of obtaining this prey type. The availability of this prey may explain the abundance of earthworm-eating rodents such as *Scapteromys* (Massoia and Fornes, 1954). The relationship between termite-eating and earthworm-eating can be seen in the echidna *Tachyglossus*, which apparently eats only ants and termites in Australia (Griffiths, 1978), but adds earthworms to its diet when it occurs in New Guinea (Van Deusen and George, 1969).

4.5. Form, Function, and Feeding

Griffiths (1968) has detailed the morphological traits that characterize specialized mammalian myrmecophages. These include (1) extensile veriform tongues with large salivary glands to lubricate them, (2) long snouts and jaws with grossly modified teeth or lacking teeth, (3) anomalous stomachs, and (4) forelimbs adapted for digging.

Not all of the mammals whose diet is composed of over 90% ants and/or termites have all of these characteristics. Friend (1982) has suggested that lack of many of these specializations, such as in the numbat and aardwolf, is correlated with a diet composed largely of termites. He suggests that this is because ants, with their heavier exoskeletons, are more difficult to digest than softer-bodied termites. This may be the case, but it seems more likely that termites are being chosen for other reasons, which, at least in the case of the aardwolf, may have to do with prey "availability."

Patterson (1975) has pointed out that irrespective of the exact pieces composing the "adaptive complex" typifying specialized myrmecophages, the function of the whole is the rapid transfer of small prey to the stomach in large quantities. The need for speed is a result of the rapidity with which ants and termites can respond to mammalian attack.

One result of the constraints imposed on such rapid ingestion is the absence of chewing. Emmons (1980) has suggested that some squirrels (genus *Funisciurus*) eat ants and termites without chewing to enable them to ingest their prey more rapidly. Absence of chewing has been documented for numbats (Calaby, 1960), striped possums (*Dactylopsila;* Smith, 1982), the mongoose *Bdeogale* (Lang, personal communication in Bequaert, 1922), nine-banded armadillos (K. H. Redford, unpublished observations), probably aardwolves, and of course the toothless anteaters and pangolins. "Chewing" or trituration is taken over by muscularized stomachs (giant anteaters), by specialized structures on the tongue (*Tachyglossus*), or by grinding action using sand or pebbles in the stomach (numbats) (Griffiths, 1968).

Lack of chewing is correlated in specialized myrmecophages with loss of teeth (anteaters and pangolins) or an apparent lack of selection on the teeth. This last fact comes from the observation of great variation in tooth number and size in such myrmecophages as numbats (Tate, 1951), aardwolves (Dobroruka, 1970), and giant armadillos (Lonnberg 1942 and Wetzel 1981 in Redford, 1985b). Loss of teeth and the concomitant lack of chewing is also correlated with ingestion of large quantities of indigestible material: sand, dirt, pebbles, sticks, leaves, and grass. All species of highly myrmecophagous mammals ingest a lot of this material: sand composed over 65% of the weight of aardwolf feces (Bothma and Nel, 1980) and has been recorded as composing 47% of material in aardvark stomachs (Smithers, 1971). Ingestion of this amount of indigestible material is a function of a long, sticky tongue rapidly flicking in and out.

Not all mammals that are highly myrmecophagous consume ants and termites without chewing. For these species the cost of chewing prey inextricably associated with dirt and sand is very worn teeth, as found in sloth bears (Pocock, 1941), a mongoose (*Rhyncogale;* W. F. H. Ansell, 1960), and a squirrel (*Rhinosciurus;* Lekagul and McNeely, 1977). These species may be embarking on the path toward loss of chewing.

However, it may be that other important components of their diet, such as fruit (sloth bears; Laurie and Seidensticker, 1977), require chewing and thus act in the opposite direction and maintain chewing and teeth. McNab (1984) notes that the amount of ingested inorganic material scales with myrmecophage body size; the smaller predators, such as mice, are able to ingest the ants and termites without associated detritus.

Two studies have suggested ways in which the indigestible material ingested by myrmecophages may benefit these animals. The first work, by Horner et al. (1965) on the highly insectivorous grasshopper mouse (*Onochomys*), suggested that sand helped process the food. Sand, found in 100% of the feces, would scour the broken pieces of arthropod chitinous exoskeleton in the gut and make available for digestion otherwise unavailable insect flesh. The second suggestion, made by Hladik (1977a) for folivorous monkeys, was that soil, particularly clay, is deliberately eaten to absorb some of the leaf tannins ingested by the monkeys. Geophagy is fairly common among primates and this absorptive ability might serve as a more general explanation for geophagy if dirt were able to absorb chemicals other than tannins. No work has been done on the digestive physiology of myrmecophages, but they apparently have ways of detoxifying the chemicals prevalent in their prey. One such way might be the use of indigestible material to absorb potentially toxic or digestion-inhibiting chemicals.

This might also explain a certain feeding behavior performed by nonspecialized myrmecophages when eating ants or termites: the deliberate ingestion of indigestible plant and/or mineral material. Chimpanzees (Hladik, 1973; Nishida, 1973), bears (Novikov, 1962), and orangutans (MacKinnon, 1974) all eat the nest material of ants, termites (including the whole nest of *Nasutitermes;* Hladik, 1973), and bees when they prey on these insects. Bark, nest carton, and inorganic nest material may all help the mammalian predator counteract the effects of the insect chemicals.

ACKNOWLEDGMENTS. I would like to thank the following organizations for funding my research: Organization of American States, National Geographic Society, Sigma Xi, Harvard University, and Friends of the National Zoo. Kay Kenyon provided unfailing bibliographic help, and V. Garber and B. L. Stanton helped me valiantly at the typewriter. P. Shaw, K. Kranz, and M. Stuwe kindly translated for me, and the following people greatly helped with unpublished data or advice: W. F. H. Ansell, P. Barreto, M. Bochert, J. Dietz, J. M. Duplantier, J. Eisenberg, L. Emmons, J. A. Friend, E. Gould, K. Johnson, B. Konstant, W. C. McGrew, D. McKey, B. McNab, R. Mittermeier, M. Nicoll, G. Prestwich,

G. Rathbun, P. Richardson, J. Robinson, G. Schaller, A. Smith, S. Srikosamatara, J. Verschuren, and P. Waser.

My social-insect support team, B. L. Thorne and J. F. A. Traniello, provided me with the usual patient answers to many questions. This paper benefitted greatly from the advice of A. W. Crompton, J. Eisenberg, B. Holldobler, Y. Lubin, K. Sebens, P. Shaw, B. Thorne, J. Traniello, and P. Waser.

REFERENCES

Adis, J., Lubin, Y. D., and Montgomery, G. G., 1984, Arthropods from the canopy of inundated and terra firme forests near Manaus, Brazil, with critical considerations on the pyrethrum-fogging technique, *Stud. Neotrop. Fauna Environ.* **19**:223–236.

Albignac, R., 1974, Observations éco-éthologiques sur le genre *Eupleres*, Viverridae de Madagascar, *Terre Vie* **28**:321–351.

Allen, G. M., 1938, *The Mammals of China and Mongolia*, American Museum of Natural History, New York.

Ansell, A. D. H., and Ansell, P. D. H., 1969, *Petrodromus tetradactylus* at Ngoma, *Puku* **5**:211–213.

Ansell, W. F. H., 1960, Contributions to the mammalogy of northern Rhodesia, *Occ. Pap. Natl. Mus. S. Rhodesia* **3**:351–390.

Ansell, W. F. H., 1964, Addenda and corrigenda to "Mammals of Northern Rhodesia," *Puku* **2**:14–52.

Ansell, W. F. H., 1965, Addenda and corrigenda to "Mammals of Northern Rhodesia," *Puku* **3**:1–14.

Arnold, G., 1915, A monograph of the Formicidae of South Africa, *Ann. S. Afr. Mus.* **24**:1–766.

Ayres, J. M., and Nessimian, J. L., 1982, Evidence for insectivory in *Chiropotes satanas*, *Primates* **23**:458–459.

Azzaroli, L., and Simonetta, A. M., 1966, Carnivori della Somalia ex-Italiana, *Monit. Zool. Ital. (Suppl.)* **74**:102–195.

Baker, R. H., 1943, May food habits of armadillos in Eastern Texas, *Am. Midl. Nat.* **19**:379–380.

Barlow, J. C., 1969, Observations on the biology of rodents in Uruguay, *Life Sci. Contrib. R. Ont. Mus.* **75**:1–59.

Beebe, W., 1922, The giant armadillo, *Zool. Soc. Bull.* **25**:117–120.

Beeman, L. E., and Pelton, M. R., 1980, Seasonal foods and feeding ecology of black bears in the Smoky Mountains, in: *Bears—Their Biology and Management* (C. J. Martinka and K. L. McArthur, eds.), Bear Biology Association USGPO, pp. 141–147.

Beng, T. K., 1965, Stomach contents of some Borneo mammals, *Sarawak Mus. J.* **12**:373–385.

Bequaert, J., 1922, The predaceous enemies of ants, *Bull. Am. Mus. Nat. Hist.* **45**:271–331.

Bodenheimer, F. S., 1951, *Insects as Human Food*, Junk, The Hague.

Bolton, B., 1973, The ant genera of West Africa: A synonymic synopsis with keys (Hymenoptera: Formicidae), *Bull. Br. Mus. (Nat. Hist.)* **27**:317–368.

Borchert, M., and Hansen, R. L., 1983, Effects of flooding and fire on valley side wet campo rodents in Central Brazil, *Rev. Bras. Biol.* **43**(3):229–240.

Bothma, J. du P., 1959, Notes on the stomach contents of certain Carnivora (Mammalia) from the Kalahari Gemsbok Park, *Koedoe* **9:**37–39.

Bothma, J. du P., 1965, Random observations on the food habits of certain Carnivora (Mammalia) in Southern Africa, *Fauna Flora* **16:**16–22.

Bothma, J. du P., 1971, Food habits of some Carnivora (Mammalia) from Southern Africa, *Ann. Transv. Mus.* **27:**15–26.

Bothma, J. du P., and Nel, J. A. J., 1980, Winter food and foraging behavior of the aardwolf *Proteles cristatus* in the Namib-Naukluft Park, *Madoqua* **12:**141–149.

Bothma, J. du P., Nel, J. A. J., and Macdonald, A., 1984, Food niche separation between four sympatric Namib Desert carnivores, *J. Zool. Lond.* **202:**327–340.

Bourliere, M. F., and Verschuren, J., 1972, Exploration du Parc National des Cirunga, Fascicule 3, Foundation pour favoriser recherche scientifique en Afrique, Brussels, Belgium.

Buttikofer, J., 1890, *Reisebilder aus Liberia,* E. J. Brill, Leiden.

Calaby, J. H., 1960, Observations on the banded ant-eater *Myrmecobius f. fasciatus* Waterhouse (Marsupialia), with particular reference to its food habits, *Proc. Zool. Soc. Lond.* **135:**183–207.

Carvalho, J. C. M., 1966, Novos dados sôbre a alimentação do tamanduá-bandeira (*Myrmecophaga tridactyla* Linnaeus, 1758), Edentata, Mammalia, *An. Acad. Bras. Cienc.* **38:**341–345.

Carvalho, J. C. M., and Kloss, G. R., 1951, Sôbre a alimentação do tamanduá-bandeira ("*Myrmecophaga tridactyla*" L., 1758), *Rev. Bras. Biol.* **11:**37–42.

Charles-Dominique, P., 1971, Éco-éthologie des prosimiens du Gabon, *Biol. Gabon.* **7:**121–228.

Charles-Dominique, P., Atramentowicz, M., Charles-Dominique, M., Gerard, H., Hladik, A., Hladik, C. M., and Prevost, M. F., 1981, Les mammifères frugivores arbicoles nocturnes d'une forêt Guyanaise: Inter-relations plantes–animaux, *Rev. Ecol.* **35:**341–435.

Chew, R. M., 1979, Mammalian predation on honey ants, *Myrmecocystus* (Formicidae), *Southwest. Nat.* **24:**677–682.

Cole, L. R., 1975, Foods and foraging places of rats (Rodentia: Muridae) in the lowland evergreen forest of Ghana, *J. Zool. Lond.* **175:**453–471.

Coles, H. R., 1980, Defensive strategies in the ecology of neotropical termites, Ph. D. thesis, University of Southampton.

Coles de Negret, H. R., and Redford, K. H., 1982, The biology of nine termite species (Termitidae: Isoptera) from the cerrado of central Brazil, *Psyche* **89:**81–106.

Crabb, W. D., 1941, Food habits of the prairie spotted skunk in southeastern Iowa, *J. Mammal.* **22:**349–364.

Davis, D. D., 1962, Mammals of the lowland rain forest of North Borneo, *Bull. Singapore Natl. Mus.* **31:**1–129.

Dean, W. R. J., 1978, Stomach contents of a specimen of *Proteles cristatus* from Angola (Carnivora: Protelidae), *Zool. Afr.* **13:**164.

De Graaf, G., 1981, *The Rodents of Southern Africa,* Butterworths, Durban.

Delany, M. J., 1964, An ecological study of the small mammals of the Queen Elizabeth Park, Uganda, *Rev. Zool. Bot. Afr.* **70:**129–147.

Deligne, J., Quennedey, A., and Blum, M. S., 1981, The enemies and defense mechanisms of termites, in: *Social Insects,* Volume II (H. R. Hermann, ed.), Academic Press, New York, pp. 2–76.

Dobroruka, L. J., 1970, Individual variation of external and cranial characters in the aardwolf, *Proteles cristatus* (Sparrman, 1783), *Rev. Zool. Bot. Afr.* **82**:163–170.

Dubost, G., 1963, Un ruminant à régime alimentaire partiellement carné: Le chevrotain aquatique (*Hyemoschus aquaticus* Ogilby), *C. R. Acad. Sci. Paris* **256**:1359–1360.

Dubost, G., 1965, Un ruminant à régime alimentaire partiellement carné: Le chevrotain aquatique (*Hyemoschus aquaticus* Ogilby), *Biol. Gabon.* **1**:21–23.

Du Toit, C. F., 1980, The yellow mongoose *Cynictis penicillata* and other small carnivores in the Mountain Zebra National Park, *Koedoe* **23**:179–184.

Eadie, W. R., 1939, A contribution to the biology of *Parascalops breweri*, *J. Mammal.* **20**:150–173.

Earle, R. A., 1981, Aspects of the social and feeding behaviour of the yellow mongoose *Cynictis penicillata* (G. Cuvier), *Mammalia* **45**:143–152.

Edwards, C. A., and Lofty, J. R., 1977, *Biology of Earthworms*, Wiley, New York.

Eisenberg, J. F., and Lockhart, M., 1972, An ecological reconnaisance of Wilpattu National Park, Ceylon, *Smith. Contrib. Zool.* **101**:1–118.

Elgmork, K., 1981, Bjorn og rev i maurtuer, *Fauna* **34**:147–149.

Emmons, L. H., 1975, Ecology and behavior of African rainforest squirrels, Ph. D. thesis, Cornell University.

Emmons, L. H., 1980, Ecology and resource partitioning among nine species of African rain forest squirrels, *Ecol. Monogr.* **50**:31–54.

Enke, B., 1978, Sieben Jahre Tamanduas (*Tamandua tetradactyla*) im Krefelder Zoo, *Zool. Gart.* **48**:19–30.

Errington, P. L., 1937, Summer food habits of the badger in northwestern Iowa, *J. Mammal.* **18**:213–216.

Erwin, T. L., 1983, Tropical forest canopies: The last biotic frontier, *Bull. Ecological Society of America* **1983** (Spring):14–19.

Fetherstonhaugh, A. H., 1940, Some notes on Malayan bears, *Malay Nat. J.* **1**:15–18.

Field, A. C., 1975, Seasonal changes in reproduction, diet and body composition of two equatorial rodents, *East Afr. Wildl. J.* **13**:221–235.

Fitch, H. S., Goodrum, P., and Newman, C., 1952, The armadillo in the southeastern United States, *J. Mammal.* **33**:21–37.

Fittkau, E. J., and Klinge, H., 1973, On biomass and trophic structure of the central Amazonian rain forest ecosystem, *Biotropica* **5**:2–14.

Fleay, D., 1942, The numbat in Victoria, *Victorian Nat.* **59**:3–7.

Fossati, L., 1937, Abitudini dell' oritteropo d'Eritrea, *Natura Milano* **28**:17–23.

Freeland, W. J., and Janzen, D. H., 1974, Strategies in herbivory by mammals: The role of plant secondary compounds, *Am. Nat.* **108**:269–289.

Freese, C. H., 1976, Predation on swollen-thorn acacia ants by white-faced monkeys, *Cebus capucinus*, *Biotropica* **8**:278–281.

Friend, T., 1982, The numbat—An endangered specialist, *Aust. Nat. Hist.* **20**(10):339–342.

Galdikas-Brindamour, B., and Brindamour, R., 1975, Orangutans, Indonesia's "people of the forest," *Nat. Geogr.* **148**:444–473.

Gautier-Hion, A., 1978, Food niches and coexistence in sympatric primates in Gabon, *Rec. Adv. Primatol.* **1**:269–286.

Gautier-Hion, A., 1980, Seasonal variations of diet related to species and sex in a community of *Cercopithecus* monkeys, *J. Anim. Ecol.* **49**:237–269.

Genest-Villard, H., 1980, Régime alimentaire des rongeurs myomorphes de forêt équatoriale (région de M'Baiki, République Centrafricaine), *Mammalia* **44**:423–484.

Gittins, S. P., and Raemaekers, J. J., 1980, Siamang, lar and agile gibbons, in: *Malayan Forest Primates* (D. J. Chivers, ed.), Plenum Press, New York, pp. 63–106.

Goodall, J., 1963, Feeding behaviour of wild chimpanzees, *Symp. Zool. Soc. Lond.* **10:**39–47.

Greegor, D. H., 1974, Comparative ecology and distribution of two species of armadillos, *Chaetophractus vellerosus* and *Dasypus novemcinctus*, Ph. D. thesis, University of Arizona.

Griffiths, M., 1968, *Echidnas*, Pergamon Press, Oxford.

Griffiths, M., 1978, *The Biology of the Monotremes*, Academic Press, New York.

Griffiths, M., and Simpson, K. G., 1966, A seasonal feeding habit of spiny anteaters, *CSIRO Wildl. Res.* **11:**137–143.

Grimm, R., 1970, Blaubocken (*Cephalophus monticola* (Thunberg, 1798); Cephalophinae, Bovidae) als Insektenfresser, *Z. Säugetierkd.* **35:**357–359.

Haddow, A. J., 1952, Field and laboratory studies on an African monkey, *Cercopithecus ascanius schmidti* Matschie, *Proc. Zool. Soc. Lond.* **122:**197–394.

Hall, K. R. L., 1965, Behaviour and ecology of the wild Patas monkey, *Erythrocebus patas* in Uganda, *J. Zool. Lond.* **148:**15–87.

Hall, S., 1980, The diets of two coexisting species of *Antechinus* (Marsupialia: Dasyuridae), *Aust. Wildl. Res.* **7:**365–378.

Hall-Martin, A. J., and Botha, B. P., 1980, A note on feeding habits, ectoparasites and measurements of the black-backed jackal *Canis mesomelas* from Addo Elephant National Park, *Koedoe* **23:**157–162.

Hamilton, W. J., III, Buskirk, R. E., and Buskirk, W. H., 1978, Omnivory and utilization of food resources by Chacma baboons, *Papio ursinus*, *Am. Nat.* **112:**911–924.

Hanney, P., 1964, The harsh furred rat in Nyasaland, *J. Mammal.* **45:**345–358.

Hanney, P., 1965, The Muridae of Malawi (Africa: Nyasaland), *J. Zool. Lond.* **146:**577–633.

Harrison, J. L., 1950, Notes on some small mammals of Malaya, *Bull. Raffles Mus.* **23:**300–309.

Harrison, J. L., 1954, The natural food of some rats and other mammals, *Bull. Raffles Mus.* **25:**157–165.

Hatt, R. T., 1940, Lagomorpha and Rodentia other than Sciuridae, Anomaluridae and Idiuridae. Collected by the American Museum Congo Expedition, *Am. Mus. Nat. Hist. Bull.* **76:**1939–1940.

Hegh, E., 1922, *Les Termites*, Desmet-Verreneuil, Brussels.

Heinsohn, G. E., 1966, Ecology and reproduction of the Tasmanian bandicoots (*Perameles gunni* and *Isoodon obesulus*), *Univ. Calif. Publ. Zool.* **80:**1–96.

Hermann, H. R., and Blum, M. S., 1981, Defensive mechanisms in the social Hymenoptera, in: *Social Insects*, Volume II (H. R. Hermann, ed.), Academic Press, New York, pp. 77–179.

Herzig-Straschil, B., 1977, Notes on the feeding habits of the yellow mongoose *Cynictis penicillata*, *Zool. Afr.* **12:**225–256.

Hladik, C. M., 1973, Alimentation et activité d'un groupe de chimpanzés reintroduits en forêt Gabonaise, *Terre Vie* **27:**343–413.

Hladik, C. M., 1977a, A comparative study of the feeding strategies of two sympatric species of leaf monkeys: *Presbytis senex* and *Presbytis entellus*, in: *Primate Ecology* (T. H. Clutton-Brock, ed.), Academic Press, New York, pp. 324–354.

Hladik, C. M., 1977b, Chimpanzees of Gabon and chimpanzees of Gombe: Some comparative data on the diets, in: *Primate Ecology* (T. H. Clutton-Brock, ed.), Academic Press, New York, pp. 481–503.

Hladik, C. M., and Hladik, A., 1972, Disponibilités alimentaires et domaines vitaux des primates a Ceylon, *Terre Vie* **1:**149–215.

Honacki, J. H., Kinman, K. E., and Koeppl, J. W., 1982, *Mammal Species of the World*, Allen Press, Lawrence.

Horner, B. B., Taylor, J. M., and Padykula, H. A., 1965, Food habits and gastric morphology of the grasshopper mouse, *J. Mammal.* **45**:513–535.

Hubert, B., Gillon, D., and Adam, F., 1981, Cycle annuel du régime alimentaire des trois principales espèces de rongeurs (Rodentia; Gerbillidae et Muridae) de Bandia (Sénégal), *Mammalia* **45**:1–20.

Husson, A. M., 1978, *The Mammals of Suriname*, E. J. Brill, Leiden.

Hutton, A. F., 1949, Notes on the Indian pangolin (*Manis crassicaudata*, Geoffer. St. Hilaire), *J. Bombay Nat. Hist. Soc.* **48**:805–806.

Iwuala, M. O. E., Braide, E. I., and Maduka, N., 1980, Observations on the food habits of some African rodents, *Rev. Biol. Trop.* **28**:227–236.

Izawa, K., 1975, Foods and feeding behavior of monkeys in the Upper Amazon Basin, *Primates* **16**:295–316.

Izawa, K., 1979, Foods and feeding behavior of wild black-capped capuchin (*Cebus apella*), *Primates* **20**:57–76.

Izawa, K., and Yoneda, M., 1981, Habitat utilization of nonhuman primates in a forest of the Western Pando, Bolivia, *Kyoto Univ. Overseas Res. Rep. New World Monkeys* **1981**:12–22.

Jewell, P. A., and Oates, J. F., 1969, Ecological observations on the lorisoid primates of African lowland forest, *Zool. Afr.* **4**:231–248.

Kalmbach, E. R., 1944, *The Armadillo: Its Relation to Agriculture and Game*, Game Fish and Oyster Commission, Austin, Texas.

Kappler, A., 1887, *Surinam, Land, Seine Natur, Bevolkeruna und seine Kultur-Verhaltnisse*, J. G. Cotta'sche, Stuttgart.

Kaufmann, J. H., 1962, Ecology and social behavior of the coati, *Nasua narica* on Barro Colorado Island, Panama, *Univ. Calif. Publ. Zool.* **60**:95–222.

Ketelhodt, H. F. V., 1966, Der erdwolf, *Proteles cristatus* (Sparrman, 1783), *Z. Säugetierkd.* **31**:300–308.

Kingdon, J., 1971, *East African Mammals*, Volume I, Academic Press, New York.

Kingdon, J., 1974a, *East African Mammals*, Volume II, (*A*) (*Insectivores and Bats*), Academic Press, New York.

Kingdon, J., 1974b, *East African Mammals*, Volume II, (*B*) (*Hares and Rodents*), Academic Press, New York.

Kingdon, J., 1977, *East African Mammals*, Volume III, (*A*) (*Carnivores*), Academic Press, New York.

Koop, K., and Velimirov, B., 1982, Field observations on activity and feeding of bat-eared foxes (*Otocyon megalotis*) at Nyai Pan, Botswana, *Afr. J. Ecol.* **20**:23–27.

Kreyenberg, M., 1907, Das chinesische schuppentier, *Zool. Beobach.* **48**:182–185.

Kruuk, H., and Sands, W. A., 1972, The aardwolf (*Proteles cristatus* Sparrman, 1783) as predator of termites, *East Afr. Wildl.* **10**:211–227.

Landry, S. O., Jr., 1970, The rodentia as omnivores, *Q. Rev. Biol.* **45**:351–372.

Langham, N. P. E., 1982, The ecology of the common tree shrew, *Tupaia glis*, in peninsular Malaysia, *J. Zool. Lond.* **197**:323–344.

Langham, N. P. E., 1983, Distribution and ecology of small mammals in three rain forest localities of peninsula Malaysia with particular reference to Kedah Peak, *Biotropica* **15**:199–206.

Laurie, A., and Seidensticker, J., 1977, Behavioural ecology of the sloth bear (*Melursus ursinus*), *J. Zool. Lond.* **182**:187–204.

Lekagul, B., and McNeely, J. A., 1977, *Mammals of Thailand*, Kurusapha Ladprao Press, Bangkok.

Levings, S. C., 1983, Seasonal, annual, and among-site variation in the ground ant community of a deciduous tropical forest: Some causes of patchy species distributions, *Ecol. Monogr.* **53**:435–455.

Lim, B. L., 1965, Food and weights of small animals from the First Division, Sarawak, *Sarawak Mus. J.* **12**:360–372.

Lim, B. L., 1967, Note on the food habits of *Ptilocercus lowii* Gray (pentail tree shrew) and *Echinosorex gymnurus* (Raffles) (moonrat) in Malaya with remarks on "ecological labelling" by parasite patterns, *J. Zool. Lond.* **152**:375–379.

Lim, B. L., and Heyneman, D., 1968, A collection of small mammals from Tuaran and the southwest face of Mt. Kinabalu, Sabah, *Sarawak Mus. J.* **16**:257–276.

Llewellyn, L. M., and Uhler, F. M., 1952, The foods of fur animals of the Patuxent Research Refuge, Maryland, *Am. Midl. Nat.* **48**:193–203.

Lubin, Y. D., and Montgomery, G. G., 1981, Defenses of *Nasutitermes* termites (Isoptera, Termitidae) against tamandua anteaters (Edentata, Myrmecophagidae), *Biotropica* **13**:66–76.

Lynch, C. D., 1980, Ecology of the suricate, *Suricata suricatta* and yellow mongoose, *Cynictis penicillata* with special reference to their reproduction, *Mem. Nas. Mus. Bloemfontein* **14**:1–145.

Macdonald, D. W., 1980, The red fox, *Vulpes vulpes*, as a predator upon earthworms, *Lumbricus terrestris*, *Z. Tierpsychol.* **52**:171–200.

MacKinnon, J., 1974, The behavior and ecology of wild orangutans, *Anim. Behav.* **22**:3–74.

Massoia, E., and Fornes, A., 1954, Notas sobre el genero *Scapteromys* (Rodentia-Cricetidae), *Physis* **24**:279–297.

Mathews, A. G. A., 1977, *Studies on termites from the Mato Grosso State, Brazil*, Academia Brasileira de Ciencias, Rio de Janeiro.

McBeath, N. M., and McGrew, W. C., 1982, Tools used by wild chimpanzees to obtain termites at Mt. Assirik, Senegal: The influence of habitat, *J. Hum. Evol.* **11**:65–72.

McGrew, W. C., 1974, Tool use by wild chimpanzees in feeding upon driver ants, *J. Hum. Evol.* **3**:501–508.

McGrew, W. C., 1983, Animal foods in the diets of wild chimpanzees (*Pan troglodytes*): Why cross-cultural variation?, *J. Ethol.* **1**:46–61.

McGrew, W. C., Tutin, C. E. G., and Baldwin, P. J., 1979, Chimpanzees, tools and termites: Cross-cultural comparisons of Senegal, Tanzania, and Rio Muni, *Man* **14**:185–214.

McIntosh, D. L., 1963, Food of the fox in the Canberra District, *CSIRO Wildl. Res.* **8**:1–20.

McNab, B. K., 1984, Physiological convergence among ant-eating and termite-eating mammals, *J. Zool. Lond.* **203**:485–510.

Melton, D. A., 1976a, The biology of aardvark (Tubulidentata-Orycteropidae), *Mammal. Rev.* **6**:75–88.

Melton, D. A., 1976b, Environmental heterogeneity produced by termitaria in western Uganda with special reference to mound usage by vertebrates, M. S. thesis, University of British Columbia.

Menzies, J. I., 1962, Feeding pangolins (*Manis* spp.) in captivity, *Int. Zoo Yearbk.* **4**:126–128.

Menzies, J. I., 1966, A note on the nutrition of the tree pangolin *Manis tricuspis* in captivity, *Int. Zoo Yearbk.* **6**:71.

Merrit, D. A., 1985, Naked-tailed armadillos, *Cabassous* sp., in: *Ecology and Evolution of Sloths, Anteaters and Armadillos (Mammalia, Xenarthra = Edentata)* (G. G. Montgomery, ed.), Smithsonian Institution Press, Washington, D.C., pp. 389–392.

Mill, A. E., 1982a, Foraging and defensive behaviour in Neotropical termites, Ph. D. thesis, University of Southampton.

Mill, A. E., 1982b, Populações de termitas (Insecta, Isoptera) em quatro habitats no baixo Rio Negro, *Acta Amazon.* **12**:53–60.

Mills, M. G. L., and Mills, M. E. J., 1978, The diet of the brown hyaena, *Hyaena brunnea* in the southern Kalahari, *Koedoe* **21**:125–149.

Moeller, E., 1975, Present-day Edentates, in: *Animal Life Encyclopedia, Mammals II* (B. Grzimek, ed.), Van Nostrand, New York, pp. 149–181.

Montgomery, G. G., 1979, El grupo alimentico (feeding guild) del oso hormiguero, *Conciencia* **6**:3–6.

Montgomery, G. G., 1983, *Cyclopes didactylus*, in: *Costa Rican Natural History* (D. H. Janzen, ed.), University of Chicago Press, Chicago, pp. 461–463.

Montgomery, G. G. (ed.), 1985, Movements, foraging, and food habits of the four extant species of Neotropical Vermilinguas (Mammalia: Myrmecophagidae) in: *Ecology and Evolution of Sloths, Anteaters and Armadillos (Mammalia, Xenarthra = Edentata)*, Smithsonian Institution Press, Washington, D.C., pp. 365–377.

Montgomery, G. G., and Lubin, Y. D., 1977, Prey influences on movements of Neotropical anteaters, in: *Proceedings of the 1975 Predator Symposium* (R. L. Phillips and C. Jonkel, eds.), University Montana, Missoula, pp. 103–131.

Moojen, J., 1965, Novo gênero de Cricetidae do Brasil central, *Rev. Bras. Biol.* **25**:281–285.

Neal, E. G., 1970, The banded mongoose, *Mungos mungo* Gmelin, *East Afr. Wildl. J.* **8**:63–71.

Neal, E. G., 1977, *Badgers*, Blanford Press, Dorset.

Nel, J. A. J., 1978, Notes on the food and foraging behavior of the bat-eared fox, *Otocyon megalotis*, *Bull. Carnegie Mus. Nat. Hist.* **6**:132–137.

Newman, H. H., 1913, The natural history of the nine-banded armadillo of Texas, *Am. Nat.* **47**:513–539.

Newsome, A. E., 1962, Rabbit-eared bandicoots or bilbies, *Aust. Nat. Hist.* **14**:97–98.

Nicoll, M. E., 1982, Reproductive ecology of *Tenrec ecaudatus* (Insectivora: Tenrecidae) in the Seychelles, Ph. D. thesis, University of Aberdeen.

Nishida, T., 1973, The ant-gathering behaviour by the use of tools among wild chimpanzees of the Mahali Mountains, *J. Hum. Evol.* **2**:357–370.

Nishida, T., and Hiraiwa, M., 1982, Natural history of a tool-using behavior by wild chimpanzees in feeding upon wood-boring ants, *J. Hum. Evol.* **11**:73–99.

Nishida, T., and Uehara, S., 1980, Chimpanzees, tools and termites: Another example from Tanzania, *Curr. Anthropol.* **21**:671–672.

Novikov, G. A., 1962, *Carnivorous Mammals of the Fauna of the U.S.S.R.*, Israel Program Scientific Translation, Jerusalem.

Pacheco, J., and Naranjo, C. J., 1978, Field ecology of *D. sabanicola* in Venezuela, in: *The Armadillo as an Experimental Model in Biomedical Research*, Pan American Health Organization Scientific Publication No. 336, pp. 13–17.

Pages, E., 1965, Notes sur les pangolins du Gabon, *Biol. Gabon.* **1**:209–238.

Pages, E., 1970, Sur l'écologie et les adaptations de l'orycterope et des pangolins sympatriques du Gabon, *Biol. Gabon.* **6**:23–92.

Pages, E., 1975, Etude éco-éthologique de *Manis tricuspis* par radiotracking, *Mammalia* **39**:613–641.

Patterson, B., 1975, The fossil aardvarks (Mammalia: Tubulidentata), *Bull. Mus. Comp. Zool.* **147**:185–237.

Pocock, R. I., 1941, *The Fauna of British India, Including Ceylon and Burma Mammalia*, Volume II, Taylor and Francis, London.

Prakash, I., 1959, Food of some Indian desert mammals, *J. Biol. Sci.* **2**:100–109.

Prater, S. H., 1965, *The Book of Indian Animals*, Diocesan Press, Madras.

Prestwich, G. D., 1984, Defense mechanisms of termites, *Annu. Rev. Entomol.* **29**:201–232.

Rahm, U., and Christiaensen, A., 1963, Les mammifères de la region occidentale du Lac Kivu, *Mus. ser IN8 (Sci. Zool.)* **118:**1–83.

Ramirez, M. F., Freese, C. H., and Revilla, J., 1977, Feeding ecology of the pygmy marmoset, *Cebuella pygmaea,* in northeastern Peru, in *The Biology and Conservation of the Callitrichidae* (D. G. Kleiman, ed.), Smithsonian Institution Press, Washington, D. C., pp. 91–104.

Rathbun, G. B., 1979, The social structure and ecology of elephant-shrews, *Z. Tierpsychol. (Suppl.)* **20:**1–76.

Redford, K. H., 1983, Mammalian myrmecophagy: Foraging, feeding and food preference, Ph. D. thesis, Harvard University.

Redford, K. H., 1984a, The termitaria of *Cornitermes cumulans* (Isoptera, Termitidae) and their role in determining a potential keystone species, *Biotropica* **16:**(2)**:**112–119.

Redford, K. H., 1984b, Mammalian predation on termites: Tests with the burrowing mouse (*Oxymycterus roberti*) and its termite prey, *Oecologia* **65:**145–152.

Redford, K. H., 1985a, Feeding and food preference in captive and wild giant anteaters (*Myrmecophaga tridactyla*), *J. Zool. Lond.* **209:**559–572.

Redford, K. H., 1985b, Food habits of armadillos (Xenarthra: Dasypodidae), in: *Ecology and Evolution of Sloths, Anteaters and Armadillos (Mammalia, Xenarthra = Edentata)* (G. G. Montgomery, ed.), Smithsonian Institution Press, Washington, D.C., pp. 429–437.

Redford, K. H., and Dorea, J. G., 1984, The nutritional value of invertebrates with emphasis on ants and termites as food for mammals, *J. Zool. Lond.* **203:**385–395.

Rengger, J. R., 1830, *Naturges chichte der Saeugethiere von Paraguay,* Schweighanserschen, Basel.

Rijksen, H. D., 1978, A field study on Sumatran orangutans (*Pongo pygmaeus* Abelii Lesson 1827), *Meded. Landbouwhogesch. Wageningen* **78:**1–420.

Roberts, T. J., 1977, *The Mammals of Pakistan,* Ernest Benn, London.

Roberts, T. J., Vielliard, T., 1971, Commentaires sur le grand pangolin Indien *Manis crassicaudata, Mammalia* **35:**610–613.

Robinson, J. G., 1986, Seasonal variation in use of time and space by the wedge-capped capuchin monkey *Cebus olivaceus:* Implications for foraging theory, *Smithson. Contrib. Zool.,* pp. 1–60.

Rood, J. P., 1975, Population dynamics and food habits of the banded mongoose, *East Afr. Wild. J.* **13:**89–111.

Roonwal, M. L., 1949, Systematics, ecology and bionomics of mammals studied in connection with Tsutsugamushi disease (scrub typhus) in the Assam-Burma war theatre during 1945, *Trans. Natl. Inst. Sci. India* **3:**67–122.

Rowe-Rowe, D. T., 1978, The small carnivores of Natal, *Lammergeyer* **25:**1–48.

Ruddle, K., 1973, The human use of insects: Examples from the Yukpa, *Biotropica* **5:**94–101.

Schaller, G. B., 1967, *The Deer and the Tiger,* University of Chicago Press, Chicago.

Schaller, G. B., 1983, Mammals and their biomass on a Brazilian ranch, *Pap. Avulsos Zool. (São Paulo)* **31**(2)**:**1–36.

Schultze, W., 1914, Notes on the Malay pangolin, *Manis javanica* Desmarest, *J. Phil. Sci. (D)* **9:**93–97.

Shaw, J. H., Carter, T. S., and Machedo-Neto, J. C., 1985, Ecology of the giant anteater (*Myrmecophaga tridactyla*) in Serra da Canastra (MG.) Brazil, in: *Ecology and Evolution of Sloths, Armadillos and Anteaters* (G. G. Montgomery, ed.), Smithsonian Institution Press, Washington, D. C., pp. 379–388.

Sheppe, W. A., 1973, Notes on Zambian rodents and shrews, *Puku* **7:**168–190.

Silveira, E. K. P., 1968, Notas sôbre a historia natural do tamanduá-mirim (*Tamandua tetradactyla chirquensis* J. A. Allen 1904, Myrmecophagidae), com referencias a fauna do Istmo do Panama, *Vellozia* **6**:9–30.

Silveira, E. K. P., 1969, Historia natural do tamanduá-bandeira, *Myrmecophaga tridactyla* Linn. 1758, Myrmecophagidae, *Vellozia* **7**:34–234.

Smith, A., 1982, Is the striped possum (*Dactylopsila trivirgata;* Marsupialia, Petauridae) an arboreal anteater?, *Aust. Mammal.* **5**:229–234.

Smithers, R. H. N., 1971, The Mammals of Botswana, *Trustees Natl. Mus. Rhodesia Mus. Mem.* **4**:1–340.

Smyth, D. R., and Philpott, C. M., 1968, Field notes on rabbit bandicoots *Macrotis lagostis* Reid (Marsupialia), from central western Australia, *Trans. R. Soc. S. Australia* **92**:3–14.

Sowerby, A. de C., 1925, The pangolin, *China J. Sci. Arts* **3**:151–153.

Stoddart, D. M., and Braithwaite, R. W., 1979, A strategy for utilization of regenerating heathland habitat by the brown bandicoot (*Isoodon obesulus;* Marsupialia, Peramelidae), *J. Anim. Ecol.* **48**:165–179.

Struhsaker, T. T., 1967, Ecology of vervet monkeys (*Cercopithecus aethiops*) in the Masai-Amboseli Game Reserve, Kenya, *Ecology* **48**:891–904.

Stuart, C. T., 1976, Diet of the black backed jackal *Canis mesomelas* in the Central Namib Desert, South West Africa, *Zool. Afr.* **11**:193–205.

Suzuki, A., 1966, On the insect-eating habits among wild chimpanzees living in the savanna woodland of western Tanzania, *Primates* **7**:481–487.

Svendsen, G. E., 1982, Weasels. *Mustela* spp., in: *Wild Mammals of North America* (J. A. Chapman and G. A. Feldhamer, eds.), Johns Hopkins University Press, Baltimore, pp. 613–628.

Sweeney, R. C. H., 1956, Some notes on the feeding habits of the ground pangolin, *Smutsia temminckii* (Smuts), *Ann. Mag. Nat. Hist. 12th Ser.* **9**:893–896.

Tate, G. H. H., 1951, The banded anteater, *Myrmecobius* Waterhouse (Marsupialia), *Am. Mus. Novit.* **1521**:1–8.

Terborgh, J., 1983, *Five New World Primates*, Princeton University Press, Princeton.

Thomas, O., and Wroughton, R. C., 1909, On mammals from the Rhio Archipelago and Malay Peninsula collected by Messrs. H. C. Robinson, C. Boden Kloss and E. Seimund, and presented to the National Museum by the Government of the Federated Malay States, *J. Fed. Malay States Mus.* **4**:99–129.

Uehara, S., 1982, Seasonal changes in the technique employed by wild chimpanzees in the Mahale Mountains, Tanzania, to feed on termites (*Pseudoacanthotermes spiniger*), *Folia Primatol.* **37**:44–76.

Urbain, A., 1954, Biologie de l'orycterope, *Mus. R. Congo. Belge Sci. Zool.* **1**:101–105.

Van Deusen, H. M., and George, G. G., 1969, Results of the Archbold Expeditions. No. 90. Notes on the Echidnas (Mammalia, Tachyglossidae) of New Guinea, *Am. Mus. Novit.* **2383**:1–23.

Verheyen, R., 1951, Contribution à l'étude éthologiques des mammifères du Parc National de l'Upemba, Institut des Parcs Nationaux du Congo Belge, Brussels, Belgium.

Verheyen, W., and Verschuren, J., 1966, Exploration du Parc National de la Garamba, Institut des Parcs Nationaux du Congo No. 50, Brussels, Belgium.

Vestgens, M. J. M., and Hall, L. S., 1977, Stomach contents of forty-two species of bats from the Australian region, *Aust. Wildl. Res.* **4**:25–35.

Vossler, J., 1907, Aus dem leben ostafrikanischer sauger, *Zool. Beobach.* **48**:193–206.

Walker, E. P., 1975, *Mammals of the World*, Johns Hopkins University Press, Baltimore.

Waser, P., 1977, Feeding, ranging and group size in the mangabey, *Cercocebus albigena*, in: *Primate Ecology* (T. H. Clutton-Brock, ed.), Academic Press, New York, pp. 183–222.

Waser, P. M., 1980, Small nocturnal carnivores: Ecological studies in the Serengeti, *Afr. J. Ecol.* **18:**167–185.

Weber, N. A., 1972, *Gardening Ants. The Attines*, American Philosophical Society, Philadelphia.

Whitaker, J. O., Jr., 1966, Food of *Mus musculus, Peromyscus maniculatus bairdi* and *Peromyscus leucopus* in Vigo County, Indiana, *J. Mammal.* **47:**473–486.

Wilson, E. O., 1971, *The Insect Societies*, Harvard University Press, Cambridge, Massachusetts.

Wilson, D. S., and Clark, A. B., 1977, Above ground predator defense in the harvester termite, *Hodotermes mossambicus* (Hagen), *J. Entomol. Soc. S. Afr.* **40:**271–282.

Zietz, A., 1872, Ameisenbar und Termiten, *Zool. Gart.* **1872:**301–304.

CHAPTER 10

A REVIEW OF DENSITY DEPENDENCE IN POPULATIONS OF LARGE MAMMALS

CHARLES W. FOWLER

1. INTRODUCTION

The study of population regulation has been central to much of the work on the population biology of both animals and plants, as seen in much of the literature (for example, Lack, 1954; Slobodkin, 1961; MacArthur and Connell, 1967; McLaren, 1971; Ricklefs, 1973; Emmel, 1976; Tamarin, 1978). Recent developments in the theoretical aspects of population dynamics have led to a growing union of the concepts of life history analysis with those of population regulation, both in an evolutionary context (for example, Boyce, 1984). Much of the literature dealing with population regulation covers research on fishes and insects, with some work devoted to the study of birds (Nikolskii, 1969; Lack, 1966; Clark et al., 1967; Weatherly, 1972; Varley et al., 1973; Hassel, 1975; Gulland, 1977). Wynne-Edwards (1962) presents information on density dependence for a number of taxonomic categories. Reviews of the literature dealing with density dependence exist for fish (McFadden, 1977) and small mammals (Christian, 1963; Krebs and Myers,

CHARLES W. FOWLER • National Marine Mammal Laboratory, Northwest and Alaska Fisheries Center, National Marine Fisheries Services, NOAA, Seattle, Washington 98115.

1974). A preliminary review for large mammals was produced by Fowler et al. (1980).

This paper is a review of the broad concept of density dependence as applied to large-mammal populations. It is an expansion of the material dealing with population regulation presented in Fowler et al. (1980) to include other density-dependent aspects of large-mammal populations. It also includes material published in the last 10 years, primarily as found in the English language. This paper has the objectives of (1) providing a list of literature dealing with density dependence in large mammals; (2) helping to bring to the attention of biologists dealing with individual species the results of research dealing with other taxonomic groups, other habitats, and trophic levels, thereby promoting interchange of information among researchers in this field; and (3) promoting comparisons both within the category of large mammals (for example, marine versus terrestrial), and between large mammals and other major categories of animals as a basis for finding important patterns in population dynamics such as those described by Fowler (1981). The utility of density-dependent information will be discussed along with brief considerations of the ways in which much of the information on large mammals relates to some of the questions that seem to be at the forefront of evolutionary population dynamics today.

The term "large mammals" is not clearly defined. The bimodality in the size distribution of mammals pointed out by Bourliere (1975) is helpful in this regard. Mammals weighing between a few kilograms and tens of kilograms occur less frequently than those above and below this range. Most large terrestrial animals are either herbivores (such as the ungulates included in this review) or carnivores (such as many of the bears, canids, and felids). All marine mammals, including sea otters, are relatively large. Based on these observations, the sizes of animals included in this review are roughly comparable to those observed in the marine environment, with the exception of a few smaller species, such as coyotes (*Canis latrans*) and monkeys (*Macaca*).

Material in this review involves both direct and indirect forms of density dependence and both internal and external mechanisms of expression (terms defined in Fowler et al., 1980, and Fowler, 1984). Thus, any change in a large-mammal population that correlates with change (or difference) in population size or density reflects density dependence as defined for this review. The main focus of this paper will be the population level parameters of mortality and reproduction along with growth. Additional material will be reviewed as it relates to the comparison of behavior, diseases, parameter variability, physiology, and population structure at varying population levels. Such comparisons may be across either time or space, each kind of comparison having its own potential problems (Fowler et al., 1980). Other potential problems are those specific to each kind of density dependence and are reviewed by Fowler et al. (1980).

The taxonomy and spellings adopted for this paper follow those of Honacki et al. (1982).

2. REPRODUCTION

Reproduction may be considered in two categories: (1) the reproductive rate of fully adult females (that is, those age classes in which most animals have or are bearing young) and (2) the reproductive rate of younger animals (that is, age classes among which those bearing young are primarily those doing so for the first time), reflective of the age at reproductive maturity.

2.1. Adult Reproduction

2.1.1. Terrestrial Mammals

2.1.1a. Ungulates. Density dependence has been observed in the reproductive rates of at least 14 ungulate species (Table I). Much of this information comes from temporal studies of separate populations. For example, density-dependent reproduction has been shown among adult female white-tailed deer (*Odocoileus virginianus*) in at least seven studies. These studies have involved populations that are widely dispersed geographically, but are generally restricted to comparisons of each population over time. A good example of intensive work showing density dependence in the reproduction of white-tailed deer is McCullough's (1979) study of the George Reserve herd.

Several species of the genus *Cervus* have been observed to exhibit density-dependent reproduction. These include sika deer (*C. nippon*) in Japan (Kaji et al., 1985), red deer (*C. elephus*) in northern Europe and New Zealand, and elk (*C. elephus*) in North America. The elk studies have involved comparisons across both space and time. Studies of this species conducted by Carbyn (1975) involved comparisons of populations whose densities were assumed to be determined largely by predation. Differences in reproductive rates were seen as density dependent based on the relationship between population and resource abundance. This parallels the influence of hunting pressure, which has been shown for elk (Knight, 1970), and would be expected to be characteristic of species in general. Intensive studies on the population

TABLE I
Large Terrestrial Mammals and Corresponding References Indicating Density Dependence in the Reproductive Rate of Adult Females[a]

Common name	Species	References
White-tailed deer	*Odocoileus virginianus*	O'Roke and Hamerstrom (1948), Cheatum and Severinghaus (1950), W. H. Adams (1960), Marburger and Thomas (1965), Hesselton et al. (1965), Teer et al. (1965), McCullough (1979), Kie et al. (1980), Newsome (1984)
Sika deer	*Cervus nippon*	Kaji et al. (1985)
Elk	*Cervus elephus*	Gross (1969), Knight (1970), Carbyn (1975), Houston (1982)
Red deer	*Cervus elephus*	Challies (1973a, 1978), Staines (1978), Clutton-Brock et al. (1982, 1983, 1985), Challies (1985)
Reindeer	*Rangifer tarandus*	Skogland (1983, 1985a)
American bison	*Bison bison*	Gross et al. (1973)
European bison	*Bison bonasus*	Krasinski (1978)
Longhorn cattle	*Bos taurus*	Gross et al. (1973)
Moose	*Alces alces*	Hatter (1950), Pimlott (1959), Markgren (1969), Priklonskii and Chervonnyi (1970), Nygren (1983)
Mountain goat	*Oreamnos americanus*	Bailey and Johnson (1977), L. G. Adams and Bailey (1982), Swenson (1985)
Bighorn sheep	*Ovis canadensis*	Woodgerd (1964), Geist (1971), Stelfox (1976)
Dall sheep	*Ovis dalli*	Murphy and Whitten (1976)
Thar	*Hemitragus jamlachicus*	Caughley (1976)
African buffalo	*Syncerus caffer*	Sinclair (1973, 1974)
Horse	*Equus caballus*	Berger (1986)
Hippopotamus	*Hippopotamus amphibius*	Laws (1968)
African elephant	*Loxodonta africana*	Buss and Savidge (1966), Buss and Smith (1966), Laws (1968, 1969a,b), Laws and Johnstone (1970), Fowler and Smith (1973), Hanks and McIntosh (1973), Laws (1973), Laws et al. (1975)
Lion	*Panthera leo*	Stevenson-Hamilton (1937), Wright (1960), Smuts (1978)

TABLE I (*Continued*)

Common name	Species	References
Grizzly bear	*Ursus arctos*	Shaffer (1978), McCullough (1981), Shaffer (1983), Stringham (1983)
Coyote	*Canis latrans*	Knowlton (1972)
Toque monkey	*Macaca sinica*	Dittus (1975)
Rhesus monkey	*Macaca mulatta*	Southwick et al. (1980)

[a]The evidence of density dependence may depend on the collective information of two or more references. A presentation of more detail is found in the preliminary review of Fowler et al. (1980).

biology of red deer are presented in detail by Clutton-Brock et al. (1982), who noted density-dependent changes in reproduction. Another intensively studied species is the reindeer (*Rangifer tarandus*), for which Skogland (1985a), for example, showed density-dependent reproduction.

Moose (*Alces alces*, Table I) have been shown to exhibit density-dependent reproduction in both European and North American populations. These studies have involved comparisons both across space and over time. Work by Nygren (1983) has shown that at least some of these changes may be due to changes in the ratios and age structure, which were also correlated with density. At least three studies involving two populations of mountain goats (*Oreamnos americanus*) have shown density-dependent reproductive rates (Table I). These results may reflect a combination of early mortality (between birth and the surveys) and age at first reproduction as well as reproduction by adults. Populations of the genus *Ovis* have been observed to show changes in rates of reproduction by adult females in several locations in North America. Reproductive rates of the European bison (*Bison bonasus*) fell as the population increased in the Bialowieza Forest (Krasinksi, 1978). The introduced thar in New Zealand also exhibit differences across space corresponding to different phases of population growth (Caughley, 1970).

Among the African ungulates, density-dependent variations in adult reproduction have been noted for the African buffalo (*Syncerus caffer*), hippopotamus (*Hippopotamus amphibius*), and the African elephant (*Loxodonta africana*). Studies of the African buffalo are presented by Sinclair (1973, 1974) and involve comparisons over time. Other detailed studies have focused on a number of populations of African elephant in which comparisons are often made across space. The results of much of the work on this species are summarized in Fowler and Smith (1973) and Laws et al. (1975).

2.1.1b. Nonungulates. Studies involving density dependence of birth rates in predators are not common in the literature. It is important to note that many predators produce relatively large litters in comparison to nonpredatory species, resulting in greater flexibility in birth rates.

Although there is some information indicating that adult reproduction exhibits density dependence for some primates, the question of whether or not such changes occur among humans is still being debated. As indicated by Dumond (1975), humans are ultimately likely to be similar to other species of large animals in showing density dependence and there are mechanisms through which such changes would probably be expressed. These issues are treated in great detail in a number of papers dealing with the biosocial mechanisms of population regulation (for example, those in Cohen et al., 1980).

2.1.2. Marine Mammals

Most marine mammals are logistically difficult to study, and for the most part, would be completely obscure were it not for the historical commercial importance placed on the species involved. References dealing with density-dependent changes in the reproductive rates of adult females are shown in Table II.

2.1.2a. Pinnipeds. The only pinniped species for which there appears to be significant information showing density-dependent changes in adult reproductive rates is the harp seal (*Phoca groenlandica*, Bowen et al., 1981). Studies on these animals have involved making comparisons of various components of a single exploited population over time.

2.1.2b. Cetaceans. A number of cetacean populations are thought to have exhibited density-dependent changes in adult reproductive rates (Fowler, 1984). Some supporting data come from studies involving comparisons across space (for example, Perrin et al., 1976, 1977). However, most information comes from historical studies associated with the long-term commercial whale fisheries for each species. To a large extent these sets of information are the results of the studies conducted by scientists in cooperation with the International Whaling Commission.

Studies of cetacean populations showing density-dependent adult reproductive rates have involved species varying in body size from the dolphins (for example, genus *Stenella*) to the blue whale (*Balaenoptera musculus*). As can be seen in Table II, such changes are thought to

TABLE II
Marine Mammals and Corresponding References Indicating Density Dependence in the Reproductive Rate of Adult Females[a]

Common name	Species	References
Harp seals	*Phoca groenlandica*	Lett and Benjaminsen (1977), Bowen et al. (1981), Lett et al. (1981), Roff and Bowen (1983)
Spotted dolphin	*Stenella attenuata*	Perrin et al. (1976, 1977), Fowler (1984), Smith (1984), Kasuya (1985)
Striped dolphin	*Stenella coeruleoalba*	Kasuya and Miyazaki (1982), Kasuya (1985)
Blue whale	*Balaenoptera musculus*	Mackintosh (1942), Gambell (1973, 1976), Gulland (1982)
Fin whale	*Balaenoptera physalus*	Laws (1961), Doi et al. (1969), Gulland (1971), Gambell (1976), Gulland (1982)
Sei whale	*Balaenoptera borealis*	Gambell (1973, 1975, 1976), Masaki (1976, 1978)
Killer whale	*Orcinus orca*	Bigg (1982), Fowler (1984)
Sperm whale	*Physeter macrocephalus*	International Whaling Commission (1971), Best (1980), Fowler (1984)

[a]The evidence of density dependence may depend on the collective information of two or more references. A presentation of more detail is found in the preliminary review of Fowler et al. (1980).

occur for both the toothed whales (Odontoceti) and the baleen whales (Mysticeti).

A special note of caution must be voiced on the interpretation and reliability of information coming from some of the studies of cetacean populations. As pointed out by Mizroch and York (1984), some of the changes observed may be the result of changing patterns in the timing and location of the commercial whale fishery or pooling data over various identifiable strata, and not in the actual population. Another aspect of cetacean research worthy of note is the competition among cetacean populations feeding on krill in the feeding grounds shared by these species. As indicated by Gambell (1975), some of the clearest information on density dependence among cetacean populations appears when looking at changes in overall whale biomass rather than in the abundance of individual species.

Detailed studies involving the monitoring of individual killer whales are reported by Bigg (1982) and summarized in Fowler (1984). This particular study parallels some of the studies conducted on ungulates in which individual animals are monitored over time.

2.2. Age at Reproductive Maturation

2.2.1. Terrestrial Mammals

Terrestrial mammals observed to exhibit density-dependent changes in age at reproductive maturation are listed in Table III. Insufficient work has been conducted among terrestrial mammals to show this form of population regulation other than for the ungulates. Much of the literature links reproductive maturation with growth, which is reviewed in a later section.

Many of the studies listed in Table III are also listed in Table I on

TABLE III
Large Terrestrial Mammals and Corresponding References Indicating Density Dependence in the Age of Reproductive Maturation[a]

Common name	Species	References
White-tailed deer	*Odocoileus virginianus*	Teer et al. (1965), Hesselton et al. (1965), McCullough (1979), Kie et al. (1980)
Mule deer	*Odocoileus hemionus*	Swank (1958)
Sika deer	*Cervus nippon*	Kaji et al. (1985)
Elk	*Cervus elephus*	Buechner and Swanson (1955), Carbyn (1975), Fowler and Barmore (1979), Houston (1982)
Red deer	*Cervus elephus*	Challies (1978), Staines (1978), Clutton-Brock et al. (1983, 1985), Challies (1985)
Reindeer	*Rangifer tarandus*	Reimers et al. (1983), Skogland (1983, 1985a,b)
Moose	*Alces alces*	Pimlott (1959)
Bighorn sheep	*Ovis canadensis*	Woodgerd (1964)
Wildebeest	*Connochaetes taurinus*	Watson (1970), Sinclair (1979)
Thar	*Hemitragus jemlachicus*	Caughley (1970)
Hippopotamus	*Hippopotamus amphibius*	Laws (1968), Sayer and Rakha (1974), Marshall and Sayer (1976)
African elephant	*Loxodonta africana*	Buss and Savidge (1966), Buss and Smith (1966), Laws (1966, 1968, 1969b, 1973), Laws and Parker (1968), Fowler and Smith (1973), Laws et al. (1975)

[a]The evidence of density dependence may depend on the collective information of two or more references. A presentation of more detail is found in the preliminary review of Fowler et al. (1980).

adult reproduction. In some cases the same populations are involved (for example, the studies of white-tailed deer, Fig. 1, by Teer et al., 1965; McCullough, 1979). Density dependence in age at maturation was also observed in another member of this genus, the mule deer (*O. hemionus;* Swank, 1958). The genus *Cervus* is again well represented by studies that are widely distributed geographically. The wildebeest (*Connochaetes taurinus*) is included in the list of African ungulates observed to exhibit density dependence by changes observed in age at first reproduction (Watson, 1970).

2.2.2. Marine Mammals

Table IV contains a list of marine mammals for which density-dependent changes in age at first reproduction have been observed. Here we note that the quality of much of the information indicating that this aspect of reproduction can exhibit density-dependent change among the pinnipeds appears to be superior to that for the cetaceans.

2.2.2a. Pinnipeds. The five species of pinnipeds shown in Table IV represent species distributed widely, occurring in both hemispheres and in most of the major oceans. Density-dependent age at first reproduction for northern fur seals is summarized by Fowler (1986) and for harp seals (Fig. 2) by Winters (1978) and Bowen et al. (1981). Studies involving the Antarctic fur seal (*Arctocephalus gazella*), as based on work by Payne (1977), are a potential basis for evidence of density dependence. However, these studies involve comparisons between species and therefore are not considered sufficiently substantive to be included in the list in Table IV. Changes in crabeater seal populations

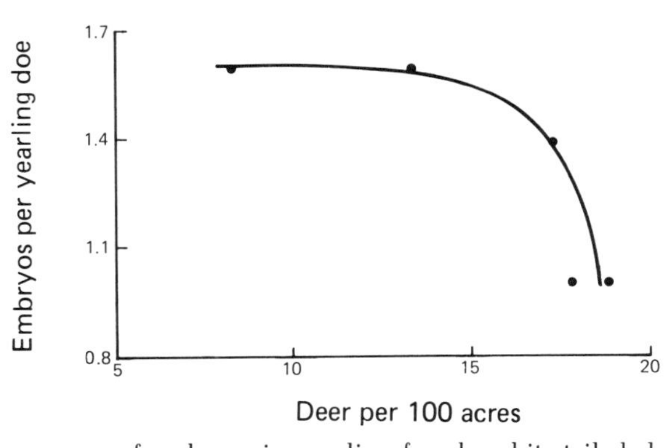

FIGURE 1. Occurrence of embryos in yearling female white-tailed deer (*Odocoileus virginianus*) at different population levels (Teer et al., 1965).

TABLE IV
Marine Mammals and Corresponding References Indicating Density Dependence in the Age of Reproductive Maturation[a]

Common name	Species	References
Northern fur seal	*Callorhinus ursinus*	Scheffer (1955), Nagasaki (1961), Chapman (1964a), Fowler (1986)
Harp seal	*Phoca groenlandica*	Mansfield (1970), Sergeant (1973), Lett and Benjaminsen (1977), Winters (1978), Bowen et al. (1981), Lett et al. (1981), Capstick and Ronald (1982)
Weddell seal	*Leptonychotes weddellii*	Stirling (1969, 1971a, 1972)
Crabeater seal	*Lobodon carcinophagus*	Bengston and Laws (1985)
Southern elephant seal	*Mirounga leonina*	Gibbney (1957), Laws (1959), Carrick and Ingham (1962), Carrick et al. (1962)
Spotted dolphin	*Stenella attenuata*	Perrin et al. (1976, 1977), Kasuya (1985)
Striped dolphin	*Stenella coeruleoalba*	Kasuya (1985)
Blue whale	*Balaenoptera musculus*	Gulland (1982)
Fin whale	*Balaenoptera physalus*	Laws (1962), Doi et al. (1969), Lockyer (1972, 1974, 1977, 1979), Gambell (1973, 1976)
Sei whale	*Balaenoptera borealis*	Lockyer (1972, 1974, 1977, 1979), Gambell (1973, 1976), Masaki (1978)
Dugong	*Dugong dugon*	Marsh (1980)

[a]The evidence of density dependence may depend on the collective information of two or more references. A presentation of more detail is found in the preliminary review of Fowler et al. (1980).

follow what is probably a decline brought on by reduced competition, which, in turn, promotes an increase in numbers and a concomitant increase in age at sexual maturity (Bengtson and Laws, 1985).

2.2.2b. Cetaceans. The studies of cetacean populations listed in Table IV follow some of the same patterns as those studies listed in Table II. Some studies involving the genus *Stenella* come from animals collected in association with fisheries and involve comparisons across the width of the Pacific Ocean; others involve single populations studied over time (Kasuya, 1985). The data for large whales come primarily

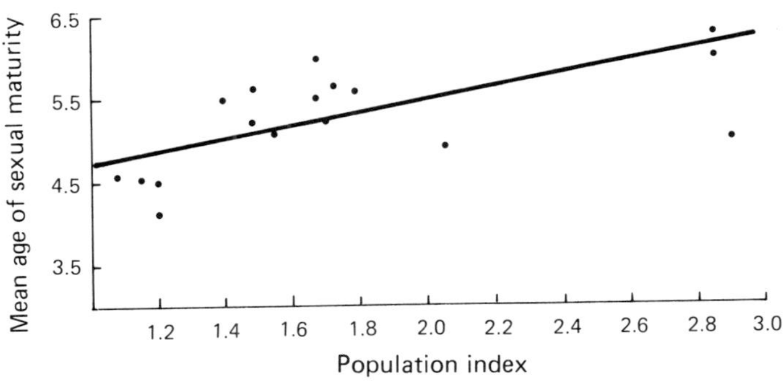

FIGURE 2. Reproductive maturity as shown by varying estimated mean ages at sexual maturity for different population levels of harp seals (*Phoca groenlandicus*) older than 1 year (Bowen et al., 1981).

from historical commercial fisheries and may be subject to some of the problems mentioned earlier. Masaki (1979) and Kato (1983) show that changes in the age at sexual maturity for minke whales are correlated with declines in overall baleen whale biomass.

2.2.2c. Sirenians. A third category of marine mammals for which density dependence has been observed includes the dugong (*Dugong dugon*), based on work by Marsh (1980). Here, however, we are dealing with change in reproductive maturation in the male portion of the population based on comparing two geographically separated populations. One, subjected to a harvest, contains males that tend to become reproductively mature earlier than in the less exploited population.

3. SURVIVAL

Although it is commonly recognized that survival rates vary continuously with age, it is nevertheless useful to consider, as in the case of reproduction, two broad categories of age for which density-dependent survival may be distinguished—juveniles and adults.

3.1. Juvenile Survival

Study of the literature reviewed in the following sections reveals that among large mammals neonate and first-winter mortality (especially in higher latitudes) are often the highest mortality rates observed among large mammals. In common with many other species, this pattern is general for mammals (Caughley, 1966). It is survival during these periods of time that is often most critical and most subject to density-dependent change.

3.1.1. Terrestrial Mammals

Table V contains a listing of terrestrial mammals for which density-dependent juvenile survival has been observed. As in the case of information for density-dependent reproduction, most of the species in this list are ungulates.

3.1.1a. Ungulates. As with other aspects of density dependence, some of the most convincing information concerning density-dependent juvenile survival comes from studies that are both extensive and intensive. Such studies for white-tailed deer are exemplified by that of McCullough (1979). Other work showing density-dependent juvenile survival in white-tailed deer is cited by Gavin (1984). Other extensive and intensive work exists for elk (Houston, 1982), red deer (Fig. 3) (Clutton-Brock et al., 1982), Soay sheep (Fig. 4) (Grubb, 1974), African buffalo (Sinclair, 1977), and African elephants (as synthesized by Fowler and Smith, 1973). The elevated juvenile mortality associated with higher populations of caribou (*R. tarandus*) by Bergerud (1978) is attributed to predation by wolves.

3.1.1b. Nonungulates. Although a number of relatively intensive studies have been conducted on various predators and primates, density-dependent juvenile survival seems less readily apparent for these groups. A clear relationship may not have been shown by these studies because density dependence for any species can (and often does) involve a delayed effect of food limitation. Resource levels at any particular time may be more a product of historic population levels of the consumer than population levels measured at the time juvenile survival is evaluated. That juvenile survival is tied to resource levels is fairly clearly demonstrated in most of the references dealing with the species listed in Table V. Changes in juvenile survival of black bears are attributed to the numbers of adult males rather than the total population (Kemp, 1976). The same may be true for other species of carnivores and primates as well (for example, Dittus, 1978).

3.1.2. Marine Mammals

Marine mammal species for which density-dependent juvenile survival has been observed are listed in Table VI. It should be noted that the lack of evidence for density-dependent juvenile survival among cetacean populations is at least partially a product of the logistic difficulty in obtaining the relevant information and the regulations prohibiting the take of juvenile animals and not necessarily an indication that density dependence does not occur. Pinnipeds are the primary

TABLE V
Large Terrestrial Mammals and Corresponding References Indicating Density Dependence in Survival among Juveniles[a]

Common name	Species	References
White-tailed deer	*Odocoileus virginianus*	McCullough (1979), Gavin (1984)
Mule deer	*Odocoileus hemionus*	Rasmussen (1941), Klein and Olson (1960)
Sika deer	*Cervus nippon*	Christian et al. (1960)
Elk	*Cervus elephus*	Fowler and Barmore (1979), Houston (1982), Sauer and Boyce (1983)
Red deer	*Cervus elephus*	Staines (1978), Clutton-Brock et al. (1982, 1985)
Reindeer and caribou	*Rangifer tarandus*	Bergerud (1978), Leader-Williams (1980), Skogland (1984a, 1985b)
Bighorn sheep	*Ovis canadensis*	Woodgerd (1964), Stelfox (1976), Douglas and Leslie (1986)
Dall sheep	*Ovis dalli*	Walters et al. (1981)
Soay sheep	*Ovis aries*	Grubb (1974)
Wildebeest	*Connochaetes taurinus*	Talbot and Talbot (1963), Watson (1969), Sinclair (1973)
African buffalo	*Syncerus caffer*	Grimsdell (1973), Sinclair (1973, 1974, 1977)
Thar	*Hemitragus jemlachicus*	Caughley (1970, 1976)
Warthog	*Phacochoerus aethiopicus*	Child et al. (1968)
Hippopotamus	*Hippopotamus amphibius*	Laws (1968)
African elephant	*Loxodonta africana*	Laws and Parker (1968), Corfield (1973), Fowler and Smith (1973)
Wolf	*Canis lupus*	Walters et al. (1981)
Coyote	*Canis latrans*	Knowlton and Stoddard (personal communication)[b]
Black bear	*Ursus americanus*	Kemp (1976)
Grizzly bear	*Ursus arctos*	McCullough (1981)
Lion	*Panthera leo*	Stevenson-Hamilton (1973), Wright (1960), Smuts (1976, 1978)
Toque monkey	*Macaca sinica*	Dittus (1975)

[a]The evidence of density dependence may depend on the collective information of two or more references. A presentation of more detail is found in the preliminary review of Fowler et al. (1980).
[b]Frederick K. Knowlton and Charles Stoddart, Predator ecology and behavior project, Utah State University, Logan, Utah.

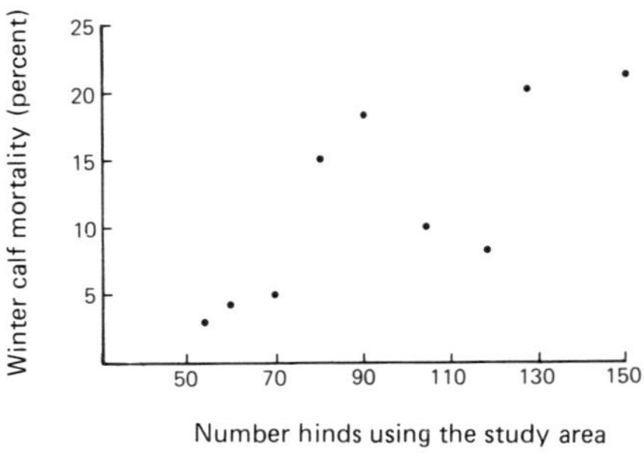

FIGURE 3. Juvenile mortality for red deer (*Cervus elephus*) as shown by varying levels of winter mortality (percent) at different population levels (measured by number of females in a study area) (Clutton-Brock et al., 1982).

group of species of marine mammals for which juvenile stages of the life cycle are available for observation. Again, the strength of information indicating that density dependence in juvenile survival comprises part of the population regulation of these species is dependent upon the degree to which the species have been studied. In the case of the extensively studied northern fur seal, density-dependent juvenile survival has been observed both during the first few weeks of life and during the first several years of life (Fig. 5) (Fowler, 1986). Among juvenile elephant seals, density-dependent survival seems to be more closely related to density on the breeding colonies than to general population size, although the two may be related. For the northern elephant seal juvenile mortality seems to be exacerbated by storms, but also involves interindividual aggression and the separation of mother–pup pairs (LeBoeuf and Briggs, 1977; Riedman and LeBoeuf, 1982). A study of harp seals (Lett et al., 1981) indicated a density-dependent change in juvenile survival, but subsequent analyses (Bowen et al., 1981) have not found such a relationship.

3.2. Density-Dependent Adult Survival

Adult survival is difficult to measure accurately. It is also observed to be less responsive to environmental variability than other age classes.

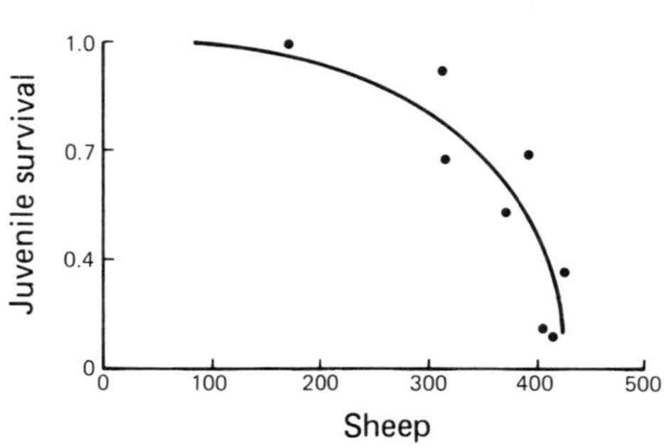

FIGURE 4. Juvenile survival of Soay sheep (*Ovis aries*) at varying population levels (based on data from Grubb, 1974).

TABLE VI
Marine Mammals and Corresponding References Indicating Density Dependence in Survival among Juveniles[a]

Common name	Species	Reference
Gray seal	*Halichoerus grypus*	Coulson and Hickling (1964), Bonner and Hickling (1971), Bonner (1975), Harwood (1981)
Northern fur seal	*Callorhinus ursinus*	Chapman (1961, 1973, 1981), Ichihara (1972), Smith (1973), Lander and Kajimura (1982), Fowler (1986)
Antarctic fur seal	*Arctocephalus gazella*	Bonner (1968), Doidge and Croxall (1984)
Harp seal	*Phoca groenlandica*	Lett et al. (1981)
Northern elephant seal	*Mirounga angustirostris*	LeBoeuf and Briggs (1977), LeBoeuf and Condit (1983)
Southern elephant seal	*Mirounga leonina*	Bester and Lenglart (1982)
Killer whale	*Orcinus orca*	Bigg (1982), Fowler (1984)

[a]The evidence of density dependence may depend on the collective information of two or more references. A presentation of more detail is found in the preliminary review of Fowler et al. (1980).

Even so, density dependence has been observed in a number of species of large mammals (Table VII). It is in the case of adult survival that it is most clear that long-term intensive studies are most likely to produce information concerning density dependence. Examples involve red deer (Clutton-Brock et al., 1982), Soay sheep (Grubb, 1974), and African

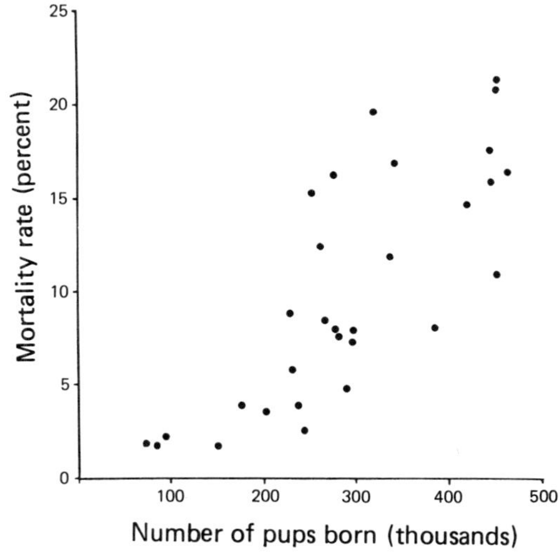

FIGURE 5. Mortality of northern fur seal pups (*Callorhinus ursinus*) at various population levels (measured by the number of pups born on St. Paul Island, Alaska) (Fowler, 1986).

TABLE VII
Large Mammals and Corresponding References Indicating Density Dependence in Adult Survival

Common name	Species	References
Mule deer	*Odocoileus hemionus*	Klein and Olson (1960)
Elk	*Cervus elephus*	Knight (1970), Carbyn (1975), Fowler and Barmore (1979)
Red deer	*Cervus elephus*	Clutton-Brock et al. (1982)
Reindeer	*Rangifer tarandus*	Leader-Williams (1982), Bonner (1984)
Wildebeest	*Connochaetes taurinus*	Sinclair (1973), Sinclair et al. (1985)
Bighorn sheep	*Ovis canadensis*	Hansen (1980)
Soay sheep	*Ovis aries*	Grubb (1974)
African buffalo	*Syncerus caffer*	Sinclair (1973, 1974, 1977)
Thar	*Hemitragus jemlachicus*	Caughley (1970)
Killer whale	*Orcinus orca*	Bigg (1982), Fowler (1984)
Human	*Homo sapiens*	Factor and Waldron (1973), Paulus et al. (1979)

buffalo (Sinclair, 1977). Hansen (1980) shows data indicating that mortality among desert bighorns increased at high population levels and that most dead animals found were adults. Adult survival is one of the density-dependent factors that seems to have been clearly demonstrated for human beings (*Homo sapiens*). One set of information for humans comes from a highly controlled study (Fig. 6) (Paulus et al., 1979), whereas other information involves a broad spectrum of cross-comparisons (Factor and Waldron, 1973).

Disease and pathological abnormalities are often seen as contributing to higher mortality at high population densities. Such is the case for the reindeer of South Georgia (Leader-Williams, 1982; Bonner, 1984). As observed by Shrauder (1984), diseases and parasitic infestations are serious for white-tailed deer only when their populations are high enough

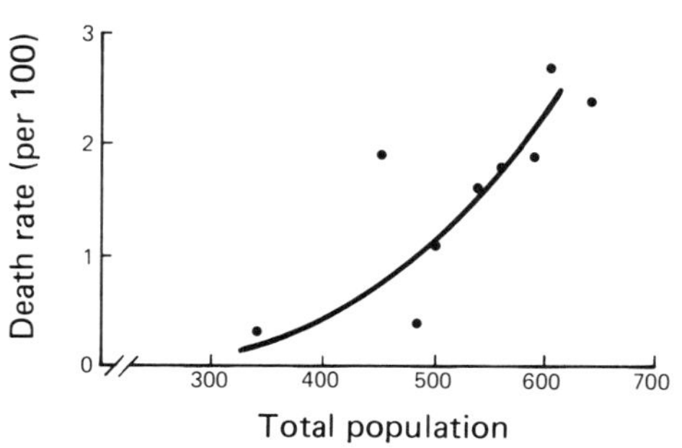

FIGURE 6. Mortality of adult male humans at different prison population levels (based on data from Paulus et al., 1979).

to have reduced their food resources, resulting in poor physical condition (see also Cook, 1984; Newsome, 1984; Whittington, 1984; Matschke et al., 1984). Torgerson and Porath (1984) cite work indicating that infestations of meningeal worms, lone star ticks, and various abdominal roundworm parasites and abomasal parasites all tend to occur at more injurious levels at high densities of white-tailed deer. The density-dependent relationship between abomasal parasites and white-tailed deer populations was reviewed by Eve and Kellogg (1977), who showed significant realtionships in which the more dense populations were more heavily parasitized. Coccidiosis and strongylatosis were found to occur in correlation to population density of roe deer by Padaiga and Marma (1970).

4. RECRUITMENT

Recruitment can be defined as the addition of young animals to a population of older animals. The population of older animals, in turn, can be defined variously as that sector of the population that is reproductively mature, those animals older than a certain specified age, or that sector of a population subjected to commercial utilization. This latter definition has been of particular importance in the case of commercially harvested species of whales. Table VIII contains a list of species of large mammals for which recruitment has been observed to be density dependent. Density-dependent recruitment, being a product of both survival and reproduction prior to the time of recruitment, is included in this review because it emphasizes the observation that both survival and reproduction are important components of population regulation for large mammals.

Measures of recruitment are often highly inferential compared to the more direct observations yielding information for reproduction and juvenile survival. Even so, there are a number of species for which density-dependent relationships are thought to occur, including a species for which density dependence has not been shown otherwise (gray whales, *Eschrichtius robustus*).

5. GROWTH AND BODY SIZE

In addition to the density dependence observed in reproduction and survival (which are both related to population regulation), there are numerous cases where density dependence has been observed in

TABLE VIII
Large Mammals and Corresponding References Indicating Density Dependence in Recruitment (Involving a Combination of Reproduction and Survival)

Common name	Species	References
Mule deer	*Odocoileus hemionis*	Kelker (1947), Dasmann and Blaisdell (1954)
Red deer	*Cervus elephus*	Bannikov and Lebedeva (1973)
Moose	*Alces alces*	Serafinski (1969), Ling (1970), Tomek (1977)
Reindeer	*Rangifer tarandus*	Klein (1968), Leader-Williams (1980)
Bighorn sheep	*Ovis canadensis*	Stelfox (1976)
Wild horse	*Equus caballus*	Berger (1986)
Lion	*Panther leo*	Bertram (1973), Smuts (1978)
Northern fur seal	*Callorhinus ursinus*	Chapman (1961, 1981), Eberhardt (1981)
Striped dolphin	*Stenella coeruleoalba*	Kasuya and Miyazaki (1982)
Blue whale	*Balaenoptera musculus*	Chapman (1964b)
Fin whale	*Balaenoptera physalus*	Allen (1966, 1972, 1973, 1978), Doi et al. (1965, 1969)
Gray whale	*Eschrichtius robustus*	Reilly (1981), Fowler (1984)

body growth and physical size. Table IX contains a list of species for which such relationships have been observed in length, weight, or girth. Among these, Stelfox (1976) indicates that the size of bighorn sheep is influenced by increasing winter-time weight loss with increasing population levels. Growth and age at first reproduction seem closely related and, as inferred by Masaki (1978) for sei whales, changes in one are often accompanied by changes in the other.

Various body parts also show density dependence in growth or size, usually correlated with general body size. Among the ungulates, antler size is density dependent, as demonstrated by Johnson (1937), Severinghaus et al. (1950), W. H. Adams (1960), Cook (1984), and McCullough (1984, 1985) for the genus *Odocoileus*. Kaji et al. (1985) showed similar trends for sika deer, as did Clutton-Brock et al. (1982) (both weight and length) and Clutton-Brock et al. (1985) for red deer. Other body parts have also been demonstrated to be density dependent in size. The jawbone and femur length of red deer has been observed to be density dependent by Klein and Strandgaard (1972). Other examples include the length of jawbones as observed for red deer by Challies (1973a–c) and Staines (1978), for sika deer by Kaji et al. (1985),

TABLE IX
Large Mammals and Corresponding References Indicating Density Dependence in Physical Growth or Body Size

Common name	Species	References
White-tailed deer	*Odocoileus virginianus*	Johnson (1937), Severinghaus (1955, 1979), W. H. Adams (1960), Teer et al. (1965), Hesselton et al. (1965), Kie et al. (1980), Cook (1984), McCullough (1984, 1985)
Mule deer	*Odocoileus hemionis*	Johnson (1937), Klein (1964)
Roe deer	*Capreolus capreolus*	Klein and Strandgaard (1972)
Sika deer	*Cervus nippon*	Christian et al. (1960), Kaji et al. (1985)
Red deer	*Cervus elephus*	Challies (1973a,b, 1978, 1985), Staines (1978), Albon et al. (1983)
Reindeer	*Rangifer tarandus*	Klein (1968), Reimers et al. (1983), Skogland (1983, 1984a, 1985a)
Bighorn sheep	*Ovis canadensis*	Stelfox (1976)
Wildebeest	*Connochaetes taurinus*	Watson (1969, 1970)
Hippopotamus	*Hippopotamus amphibius*	Laws (1968)
African elephant	*Loxodonta africana*	Laws (1966, 1968, 1969a,b, 1973), Laws and Parker (1968)
Northern fur seal	*Callorhinus ursinus*	Scheffer (1955), Nagasaki (1961), Fowler (1986)
Grey seal	*Halichoerus grypus*	Coulson and Hickling (1964), Bonner (1975)
Weddell seal	*Leptonychotes weddellii*	Stirling (1971a,b)
Southern elephant seal	*Mirounga leonina*	Carrick et al. (1962)
Sei whale	*Balaenoptera borealis*	Masaki (1978)

and for reindeer (Fig. 7) by Skogland (1983). Hoof size and number of points on the antlers of sika deer have also been observed to be density dependent by Kaji et al. (1985). Tooth size in northern fur seals has been determined to be density dependent (Fig. 8), as reported by Fowler (1982, 1986).

Fetal growth and size can also show density-dependent change, as shown for reindeer by Skogland (1983, 1984a, 1985b). Similar trends seem to exist for northern fur seals (Fowler, 1986), based on size at birth.

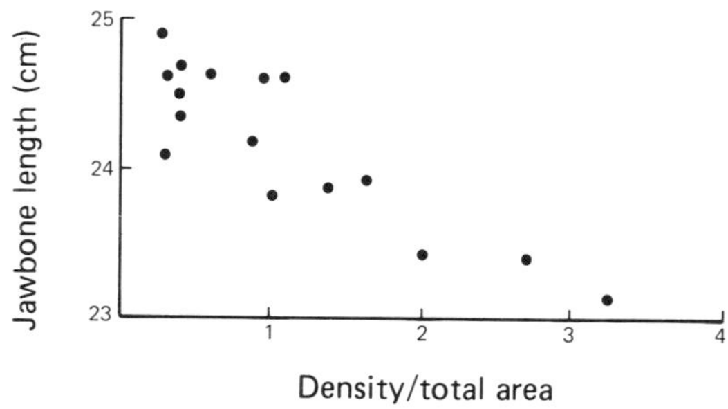

FIGURE 7. Size of jawbones (mean length) for reindeer (*Rangifer tarandus*) at various population levels (Skogland, 1983).

6. OTHER DENSITY-DEPENDENT TRAITS

With the possible exception of diseases and predation as factors contributing to mortality, the categories of density dependence reviewed above are those best represented by published literature. However, many other factors have also been observed to change with population density in parallel with regulatory factors and growth.

Some aspects of the behavior of several different species have been observed to be density dependent. Stranding in toothed whales has been suggested as being more frequent at high population levels (Sergeant, 1982). Houston (1979) notes that the movement of elk in and out of the northern Yellowstone National Park is different at varying levels of population density. The same is noted by Berger (1986) for horses (*Equus caballus*). Emigration has been observed to increase with density for northern elephant seals (Reiter et al., 1981) and moose (Tomek, 1977). Emigration was also found to show signs of being correlated with density for red deer by Clutton-Brock et al. (1985). Skogland (1985b) indicates that both emigration and group size in reindeer are density dependent, small groups being associated with higher densities. Group

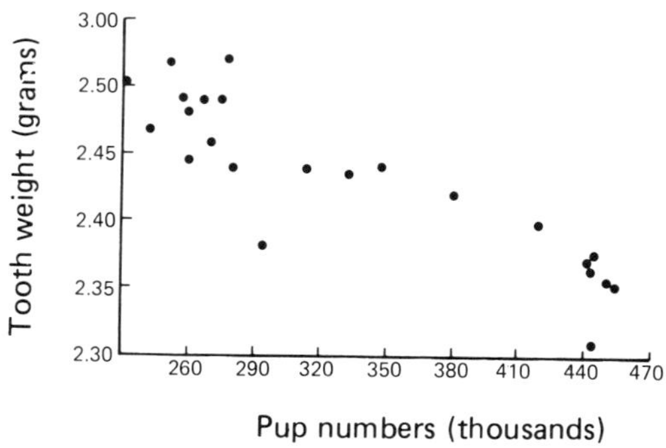

FIGURE 8. Mean maxillary canine tooth weight of northern fur seals at various population levels (pup numbers) (Fowler, 1986).

size for European bison (*B. bonasus*) has also been observed to increase with density (Krasinksi, 1978). Various aspects of the behavior of red deer have been noted to be density dependent (Clutton-Brock et al., 1985), including the duration of suckling bouts (possibly related to milk production), rut-related behavior, and aggressive interactions. Daily suckling time has also been shown to vary according to population density for reindeer (Skogland, 1985b). Similar relationships may occur in musk oxen (*Ovibos moschatus*), as observed by Jingfors (1984). Territorial size in red deer declines with density (Klein and Strandgaard, 1972). Berger (1979a,b, 1980) has shown differences in behavioral repertoires that are related to population density for bighorn sheep (*O. canadensis*). In this work lambs raised in populations with higher densities showed more complex behavior, were more active in social development, and played more than sheep in less dense populations. The mating strategy and success of northern elephant seals has been observed to change with density (Reiter et al., 1981; LeBoeuf and Reiter, 1987; LeBoeuf, 1974).

Infanticide is thought to change with density in several species. Shaffer (1978), McCullough (1981), and Stringham (1983) present information suggesting the possibility of density-dependent infanticide for grizzly bears. It is thought that infanticide may increase with density among lions (Smuts, 1978). Divale (1972) reviews information related to infanticide among humans.

Other aspects of human behavior have also been shown to be density dependent. McCain et al. (1976) have shown that the level of complaints increases among people under crowded conditions. Nacci et al. (1977) maintain that misconduct is also observed at higher frequency among more crowded populations. In a more positive vein, the organizational traits of Plains Indians was observed to increase with population size in the anthropological studies of Carneiro (1967). This is an example of the potential for a stimulating (rather than inhibitory) effect resulting from increasing density.

The feeding habits of individuals within populations may change in reaction to the changing composition of available forage in the environment (itself a response to changing consumer population levels). Such changes have been observed by Kie et al. (1980), who have shown that the composition of the diet of deer changed in parallel with changes in population size. Similar information is presented by Dasmann and Blaisdell (1954) and McCullough (1985) for deer. Skogland (1984b, 1985b) indicates that the diet and feeding strategy of reindeer change with food limitation at higher densities, as is also shown by Stelfox (1976) for bighorn sheep.

Sex ratios at particular stages of the life cycle have also been observed to be density dependent. The sex ratio of reindeer at birth is density dependent, according to Skogland (1986a). The sex ratio of fawns in white-tailed deer was found to vary (McCullough, 1979). The sex ratio of young wolves has been observed to show changes in response to population size (Mech, 1975). The sex ratio of older animals in red deer populations has been observed to be density dependent by Clutton-Brock et al. (1982), as it has for northern elephant seals by LeBoeuf (1974). This may reflect what appears to be a greater tendency for males than females to respond in a density-dependent manner in many populations. For instance, in wild horses males are more apt to disperse farther distances as population size increases (Berger, 1986).

The timing of various events in the annual cycle of large mammals has also been shown to be density dependent. Kie et al. (1980) indicate that conception occurs later in the year for white-tailed deer at high population levels. Teer (1984) finds that ovulation is delayed, and both McCullough (1979) and Teer (1984) report delayed breeding seasons at higher densities for the species. Kie et al. (1980) also indicate that under similar conditions, velvet is retained longer and the shedding of antlers is delayed. The Tule elk also shows changes in breeding season associated with density (McCullough, 1969). Clutton-Brock et al. (1982, 1985) indicate that the timing of conception, molting, antler loss, rut, and parturition have been observed to be density dependent for red deer. Reimers et al. (1983) and Skogland (1985b) indicate that birth is delayed in reindeer at high density. The length of the lactation and resting phases of dolphin reproductive cycles have been shown to change as a response to exploitation and the resulting changes in population levels (Kasuya, 1985).

Physiological aspects of individuals within populations are also known to show density-dependent responses. For various levels of malnutrition in white-tailed deer populations, Kie et al. (1980) observed that kidney fat, femur marrow fat, and various blood parameters exhibited parallel changes. Fat reserves observed for the thar are thought to follow similar patterns (Caughley, 1976). Other aspects of body condition (see review by Hanks, 1981) may be reflected in the observed relationship between weight at median fertility and population size for red deer (Albon et al., 1983).

The number of falls suffered by young reindeer searching for food (scarce at high densities) has been observed to be density dependent by Leader-Williams (1980) based on comparing separate populations. As mentioned above, diseases are often thought to be among the factors contributing to mortality at high population levels. This was seen as significant for wildebeest by Talbot and Talbot (1963). Ill health was also observed to be more of a problem at higher population levels for bighorn sheep by Stelfox (1976) and for reindeer (Leader-Williams, 1982).

Density dependence can also involve the variance of a particular parameter rather than (or in addition to) its actual level. Many of the regulatory factors reviewed earlier in this paper exhibit more variability (or density vagueness, to use the term of Strong, 1984) at maximum population levels sustainable by the natural environment. At these levels the population is most sensitive to temporal variability of the physical (density-independent) aspects of the environment. This change in variability within the population, however, can be somewhat difficult to separate from the nonlinear changes often observed in population parameters (Fowler, 1981). For example, increased variability in the birth rate of longhorn cattle at high population levels is suggested by the data presented by Gross et al. (1973). Similar observations apply to lamb production by bighorn sheep (Woodgerd, 1964), for both juvenile and adult survival among Soay sheep (Grubb, 1974), and for recruitment among large whales (for example, see Chapman, 1964b). Such density dependence is explicitly presented by Fowler (1986) for the variability of juvenile survival among northern fur seals. Variance in population numbers themselves was observed to increase with population numbers for the southern elephant seal by Van Aarde (1980, 1984).

7. DISCUSSION

Biologists studying populations of large mammals find themselves faced with extensive information regarding density dependence in a wide variety of species. The task is to interpret such information as it relates to current questions of both theoretical and practical importance. Using the published literature as a resource, information has been assembled and very briefly presented. At this point we must find what can be said about this collection of information that may be helpful in its interpretation, its application to future studies, and its use in the management of large-mammal populations. The data reviewed show that, in general, factors limiting populations become more pronounced as the populations increase. Mortality and emigration increase, whereas reproduction and growth decrease with higher population levels. The general trend is one in which, regardless of mechanism (for example, predation, food limitation, disease, or behavior), density-dependent reactions usually resist increased population density.

One question regarding the evolutionary aspects of population dynamics is whether or not one mode of population regulation predominates as a regulatory factor in the populations exhibiting life history strategies characteristic of large mammals (delayed maturation, small litters, high survival, etc.). The data (Table X) (Fowler, 1984) support the view that population regulation among large mammals is a set of phenomena involving the stages of life just before, during, and just after parturition. This is seen in two ways. First, there is the information indicating that reproductive rates of both young and older animals exhibit density dependence among a large number of species. Second, juvenile survival often responds to density. It must be remembered, however, that the process of measuring adult survival is logistically difficult and we lack the tools to detect the small changes often characteristic of the variability of this parameter. In view of the great sensitivity of a population's rate of increase (or decrease) to changes in adult survival, it seems premature to conclude that the adult survival of large-mammal populations is not an important density-dependent factor. On theoretical grounds it can be argued that evolutionary forces may be expected to operate to change any imbalance in the expression of density dependence so as to produce relative uniformity among the effects of the various factors involved in population regulation. The relative strength of any particular factor (strength measured as contribution to long-term mean rates of change in population) would be expected to be proportional to the degree to which it would be sensitive

TABLE X
Comparison of the Mode of Expression of Density Dependence in Population Regulation of Large Mammals[a]

	Number of species		
	Marine mammals	Terrestrial mammals	Total
Birth rate	8	21	29
Age at reproductive maturation	11	11	22
Juvenile survival	7	20	27
Adult survival	1	9	10
Total[b]	17	28	45

[a]A count of species for which density dependence has been reported in the literature as based on Tables I–VII.
[b]Some species fall into more than one category, so the total is not the sum of the column. Thus, for example, 8 of 17 species of marine mammals show density dependence in adult birth rate.

to evolutionary changes. On these grounds no single factor would be expected to predominate as the most important element in population regulation.

Generally, food and food resources seem to be the main factor behind the expression of density dependence in large mammals (for example, Klein, 1970, 1985). The bulk of density-dependent changes presented in the literature are attributed to the lack of food resources at high populations compared to their availability at lower population levels (Fowler et al., 1980; Fowler, 1984). Even when other factors such as disease or predation are mentioned, these factors are often observed to occur in response to the weakened condition of animals faced with limited resources (for example, see the various papers in Halls, 1984). Even the behavioral and social factors thought to operate in population regulation of many higher vertebrates are thought to operate in response to food levels available in the environment. An example of this is seen in the work of Dittus (1975, 1978) regarding the population dynamics of the toque monkey (*Macaca sinica*). The general importance of food resources is emphasized by the energetics-based expression of density dependence in body size, growth, and the miscellaneous density-dependent factors other than population regulation or growth. Many of these factors are seen as being a response to decreased levels of food resources available to individuals at high population levels due to increased consumption by the total population.

The emphasis on food limitation apparent in the literature should not be overstated, however. Other mechanisms, especially diseases and predation, can be involved. In this regard it must be pointed out that many of the studies of terrestrial ungulates involve populations for which there are no longer natural predators. It is apparent that studies of populations with and without predation tend to produce dissimilar results. This is the case with reindeer populations subjected to predatory population regulation (Bergerud, 1980) as compared to populations that suffer less predation (Skogland, 1983). Moose populations subjected to predator-caused mortality do not seem to exhibit the same kind of density dependence observed in other populations (Messier and Crete, 1984, 1985). Carbyn (1975) suggests that populations of elk subjected to predation exhibit less food-related density-dependent regulation than predator-free populations. Even so, populations subjected to natural predation, diseases, and parasites often show signs of food limitation (for example, see Sinclair and Norton-Griffiths, 1982, and the various papers in Halls, 1984).

The relative roles of various limiting factors in human population dynamics are not clear. It is probable that a combination of factors operated to regulate prehistoric human populations (Cohen, 1980). Evidence for the primary importance of food-based causes in prehistoric populations, with secondary roles for predation and disease, is strengthened by extrapolation from natural wild populations of nonhuman large mammals as reviewed in this paper. It is thought that human populations without the benefit of modern technology respond to food levels in the manner such extrapolations would predict (for example, Hayden, 1972; Hassen, 1981). The influence of food may be mediated by or expressed through other sociobiological factors, as emphasized by the review of Dumond (1975), much of the work in Cohen et al. (1980), and Hassen (1981).

Another pattern in the population dynamics of large mammals is the nonlinearity in density dependence—populations exhibit most population regulation at levels close to the carrying capacity (see Figs. 1 and 3–6 and Fowler, 1981). Much of the work on large mammals yields evidence for this pattern (for example, Christian, 1971). Examples of nonlinearity in the per capita rates behind this phenomena are increasingly prevalent. Nygren's (1983) data for moose show such nonlinearity in reproductive rates (including a reduced rate at the lowest population levels). The juvenile mortality for red deer shown by Clutton-Brock et al. (1985) appears nonlinear, with most change occurring at higher densities. In Swenson's (1985) work on mountain goats maximum production occurred at levels over 50% of those sustainable by the environment. Direct tests of this pattern were undertaken by Skogland (1985a, 1986b), who found that populations of reindeer show most density-dependent food limitation at levels above 50% of the carrying capacity. By contrast, the evidence for density-dependent growth seems to show less nonlinearity (for example, Figs. 7 and 8). If growth is more linear in its density-dependent relationships than are the vital rates associated with reproduction and mortality, it would support the view that the vital rates may exhibit most change around physiological thresholds or thresholdlike phenomena. This aspect of density dependence in large mammals should be studied in greater detail.

Just how this nonlinearity in density dependence relates to life history strategy is not totally clear. It may involve the degree to which a species acts as a consumer (regulating its resource) versus the degree to which it is consumed and thereby regulated by other populations (Fowler, 1980a, 1981). Another possibility is that nonlinearity is a result of evolutionary pressures involved in the production of life history traits, measured by the rate of increase per generation. A preliminary analysis (Fowler, 1984) indicates that species with relatively low rates of population increase per generation exhibit more regulation at high population levels compared to species with high rates of increase per generation. Considerable effort is being spent on the analysis of the evolutionary aspects of population dynamics and the results should shed light on these apparent patterns.

One of the observations made possible by this review is that information on density dependence can be a useful interpretive tool in describing the status of populations. Not only are altered population levels observed to show changes in their characteristics, but, reversing the relationship, such changes can serve as indices of population levels. Severinghaus (1955, 1979), for example, showed how antler and body size serve as assessments of the condition of a population in relation to the carrying capacity of its habitat. Jawbone length is used in the management of wild reindeer. The use of density-dependent information for determining status is also exemplified by the work of Challies (1973b, 1978) for red deer in New Zealand. The strong tendency for diseases and parasite loads to be density dependent has led managers of white-tailed deer to use diseases and parasites as indicators of deer densities in relation to their carrying capacities (especially as related to nutritional status, Eve and Kellogg, 1977; Matschke et al., 1984). The general utility of such indices are described by Eberhardt and Siniff (1977) and Eberhardt (1977), who also present arguments for increased utilization of such indices. This approach was applied to northern fur seals by Fowler (1985). Such approaches are particularly helpful where specific guidelines are established for population management (for example, levels needed to achieve maximum sustained yield or required as management criteria through legislation such as the U. S. Marine Mammal Protection Act). A review of the utility of various indices of population condition is found in Hanks (1981).

Whether or not there are taxonomic patterns in population regulation remains unclear. The same holds true for whether or not there are clearly discernible differences between species within separate trophic levels. Further work is needed to determine whether or not marine mammals are more regulated by age at first reproduction than are terrestrial mammals, for example. Eberhardt (1977) expected a predictable order in the expression of density-dependent responses along the spectrum of population size. Whether or not this happens for specific populations, or for large mammals in general, remains to be clearly shown (Fowler, 1980b). Such questions are of importance both theoretically and practically, but await further research for clear answers. The discovery and description of any such patterns, however, require both ongoing and future research in conjunction with the base of published information as presented in this paper.

As with other taxonomic categories, it appears that among large mammals males are often more sensitive than females to density-dependent change and their responses are often more pronounced (for example, Peek et al., 1967; Clutton-Brock et al., 1985). Such differences seem to include fetal age classes as well (Robinette et al., 1957). The ecological implications of these differences between the sexes are not well understood, but may involve body size, metabolic requirements, and reproductive behavior.

8. SUMMARY

It is clear that large-mammal populations exhibit density dependence in various forms. Some result in population regulation; others appear as changes in characteristics of individual animals that often relate to reproduction or maturity. No one mode of expression is clearly defined in large mammals, but there is a strong indication that the reproductive process is involved. Food resources seem to be the most common causative agents behind the expression of density dependence. The density-dependent aspects of large-mammal populations serve as a very useful tool for describing population status, especially in relationship to levels sustainable by the environment. This is particularly true for males in view of their more pronounced responses to population levels. Most large mammals appear to exhibit population regulation most prominently at population levels close to those sustainable by the environment. The population level at which such regulation occurs may be related to its rate of increase per generation as a quantitative measure of the life history strategy of all species. In view of the current interest in the evolutionary aspect of population dynamics (especially as related to life history strategy), it is hoped that by making the literature reviewed in this paper more available to the researchers involved in this field, progress of both practical and theoretical importance will be enhanced.

Finally, there are undoubtedly a significant number of papers (especially non-English works) that have been overlooked in compiling this review. I would be deeply grateful to readers who could provide reprints or lists of relevant publications to be used in any future update of this work or for further synthesis.

ACKNOWLEDGMENTS. I am especially grateful to W. Trent Bunderson, Marion B. Cherry, Ronald J. Ryel, and Benjamin B. Steele, who helped in the compilation of material used to produce the preliminary review of density dependence in large mammals (Fowler et al., 1980). Bill Barmore kindly provided access to his reference files during this work. Jason Baker was very helpful in locating more recent literature published since the first review. Earlier versions of this paper were reviewed by John Bengtson, Joel Berger, Marthan Bester, Nigol Bonner, Don Bowen, Howard Braham, Jeff Breiwick, Chris Challies, Susan Chivers, Tim Clutton-Brock, Dennis Crocker-Bedford, Don Dumond, John Eisenberg, Jerran Flinders, Sharon Giese, Anne Gunn, Aleta Hohn, Doug Houston, Koichi Kaji, David Klein, Burney LeBoeuf, Dale McCullough, Tuire Nyrgren, Nancy Pagh, Roger Pearson, William Perrin, Bernt-Erik Saether, David Sergeant, John Skinner (and several others of the staff at the University of Pretoria), Terje Skogland, and Pamela Wilder. All provided useful comments and the suggestions from these reviews added significantly to the quality and scope of the paper. I thank Leola Hietala for her persistence and patience in typing and correcting the numerous revisions of this paper.

The preliminary review in Fowler et al. (1980) was work conducted under contract MM7AC013 from the Marine Mammal Commission to Utah State University. I would like to thank the staff of both for their support, cooperation, and reviews.

REFERENCES

Adams, L. G., and Bailey, J. A., 1982, Population dynamics of mountain goats in the Sawatch Range, Colorado, *J. Wildl. Manage.* **46**:1003–1009.

Adams, W. H., 1960, Population ecology of white-tailed deer in northeastern Alabama, *Ecology* **41**:706–715.

Albon, S. D., Mitchell, B., and Staines, B. W., 1983, Fertility and body weight in female red deer: A density dependent relationship, *J. Anim. Ecol.* **52**:969–980.

Allen, K. R., 1966, Some methods for estimating exploited populations, *J. Fish. Res. Board Can.* **23**(10):1553–74.

Allen, K. R., 1972, Annex F, Further notes on the assessment of Antarctic fin whale stocks, *Rep. Int. Whaling Comm.* **22**:43–53.

Allen, K. R., 1973, Analysis of the stock-recruitment relation in Antarctic fin whales, *Rapp. P.-V. Reun. Cons. Int. Explor. Mer* **164**:132–137.

Allen, K. R., 1978, Notes on the calculation of recruitment rates (R II) from age composition, *Rep. Int. Whaling Comm.* **28**:147–149.

Bailey, J. A., and Johnson, B. K., 1977, Status of introduced mountain goats in the Sawatch Range of Colorado, *Proc. Int. Mountain Goat Symp.* **1**:54–63.

Bannikov, A. G., and Lebedeva, L. S., 1973, Deer population of Biryuchii Island, *Sov. J. Ecol.* (*Engl. Transl. Ekologiya*) **2**:336–338.

Bengtson, J. L., and Laws, R. M., 1985, Trends in crabeater seal age at maturity: An insight into Antarctic marine interactions, in: *Antarctic Nutrient Cycles and Food Webs* (L. R. Siegfried, P. R. Condy, and R. M. Laws, eds.), Springer, Berlin, pp. 669–676.

Berger, J., 1979a, Weaning, social environments and the ontogeny of spatial associations in bighorn sheep, *Biol. Behav.* **4**:463–472.
Berger, J., 1979b, Social ontogeny and behavioral diversity: Consequences for bighorn sheep *Ovis canadensis* inhabiting desert and mountain environments, *J. Zool. Lond.* **188**:251–266.
Berger, J., 1980, The ecology, structure and functions of special play in bighorn sheep (*Ovis canadensis*), *J. Ecol.* **192**:531–542.
Berger, J., 1986, *Wild Horses of the Great Basin, Social Competition and Population Size*, University of Chicago Press, Chicago.
Bergerud, A. T., 1978, Caribou, in: *Big Game of North America, Ecology and Management* (J. L. Schmidt, and D. L. Gilbert, eds.), Stackpole Books, Harrisburg, Pennsylvania, pp. 83–101.
Bergerud, A. T., 1980, A review of the population dynamics of caribou and wild reindeer in North America, in: *Proceedings of 2nd International Reindeer/Caribou Symposium* (E. Reimers, F. Gaare, and S. Skjenneberg, eds.), Direktoratet for vilt og ferskvannfisk, Trondheim, Norway, pp. 556–581.
Bertram, B. C. R., 1973, Lion population regulation, *East Afr. Wildl. J.* **11**(2):215–225.
Best, P. B., 1980, Pregnancy rates in sperm whales off Durban, *Rep. Int. Whaling Comm. Spec. Issue* **2**:93–77.
Bester, M. N., and Lenglart, P.-Y., 1982, An analysis of the southern elephant seal *Mirounga leonina* breeding population at Kerguelen, *S. Afr. J. Antarct. Res.* **12**:11–16.
Bigg, M., 1982, An assessment of killer whale (*Orcinus orca*) stocks off Vancouver Island, British Columbia, *Rep. Int. Whaling Comm.* **32**:655–66.
Bonner, W. N., 1968, The Fur Seal of South Georgia, *British Antarctic Survey Scientific Report* No. 56.
Bonner, W. N., 1975, Population increase of grey seals at the Farne Islands, *Rapp. P.V. Reun. Cons. Int. Explor. Mer* **169**:336–370.
Bonner, W. N., 1984, Introduced mammals, in: *Antarctic Ecology* (R. M. Laws, ed.), Academic Press, London, pp. 237–278.
Bonner, W. N., and Hickling, G., 1971, The grey seals of the Farne Islands, report for the period October, 1969 to July, 1971, *Trans. Nat. Hist. Soc. Northumber. N. S.* **17**:141–162.
Bourliere, F., 1975, Mammals, small and large: The ecological implications of size, in: *Small Mammals—Their Productivity and Population Dynamics* (F. B. Golly and K. Petrusewicz, eds.), Cambridge University Press, Cambridge, pp. 1–8.
Bowen, W. D., Capstick, C. K., and Sergeant, D. E., 1981, Temporal changes in the reproductive potential of female harp seals (*Pagophilus groenlandicus*), *Can. J. Fish. Aquat. Sci.* **38**(5):495–503.
Boyce, M. S., 1984, Restitution of r- and *K*-selection as a model of density dependent natural selection, *Annu. Rev. Ecol. Syst.* **15**:427–47.
Buechner, H. K., and Swanson, C. V., 1955, Increased natality resulting from lowered population density among elk in southeastern Washington, *Trans. N. Am. Wildl. Nat. Resour. Conf.* **20**:560–567.
Buss, I. O., and Savidge, J. M., 1966, Change in population number and reproductive rate of elephants in Uganda, *J. Wildl. Manage.* **30**(4):791–809.
Buss, I. O., and Smith, N. S., 1966, Observations on reproduction and breeding behavior of the African elephant, *J. Wildl. Manage.* **30**(2):375–388.
Capstick, C. K., and Ronald, K., 1982, Modeling seal populations for herd management, in: *Mammals in the Seas: Small Cetaceans, Seals, Sirenians, and Otters*, FAO Fisheries Series 5, Volume 4, pp. 177–190.
Carbyn, L. N., 1975, Wolf predation and behavioral interactions with elk and other ungulates in an area of high prey diversity, Ph. D. thesis, University of Toronto.

Carneiro, R. L., 1967, On the relationship between size of population and complexity of social organization, *Southwest. J. Anthropol.* **23**:234–243.

Carrick, R., and Ingham, S. E., 1962, Studies on the southern elephant seal, *Mirounga leonina* (L.), V., Population dynamics and utilization, *CSIRO Wildl. Res.* **7**(2):198–206.

Carrick, R., Csordas, S. E., and Ingham, S. E., 1962, Studies on the southern elephant seal, *Mirounga leonina* (L.), IV., Breeding and development, *CSIRO Wildl. Res.* **7**(2):161–197.

Caughley, G., 1966, Mortality patterns in mammals, *Ecology* **47**(6):906–918.

Caughley, G., 1970, Eruption of ungulate populations, with emphasis on Himalayan thar in New Zealand, *Ecology* **51**(1):53–72.

Caughley, G., 1976, Wildlife management and the dynamics of ungulate populations, in: *Applied Biology* (T. H. Coaker, ed.), Academic Press, London, pp. 183–246.

Challies, C. N., 1973a, Trends in red deer (*Cervus elaphus*) populations in Westland forests, *Proc. N. Z. Ecol. Soc.* **21**:45–50.

Challies, C. N., 1973b, The use of physical size and growth rate for deer management, *N. Z. For. Serv. Res. Inst. FRI Symp.* **14**:44–53.

Challies, C. N., 1973c, The effect of the commercial venison industry on deer populations, *N. Z. For. Serv. Res. Inst. FRI Symp.* **14**:164–171.

Challies, C. N., 1978, Assessment of the physical well-being of red deer (*Cervus elephus*) populations in South-Westland, *N. Z. For. Serv. Rep. For. Res. Inst.* **163**:1–104.

Challies, C. N., 1985, Establishment, control, and commercial exploitation of wild deer in New Zealand, *R. Soc. N. Z. Bull.* **22**:23–26.

Chapman, D. G., 1961, Population dynamics of the Alaska fur seal herd, *Trans. N. Am. Wildl. Nat. Resour. Conf.* **26**:356–369.

Chapman, D. G., 1964a, Critical study of Pribilof fur seal population estimates, *U. S. Fish Wildl. Serv. Fish. Bull.* **63**(3):657–669.

Chapman, D. G., 1964b, Final report, Special Committee of Three Scientists, *Rep. Int. Whaling Comm.* **14**:39–92.

Chapman, D. G., 1973, Spawner-recruit models and estimation of the level of maximum sustainable catch, *Rapp. P. V. Reun. Cons. Int. Explor. Mer* **164**:325–332.

Chapman, D. G., 1981, Evaluation of marine mammal population models, in: *Dynamics of Large Mammal Populations* (C. W. Fowler and T. D. Smith, eds.), Wiley, New York, pp. 277–296.

Cheatum, E. L., and Severinghaus, C. W., 1950, Variations in fertility of white-tailed deer related to range conditions, *Trans. N. Am. Wildl. Nat. Resour. Conf.* **15**:170–190.

Child, G., Roth, H. H., and Kerr, M., 1968, Reproduction and recruitment patterns in warthog (*Phacochoerus aethiopicus*) populations, *Mammalia* **32**(1):16–29.

Christian, J. J., 1963, Endocrine adaptive mechanisms and the physiologic regulation of population growth, in: *Physiological Mammalogy* (W. V. Mayer and R. G. Van Gelder, eds.), Academic Press, New York, pp. 189–353.

Christian, J. J., 1971, Population density and fertility in mammals, in: *The Action of Hormones: Genes to Populations* (P. Foa, ed.), C. C. Thomas, Springfield, Illinois, pp. 471–510.

Christian, J. J., Flyger, V., and Davis, D. E., 1960, Factors in the mass mortality of a herd of sika deer, *Cervus nippon*, *Chesapeake Sci.* **1**(2):79–95.

Clark, L. R., Geier, P. W., Hughes, R. D., and Morris, R. F., 1967, *The Ecology of Insect Populations in Theory and Practice*, Methuen, London.

Clutton-Brock, T. H., Guinness, F. E., and Albon, S. D., 1982, *Red Deer*, University of Chicago Press, Chicago, Illinois.

Clutton-Brock, T. H., Guinness, F. E., and Albon, S. D., 1983, The costs of reproduction to red deer hinds, *J. Anim. Ecol.* **52**:367–383.

Clutton-Brock, T. H., Major, M., and Guinness, F. E., 1985, Population regulation in male and female red deer, *J. Anim. Ecol.* **54**:831–846.

Cohen, M. N., 1980, Speculations on the evolution of density measurement and population regulation in *Homo sapiens*, in: *Biosocial Mechanisms of Population Regulation* (M. N. Cohen, R. S. Malpass, and H. G. Klein, eds.), Yale University Press, New Haven, Connecticut.

Cohen, M. N., Malpass, R. S., and Klein, H. G. (eds.), 1980, *Biosocial Mechanisms of Population Regulation*, Yale University Press, New Haven, Connecticut.

Cook, R. L., 1984, Texas, in: *White-tailed Deer: Ecology and Management* (L. K. Halls, ed.), Stackpole Books, Harrisburg, Pennsylvania, pp. 457–474.

Corfield, T. F., 1973, Elephant mortality in Tsavo National Park, Kenya, *East Afr. Wildl. J.* **11**:339–368.

Coulson, J. C., and Hickling, G., 1964, The breeding biology of the grey seal, *Halichoerus grypus* (Fab.), on the Farne Islands, Northumberland, *J. Anim. Ecol.* **33**:485–512.

Dasmann, W. P., and Blaisdell, J. A., 1954, Deer and forage relationship on the Lassen-Washoe interstate winter deer range, *Calif. Dep. Fish Game Bull.* **40**:215–234.

Dittus, W. P. J., 1975, Population dynamics of the toque monkey, *Macaca sinica*, in: *Socioecology and Psychology of Primates* (R. H. Tuttle, ed.), Mouton, The Hague, pp. 125–151.

Dittus, W. P. J., 1978, The social regulation of population density and age–sex distribution in the toque monkey, *Behaviour* **63**:281–322.

Divale, W. T., 1972, Systematic population control in the middle and upper Palaeolithic: Inferences based on contemporary hunter-gatherers, *World Archaeol.* **4**:222–243.

Doi, T., Ohsumi, S., and Nemoto, T., 1965, On diagnosis of the fin whale population in the Antarctic and optimum level of catching, *Bull. Tokai Reg. Fish. Res. Lab.* **41**:1–21.

Doi, T., Ohsumi, S., Nasu, K., and Shimadzu, Y., 1969, Advanced assessment of fin whale stock in the Antarctic, *Bull. Tokai Reg. Fish. Res. Lab.* **60**:95–126.

Doidge, D. W., and Croxall, J. P., 1984, Density-dependent pup mortality in the Antarctic fur seal *Artocephalus gazella* at South Georgia, *J. Zool. Lond.* **202**:449–460.

Douglas, C. L., and Leslie, P. M., Jr., 1986, Influence of weather and density on lamb survival of desert mountain sheep, *J. Wildl. Manage.* **50**:153–156.

Dumond, D. E., 1975, The limitation of human population: A natural history, *Science* **187**:713–721.

Eberhardt, L. L., 1977, Optimal management policies for marine mammals, *Wildl. Soc. Bull.* **5**(4):162–169.

Eberhardt, L. L., 1981, Population dynamics of the Pribilof fur seals, in: *Dynamics of Large Mammal Populations* (C. W. Fowler and R. D. Smith, eds.), Wiley, New York, pp. 197–220.

Eberhardt, L. L., and Siniff, D. B., 1977, Population dynamics and marine mammal management policies, *J. Fish. Res. Board Can.* **34**:183–190.

Emmel, T. C., 1976, *Population Biology*, Harper and Row, New York.

Eve, J. H., and Kellog, F. E., 1977, Management implications of abomasal parasites in southeastern white-tailed deer, *J. Wildl. Manage.* **41**:169–177.

Factor, R. M., and Waldron, I., 1973, Contemporary population densities and human health, *Nature* **243**:381–384.

Fowler, C. W., 1980a, Exploited populations of predator and prey: Implications of a model, in: *Comparative Population Dynamics of Large Mammals: A Search for Management Criteria* (C. W. Fowler, W. T. Bunderson, M. B. Cherry, R. J. Ryel, and B. B. Steele, eds.), U. S. Department of Commerce, National Technical Information Service, Springfield, Virginia, pp. 260–281.

Fowler, C. W., 1980b, Indices of population status, in: *Comparative Population Dynamics of Large Mammals: A Search for Management Criteria* (C. W. Fowler, W. T. Bunderson, M. B. Cherry, R. J. Ryel, and B. B. Steele, eds.), U. S. Department of Commerce, National Technical Information Service, Springfield, Virginia, pp. 282–295.

Fowler, C. W., 1981, Density dependence as related to life history strategy, *Ecology* **62**(3):602–610.

Fowler, C. W., 1982, Interactions of northern fur seals and commercial fisheries, *Trans. N. Am. Wildl. Nat. Resour. Conf.* **47**:278–292.

Fowler, C. W., 1984, Density dependence in cetacean populations, *Rep. Int. Whaling Comm. Spec. Issue* **6**:373–379.

Fowler, C. W., 1985, Status review: Northern fur seals (*Callorhinus ursinus*) of the Pribilof Islands, Alaska, National Marine Mammal Laboratory, Northwest and Alaska Fisheries Center, Seattle, Washington; background paper submitted to the 28th Annual Meeting of the North Pacific Fur Seal Commission, Tokyo, Japan.

Fowler, C. W., 1986, Density dependence in northern fur seals (*Callorhinus ursinus*), in *Proceedings of a Conference on Fur Seals of the World, Cambridge, England, April 1984*, in press.

Fowler, C. W., and Barmore, W. J., 1979, A population model of the northern Yellowstone elk herd, in: *Proceedings of the First Conference on Scientific Research in National Parks*, Volume 1 (R. M. Linn, ed.), U. S. National Park Service, pp. 427–434.

Fowler, C. W., and Smith, T., 1973, Characterizing stable populations: An application to the African elephant population, *J. Wildl. Manage.* **37**(4):513–523.

Fowler, C. W., Bunderson, W. T., Ryel, R. J., and Steel, B. B., 1980, A preliminary review of density dependent reproduction and survival in large mammals, in: *Comparative Population Dynamics of Large Mammals: A Search for Management Criteria* (C. W. Fowler, W. T. Bunderson, M. B. Cherry, R. J. Ryel, and B. B. Steele, eds.), U. S. Department of Commerce, National Technical Information Service, Springfield, Virginia, pp. 48–173.

Gambell, R., 1973, Some effects of exploitation on reproduction in whales, *J. Reprod. Fertil. Suppl.* **19**:533–553.

Gambell, R., 1975, Variations in reproduction parameters associated with whale stock sizes, *Rep. Int. Whaling Comm.* **25**:182–189.

Gambell, R., 1976, Population biology and the management of whales, *Appl. Biol.* **1**:247–343.

Gavin, T. A., 1984, Pacific Northwest, in: *White-tailed Deer: Ecology and Management* (L. K. Halls, ed.), Stackpole Books, Harrisburg, Pennsylvania, pp. 487–496.

Geist, V., 1971, *Mountain Sheep: A Study in Behavior and Evolution*, University of Chicago Press, Chicago, Illinois.

Gibbney, L. F., 1957, The seasonal reproductive cycle of the female elephant seal—*Mirounga leonina* L.—at Heard Island, *Aust. Nat. Antarct. Res. Exp. Rep. B* **1**(*Zool.*):1–26.

Grimsdell, J. J. R., 1973, Reproduction in the African buffalo, *Syncerus caffer*, in Western Uganda, *J. Reprod. Fertil. Suppl.* **19**:303–318.

Gross, J. E., 1969, Optimum yield in deer and elk populations, *Trans. N. Am. Wildl. Nat. Resour. Conf.* **34**:372–386.

Gross, J. E., Roelle, J. E., and Williams, G. L., 1973, Program ONEPOP an Information processor: A Systems Modeling and Communications Project, Progress Report, Colorado Cooperative Wildlife Research Unit, Colorado State University, Fort Collins, Colorado.

Grubb, P., 1974, Population dynamics of the Soay sheep, in: *Island Survivors: The Ecology of the Soay Sheep of St. Kilda* (P. A. Jewell, C. Milner, and J. M. Boyd, eds.), Athlone Press, London, pp. 242–72.

Gulland, J. A., 1971, The effect of exploitation on the numbers of marine animals, in:

Proceedings: Advanced Study Institute on "Dynamics of Numbers in Populations" (P. J. Den Boer and G. R. Gradwell, eds.), Centre for Agriculture Publication and Documentation, Wageningen, The Netherlands, pp. 450–468.

Gulland, J. A., 1977, *Fish Population Dynamics*, Wiley, London.

Gulland, J. A., 1982, A note on the abundance of Antarctic blue whales, in: *Mammals in the Seas: General Papers and Large Cetaceans*, FAO Fisheries Series 5, Volume 3, pp. 219–228.

Halls, L. K., (ed.), 1984, *White-tailed Deer: Ecology and Management*, Stackpole Books, Harrisburg, Pennsylvania.

Hanks, J., 1981, Characterization of population condition, in: *Dynamics of Large Mammal Populations* (C. W. Fowler and T. D. Smith, eds.), Wiley, New York, pp. 47–73.

Hanks, J., and McIntosh, J. E. A., 1973, Population dynamics of the African elephant (*Loxodonta africana*), *J. Zool. Lond.* **169**:29–38.

Hansen, C. G., 1980, Population dynamics, in: *The Desert Bighorn* (G. Monson and L. Summer, eds.), University of Arizona Press, Tucson, pp. 217–235.

Harwood, J., 1981, Managing gray seal populations for optimal stability, in: *Dynamics of Large Mammal Populations* (C. W. Fowler and T. D. Smith, eds.), Wiley, New York, pp. 159–172.

Hassell, M. P., 1975, Density-dependence in single-species populations, *J. Anim. Ecol.* **44**(1):283–295.

Hassen, F. A., 1981, *Demographic Archaeology*, Academic Press, New York.

Hatter, J., 1950, The moose of Central British Columbia, Ph. D. thesis, State College of Washington, Pullman, Washington.

Hayden, B., 1972, Population control among hunter/gatherers, *World Archael.* **4**:205–221.

Hesselton, W. T., Severinghaus, C. W., and Tanck, J. E., 1965, Population dynamics of deer at the Seneca Army Depot, *N. Y. Fish Game. J.* **12**(1):17–30.

Honacki, J. H., Kinman, K. E., and Koepple, J. W., 1982, *Mammal Species of the World*, Allen Press, Lawrence, Kansas.

Houston, D. B., 1979, The northern Yellowstone elk—Winter distribution and management, in: *North American Elk: Ecology, Behavior and Management* (M. S. Boyce and L. D. Hayden-Wing, eds.), University of Wyoming, pp. 263–272.

Houston, D. B., 1982, *The Northern Yellowstone Elk: Ecology and Management*, MacMillan, New York.

Ichihara, T., 1972, Maximum sustainable yield from the Robben fur seal herd, *Bull. Far Seas Fish. Res. Lab. (Shimizu)* **6**:77–94.

International Whaling Commission, 1971, Report of the special meeting on sperm whale biology and stock assessment, Honolulu, Hawaii, 13th–24th March, 1970, *Rep. Int. Whaling Comm.* **21**:40–50.

Jingfors, K., 1984, Observations of cow–calf behavior in free-ranging musk oxen, *Biol. Pap. Univ. Alaska Spec. Rep.* **4**:105–109.

Johnson, F. W., 1937, Deer weights and antler measurements in relation to population density and hunting effort, *Trans. N. Am. Wildl. Nat. Resour. Conf.* **2**:446–457.

Kaji, K., Koizumi, T., and Ohtaishi, N., 1985, Population growth of sika deer introduced into Nakanoshima Island in Lake Toya, Hokkaido, Presentation at the IV International Theriological Congress, Edmonton, Alberta, 22–25 August 1985.

Kasuya, T., 1985, Effect of exploitation on reproductive parameters of the spotted and striped dolphins off the Pacific coast of Japan, *Sci. Rep. Whales Res. Inst. Tokyo* **36**:107–138.

Kasuya, T., and Miyazaki, N., 1982, The stock of *Stenella coeruleoalba* off the Pacific coast of Japan, in: *Mammals in the Seas: Small Cetaceans, Seals, Sirenians, and Otters*, FAO Fisheries Series 5, Volume 4, pp. 21–37.

Kato, H., 1983, Some considerations on the decline in the age at sexual maturity of the Antarctic minke whale, *Rep. Int. Whaling Comm.* **33**:393–399.

Kelker, G. H., 1947, Computing the rate of increase for deer, *J. Wildl. Manage.* **11**(2):177–183.

Kemp, G. A., 1976, The dynamics and regulation of black bear *Ursus americanus* populations in northern Alberta, in: *Bears—Their Biology and Management* (M. R. Pelton, J. W. Lentfer, and G. E. Folk, eds.), IUCN, Morges, Switzerland, pp. 191–197.

Kie, J. G., Drawe, D. L., and Scott, G., 1980, Changes in diet and nutrition with increased herd size in Texas white-tailed deer, *J. Range Manage.* **33**(1):28–34.

Klein, D. R., 1964, Range-related differences in growth of deer reflected in skeletal ratios, *J. Mammal.* **45**:226–35.

Klein, D., 1968, The introduction, increase and crash of reindeer on St. Matthew Island, *J. Wildl. Manage.* **32**(2):350–367.

Klein, D. R., 1970, Food selection by North American deer and the response to overutilization of preferred plant species, in: *Animal Populations in Relation to Their Food Resources* (A. Watson, ed.), Blackwell, Oxford, pp. 25–46.

Klein, D. R., 1985, Population ecology: The interaction between deer and their food supply, *R. Soc. N. Z. Bull.* **22**:13–22.

Klein, D., and Olson, S. T., 1960, Natural mortality patterns of deer in southeast Alaska, *J. Wildl. Manage.* **24**(1):80–88.

Klein, D. R., and Strandgaard, H., 1972, Factors affecting growth and body size of roe deer, *J. Wildl. Manage.* **36**:64–79.

Knight, R. R., 1970, The Sun River elk herd, *Wildl. Monogr.* **23**:1–66.

Knowlton, F. F., 1972, Preliminary interpretations of coyote population mechanics with some management implications, *J. Wildl. Manage.* **36**(2):369–382.

Krasinksi, Z. A., 1978, Dynamics and structure of the European bison population in the Bialowieza Primeval Forest, *Acta Theriol.* **23**:3–48.

Krebs, C. J., and Myers, J. M., 1974, Population cycles in small mammals, in: *Advances in Ecological Research*, Volume 8 (A. MacFadyen, ed.), Academic Press, New York, pp. 267–399.

Lack, D., 1954, *The Natural Regulation of Animal Numbers*, Oxford University Press, London.

Lack, D., 1966, *Population Studies of Birds*, Clarendon Press, Oxford.

Lander, R. H., and Kajimura, H., 1982, Status of northern fur seals, in: *Mammals in the Seas: Small Cetaceans, Seals, Sirenians, and Otters*, FAO Fisheries Series 5, Volume 4, pp. 319–346.

Laws, R. M., 1959, Accelerated growth in seals, with special reference to the Phocidae, *Nor. Hvalfangst-Tid.* **9**:425–452.

Laws, R. M., 1961, Reproduction, growth and age of Southern Hemisphere whales, *Discovery Rep.* **31**:327–485.

Laws, R. M., 1962, Some effects of whaling on the southern stocks of baleen whales, in: *The Exploitation of Natural Animal Populations* (E. D. Le Cren and M. W. Holgate, eds.), Wiley, New York, pp. 137–158.

Laws, R. M., 1966, Age criteria for the African elephant, *Loxodonta a. africana*, *East Afr. Wildl. J.* **4**:1–37.

Laws, R. M., 1968, Interactions between elephant and hippopotamus populations and their environments, *East Afr. Agric. For. J.* **33**:140–147.

Laws, R. M., 1969a, The Tsavo research project, *J. Reprod. Fertil. Suppl.* **6**:495–531.

Laws, R. M., 1969b, Aspects of reproduction in the African elephant, *Loxodonta africana*, *J. Reprod. Fertil. Suppl.* **6**:193–217.

Laws, R. M., 1973, Effects of human activities on reproduction in the wild, *J. Reprod. Fertil. Suppl.* **19**:523–532.

Laws, R. M., and Johnstone, R. C. B., 1970, Elephants and habitats in North Bunyoro, Uganda, *East Afr. Wildl. J.* **8:**163–180.

Laws, R. M., and Parker, I. S. C., 1968, Recent studies on elephant populations in East Africa, *Symp. Zool. Soc. Lond.* **21:**319–359.

Laws, R. M., Parker, I. S. C., and Johnson, R. C. B., 1975, *Elephants and Their Habitats: The Ecology of Elephants in North Bunyoro, Uganda,* Clarendon Press, Oxford.

Leader-Williams, N., 1980, Population dynamics and mortality of reindeer introduced into South Georgia, *J. Wildl. Manage.* **44**(3)**:**640–657.

Leader-Williams, N., 1982, Relationship between a disease, host density and mortality in a free living deer population, *J. Anim. Ecol.* **51:**235–240.

LeBoeuf, B. J., 1974, Male–male competition and reproductive success in elephant seals, *Am. Zool.* **14:**163–176.

LeBoeuf, B. J., and Briggs, K. T., 1977, The cost of living in a seal harem, *Mammalia* **41:**167–195.

LeBoeuf, B. J., and Condit, R. S., 1983, The high cost of living on the beach, *Pac. Discovery* **36:**12–14.

LeBoeuf, B. J., and Reiter, J., 1987, Lifetime reproductive success in northern elephant seals in: *Reproductive Success* (T. H. Clutton-Brock, ed.), University of Chicago Press, Chicago, Illinois, in press.

Lett, P. F., and Benjaminsen, T., 1977, A stochastic model for the management of the northwestern Atlantic harp seal (*Pagophilus groenlandicus*) population, *J. Fish. Res. Board Can.* **34:**1155–1187.

Lett, P. F., Mohn, R. K., and Gray, D. F., 1981, The impact of the determination of density-dependent processes on the management strategy for the northwestern Atlantic harp seal, in: *Dynamics of Large Mammal Populations* (C. W. Fowler and T. D. Smith, eds.), Wiley, New York, pp. 135–158.

Ling, H., 1970, Hirvikannan rakenteesta ja tuotosta eestissa [On the composition and productivity of the moose (*Alces alces*) population in the Estonian SSR]. *Suom. Riista* **22:**14–23 [in Finnish with English summary].

Lockyer, C. A., 1972, The age at sexual maturity of the southern fin whale (*Balaenoptera physalus*) using annual layer counts in the ear plug, *J. Cons. Cons. Int. Explor. Mer* **34**(2)**:**276–294.

Lockyer, C. A., 1974, Investigation of the ear plug of the southern sei whale, *Balaenoptera borealis,* as a valid means of determining age, *J. Cons. Cons. Int. Explor. Mer* **36**(1)**:**71–81.

Lockyer, C. A., 1977, A preliminary study of variations in age at sexual maturity of the fin whale with year class, in six areas of the Southern Hemisphere, *Rep. Int. Whaling Comm.* **27:**141–147.

Lockyer, C. A., 1979, Changes in a growth parameter associated with exploitation of southern fin and sei whales, *Rep. Int. Whaling Comm.* **29:**191–196.

MacArthur, R., and Connell, J., 1967, *The Biology of Populations,* Wiley, New York.

Mackintosh, N. A., 1942, The southern stocks of whalebone whales, *Discovery Rep.* **22:**197–300.

Mansfield, A. W., 1970, Population dynamics of some arctic seals, in: *Antarctic Ecology,* Volume I (M. W. Holgate, ed.), Academic Press, New York, pp. 429–446.

Marburger, R. G., and Thomas, J. W., 1965, A die off in white-tailed deer of the central mineral region of Texas, *J. Wildl. Manage.* **29**(4)**:**706–716.

Markgren, G., 1969, Reproduction of moose in Sweden, *Viltrevy* **6**(3)**:**127–299.

Marsh, H., 1980, Age determination of the dugong (*Dugong dugon* (Muller)) in Northern Australia, *Rep. Int. Whaling Comm. Spec. Issue* **3:**181–202.

Marshall, P. J., and Sayer, J. A., 1976, Population ecology and response to cropping of a hippopotamus population in Eastern Zambia, *J. Appl. Ecol.* **13**(2)**:**391–403.

Masaki, Y., 1976, Biological studies on the North Pacific sei whale, *Bull. Far Seas Fish. Res. Lab. (Shimizu)* **14**:1–104.

Masaki, Y., 1978, Yearly change in the biological parameters of the Antarctic sei whale, *Rep. Int. Whaling Comm.* **28**:421–429.

Masaki, Y., 1979, Yearly changes in biological parameters of the Antarctic minke whale, *Rep. Int. Whaling Comm.* **29**:375–395.

Matschke, G. H., Fagerstone, K. A., Harlow, R. F., Hayes, F. A., Nettles, V. F., Parker, W., and Trainer, D. O., 1984, Population influences, in: *White-tailed Deer: Ecology and Management* (L. K. Halls, ed.), Stackpole Books, Harrisburg, Pennsylvania, pp. 169–200.

McCain, G., Cox, V. C., and Paulus, P. B., 1976, The relationship between illness complaints and degree of crowding in a prison environment, *Environ. Behav.* **8**(2):283–291.

McCullough, D. R., 1969, The Tule elk, *Univ. Calif. Publ. Zool.* **88**:1–209.

McCullough, D. R., 1979, *The George Reserve Deer Herd: Population Ecology of a K-Selected Species*, University of Michigan Press, Ann Arbor, Michigan.

McCullough, D. R., 1981, Population dynamics of the Yellowstone grizzly bear, in: *Dynamics of Large Mammal Populations* (C. W. Fowler and T. D. Smith, eds.), Wiley, New York, pp. 173–196.

McCullough, D. R., 1984, Lessons from the George Reserve, Michigan, in: *White-tailed Deer: Ecology and Management* (L. K. Halls, ed.), Stackpole Books, Harrisburg, Pennsylvania, pp. 211–242.

McCullough, D. R., 1985, Variables influencing food habits of white-tailed deer in the George Reserve, *J. Mammal.* **66**(4):682–692.

McFadden, J. T., 1977, An argument supporting the reality of compensation in fish populations and a place to let them exercise it, in: *Proceedings of the Conference on Assessing the Effects of Power-Plant-Induced Mortality on Fish Population* (W. V. Winkle, ed.), Pergamon Press, Fairview Park, New York, pp. 153–183.

McLaren, I. A. (ed.), 1971, *Natural Regulation of Animal Populations*, Atherton Press, New York.

Mech, L. D., 1975, Disproportionate sex ratio of wolf pups, *J. Wildl. Manage.* **39**(4):737–740.

Messier, F., and Crete, M., 1984, Body condition and population regulation by food resources in moose, *Oecologia (Berl.)* **65**:44–50.

Messier, F., and Crete, M., 1985, Moose–wolf dynamics and the natural regulation of moose populations, *Oecologia (Berl.)* **65**:503–512.

Mizroch, S. A., and York, A. E., 1984, Have pregnancy rates of Southern Hemisphere fin whales, *Balaenoptera physalus*, increased?, *Rep. Int. Whaling Comm. Spec. Issue* **6**:401–410.

Murphy, E. C., and Whitten, K. R., 1976, Dall sheep demography in McKinley Park and a reevaluation of Murie's data, *J. Wildl. Manage.* **40**(4):597–609.

Nacci, P. L., Teitelbaum, H. E., and Prather, J., 1977, Population density and inmate misconduct rates in the federal prison system, *Fed. Proc.* **151**:6–31.

Nagasaki, F., 1961, Population study of the fur seal herd, *Tokai Reg. Fish. Res. Lab. Spec. Publ.* **7**:1–60.

Newsome, J. D., 1984, Coastal Plain, in: *White-tailed Deer: Ecology and Management* (L. H. Halls, ed.), Stackpole Books, Harrisburg, Pennsylvania, pp. 367–380.

Nikolskii, G. V., 1969, *Theory of Fish Population Dynamics as the Biological Background for National Exploitation and Management of Fishery Resources*, Oliver and Boyd, Edinburgh, Scotland.

Nygren, T., 1983, The relationship between reproduction rate and age structure, sex ratio and density in the Finnish moose population, Presentation at the XVI International Congress of Game Biologists, High Tatras, Czechoslovakia, 25 September–2 October 1983.

O'Roke, E. C., and Hamerstrom, F. N., Jr., 1948, Productivity and yield of the George Reserve deer herd, *J. Wildl. Manage.* **12**(1):78–86.

Padaiga, V. I., and Marma, B. B., 1970, Invasion of the roe deer with parasites depending on its population density and conditions of life, *Zool. Zh.* **49**:238–287.

Paulus, P. B., McCain, G., and Cox, V. C., 1979, Death rates, psychiatric commitments, blood pressure, and perceived crowding as a function of institutional crowding, *Environ. Psychol. Nonverbal Behav.* **3**(2):107–116.

Payne, M. R., 1977, Growth of a fur seal population, *Philos. Trans. R. Soc. Lond. B Biol. Sci.* **279**:67–79.

Peek, J. M., Lovaas, A. L., and Rouse, R. A., 1967, Population changes within the Gallatin elk herd, 1932–65, *J. Wildl. Manage.* **31**:304–316.

Perrin, W. F., Coe, J. M., and Zweifel, J. R., 1976, Growth and reproduction of the spotted porpoise, *Stenella attenuata,* in the offshore eastern tropical Pacific, *U. S. Natl. Mar. Fish. Serv. Fish. Bull.* **74**(2):229–269.

Perrin, W. F., Miller, R. B., and Sloan, P. A., 1977, Reproductive parameters of the offshore spotted dolphin, a geographical form of *Stenella attenuata,* in the eastern tropical Pacific, 1973–1975, *U. S. Natl. Mar. Fish. Serv. Fish. Bull.* **75**(3):629–633.

Pimlott, D. H., 1959, Reproduction and productivity of Newfoundland moose, *J. Wildl. Manage.* **23**(4):381–401.

Priklonskii, S. G., and Chervonnyi, V. V., 1970, The fertility of elk in the central regions of the European RSFSR, *Sov. J. Ecol. (Engl. Transl. Ekologiya)* **1**(2):171–172.

Rasmussen, D. I., 1941, Biotic communities of the Kaibab Plateau, Arizona, *Ecol. Monogr.* **11**(3):229–275.

Reilly, S. B., 1981, Population assessment and population dynamics of the California gray whale (*Eschrichtius robustus*), Ph. D. thesis, University of Washington, Seattle, Washington.

Reimers, E., Klein, D. R., and Sorumgard, R., 1983, Calving time, growth rate and body size of Norwegian reindeer on different ranges, *Arct. Alp. Res.* **15**(1):107–118.

Reiter, J., Pankin, K. J., and LeBoeuf, B. J., 1981, Female competition and reproductive success in northern elephant seals, *Anim. Behav.* **29**:670–687.

Ricklefs, R. E., 1973, *Ecology*, Chiron Press, Portland, Oregon.

Riedman, M. L., and LeBoeuf, B. J., 1982, Mother–pup separation and adoption in northern elephant seals, *Behav. Ecol. Sociobiol.* **11**:203–215.

Robinette, W. L., Gashwiler, J. S., Low, J. B., and Jones, D. A., 1957, Differential mortality by sex and age among mule deer, *J. Wildl. Manage.* **21**:1–16.

Roff, D. A., and Bowen, W. D., 1983, Population dynamics and management of the northwest Atlantic harp seal (*Phoca groenlandica*), *Can. J. Fish. Aquat. Sci.* **40**(7):919–932.

Sauer, J. R., and Boyce, M. S., 1983, Density dependence and survival of elk in northwestern Wyoming, *J. Wild. Manage.* **47**(1):31–37.

Sayer, J. A., and Rakha, A. M., 1974, The age of puberty of the hippopotamus (*Hippopotamus amphibius* Linn.) in the Luangwa River in Eastern Zambia, *East Afr. Wildl. J.* **12**(3):227–232.

Scheffer, V. B., 1955, Body size with relation to population density in mammals, *J. Mammal.* **36**(4):493–515.

Serafinski, W., 1969, Reproduction and dynamics of moose (*Alces alces* L.) population in the Kampinos National Park, *Ekol. Pol. A* **17**(37):709–718.

Sergeant, D. E., 1973, Environment and reproduction in seals, *J. Reprod. Fertil. Suppl.* **19**:555–561.

Sergeant, D. E., 1982, Mass strandings of toothed whales (Odontoceti) as a population phenomenon, *Sci. Rep. Whales Res. Inst. Tokyo* **34**:1–47.

Severinghaus, C. W., 1955, Deer weights as an index of range conditions in two wilderness areas in the Adirondack region, *N. Y. Fish Game J.* **2**(27):154–160.

Severinghaus, C. W., 1979, Weights of white-tailed deer in relation to range conditions in New York, *N. Y. Fish Game J.* **26**(2):163–187.

Severinghaus, C. W., Maguire, H. F., Cookinham, R. A., and Tanck, J. E., 1950, Variations by age class in the antler beam diameter of white-tailed deer related to range conditions, *Trans. N. Am. Nat. Resour. Wildl. Conf.* **15**:551–570.

Shaffer, M. L., 1978, Determining minimum viable population sizes: A case study of the grizzly bear (*Ursus arctos* L.), Ph. D. thesis, Duke University, Durham, North Carolina.

Shaffer, M. L., 1983, Determining minimum viable population sizes for the grizzly bear, in: *International Conference on Bear Research and Management*, Volume 5, pp. 135–141.

Shrauder, P. A., 1984, Appalachian mountains, in: *White-tailed Deer: Ecology and Management* (L. K. Halls, ed.), Stackpole Books, Harrisburg, Pennsylvania, pp. 331–344.

Sinclair, A. R. E., 1973, Regulation, and population models for a tropical ruminant, *East Afr. Wildl. J.* **11**:307–316.

Sinclair, A. R. E., 1974, The natural regulation of buffalo populations in East Africa, III: Population trends and mortality, *East Afr. Wildl. J.* **12**:185–200.

Sinclair, A. R. E., 1977, *The African Buffalo: A Study of Resource Limitation of Populations*, University of Chicago Press, Chicago, Illinois.

Sinclair, A. R. E., 1979, The eruption of the ruminants, in: *Serengeti, Dynamics of an Ecosystem* (A. R. E. Sinclair and M. Norton-Griffiths, eds.), University of Chicago Press, Chicago, Illinois, pp. 81–103.

Sinclair, A. R. E., and Norton-Griffiths, M., 1982, Does competition or facilitation regulate migrant ungulate populations in the Serengeti? A test of hypotheses, *Oecologia (Berl.)* **53**:364–369.

Sinclair, A. R. E., Dublin, H., and Borner, M., 1985, Population regulation of Serengeti wildebeest: A test of the food hypothesis, *Oecologia (Berl.)* **65**:266–268.

Skogland, T., 1983, The effects of density dependent limitation on size of wild reindeer, *Oecologia (Berl.)* **60**:156–168.

Skogland, T., 1984a, The effects of food and maternal conditions on fetal growth and size in wild reindeer, *Rangifer* **4**(2):39–46.

Skogland, T., 1984b, Wild reindeer foraging-niche organization, *Holarct. Ecol.* **7**:345–379.

Skogland, T., 1985a, The effects of density-dependent resource limitation on the demography of wild reindeer, *J. Anim. Ecol.* **54**:359–374.

Skogland, T., 1985b, Life History Characteristics of Wild Reindeer (*Rangifer tarandus tarandus* L.) in Relation to Their Food Resources; Ecological Effect and Behavioral Adaptations, Meddlser fra Norsk Viltforskning, Series 3, Number 4, Directorate for Wildlife and Freshwater Fish, Trondheim, Norway.

Skogland, T., 1986a, Sex ratio variation in relation to maternal condition and prenatal investment in wild reindeer (*Rangifer tarandus tarandus* L.), *Oikos*, in press.

Skogland, T., 1986b, Density dependent food limitation and maximal production in wild reindeer herds, *J. Wildl. Manage.* **50**(2):314–319.

Slobodkin, L. B., 1961, *Growth and Regulation of Animal Populations*, Holt, Rinehart and Winston, New York.

Smith, T. D., 1973, Variable population projection matrices: Theory and application to the evaluation of harvesting strategy, Ph. D. thesis, University of Washington, Seattle.

Smith, T. D., 1984, Estimating the dolphin population size yielding maximum production, *Rep. Int. Whaling Comm. Spec. Issue* **6**:187–190.

Smuts, G. L., 1976, Population characteristics and recent history of lions in two parts of the Kruger National Park, *Koedoe* **19**:153–164.

Smuts, G. L., 1978, Effects of population reduction on the travels and reproduction of lions in Kruger National Park, *Carnivore (Seattle)* **1**(2):61–72.

Southwick, C. H., Richie, T., Taylor, H., Teas, H. J., and Siddiqi, M. F., 1980, Rhesus monkey populations in India and Nepal: Patterns of growth, decline, and natural regulation, in: *Biosocial Mechanisms of Population Regulation* (M. N. Cohen, R. S. Malpass, and H. G. Klein, eds.), Yale University Press, New Haven, Connecticut, pp. 151–170.

Staines, B. W., 1978, The dynamics and performance of a declining population of red deer (*Cervus elephus*), *J. Zool. Lond.* **184**:403–419.

Stelfox, J. G., 1976, Range ecology of Rocky Mountain bighorn sheep in Canadian National Parks, *Can. Wildl. Serv. Rep. Ser.* **39**:1–50.

Stevenson-Hamilton, J., 1937, *South African Eden*, Casell, London.

Stirling, I., 1969, Ecology of the Weddell seal in McMurdo Sound, Antarctica, *Ecology* **50**:573–586.

Stirling, I., 1971a, Population dynamics of the Weddell seal (*Leptonychotes weddelli*) in McMurdo Sound, Antarctica, 1966–68, in: *Antarctic Pinnipedia* (W. H. Burt, ed.), Antarctic Research Series 18, pp. 141–167.

Stirling, I., 1971b, Population aspects of Weddell seal harvesting at McMurdo Sound, Antarctica, *Polar Record* **15**(98):653–667.

Stirling, I., 1972, Regulation of numbers of an apparently isolated population of Weddell seals (*Leptonychotes weddelli*), *J. Mammal.* **53**:107–115.

Stringham, S. F., 1983, Roles of adult males in grizzly bear population biology, in: *International Conference on Bear Research and Management*, Volume 5, pp. 140–151.

Strong, D. R., 1984, Density-vague ecology and liberal population regulation in insects, in: *A New Ecology, Novel Approaches to Interactive Systems* (P. W. Price, C. N. Slobodchikoff, and W. S. Gaud, eds.), Wiley, New York, pp. 313–327.

Swank, W. G., 1958, The Mule Deer in Arizona Chaparral, Wildlife Bulletin No. 3, Arizona Game and Fish Department.

Swenson, J. E., 1985, Compensatory reproduction in an introduced mountain goat population in the Absaroka Mountains, Montana, *J. Wildl. Manage.* **49**(4):837–843.

Talbot, L. M., and Talbot, M. H., 1963, The wildebeest in western Masailand, *Wildl. Monogr.* **12**:1–88.

Tamarin, R. H. (ed.), 1978, *Population Regulation*, Dowden, Hutchinson, and Ross, Stroudsburg, Pennsylvania.

Teer, J. G., 1984, Lessons from the Llano Basin, Texas, in: *White-tailed Deer: Ecology and Management* (L. K. Halls, ed.), Stackpole Books, Harrisburg, Pennsylvania, pp. 261–290.

Teer, J. G., Thomes, J. W., and Walker, E. A., 1965, Ecology and management of white-tailed deer in the Llano Basin of Texas, *Wildl. Monogr.* **15**:62.

Tomek, A., 1977, The occurrence and some ecological parameters of the moose in Poland, *Acta Theriol.* **22**:485–508.

Torgerson, O., and Porath, W. R., 1984, Midwest oak/hickory forest, in: *White-tailed deer, Ecology and Management* (L. K. Halls, ed.), Stackpole Books, Harrisburg, Pennsylvania, pp. 411–426.

Van Aarde, R. J., 1980, Fluctuation in the population of southern elephant seals *Mirounga leonina* at Kerguelen Islands, *S. Afr. J. Zool.* **15**:99–106.

Van Aarde, R. J., 1984, Aspects of the population biology of the southern elephant seal, *Mirounga leonina,* at Iles Kerguelen, *S. Afr. J. Sci.* **80:**31–32.

Varley, G. C., Gradwell, G. R., and Hassell, M. P., 1973, *Insect Population Ecology: An Analytical Approach,* University of California Press, Berkeley.

Walters, C. J., Stocker, M., and Haber, C. G., 1981, Simulation and optimization models for a wolf–ungulate system in: *Dynamics of Large Mammal Populations* (C. W. Fowler and T. D. Smith, eds.), Wiley, New York, pp. 317–338.

Watson, R. M., 1969, Reproduction of wildebeest, *Connochaetes taurinus albojubatus* Thomas, in the Serengeti region, and its significance to conservation, *J. Reprod. Fertil. Suppl.* **6:**287–310.

Watson, R. M., 1970, Generation time and intrinsic rates of natural increase in wildebeest (*Connochaetes taurinus albojubatus* Thomas), *J. Reprod. Fertil.* **22**(3):557–561.

Weatherly, A. H., 1972, *Growth and Ecology of Fish Populations,* Academic Press, London.

Whittington, R. W., 1984, Piedmont Plateau, in: *White-tailed Deer: Ecology and Management* (L. K. Halls, ed.), Stackpole Books, Harrisburg, Pennsylvania, pp. 211–242.

Winters, G. H., 1978, Production, mortality, and sustainable yield of Northwest Atlantic harp seals (*Pagophilus groenlandicus*), *J. Fish. Res. Board Can.* **35:**1247–1261.

Woodgerd, W., 1964, Population dynamics of bighorn sheep on Wildhorse Island, *J. Wildl. Manage.* **28**(2):381–391.

Wright, B. S., 1960, Predation on big game in East Africa, *J. Wildl. Manage.* **24**(1):1–15.

Wynne-Edwards, V. C., 1962, *Animal Dispersion in Relation to Social Behaviour,* Hafner, New York.

CHAPTER 11

BEHAVIORAL ECOLOGY OF THE NEVADA KIT FOX (*VULPES MACROTIS NEVADENSIS*) ON A MANAGED DESERT RANGELAND

G. TIMOTHY O'NEAL, JERRAN T. FLINDERS, and WARREN P. CLARY

1. INTRODUCTION

The kit fox (*Vulpes macrotis* Merriam) inhabits desert and semiarid areas of western North America. Recent research and evaluations consider the kit fox distinct from the swift fox *V. velox* (Say) (Rohwer and Kilgore, 1973; Egoscue, 1979; McGrew, 1979; Honacki et al., 1982; Nowack and Paradiso, 1983). Earlier accounts (Hall and Kelson, 1959) recognized eight subspecies of *V. macrotis*. McGrew's (1979) thorough review of research dealing with subspecific categories for the kit fox reduced the eight subspecies to five: *V. m. macrotis* Merriam, *V. m. mutica* Merriam, *V. m. nevadensis* Goldman, and *V. m. zinseri* Benson. The San Joaquin kit fox (*V. m. mutica*) is considered an endangered

G. Timothy O'Neal was killed in an automobile accident 2 November 1985. This report on kit foxes is dedicated to his memory as a scientist and friend.

G. TIMOTHY O'NEAL and JERRAN T. FLINDERS • Wildlife and Range Resources Program, Brigham Young University, Provo, Utah 84602. WARREN P. CLARY • U. S. Forest Service Forestry Science Laboratory, Boise, Idaho 83704.

species by the U. S. Department of Interior and receives appropriate protection in California (Morrell, 1972; U. S. Department of Interior, Fish and Wildlife Service, 1976).

The Nevada kit fox (*V. m. nevadensis*) occurs farther north than other subspecies and is largely identified with the Great Basin and adjacent cold desert habitats (McGrew, 1979). Egoscue (1962) did the first thorough study of the Nevada kit fox, but lacked the benefit of radiotelemetry to document day-to-day movements of foxes. McGrew (1977) delineated the geographic distribution and also documented aspects of steppe and desert climates associated with the occurrence of the Nevada kit fox.

A 1-year study (January–December 1983) of *V. m. nevadensis* was completed at the Desert Experimental Range in western Utah under joint cooperation with Brigham Young University, the U. S. Forest Service, and the Utah Division of Wildlife Resources.

The overall objective was to document adaptive habitat selection, reproduction, food habits, dispersal, and other behavioral attributes of kit foxes on a desert rangeland managed for winter grazing by domestic sheep and cattle.

2. DESCRIPTION OF STUDY AREA

2.1. Topography and Climate

The Desert Experimental Range (DER) (Fig. 1) is a 225 km^2 U. S. Forest Service facility located near the southeastern edge of the known distribution of *V. m. nevadensis* (Hall, 1981). The topography of this facility is typical of the cold-desert Great Basin. About 75% of the study area is alluvial or flat valley bottom and the remainder is mountainous. Average annual precipitation is 157 mm, much of which falls as snow during winter months. Daily minimum and maximum temperatures range from -11 and 4.2°C in January to 13 and 33°C in July (Holmgren, 1973). Elevation of the study area ranges from 1600 to 1700 m.

2.2. Vegetation

Dominant shrub species are common winterfat (*Ceratoides lanata*), budsage (*Artemisia spinescens*), black sagebrush (*Artemisia nova*), shadscale saltbush (*Atriplex confertifolia*), and low rabbitbrush (*Chrysothamnus viscidiflorus* sp.). Perennial grasses include Indian ricegrass (*Oryzopsis hymenoides*), galleta grass (*Hilaria jamesii*), and sand drop-

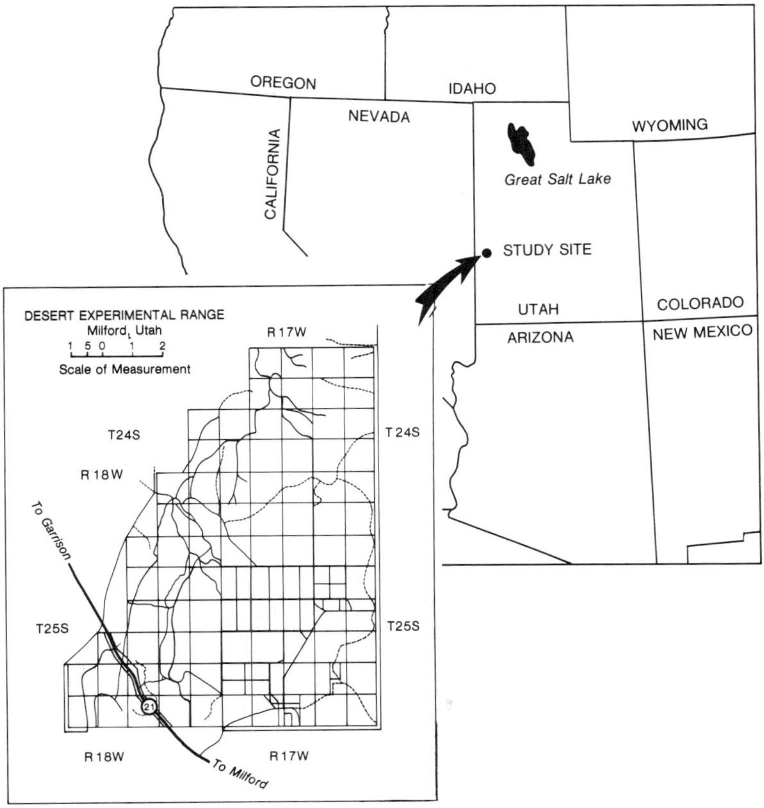

FIGURE 1. The Desert Experimental Range (U. S. Forest Service), a 225 km^2 desert rangeland located in the Great Basin of southwestern Utah.

seed (*Sporobolus cryptandrus*) (Holmgren, 1973). The Desert Experimental Range has been involved continuously for 50 years in winter grazing studies of domestic sheep and their relationship to desert plant communities. Cattle are also grazed during winter months within some areas on the site.

3. METHODS AND PROCEDURES

3.1. Kit Fox Capture

The study area was searched extensively on foot, from fixed-wing aircraft, and from automobiles for active den sites. Kit foxes were captured in two different sizes of Tomahawk traps (38 by 38 by 107 cm double-door for adults and 26 by 26 by 66 cm single-door for pups) baited with pieces of black-tailed jackrabbit (*Lepus californicus*). Adults were trapped at active dens, inactive dens, and at sites used as scent posts. Pups were trapped at active dens only. Dens selected for study were established by foxes in a variety of grazed paddocks as well as ungrazed areas.

3.2. Body Measurements

At each of six active dens, a pair of adult foxes was anesthetized by injection of ketamine hydrochloride (0.3 cm^3/kg dosage), sexed, weighed, measured, marked with ear tags, checked for external parasites, radiocollared, and released. Measurements included total length, length of tail, ear, hind foot, zygomatic breadth, and head length (nose pad to back of head). Upon emergence from dens, pups were processed on a weekly basis in the same manner as adults, and one male and female from each den were radiocollared later in the summer. In addition, tooth eruption and development were noted. Tooth measurements, including distal edge of canine to proximal edge of PM 4 (upper), distal edge of incisors to proximal edge of PM 4 (upper), upper canine length, and distal edge of canine to proximal edge of PM 1, were taken. Approximate whelping dates were determined by comparing tooth eruptions and body weights of pups with those of two pups reared in the laboratory during the study.

3.3. Blood Collection

Blood samples (obtained from the jugular vein) were taken from 33 of the 42 foxes, and a white cell differential was determined for each sample to help determine physical condition. Blood samples were obtained from laboratory-reared kit foxes for comparison.

3.4. Radio Tracking

In order to monitor effectively the activity of each fox, a lightweight transmitter was attached around the neck. The collar, transmitter, and battery weighed 122–125 g and the cable antenna extended approximately 36 cm from the collar. Each transmitter operated on a different frequency, allowing for identification of individual foxes. Tracking of radiocollared foxes was accomplished by use of a Telonics TR-2 receiver/scanner using both a truck-mounted and hand-held antenna. Locations were plotted from angles determined by triangulation at varying times during the night. Considerable effort was made to locate instrumented foxes at least every other night to document movement and every day to document use of dens. An average of 110 plots (from triangulation)/adult fox was used in determining home range size. By plotting coordinates onto a topographic map, approximate locations on any given night were determined for each adult fox. Shape and size of home range were determined for each fox by connecting outermost radio contact points (Mohr, 1947).

3.5. Scat Collection and Processing

Scats were collected from around each of the six dens approximately every 2 weeks. Dried scats were prepared for analysis following the methods of Green and Flinders (1981). Scats were dried, washed, and separated from insoluble material. Hair, bones, feathers, as well as arthropod remains from scat were compared to reference specimens. One hundred random points were used to locate food items in pooled scats from each den during each collection period. Scats of young were collected separately until early July, when they could no longer be distinguished from adult scat.

3.6. Prey Base Assessment

To assess abundance of small mammals, 100 Sherman traps at 10-m intervals were established randomly within known hunting areas of each pair of foxes. Museum specials (kill traps) were not used, because of the likelihood of rodents being removed from traps by foxes. Small mammals were trapped, marked, and released for five consecutive nights, from 26 May to 4 June 1983. Additional trapping near Dens 3 and 6 took place between 5 and 9 September 1983. Density/ha was calculated using original captures within a 10 by 1000-m belt transect (Caughley, 1977). Leporid abundance was assessed in spring, summer, and autumn periods by counting number of leporids considered within the 9.2-m (30-ft)-belt transect along selected sections of dirt roadways on the study site (Flinders and Hansen, 1973). The average of three nights of counting was used as an estimate of abundance. Birds were censused late in July 1983 for three early mornings along the same roadways used to assess leporids. Density (number/km^2) estimates for birds and hares were calculated from numbers seen/area of transect (Caughley, 1977).

3.7. Vegetative Assessments

Vegetative characteristics near natal dens were assessed in two ways. The point-centered quarter method (Mueller-Dombois and Ellenberg, 1974) was used to determine height, cover, and density of woody plant species at 25 points on a 40 by 40-m grid centered over each natal den site. Percent cover of grasses and forbs was assessed within 25 plots (20 by 50 cm) located at the same points as above, according to methods outlined by Daubenmire (1959). Soil samples were taken from each natal den site at the surface and 15 cm below the surface to determine soil texture. Separate similarity indices (Ruzicka, 1958) were computed for relative dominance and average shrub height for natal dens, home range vegetation, relative percent frequency of adult and juvenile diets for each den relative to other dens, similarity between winter–spring and summer–autumn adult diets, and between prey availability and relative percent frequency of prey in adult and juvenile diets. Statistical cluster analyses (Sneath and Sokal, 1973) were applied to those similarity indices that compared each den or home range relative to other dens or home ranges. All statistical tests, including correlation, regression, linear regression, and χ^2, were considered significant at the 0.05 level.

4. RESULTS

4.1. Kit Fox Trapping Success

During the course of the study, 65 different kit foxes (35 males and 30 females) were trapped and ear-tagged. Thirty-eight foxes were radiocollared for detailed tracking. The radio collars did not appear to have any adverse effect on kit fox behavior. Radiocollared foxes appeared to adjust to collars within 24 h. There was little difficulty in initially trapping adult and young foxes. Most adults, however, avoided traps after the first or second time they were captured and only a few were recaptured during the course of the study. Consequently, there was difficulty in recapturing adults at the end of the study for collar removal. Juveniles began avoiding the smaller traps in midsummer and larger adult traps were needed to capture them. By early autumn, most juveniles had died, dispersed, moved to other dens away from their parents, or simply avoided the traps.

Pieces of black-tailed jackrabbit either fresh, putrid, or scented with fox lure proved to be the most successful bait. In late July, water was used with some success for foxes not near permanent water, but rains in August rendered this bait ineffective. Near the end of the study, smoked salmon worked well as bait to recapture some of the more wary foxes for radio collar removal. Live deer mice (*Peromyscus manicu-*

latus) were used successfully in recapturing a transient male that could not be retrapped with jackrabbit bait.

4.2. Mortality

Twenty-three known mortalities occurred in the 42 kit fox pairs and whelps at the six den sites. Data in Table I include the known information regarding natality and mortality of adult pairs and resultant whelps at each of the six den sites studied. Several things concerning the mortalities are noteworthy. For example, a yearling female helper at Den 1 died along with perhaps all of the foxes at that den. It is not known whether the possible drowning was actually a drowning, since the lungs floated when placed in water during a necropsy (water-filled lungs usually sink), and kit foxes are known to swim (Reeder, 1949). The kit fox killed by a golden eagle represented a rare occurrence. In a study of golden eagle nest contents in western Texas and in New Mexico, only one nest was found to contain kit fox remains (Mollhagen et al., 1972). The apparent kills by coyotes also appear unique, in that such mortalities were not reported by Egoscue (1962) or Morrell (1972), although Seton (1925) observed coyotes attempting to capture kit foxes. There is a possibility that predation on kit foxes in the study area reflected prey switching on the part of coyotes as a result of a cyclic low in the black-tailed jackrabbit population at the time of the study. A radiocollared juvenile female from Den 5 was found to have a severe elbow injury on her left front leg. Treatment, unfortunately, was not in time and she was found dead 2 days later, apparently unable to successfully hunt in her injured condition. The Den 4 adult male was caught by a trapper just south of the DER headquarters 1 month after the study ended. With the exception of the Den 5 adult male and a Den 6 juvenile female, who were found inside dens, all other deaths occurred away from dens. Because of the relative isolation and nature of the study area, varmint hunting and road kills apparently are not major mortality factors in the study area. Only one juvenile female fox (whose ear tags were missing) was found shot.

4.3. Den Sites

4.3.1. External Characteristics

The overall number of entrances at studied den sites varied from one to 25. One or two entrances at each den were apparently used for waste disposal, judging from the large amount of scat deposited outside

TABLE I
Known Aspects of Natality and Mortality of Male (M) and Female (F) Kit Fox Adults and Young Associated with Six Den Sites

	Natality			Mortality							
	Pups born			Pups		Julian date		Adults		Julian date	
Den	M	F	Julian date	M	F	M	F	M	F	M	F
1	3	1	080	1[a]	1[a]	249	270	1[b]	1[b]	121	306
					1[b]		203				
					1[b]		Unknown				
2	2	3	071	1[b]	1[c]	262	203	1[i]	0	227	
3	2	2	105	1[d]	1[a]	139	241	0	0		
				1[e]		342					
4	3	4	078	1[f]	1[d]	364	183	0	0		
					1[a]		304				
5	3	3	073	1[b]	1[g]	241	173	1[b]	0	161	
				1[b]		338					
6	2	1	072	1[h]	1[b]	330	182	1[i]	1[i]	340	340

[a]Killed by coyote.
[b]Cause unknown.
[c]Killed by golden eagle.
[c]Trap-related.
[e]Possible starvation.
[f]Caught by trapper.
[g]Injury-related.
[h]Possible drowning.
[i]Sacrificed for parasite study.

these entrances. The cleaning out of scat occurred every 1–2 weeks. Some entrances, particularly in older, multiple-entrance dens, were in poor condition and apparently not used. Entrances were usually taller than wide and the average diameter of natal den entrances was 23 by 15 cm. Each entrance had a mound of excavated dirt in front and many had a long ramp of excavated dirt extending as far as 3 m from the entrance. Egoscue (1956) reported dens typically had at least one ramp extending 1.5–2.5 m from the mouth of the entrance. Most dens had one or two shallow basins on the surface where adults were often seen napping in the daytime. The Den 4 family moved from the natal den to an already established den beneath a group of four juniper post piles when their pups were approximately 2 weeks old. This site was shared with a family of striped skunks (*Mephitis mephitis*). The post piles were approximately 12 m apart, three of which were connected by fox tunnels. These same foxes later moved into an old den beneath an abandoned building and remained there for approximately 1 month. This was the only family found to have lived in association with man-made structures.

The larger dens containing numerous tunnels were probably the result of many years of continuous use (Egoscue, 1956, 1962; Morrell, 1972). Some of the smaller, one-entrance dens appeared to have originally been badger (*Taxidea taxis*) diggings. A few were located on old kangaroo rat (*Dipodomys*) mounds. McGrew (1977) suggested kit foxes in the western deserts depended on other burrowing animals (particularly badgers) as a start for new dens, and if soils could not be burrowed, alternate shelters were used. Kit foxes have been described as both poor diggers (Cahalane, 1947) and excellent diggers (Lechleitner, 1969). Observations made during this study tend to support the latter.

4.3.2. Internal Characteristics

Two natal kit fox dens, not otherwise used in this study, were entirely excavated in order to obtain pups for laboratory studies, and this provided an opportunity to examine the internal structure of a kit fox den. The first den consisted of a series of spiraling, interconnecting tunnels that led to two grass-lined chambers. The first chamber (approximately 0.6 m below the surface) was apparently used for food storage, since it contained a little pocketmouse (*Perognathus longimembris*), an Ord's kangaroo rat (*Dipodyms ordii*), and the head and torso of a young black-tailed jackrabbit. The second chamber, approximately 2.5 m below the surface, was larger and probably used for sleeping. The second den (Fig. 2), which was not as deep (approximately 1.3 m below the surface), also contained two grass-lined chambers. The food chamber contained two Ord's kangaroo rats and a northern grasshopper mouse (*Onychomys leucogaster*). Approximately six nonnatal dens were partially excavated in order to retrieve collars from living or dead foxes. Most of these dens had three to ten entrances and extended at least 1.3 m below the surface. Seton (1925) described dens in the Mojave Desert in California extending 2.5–3.0 m below the surface. Morrell (1972) described a den extending as far as 1.3 m below the surface. If dens were otherwise disturbed (for example, if one or more entrances were blocked or caved in), foxes would move to another den during the night. Most of these were older, established dens, but new dens were occasionally dug. For example, a striped skunk was caught in a trap outside a known kit fox den. The former resident and her juvenile offspring were found the next morning in a freshly dug two-entrance den some 100 m away. A radiocollared pair from an excavated den were tracked the following morning to a ten-entrance den, which appeared to have been dug overnight. Such activity would support the idea that kit foxes are excellent diggers. Often, a den change was accompanied by a definite shift in home range to the area surrounding the new den. This suggests that one reason for den changes may be to move closer to an area of more available prey. External parasite loads (particularly fleas) often decreased temporarily after a

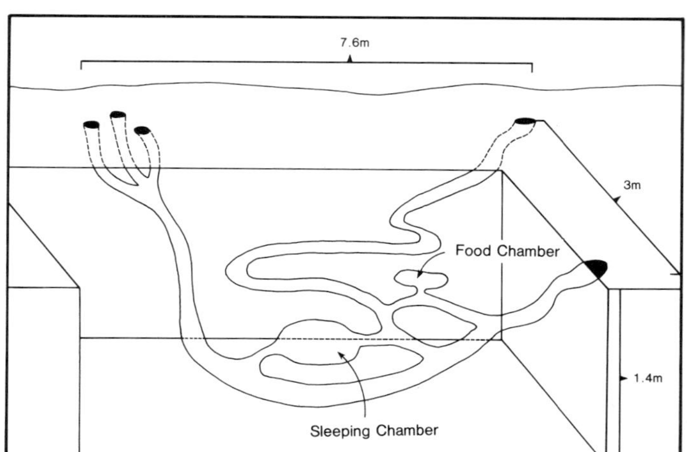

FIGURE 2. Drawing of an excavated kit fox natal den showing entrances and tunnels to the sleeping and food chambers.

den change, suggesting another reason for changes. Total number of den changes was as high as 33 for males and 19 for females during the course of the study. Dens occurred in groups or clusters, with 7–17 dens in each cluster ($\overline{X}$ = 9.5, S = 4.0), and were not randomly scattered (Table II). All dens used by a particular family of foxes were considered to be in that family's cluster. Each cluster of dens was no more than 2 km^2 in size and most dens in a cluster were never more than 100 m apart.

4.3.3. Vegetational Profiles and Similarities of Natal Dens

All natal dens were located in raised, well-drained areas as close as 1.5 km to each other. The number of entrances ranged from three to ten. Seven shrub species dominated vegetational cover at natal den sites (Table III). Shadscale with its spiny protection resists grazing pressure by sheep and cattle. As a result, relative dominance (85.2%) was highest for this species in paddocks under heavy winter–spring grazing by sheep (Table III). Low rabbitbrush was the dominant shrub for Dens 4 and 6 ($\overline{X}$ height = 14.7 and 14.2 cm, respectively). Common winterfat was the dominant shrub for Den 1 ($\overline{X}$ height = 17.9 cm), desert molly (*Kochia americana*) dominated the Den 2 site ($\overline{X}$ height = 10.1 cm), and fourwing saltbush (*Atriplex canescens*) dominated the Den 5 site ($\overline{X}$ height = 31.9 cm). Cluster analysis of natal den sites based on similarities of relative dominance for shrub species showed that Dens

TABLE II
Total Number of Dens Used (Clusters) by Each Fox Family along with the Total Number of Times Each Adult Male (M) and Female (F) Changed Dens

Den	Total dens used	Total den changes	
		M	F
1	7	2	14[b]
2	11	15[a]	14
3	17	33	19[c]
4	8	13	12
5	6	2[b]	18
6	8	14	10

[a]Died in late summer.
[b]Died early in the study.
[c]Could not determine all den changes, due to radio failure.

TABLE III
Kit Fox Den Sites (1–6) in Relationship to Grazing History of Sheep and Cattle As Well As Aspects of Vegetation

Den	Grazing status	Shrub species	Relative dominance (%)	Average shrub height (cm)	Canopy cover: Grasses (%)	Canopy cover: Forbs (%)
1	Ungrazed	Common winterfat	50.0	17.9	5.9	7.7
		Low rabbitbrush	49.3	9.4		
		Broom snakeweed	0.5	12.5		
		Bud sagebrush	0.2	5.0		
2	Sheep (every other spring)	Common winterfat	24.2	19.8	11.6	4.4
		Desert molly	75.8	10.1		
3	Sheep (heavy winter–spring)	Common winterfat	13.9	14.6	12.1	4.0
		Shadscale	85.2	14.2		
		Bud sagebrush	0.8	37.7		
		Plains prickleypear	0.1	9.7		
4	Cattle (winter)	Common winterfat	30.7	14.6	14.5	2.5
		Low rabbitbrush	46.5	14.7		
		Shadscale	2.4	27.0		
		Fourwing saltbush	20.4	29.7		
5	Sheep (autumn–spring)	Common winterfat	14.9	11.8	15.7	2.5
		Low rabbitbrush	17.4	19.6		
		Fourwing saltbush	67.7	31.9		
6	Cattle (winter)	Common winterfat	39.8	11.6	16.8	2.5
		Low rabbitbrush	60.2	14.2		

1 and 6 were most similar, with a similarity index of 80%, followed by Dens 1 and 4, with 63% similarity. Dens 1 and 3 were the least similar, at only 8%. Cluster analysis of natal den sites based on average shrub height showed Dens 4 and 5 to have the highest similarity index (60%), followed by Dens 1 and 6 (42%). Dens 1 and 3 were least similar at only 9%.

According to McGrew (1977), most kit fox sightings in Utah occur in salt desert shrub habitats (shadscale, greasewood—*Sarcobatus vericulatus,* and winterfat), followed by the southern desert shrub habitats (creosote—*Larrea divaricata,* and blackbrush—*Coleogyne ramosissima*). On the DER, grass cover at natal dens ranged from 5.9% (Den 1) to 15.7% (Den 5). Forb cover was even lower, ranging from 2.5% at Dens 4–6 to 7.7% at Den 1 (Table III).

The most important characteristic of kit fox den habitat appears to be the percent ground cover and average height of vegetation (McGrew, 1977). Grasses and forbs at unused dens appeared to be denser than in the surrounding area, perhaps as a result of increased nutrients from fox scat (Egoscue, 1962), but vegetative cover at active dens appeared to be less than in the surrounding area, probably due to constant trampling by foxes. Soil texture at Den 2 was a sandy clay loam, whereas textures at the other five natal dens were all rated as sandy loams. Similar soil textures were reported by Egoscue (1962) and Morrell (1972) at their study sites. Average year-round temperature for average maximum den depth (130 cm) was 14.0°C (R. C. Holmgren, personal communication), providing a comfortable year-round temperature. The amount of shrub cover apparently affects soil temperature up to a depth of approximately 40 cm (R. C. Holmgren, personal communication). Since shrubs are less abundant at den sites, upper soil temperature would be warmer in the summer and colder in the winter at these sites compared to surrounding sites.

4.4. Body Development and Condition Assessment

4.4.1. Natality and Sex Ratio of Young

Kit foxes breed near the first of January (Egoscue, 1956). All vixens that whelped had the distinctive reddish fur around their mammary nipples, an indication of lactation (Egoscue, 1975). The coloration did not completely disappear until acquisition of winter pelage in late summer and was not present on the female helper at Den 1. Whelping occurred from mid-March to mid-April. Approximate whelping dates were 22 March for Den 1, 12 March for Den 2, 15 April for Den 3, 19 March for Den 4, 14 March for Den 5, and 13 March for Den 6. Litter

TABLE IV
Body Measurements of Male and Female Adult Kit Foxes and Male and Female Juveniles and the Coefficient of Variation (CV) of Each Body Measurement[a]

	Weight (kg)	Total length (mm)	Tail length (mm)	Hind foot length (mm)	Ear length (mm)	Zygomatic breadth (mm)	Head length (mm)
Adult							
Male	2.1	814.0	311.3	122.8	86.5	59.2	112.7
CV	5.3	7.1	7.1	1.7	2.7	2.0	0.4
Female	2.1	750.2	282.3	118.0	84.3	57.8	112.3
CV	12.7	5.3	15.3	3.3	2.5	3.4	0.7
Juvenile							
Male	1.5	708.4	277.9	110.9	78.2	52.1	102.9
CV	13.8	8.7	8.2	5.0	6.2	3.1	4.9
Female	1.4	677.9	251.6	107.2	75.7	50.4	100.3
CV	13.5	6.9	8.1	3.9	4.2	4.8	3.4

[a]Measurements were taken in June for juveniles in Dens 1, 2, and 4–6 and in July in Den 3.

sizes ranged from three to seven pups, with an average of 4.6 pups/litter. The sex ratio of pups was essentially 1:1 (15 males:14 females), indicating a fairly stable population. Egoscue (1975) suggested a high male:female ratio would indicate an overcrowded situation, whereas a high female:male ratio would speed population recovery. The typical litter sizes of three to seven pups and the fact that all adult females trapped were found to be lactating gives further evidence of a fairly stable population (Egoscue, 1975).

4.4.2. Body Measurement Comparisons

Average weight gain/month for young was 462 g for males and 353 g for females, compared to 464 g for the laboratory-reared male and 456 g for the laboratory-reared female (both lab-reared foxes were fed dry dog food ad lib). A weight gain ranging from 340 to 456 g/month for San Joaquin kit fox young was reported by Morrell (1972). Measurements taken in June (Dens 1, 2, and 4–6) and July (Den 3, because of the late whelping date) were used for comparison between juveniles (Table IV). The juveniles were still growing at this time, most were still alive, and none had yet dispersed.

In order for juveniles to be compared at approximately the same age (3 months), the June or July date of each measurement used for between-den comparisons, depending upon the whelping date of each den of juveniles. With the exception of ear length, juvenile males surpassed females in all body measurements. The only significant differences between mean body measurements of male and female juveniles were weight, tail length, and zygomatic breadth (Figs. 3–5). Only weight and head length (nose pad to back of head) were significantly different in male juveniles of the six dens and there were no significant differences in mean body measurements in female juveniles of the six dens. While most body measurements began leveling off by mid-August, weight continued to increase for nondispersing juveniles until around mid-October. Adult males surpassed females in all body measurements (Table IV), though only hind foot length was significantly different. Average weight of the two sexes was close, probably because two females were weighed while still pregnant.

4.4.3. Condition Index

A body condition index (CI) was calculated for each adult male fox (Table V) and juvenile (Table VI) by using the following formula:

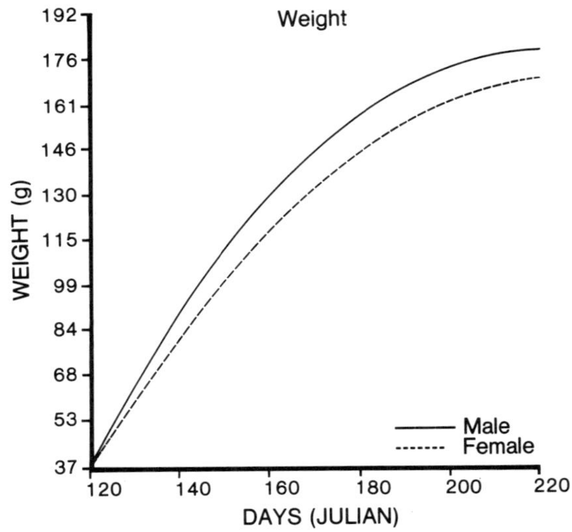

FIGURE 3. Weights (kg) of young male and female kit foxes plotted against days (Julian) from late April until early August 1983.

$$\mathrm{CI} = \frac{W_i T_i C_i Z_i}{W_m T_m C_m Z_m} \times 100$$

where i indicates the measurements for individual foxes, m denotes the maximum value for a fox of appropriate age and sex, W is body weight (kg), T is total length (mm), C is head length (mm; nose pad to back of head), and Z is zygomatic arch width (mm).

Condition indices were not calculated for adult females, because of weight discrepancies between females trapped only before whelping and those trapped only after whelping. Size of home range can be incorporated directly into the CI for adult males by using a multiplier ratio derived as follows:

$$\mathrm{HR} = 1 - \frac{\mathrm{HR}_i}{\mathrm{HR}_{\max}} \times 100$$

where HR is the inverse ratio of proportional home range size, HR_i is the size of the home range (km^2) of the ith male fox, and $\mathrm{HR}_{\max}$ is the largest home range (km^2) of all male foxes monitored.

A condition index that reflects the negative aspects of home range size is calculated by HR × CI/100 (Table V). Kit fox males with high

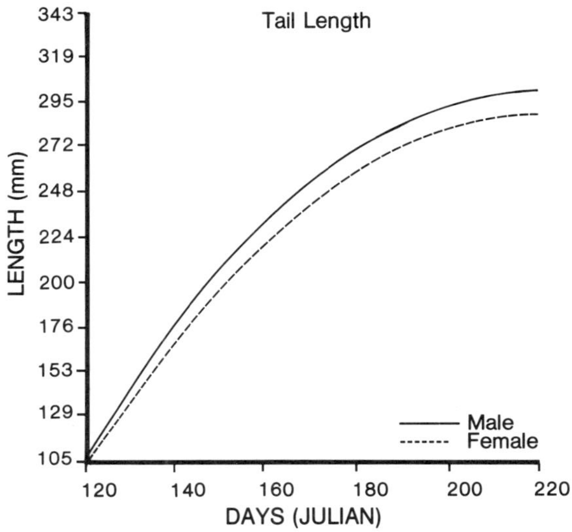

FIGURE 4. Measurements of tail length (mm) of young male and female kit foxes plotted against days (Julian) from late April until early August 1983.

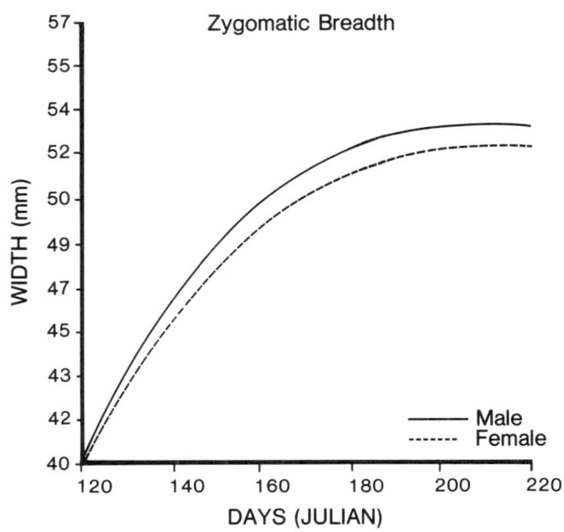

FIGURE 5. Measurements of zygomatic breadth (mm) of young male and female kit foxes plotted against days (Julian) from late April until early August 1983.

TABLE V
Body Condition Index (CI) for Six Adult Fox Males along with an Inverse Percent of Proportional Home Range Size (HR)[a]

Den	CI (%)	HR (%)	Condition index with home range factor (HRCI)
1	75.4	00.0[b]	00.0
2	81.2	20.0	16.2
3	88.5	30.0	26.5
4	100.0	33.0	33.0
5	94.6	00.0[b]	00.0
6	91.1	36.0	32.8

[a]These values are combined (HR × CI/100) into an overall condition index (HRCI) that reflects the negative aspects of large home range size for foraging male foxes.
[b]Adult male foxes at Dens 1 and 5 died (Table I) prior to adequate sampling for size of respective home ranges.

CI values were not recorded as mortalities. Den 1 and 2 adult males, who had the two lowest condition indices (16.2 and 26.5%, respectively), both died of unknown causes. No strong relationship appeared evident between CI values (Table VI) and unknown factors causing mortality among kit fox juveniles prior to dispersal. Juvenile females #219 (Den 2) and #228 (Den 6) both had rather low CI values (50.1 and 69.7%, respectively) and were the only two females to die of unknown causes prior to dispersal. None of the male young died prior to dispersal of causes that could be related to low CI scores. No significant correlation was found either between numbers of nucleated red blood cells or lymphocytes and CI values in either adult or young foxes. In fact, some of the pups and adults with higher numbers of nucleated red blood cells also had higher CI scores. There did not appear to be any significant correlation between external parasite loads and CI values. There was a strong negative relationship between condition indices of adult males and June litter size of young, apparently because larger litters place increased demands upon parents for food and other provisions.

4.4.4. External and Internal Parasite Assessments

Assessments of external and internal parasites of kit foxes indicated these organisms could influence the observed rate of mortality with or without directly causing death. External parasites included fleas (representing three families), small biting lice (*Trichodectes canis*), and ticks (*Haemaphysalis* sp.). Parasite loads increased in early summer

TABLE VI
Body Condition Index (CI) for Male (M) and Female (F) Kit for Juveniles Calculated from Body Measurements Taken in June and July 1983

Den	Pup number	Sex	CI (%)	Den	Pup number	Sex	CI (%)
1	21	M	87.0	4	26	M	98.7
	20	M	43.5		27	M	82.3
	211	F	71.3		23	M	78.2
2	29	M	71.3		213	F	74.9
	38	M	79.2		212	F	94.6
	221	F	89.0		215	F	85.9
	219	F	50.1	5	30	M	100.0
	220	F	78.6		39	M	100.0
3	37	M	79.2		28	M	97.7
	32	M	94.7		226	F	100.0
	224	F	65.5		218	F	89.1
	223	F	59.3		216	F	86.7
				6	35	M	93.4
					40	M	87.3
					228	F	69.7

and remained at medium to high levels until late summer. There were significant ($P \leq 0.5$, χ^2) differences between dens regarding degree of infestation of resident foxes by fleas, ticks, and lice. Foxes in Den 2 had the highest overall mean external parasite load for all three external parasites (1.3 on a scale from 0 to 3) and Den 4 had the lowest (1.1).

Test samples from 24 foxes (12 adults and 12 juveniles) indicated fecal material contained many detached proglottids from tapeworms, as well as eggs of nematodes. Three adult foxes (two males and one female) were sacrificed (December 1983) for examination for internal parasites. Eight specimens of the nematode *Physaloptera rara* were found in stomachs of these foxes. The large intestines of all three foxes supported large infections of tapeworm cestodes (*Taenia* sp.). In two foxes, the intestines were nearly blocked with adult cestodes. The observed rate of infection of the three sample foxes would cause problems from ulceration of the epithelial lining of the intestines, as well as loss of nutrients through decreased functional effectiveness of the intestines.

4.4.5. Analyses of White Blood Cell Differentials

Analyses of white blood cell differentials were performed on 33 of the 42 study foxes and on two laboratory-reared foxes (Table VII). Averages for white cell differentials for all wild foxes were as follows: polysegmented (mature) neutrophils 79.8%, immature neutrophils 0.3%, lymphocytes 15.9%, atypical lymphocytes 0.4%, monocytes 3.2%, eosinophils 0.03%, and basophils 0.03%. An average of 2.5 nucleated red blood cells/specimen was also present (Table VII). Premature release of these immature red blood cells into the blood stream is often indicative of hemolytic anemia and may be attributed to poor diet, heavy parasite loads, or both (Bauer et al., 1974).

Results from blood analyses of two male and female foxes reared in the laboratory differed considerably from those obtained from foxes in the field. Average results for these foxes were as follows (Table VII): polysegmented neutrophils 62.5%, lymphocytes 36.5%, and monocytes 1%. No nucleated red blood cells were seen (Table VII). This difference in white blood cell differentials was probably due to the two different environments in which the pups were reared. Laboratory-reared foxes were dusted for external parasites and treated for internal worms. Normal values for dogs (Schlam et al., 1975) are as follows: segmented neutrophils 60–77%, immature neutrophils 0–3%, lymphocytes 12–30%, monocytes 3–10%, eosinophils 2–10%, and basophils rare. The cause for the higher than average percent of segmented neutrophils (and therefore a higher neutrophil to lymphocyte ratio) in the wild foxes is un-

TABLE VII
Results of Analyses of White Blood Cell Differentials in Blood from Two Laboratory-Reared Kit Foxes (Male and Female) and Adult and Young Foxes in the Six Study Dens[a]

Fox location	Poly	IN	Lym	Aty lym	Mono	Eos	Baso	NRBC
Laboratory	62.5	0.0	36.5	0.0	1.0	0.0	0.0	0.0
Den 1	81.0	1.8	11.6	0.4	4.2	0.0	0.2	4.4
Den 2	78.7	0.2	18.1	1.6	1.6	0.0	0.0	1.3
Den 3	78.4	0.0	19.8	0.0	0.4	0.2	0.0	1.6
Den 4	81.8	0.4	12.3	0.6	4.3	0.0	0.0	2.4
Den 5	79.0	0.0	14.7	0.2	2.3	0.0	0.0	4.0
Den 6	74.5	0.0	12.5	2.0	10.0	0.0	0.0	0.0

[a]Shown are the average percent polysegmented (mature) neutrophils (Poly), immature neutrophils (IN), lymphocytes (Lymph), atypical lymphocytes (Aty lym), monocytes (Mono), eosinophils (Eos), and basophils (Baso), and the number of nucleated red blood cells (NRBC).

known. Bacterial infections are known to cause an increased percentage of immature neutrophils (Bauer et al., 1974; Schlam et al., 1975), though there were no signs of infection in any fox when blood samples were taken. Although lymphocyte percentages for most foxes were within the normal range for dogs, ten foxes were below this range. Distemper may cause low lymphocyte levels in canids (Schlam et al., 1975). It is not known if any foxes had this disease, although a necropsy on the Den 5 adult male revealed discoloration of the lungs, sometimes an indication of distemper.

4.5. Home Range and Prey Assessment

4.5.1. Home Range Size and Fox Movement

The average home range size for adult male foxes was 3.4 km^2, compared to 3.0 km^2 for adult females (Table VIII). Morrell (1972) reported the San Joaquin kit fox spends its entire life in an area of 2.6–5.2 km^2 with much overlap in home ranges. Four of the six study dens had at least some overlap in home ranges at different times during the study. Figure 6 shows home range shape and size of the Den 4 adults over the course of 11 months.

Females remained close to their natal dens during whelping and for approximately 2 weeks thereafter. The female helper at Den 1 remained even closer to the den than did the mother. Although home range area increased or decreased seasonally, such changes varied from den to den and no seasonal trend could be detected. Home range size

TABLE VIII
Average Home Range Size of Each Adult Male and Adult Female Kit Fox from the Six Study Dens

	Home-range size (km^2)	
Den	Male	Female
1[a]	—	3.9
		2.6 (helper)
2	3.9	3.7
3	3.4	2.3
4	3.3	2.8
5[a]	—	2.3
6	3.1	2.9

[a]Adult male foxes at Dens 1 and 5 both died (Table I) prior to adequate sampling for sizes of respective home ranges.

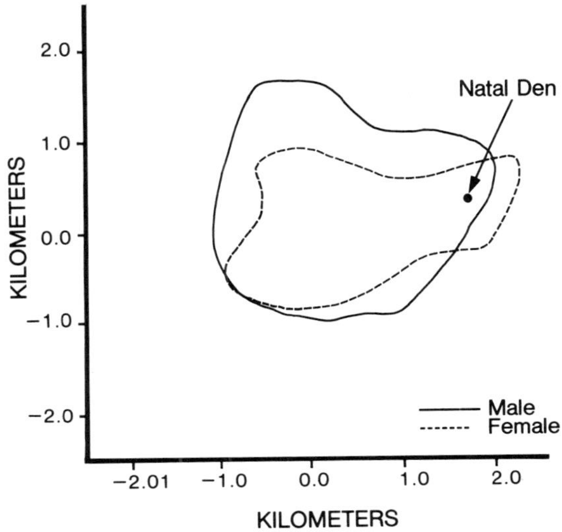

FIGURE 6. Size and shape of home ranges for the adult male and female kit foxes of Den 4 over an 11-month period of time (February–December 1983); these are representative of similar measurements taken for adults at the other five dens.

may be partly influenced by ravines or gullies within the home range. A strong negative relationship existed between total length of all ravines within a home range and home range size. One possible explanation is that if prey were more abundant in these ravines, then a larger foraging area would not be needed. Females that lost mates did not appear to range any further than females with mates. Paired foxes seldom hunted together or in close proximity to each other. Some foxes hunted in the same area for several nights before changing to another area of their home range, apparently responding to success at capturing prey.

Juveniles began leaving the immediate area of their dens, accompanied by their parents, in early and mid-July. Occasionally, the entire family hunted together, but parents usually hunted separately, each taking some of the pups with them. After a trial period lasting no more than 1 week, whelps began hunting on their own. Home ranges from July to November averaged 1.9 km^2 for nondispersing and late-dispersing juveniles.

4.5.2. Vegetational Profile of Home Ranges

Clustering of home ranges based on similarities between percent frequency of major grass, forb, and woody species showed the home ranges for foxes in Dens 1 and 4 to have a similarity index of 68%, followed by Dens 2 and 5 (61%). Dens 1 and 4 were 54% similar to Den 6. All dens showed 40% similarity in major plant species. One possible explanation of such a high correlation between Den 1 and Den 4 home ranges is their close proximity to each other and their partial overlap. In addition, the vegetation of the DER is quite homogeneous (Fig. 7).

4.5.3. Available Prey Analyses

Leporid transects were assessed in May, July, and October, and estimated average population densities are shown in Table IX. Transects for small rodents were assessed in late May and early June. Transects for Dens 4 and 6 were repeated in mid-September due to unsatisfactory results obtained in June. The number of small rodents trapped was low (usually only a 1–8% success rate), resulting in low population estimates of density. Part of the low success rate may have resulted from foxes defecating and urinating on some of the traps. Small rodents trapped included the Ord's kangaroo rat, little pocket mouse, deer mouse, northern grasshopper mouse, and antelope ground squirrel (*Ammospermophilus leucurus*). Despite known abundance in years past, no chisel-tooth kangaroo rats (*Dipodomys microps*) were trapped, found

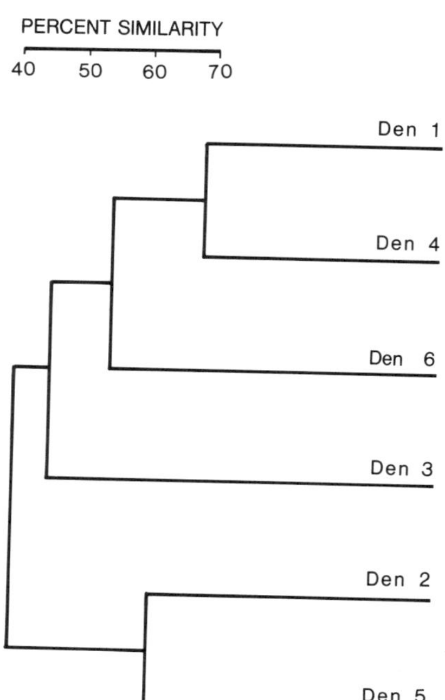

FIGURE 7. Dendrogram showing percent similarities within a cluster analysis based on relative frequency of major grass, forb, and woody species in home ranges of male and female kit foxes associated with specific dens.

TABLE IX

Estimates of Abundance for Potential Prey of Kit Foxes Generated from Periodic Sampling (1983) of Transects on the Desert Experimental Range in Southwestern Utah

Taxon	Potential prey abundance for kit fox den closest to prey sampling transect					
	Den 1	Den 2	Den 3	Den 4	Den 5	Den 6
Jackrabbit[a] (n/km^2)	26.3	29.9	19.7	23.9	29.9	21.7
Horned lark[b] (n/km^2)	167.9	324.9	157.3	252.7	324.9	337.5
Common raven[b] (n/km^2)	27.1	5.2	3.8	16.9	5.2	6.6
Swallow sp.[b] (n/km^2)	39.0	—	—	39.0	6.6	—
House wren sp.[b] (n/km^2)	—	7.8	—	—	7.8	—
Ord's kangaroo rat[c] (n/ha)	1.0	5.4	—	—	—	1.0
Little pocket mouse[c] (n/ha)	2.0	—	1.2	3.0	0.2	1.0
Deer mouse[c] (n/ha)	2.0	—	—	—	—	—
Antelope squirrel[c] (n/ha)	0.6	—	—	—	—	—

[a]Nearest leporid transect (vehicle) for night census (29 April–3 May, 3–8 July, and 19–21 October).
[b]Nearest avian transect (vehicle) for early morning census (28–30 July).
[c]Live-trap and release census (26 May–4 June). The census for Dens 3 and 6 was repeated 5 September–9 September.

in fox scat, or otherwise observed, apparently because of a naturally induced low point in the population. This kangaroo rat is known to occur in kit fox diets in other areas (Egoscue, 1962). Assessment of bird transects in late July revealed mainly the horned lark (*Eremophilia alpestris*), followed by the common raven (*Corvus brachyrhynchus*), house wren (*Troglodytes aedon*), and swallow sp. (*Hirunidae* sp.). Horned lark densities were greater than expected and may be a result of the horned lark's tendency to congregate at roadsides. Other prey species, such as the mourning dove (*Zenaida macroura*) and the black-billed magpie (*Pica pica*), were not found in the census, probably because the census was conducted after their local migration. A summary of estimates of potential avian prey populations is shown in Table IX. The transect for Den 3 contained the highest diversity (MacArthur and Wilson, 1963) of potential prey species compared to the other transects (39%), followed by Den 1 (33%) and Den 4 (15%). Transects for Dens 2, 5, and 6 all had a prey species diversity of 9%.

4.6. Food Habits

4.6.1. Adult Diets

Diets of adults and larger juvenile kit foxes on the DER appear unique in that strong dependence on a single species is not shown (Table X). Dominant food items in diets varied considerably between dens of foxes in winter–spring (W–SP) as well as summer–autumn (S–A) seasonal periods. Black-tailed jackrabbits ranked first in average percent relative frequency (21.1%, CV = 49.3%) in W–SP diets, but ranked second (18.3%, CV = 33.7%) in average S–A diets. The Ord's kangaroo rat ranked first in average percent relative frequency (21.9%, CV = 52.5%) during S–A and dropped to third ranking (17.1%, CV = 43.8%) in W–SP diets. This is the first study to document horned larks as a major prey item for kit foxes. Horned larks ranked second in percent relative frequency (18.5%, CV = 64.9%) in W–SP diets, but dropped to fifth ranking (9.5%, CV = 96.8%) in S–A diets. Foxes in Den 5 showed distinct variability regarding dietary dependence on horned larks. Diets of these foxes showed horned larks ranked first in percent relative frequency (39.5%) in W–SP diets, but were not detected in diets sampled in the S–A period, when these foxes switched to black-tailed jackrabbits and Ord's kangaroo rats as major prey items (Table X). Jerusalem crickets (*Stenopelmatus fuscus*) were surprisingly strong year-round components of kit fox diets. These insects ranked fourth in percent relative frequency (14.4%, CV = 67.4%) in W–SP diets and in-

TABLE X
Relative Percent Frequencies of Food Items in Winter–Spring (W–SP) and Summer–Autumn (S–A) Diets of Adult Kit Foxes during 1983 on the Desert Experimental Range, Utah

Food	Den 1		Den 2		Den 3		Den 4		Den 5		Den 6	
	W–SP	S–A	W–SP	S–A	W–SP	S–A	W–SP	S–A	W–SP	S–A	W–SP	S–A
Desert cottontail rabbit	1.2	0.0	0.0	0.0	0.0	0.0	0.0	0.3	3.0	0.0	0.0	0.0
Black-tailed jackrabbit	12.5	18.0	28.3	26.6	20.2	9.0	20.8	23.3	8.2	15.3	36.9	17.7
Botta's pocket gopher	5.6	0.0	5.7	1.1	3.0	1.3	2.7	4.6	1.3	0.4	3.4	2.1
Little pocket mouse	13.3	17.1	7.7	9.3	22.0	15.6	14.6	15.5	7.4	7.5	23.3	21.1
Ord's kangaroo rat	18.2	2.4	28.1	27.6	15.2	36.0	13.8	27.3	21.5	17.5	5.9	20.3
Deer mouse	0.5	1.6	0.0	0.1	0.1	0.0	0.2	0.0	0.0	14.0	2.6	0.3
Northern grasshopper mouse	0.3	0.0	0.0	0.0	0.0	0.0	0.0	0.0	0.0	0.0	0.0	0.0
Pronghorn antelope	0.3	0.0	0.0	0.0	0.0	0.0	0.0	0.0	3.0	0.0	0.0	0.0
Horned lark	23.6	14.2	5.2	5.5	13.4	11.6	10.7	24.4	39.5	0.0	18.7	1.1
Black-billed magpie	0.0	4.2	0.0	0.3	0.0	0.0	0.0	0.3	0.0	0.0	0.0	0.7
Common raven	0.3	0.2	0.0	0.0	0.0	0.0	0.0	3.1	0.0	0.0	0.0	0.0
Lizard spp.	0.0	0.0	0.0	0.0	1.2	0.0	1.5	0.0	0.0	0.0	0.0	0.7
Scorpion	10.4	9.7	4.0	6.7	3.2	0.0	0.0	0.3	0.0	8.2	0.0	9.5
Jerusalem cricket	8.8	25.4	14.8	19.6	14.3	18.3	33.0	0.0	8.7	36.1	6.6	24.4
Grasshopper spp.	0.0	0.0	0.0	0.0	0.0	0.6	0.0	0.0	0.3	0.0	0.0	0.0
Unknown arthropods	0.0	4.4	1.8	1.0	0.0	2.0	0.0	0.0	3.4	0.5	0.0	0.0
Vegetation	5.0	2.8	4.4	2.2	7.4	5.6	2.7	0.9	4.0	0.5	2.6	2.0

creased to third ranking (20.6%, $CV = 57.8\%$) in S–A diets. Evidently, the fossorial habits of these crickets, as well as scorpions (*Centruroides* spp.), allow kit foxes to locate them either by digging or within the subterranean chambers of active and inactive dens. Young pronghorn (*Antilocapra americana*) remains found around the dens were probably scavengered carcasses resulting from many forms of mortality.

Cluster analysis of similarity indices on adult W–SP diets showed highest similarity between Dens 3 and 4, with a similarity index of 67%, followed by Dens 1 and 5 (57%). All dens showed 44% similarity in diet. Cluster analysis of similarity indices on adult S–A diets combined Dens 2 and 6 at the highest similarity of 64%, followed by Dens 2 and 6 with Den 5 (55%). All dens showed 43% similarity in diet. Den 5 had the highest similarity index between W–SP and S–A diets (62%), followed by Den 2 (49.3%), Den 1 (38.8%), Den 6 (34.2%), Den 3 (32.6%), and Den 4 (30.3%). If adult male body condition (CI) is representative for a pair at each den, it is notable that Den 4 and Den 5 foxes, with the highest indices of body condition (Table V), followed quite different strategies in dietary selection during W–SP and S–A seasons. Apparently, adult kit foxes can select optimal diets from a wide array of prey items present in far different levels of abundance in their home ranges. Results of similarity indices between prey availability and relative percent frequency of prey in adult diets revealed Den 1 as having the highest similarity indices (40.4%), followed by Den 6 (36.4%), Den 2 (28.7%), Den 4 (27.3%), Den 3 (22.5%), and Den 5 (8.8%).

4.6.2. Juvenile Diets

Analysis of scat of kit fox young showed similar results in average percent relative frequency of prey items. There was a heavy reliance on Ord's kangaroo rats as well as horned larks and Jerusalem crickets (Table XI). The Ord's kangaroo rat ranked highest in mean percent (16.4%, $CV = 53\%$), followed by the horned lark (15.3%, $CV = 14.8\%$), and the little pocketmouse (14.1%, $CV = 83.6\%$). The majority of black-tailed jackrabbits were probably brought to dens by parents, because juveniles were not large enough at the time to kill these hares on their own. This may explain why black-tailed jackrabbits constituted an average of only 8.2% of juvenile diets. Cluster analysis of similarity indices on juvenile diets showed the young of Dens 2 and 6 had a diet that was 61% similar, followed by Dens 1 and 3 (46%). The average similarity in diets of young of all dens was 25%. Similarity indices between prey availability and relative percent frequency of prey in the diet was somewhat higher for juveniles. Den 6 had the highest similarity

TABLE XI
Relative Percent Frequencies of Food Items in Diets (until Early July) of Kit Fox Whelps during 1983 at the Desert Experimental Range, Utah

Food	Den 1	Den 2	Den 3	Den 4	Den 5	Den 6
Black-tailed jackrabbit	12.4	22.8	36.3	13.4	13.0	21.9
Botta's pocket gopher	4.7	4.2	2.8	12.3	23.9	1.4
Little pocket mouse	26.3	2.4	32.0	15.6	0.0	8.3
Ord's kangaroo rat	16.9	31.7	13.4	10.5	4.3	21.6
Deer mouse	2.8	10.2	0.0	0.0	0.0	4.2
Northern grasshopper mouse	1.0	1.2	0.0	0.0	0.0	0.0
Horned lark	0.0	24.0	2.3	0.0	32.6	33.0
Scorpion	16.6	3.6	2.8	7.7	0.0	0.0
Jerusalem cricket	9.2	0.0	0.0	30.7	26.1	0.0
Unknown arthropods	1.0	0.0	6.4	9.5	0.0	0.0
Vegetation	9.3	0.0	4.2	0.0	0.0	9.8

index (53.2%), followed by Den 2 (46.0%), Den 1 (43.4%), Den 3 (29.0%), Den 5 (24.0%), and Den 4 (10.1%).

Most hare remains found in and around dens appeared to be those of immature individuals. This was also found by Egoscue (1962), who noted that kit foxes have considerable difficulty killing adult jackrabbits. Cutter (1958) also reported that jackrabbit and cottontail remains found in stomachs and scat remains of the swift fox (*Vulpes velox*) were all immature.

The only item not found in the scat analysis, but observed at a den site, was the remains of a great horned owl (*Bubo virginianus*) at Den 2. The den was not near power poles or heavily traveled roads, and thus the owl was probably sick or dead when encountered by the fox. Pups at Den 4 were observed chasing and eating moths shortly before sunset and were also observed scavenging two striped skunk carcasses. Additionally, scat not belonging to any of the study foxes was found to contain pieces of horned lark egg shells.

Kit foxes are considered opportunistic foragers (Egoscue, 1962; Laughrin, 1970; Morrell, 1972), but there is also evidence of dietary specialization. Laughrin (1970) and Morrell (1972) observed that dietary habits of *V. m. mutica* were dependent on distribution and abundance of kangaroo rats. Egoscue's (1962) studies of *V. m. nevadensis* in Utah reported kangaroo rats abundant in areas occupied by foxes, but utilized for food far less than expected. Egoscue's more recent studies (1975) support the idea that dietary dependence of *V. m. nevadensis* on black-tailed jackrabbits is significant and reproductive success of kit foxes is correlated to the fluctuating abundance of these hares.

4.7. Dispersal

4.7.1. Emigration of Young

Dispersal of juveniles began in mid-August and continued through early November. Straight-line distances traveled from natal dens for those foxes located by radiotelemetry varied from 24 to 64 km. With the exception of one male who traveled north, all dispersing radiocollared juveniles that were located traveled in a southeast to southwest direction from the point of origin. In almost every instance, each dispersing fox traveled to a mountainous area for at least a period of time. One female remained in the foothills of the west slope of the Wah Wah mountains before crossing the range and entering Wah Wah valley for a total of 64 km before being killed by a coyote. A dispersing male crossed the Mountain Home Range approximately 27 km southwest of his den and was found dead on the west slope of the range, apparently having been caught in a snowstorm. Yellowish bone marrow suggested malnutrition may have been a contributing cause of death. Another male traveled to Middle Mountain approximately 24 km northeast of his den and remained there 2 days before traveling to the Tunnel Spring mountains, 9.6 km to the west. He was found 3 days later, only 1.6 km south of his original den, apparently killed by a coyote. This was the only dispersing juvenile known to have returned to the area from which it originally left. Both this male and the previously mentioned female were living in rock crevices during their dispersal. The only fox (a juvenile male) to inhabit a desert floor area during dispersal did so for approximately 2 weeks. During this time, he occupied three different sites (which appeared to be old badger diggings) approximately 26 km southeast of his original den before disappearing. He was found later, in early December, in a cattle watering tank under approximately 30 cm of ice. This site was located in a pinyon-juniper area in the south end of the valley approximately 45 km southeast of his original den. Egoscue (1956) reported a distance of 32 km traveled both by an adult female kept as a pet and another young fox. Of 12 dispersing radiocollared juveniles, the eight eventually located were found dead, suggesting that dispersal carries a high mortality rate. Death may result from increased energy demands from travel, inability to capture prey in various new habitats, increased exposure to predators (Errington, 1946; e.g., coyotes), lack of shelter, or competition from other foxes for food and shelter. Dispersing foxes may also be captured in traps (Phillips et al., 1972), because trapping season and dispersal tend to coincide, though none were reported by trappers.

4.7.2. Immigration

At the same time juveniles were dispersing from the study area, four new male and two new female foxes were trapped in the study area. One juvenile male alternated between a den occupied by the Den 2 female (who had lost her mate in August) and a den occupied by her juvenile daughter. Another juvenile male was found with a newly arrived juvenile female at the Den 5 natal den. A third newly arrived juvenile male was radiocollared and remained with a Den 3 juvenile female for 2 weeks before moving on to an unknown location. A fourth male (adult) was found with the Den 1 adult female who had lost her mate in May. Whether this association was to last is not known, since the female was found dead 1 week later of unknown causes, and the male was never seen again. A second juvenile female was trapped and radiocollared at a den site occasionally occupied by Den 4 foxes. This female was found to alternate between this den and 1 km to the west in a rock crevice near the top of a steep ridge. The reasons for occupation of seemingly uncharacteristic habitats such as rocky, mountainous areas by emigrating and immigrating individuals could be avoidance of other foxes or greater availability of prey in these areas. Less energy would be required to live in an already established shelter than to dig a new den or clean out an old one. For example, a transient adult male was trapped in April near an abandoned den, but continued to inhabit a rock crevice on a nearby steep ridge. *Vulpes m. arsipus* Elliot has similarly been found in the high, rocky, juniper-covered hills of the Mojave desert (Grinnell et al., 1937). No mortalities were observed for any of the newcomers. Egoscue (1975) did not detect an influx of newcomers and attributed it to the possibility that newcomers were either avoiding areas occupied by adults or were avoiding the traps.

5. DISCUSSION AND CONCLUSIONS

5.1. Social Organization

A prenatal shuffle occurring up to 1 week before March whelping was evident for fox pairs in Dens 1 and 3 and involved a series of movements that separated and reunited members of each mated pair. One possible reason for such shuffling may have been a search for a natal den, because kit foxes are known to clean out and inspect most usable dens before selecting a natal den (Egoscue, 1962). In early May, another type of separation and reuniting behavior occurred at all dens. Typically, a male would leave the female and pups and move into a nearby den for up to 1 week at a time. This was not abandonment, for the male still helped in providing food for the female and young. Perhaps this behavior allowed the male to rest better in the daytime without the annoyance of the begging whelps and/or female. Morrell (1972) also reported this behavior in the San Joaquin kit fox. As the pups grew older, the male and female still occasionally separated, but each would take some of the pups to their respective dens. Family units did not really break up until juveniles began dispersing in August. Nondispersing juvenile females continued to live periodically in the same den or nearby their parents and/or their siblings at least until the study ended in December. Female inclusive fitness may best be served by juvenile females remaining in the natal home range as helpers for 1 year before reaching sexual maturity. By using this strategy, these females may eventually take over portions or all of the natal home range, a form of territorial budding (Woolfenden and Fitzpatrick, 1978; Moehlman, 1983). Egoscue (1975) noted that more juveniles tend to remain instead of dispersing when adult numbers are low. Apparently, enough adult males entered the study area to replace adult males that died. Our data suggest a high mortality rate for males.

Visitation between foxes was observed on several occasions. In early June, an adult female was trapped at a den being used by the Den 2 foxes, indicating that female visitation to a den occupied by other foxes occurred at least on one occasion. An adult male (ear-tagged in March along with his mate at their nearby natal den) was trapped in November at a den periodically used by the Den 4 foxes. This was the first time since March this male had been trapped, though weekly trappings took place in this area for Den 4 juveniles, suggesting that this male may have been expanding or shifting his home range in search of food and/or a new mate. On two occasions, during postwhelping separation of pairs, male foxes were found with foxes other than their mates. In early June, the adult male from Den 1 was found with a fox whose sex was unknown. In early September, the Den 3 adult male was found with the Den 1 adult female (who had lost her mate) and her juvenile daughter. The adult female moved the next day, but the male remained with the juvenile female for 2 more days before moving back with his own mate and whelps. Such visitations suggest that nonaggressive contact with other foxes is made by members of mated pairs. Spatial-temporal scent marking (Mech, 1970) occurred but apparently did not serve to establish or maintain exclusive hunting territories for mated pairs. Perhaps such an expanded social structure serves to allow contact between male and female foxes who have lost mates during breeding season. No evidence was found that visitations by males or females took place immediately before, during, or after whelping. Nearby dens not being used at the time were also visited and marked with feces and urine. Visitation and marking of unused dens could be a survey of available dens or a search for prospective mates by unpaired individuals (Egoscue, 1956).

In light of these data, kit foxes do not appear to be strongly territorial. Our observations suggest these animals have an expanded social structure with a dominance hierarchy between foxes that parcels out best den sites and determines those foxes allowed to remain in occupied areas. The expanded social structure may indicate a more ritualized system of competition between kit foxes. All surviving pairs were still found together when the study ended in December, suggesting they were again continuing the pair bond for the following year. Kit foxes are said to be monogamous (Egoscue, 1956, 1962; Ingles, 1965). They do not breed successfully their first year and nonbreeding males are solitary (Morrell, 1972).

5.2. Fitness Assessment

For adult males in Dens 2–4 and 6 there was a nonsignificant negative Spearman's rank correlation coefficient (Siegal, 1956) of $r_s = -0.80$ between body condition index (CI) and size of home range. Although an r_s value of -0.80 substantially implies a negative cost of foraging within a large home range, the low sample size ($n = 4$) indicates that this relationship needs further study. The early deaths (Table I) of adult males at Dens 1 and 5, before adequate sampling of home range sizes, prevented a better statistical test of this proposed interaction. The consequential error of exceeding optimal limits of foraging space appears to be reflected in reduced CI scores. The Den 5 male, prior to his death of unknown cause, had a CI value of 94.6% (Table I) and apparently occupied a very good denning site and home range. Condition indices for Den 5 juveniles were high (Table VI) and the home range for the mother at this den was small (2.3 km^2) in spite of the fact that she had to rear all six pups on her own. The Den 5 female had the highest linear distance (9 km/2.3 km^2) of ravines and gullies in her home range and the resultant edge effect with its associated potential prey could be related to higher quality foraging habitat. The Den 1 male had the lowest CI score (75.4%) and died on 1 May 1983 of unknown causes. Eventually all pups and the mother at Den 1 died, or were killed, along with a female helper. The high CI scores of foxes at Den 4 may be partially explained by the fact that these foxes frequently hunted around both inhabitated and uninhabited buildings where more rodents might be expected to live, resulting in reduced time and energy spent in hunting.

5.3. Foraging Behavior

Kit foxes began hunting and related activities as early as 1 h before sunset but always ended at daybreak. Tracks left in snow tended to meander back and forth in an optimal foraging pattern that minimized path crossing, but effectively explored the foraging area (Cody, 1971). Judging from position of tracks in the snow, it appeared that hares were generally stalked and killed rather than caught during a chase. Prey caught by kit foxes may be entirely eaten on the spot or parts of it may be cached or taken back to the den (Grinnell et al., 1937; Egoscue, 1962).

Shrub patches were apparently selected and searched in an attempt to locate available prey. Both heavy snow and rain appeared to limit kit fox hunting. Hunting would take place for only a short portion of the night or not at all during such inclement weather. Both kit fox and coyote tracks were at times found on the same trails, and it did not appear that foxes were avoiding coyotes. When snow or rain is not present, kit foxes apparently obtain all the water they need from their prey (Egoscue, 1956). The cool, moist, microhabitat of their dens probably keeps water loss to a minimum (Golightly and Ohmart, 1983, 1984). Reliance on evaporation for heat dissipation is minimal, probably as an adaptation to water scarcity in the desert environment (Golightly and Ohmart, 1983), thus further reducing the need for water. They will, however, drink free water if available. Dens 2 and 4 were located near permanent water supplies (antelope water troughs) and foxes from these dens were seen drinking from them, although their foraging patterns did not appear to center around these water sources. The laboratory-reared foxes also drank free water regularly.

5.4. Relationship of Grazing to Kit Fox Activity

Approximately 2200 sheep and 300 cattle graze portions of the DER from 29 November to 5 April. Kit foxes did not appear to react to the presence or absence of sheep or cattle. Foxes would leave dens in a normal manner, even in close proximity to sheep, and hunt in their established areas with no apparent reaction to bedded or grazing sheep. They seemed to react to cattle and pronghorn in much the same way. This relationship is important, since over 75% of kit fox habitats in Utah are used for grazing by livestock (McGrew, 1977). Selection of habitat by kit foxes appears to be positively influenced by the presence of low shrub vegetation, although there does not appear to be preference for one vegetative type over another. Likewise, grazing intensity within low shrub vegetation types does not appear to curtail fox usage.

5.5. Optimal Population Model of an Evolutionarily Stable Strategy for Kit Foxes

An optimal population model for an evolutionarily stable strategy (ESS) (Smith and Price, 1973) for kit foxes appears to include four major factors. First, a male or female kit fox must secure an upper level of dominance within the expanded social structure. Immigrant individuals arriving must find a position in this social dominance hierarchy in order to be allowed to stay. Helpers at dens with whelps may be related or unrelated to the mated pair they are helping. Helpers are probably immature females that were unable to find an unoccupied area. We do not know if male helpers are allowed. Apart from maximizing fitness by aiding relatives, there are very good selfish reasons for helping. Related helpers and nonrelated helpers may eventually inherit all or part of the natal home range (Woolfenden and Fitzpatrick, 1978). Second, a large traditional den would be optimal in that little effort would be required to make necessary modifications (usually enlargements). Third, a smaller home range, preferably containing a large vegetated ravine or numerous smaller ones, is an advantage in that less energy is expended in locating prey. Individuals with smaller home ranges appear to be in better condition (that is, have a higher CI score) than those covering larger home ranges. Because individuals with larger home ranges suffered greater mortalities, the consequential error of foraging within too large an area appears to be death. A fourth major factor is that dispersal of young away from a natal area as well as the arrival of new individuals into the area apparently aids in mate replacement for males and females that lose mates during or after the whelping season. Individuals not finding mates or open spaces in mosaics of occupied habitats may or may not be accepted as helpers. Patterns of dispersal are still largely unknown, but to be optimal, must minimize crossing of one's own path and maximize contact with some prime habitat and other foxes.

6. SUMMARY

Habitat selection, reproduction, food habits, dispersal, and other behavioral expressions of kit foxes on a managed desert rangeland were investigated from January 1983 through December 1983 at the Desert Experimental Range in western Utah. Six adult kit fox pairs along with a juvenile helper from six natal dens were captured and equipped with radio transmitters to document daily and seasonal movements. Twenty-nine male and female pups from each of the six dens were captured on a weekly basis to determine condition, and a male and female pup from each den were radiocollared in midsummer to document daily movements and, later, dispersal. Major findings were:

1. Coyote predation was the largest known cause of death in juveniles. However, most causes of death were unknown in both adults and juveniles and may have been related to low indexes of body condition.

2. The 1:1 sex ratio of pups and the presence of lactating females suggests that a fairly stable kit fox population occupied the study area during this study, although adult male mortality seemed high.

3. There was a negative correlation between an index of body condition and home range size of adult male kit foxes. There appeared to be a strong relationship between indexes of body condition and mortality in juveniles prior to dispersal.

4. Home ranges averaged 3.7 km^2 for males and 3.0 km^2 for females. Shifts in home ranges occurred, but did not appear to be seasonal. There was overlap of home ranges between foxes in four of the six dens.

5. The diets of both adults and pups appeared unique in that a dependence on a single species (such as the black-tailed jackrabbit) was not shown. For the first time, horned larks were documented as a major prey item for kit foxes. Jerusalem crickets were surprisingly strong year-round components of diets.

6. Straight-line dispersal distances were large, up to 64 km. None of the dispersing juveniles located were found alive. Utilization of seemingly uncharacteristic habitats, such as rock crevices on mountainsides, was observed for all but one of the dispersing juveniles.

7. An influx of immigrants to the study site (mostly juveniles) was observed during and after the resident juveniles had begun dispersing.

8. An expanded social system appears to exist among these kit foxes. Social visitations were regularly documented between nonmated individual kit foxes. Territoriality seemed poorly expressed, although scent markings were used.

9. Kit foxes did not react to the presence or absence of sheep or cattle, even when these animals were in close proximity to dens or established hunting areas of the foxes. Likewise, they do not appear to react to the presence or absence of pronghorn. Habitat selection did not appear to be influenced by grazing intensity of the habitat.

REFERENCES

Bauer, J. D., Ackerman, G., and Gelson, T., 1974, *Clinical Laboratory Methods*, C. V. Mosby, St. Louis, Missouri.

Cahalane, V. H., 1947, *Mammals of North America*, Macmillan Co., New York.

Caughley, G., 1977, *Analysis of Vertebrate Populations*, Wiley, New York.

Cody, M. L., 1971, Finch flocks in the Mohave Desert, *Theor. Popul. Biol.* **2**:142–158.

Cutter, W. L., 1958, Denning of the swift fox in Northern Texas, *J. Mammal.* **39**(1):70–74.

Daubenmire, R., 1959, A canopy-coverage method of vegetational analysis, *Northwest. Sci.* **30**:43–64.

Egoscue, H. R., 1956, Preliminary studies of the kit fox in Utah, *J. Mammal.* **37**:351–357.

Egoscue, H. R., 1962, Ecology and life history of the kit fox in Tooele County Utah, *Ecology* **43**:481–497.

Egoscue, H. R., 1975, Population dynamics of the kit fox in western Utah, *Bull. S. Calif. Acad. Sci.* **74**:122–127.

Egoscue, H. R., 1979, Vulpes Velox, *Mammal. Species* **122**:1–5.

Errington, P. L., 1946, Predation and vertebrae populations, *Q. Rev. Biol.* **21**:144–177, 221–245.

Flinders, J. T., and Hansen, R. M., 1973, Abundance and dispersion of leporids within a shortgrass ecosystem, *J. Mammal.* **54**:287–291.

Golightly, R. T., and Ohmart, R. D., 1983, Metabolism and body temperature of two desert canids: Coyotes and kit foxes, *J. Mammal.* **64**(4):624–635.

Golightly, R. T., Jr., and Ohmart, R. D., 1984, Water economy of two desert canids: Coyote and kit fox, *J. Mammal.* **65**(1):51–58.

Green, J. S., and Flinders, J. T., 1981, Diets of sympatric red foxes and coyotes in southeastern Idaho, *Great Basin Nat.* **41**:251–254.

Grinnell, J., Dixon, J. S., and Lindsale, J. M., 1937, *Fur-bearing Mammals of California, Their Natural History, Systemic Status, and Relation to Man,* Volume II, University of California Press, Berkeley, California.

Hall, E. R., 1981, *The Mammals of North America,* 2nd ed., Wiley, New York.

Hall, E. R., and Kelson, K. R., 1959, *The Mammals of North America,* Ronald Press, New York.

Holmgren, R. C., 1973, The Desert Experimental Range: Description, history, and program, in: *Arid Shrublands—Proceedings of the Third Workshop of the United States/Australia Rangeland Panel* (D. N. Hyder, ed.), Society of Range Management, Denver, Colorado, pp. 18–22.

Honacki, J. H., Kinman, K. E., and Koeppl, J. W., 1982, *Mammal Species of the World. A Taxonomic and Geographic Reference,* Allen Press, Lawrence, Kansas.

Ingles, L., 1965, *Mammals of the Pacific States,* Stanford University Press, Stanford, California.

Laughrin, L., 1970, San Joaquin Kit Fox, Its Distribution and Abundance, California Department of Fish and Game, Wildlife Management Branch Administrative Report 70-2, Sacramento, California.

Lechleitner, R. R., 1969, *Wild Mammals of Colorado,* Ptuett Publishing Co., Boulder, Colorado.

MacArthur, R. H., and Wilson, E. O., 1963, An equilibrium theory of insular zoogeography, *Evolution* **17**:373–387.

McGrew, J. C., 1977, Distribution and habitat characteristics of the kit fox (*Vulpes macrotis*) in Utah, M. S. thesis, Utah State University.

McGrew, J. C., 1979, *Vulpes macrotis, Mammal. Species* **123**:1–6.

Mech, L. D., 1970, *The Wolf: The Ecology and Behavior of an Endangered Species,* Natural Press, Garden City, New York.

Moehlman, P. D., 1983, Socioecology of silverbacked and golden jackles (*Canis mesomelas* and *Canis aureus*), in: *Advances in the Study of Mammalian Behavior* (J. F. Eisenberg and D. G. Kleiman, eds.), American Society of Mammalogists Special Publication No. 7, pp. 423–453.

Mohr, C. O., 1947, Table of equivalent populations of North American small mammals, *Am. Midl. Nat.* **37**:223–249.

Mollhagen, T. R., Riley, R. W., and Packard, R., 1972, Prey remains in golden eagle nests: Texas and New Mexico, *J. Mammal.* **36**(3):784–792.

Morrell, S., 1972, Life history of the San Joaquin kit fox, *Calif. Fish Game* **58**:162–174.

Mueller-Dombois, D., and Ellenburg, H., 1974, *Aims and Methods of Vegetation Ecology*, Wiley, New York.

Nowack, R. M., and Paradiso, J. L., 1983, *Walker's Mammals of the World*, 4th ed., Johns Hopkins University Press, Baltimore.

Phillips, R. L., Andrews, R. D., Storm, G. L., and Bishop, R. H., 1972, Dispersal and mortality of red foxes, *J. Wildl. Manage.* **36**(2):237–248.

Reeder, W. G., 1949, Aquatic activity of a desert kit fox, *J. Mammal.* **42**:413–414.

Rohwer, S. A., and Kilgore, D. L., Jr., 1973, Interbreeding in the arid-land foxes, *Vulpes velox* and *V. macrotis*, *Syst. Zool.* **22**:157–165.

Ruzicka, M., 1958, Awendung Mathematisch-Statisticher methoden in der geobotanik (Synthetische bearbeitung von aufnahmen), *Biol. Batisl.* **13**:647–661.

Seton, E. T., 1925, *Lives of Game Animals, Cats, Wolves and Foxes*, Doubleday, Garden City, New York.

Schlam, O. W., Jain, N. C., and Carroll, E. J., 1975, *Veterinary Hematology*, Lea and Febiger, Philadelphia.

Siegel, S., 1956, *Nonparametric Statistics for the Behavioral Sciences*, McGraw-Hill, New York.

Smith, M. J., and Price, G. R., 1973, The logic of animal conflicts, *Nature* **246**:15–18.

Sneath, R. H. A., and Sokal, R. R., 1973, *Numerical Taxonomy: Principles and Practices of Numerical Classification*, Freeman, San Francisco.

U. S. Department of Interior, Fish and Wildlife Service, 1976, Endangered and threatened wildlife and plants. Republication of the list of species, *Fed. Reg.* **41**(208):47180–47198.

Woolfenden, G. E., and Fitzpatrick, J., 1978, The inheritance of territory in group-breeding birds, *BioSci.* **28**:104–108.

CHAPTER 12

SOMATIC MUTATION IN THE POLYNESIAN RAT (*RATTUS EXULANS*) AT ENEWETAK NUCLEAR TEST SITE

MANFRED TEMME

1. INTRODUCTION

Palatal ridge patterns in about 530 mammal species have been described and depicted by Eisentraut (1976). The intermolar configurations of the palatal ridges in the genus *Rattus* have been used as an epigenetic marker to describe variation of distinctiveness within and among rodent populations using a nonmetrical statistic (Temme and Jackson, 1978). Diastemal or antemolar ridge patterns have not yet been the subject of an extensive comparative population study.

Antemolar ridges in the diastema area of rats (genus *Rattus*) normally are two humplike, relatively straight bars, each possessing a thin, solid crest, which may be smooth and entire or more or less verrucosed (Fig. 1). In the variant form, the humps often are centrally divided, may point anteriorly or posteriorly, and often extend between each other (Fig. 2).

MANFRED TEMME • Department of Biological Sciences, Bowling Green State University, Bowling Green, Ohio 43403. *Present address:* Alter Horst 18, 2982 Norderney, West Germany.

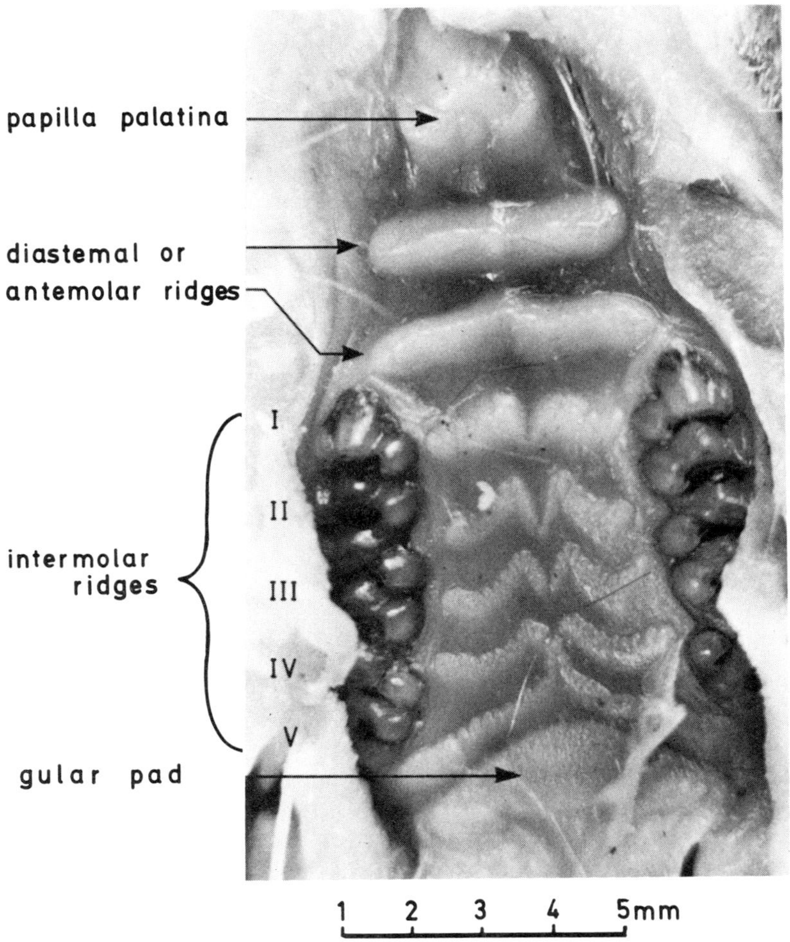

FIGURE 1. Nomenclature of palatal ridges. This example shows a typical and regular pattern in a Polynesian rat (*R. exulans*).

Such aberrations of antemolar ridges found in Polynesian rat (*Rattus exulans*) populations on several islets of Enewetak Atoll (a former nuclear weapons test site) during a collecting visit in March 1977 made further examination of this phenomenon desirable. A working hypothesis stated that these aberrant antemolar palatal ridge patterns resulted from radiation-induced mutations (Temme and Jackson, 1978). Therefore, such configurations should be absent in rats on other atolls or countries where *R. exulans* is distributed, except probably for Bikini Atoll, which also was a nuclear weapons test site.

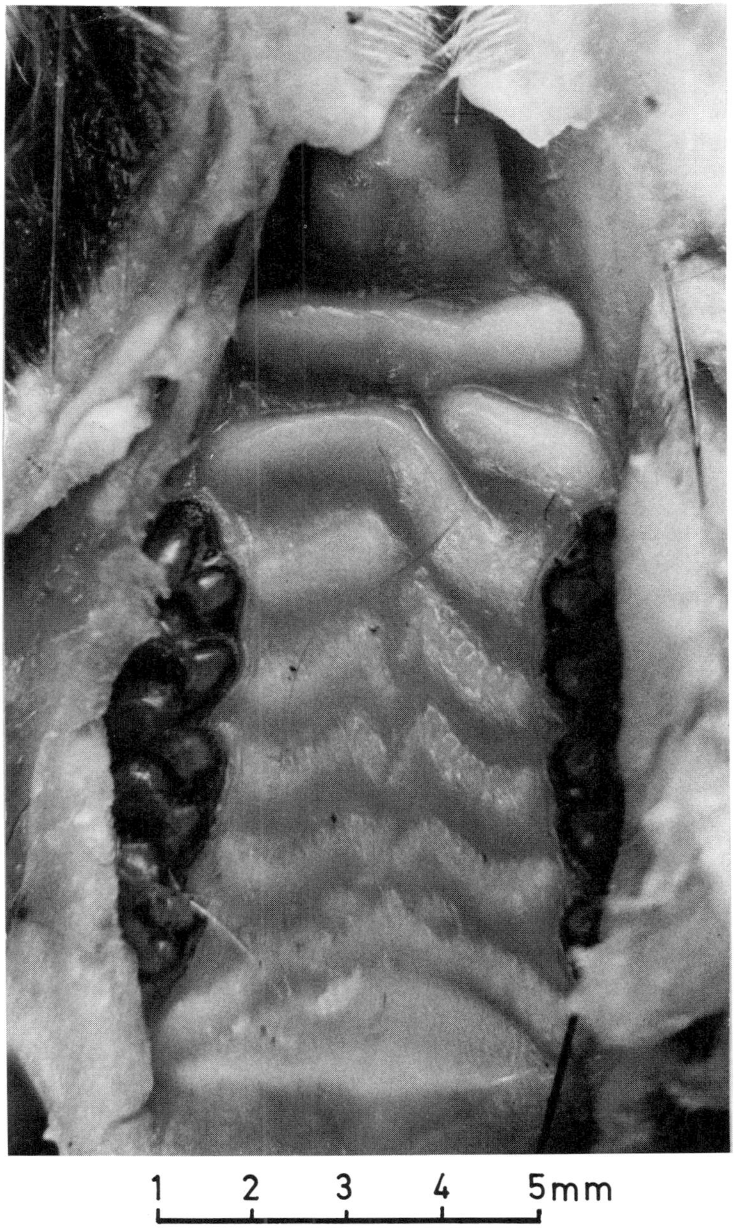

FIGURE 2. Example of an aberrant palatal ridge pattern in the Polynesian rat (*Rattus exulans*) from Lujor islet, Enewetak Atoll. The second antemolar hump is divided and partially bent downward into the area of the first (I) intermolar ridge, which also shows a humplike appearance.

An additional and larger collection of Polynesian rats was made during an expedition to several atolls in the Northern Marshall Islands from 11 October to 21 November 1978 (Temme, 1981). These specimens permitted further testing of the hypothesis and enlargement of the earlier preliminary analysis, which had utilized 236 Polynesian rats.

The islets in the northern half of Enewetak Atoll were primary test sites during the nuclear weapons testing program between 1948 and 1958, and these nuclear events caused heavy physical destruction and loss of biota. (See Anonymous, 1981, for summary of test programs and cleanup operations.) Four of these islets, Lujor, Aomon, Bijire, and Lojwa, are inhabited by *R. exulans* populations.

2. METHODS

Antemolar ridges from 719 Polynesian rats taken in 1977 and 1978 from Enewetak Atoll were analyzed. Rats, collected with snap traps, were decapitated; heads were preserved in 10% formalin. Later, after placement in boiling water, the flesh was removed, and the skulls were placed in ethyl alcohol. The palatal ridges remained with the skull and were examined under a dissecting scope; the configuration of the intermolar ridges was scored five times independently, with uniformity in palate ridge classification being requisite for use in statistical analyses (Temme, 1979; Temme and Jackson, 1978). In this study on diastemal ridges the appearance was scored dichotomously to obtain statistical attributes. Any irregular shape, especially divided, missing, or additional ridge bars, was classified as "abnormal." Calculations of correlation coefficients and significance tests followed Sokal and Rohlf (1969).

Estimates of total radiation exposure at the time nuclear devices were detonated were obtained from tabular data presented by Lynch and Gudiksen (1973). Measurements in 1972 of residual γ radiation and levels of ^{137}Cs in biota of these islets were taken from Stuart and Meibaum (1973).

3. RESULTS

After viewing palatal ridge patterns in *R. exulans* from the Philippines (n = 40), Hawaiian Islands (n = 73), Thailand (n = 20), and some atolls in the Northern Marshalls (n = 483), diastemal ridge var-

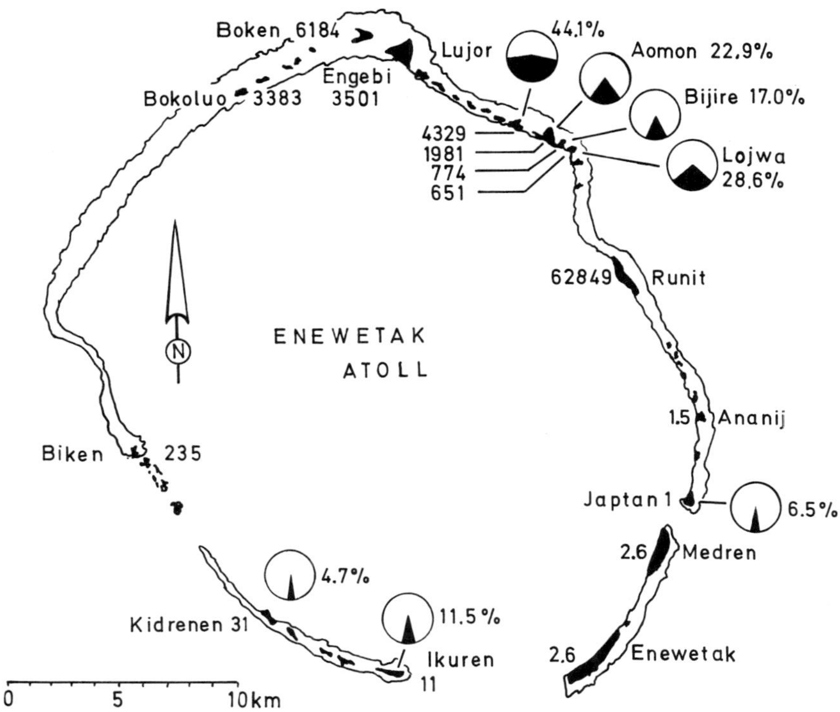

FIGURE 3. The sections in the circles represent the percentage of abnormal antemolar palatal ridge pattern in the Polynesian rat populations. The four islets with the highest percentages are centered between the islets with the highest contamination data. The values given for several of the larger islets are average fallout contamination levels in R/h, corrected to $H + 1$ h past detonation of the atomic testing devices (taken from Lynch and Gudiksen, 1973).

iants found at Enewetak Atoll are considered distinctive and abnormal and thus are termed aberrations.

In the present study at Enewetak Atoll, Polynesian rats from Aomon, Bijire, Lojwa, Japtan, Lujor, Ikuren, and Kidrenen islets were analyzed. (Polynesian rats were not found on other islets in extensive trapping at different times and in nocturnal observations.) The highest incidence of abnormal antemolar ridges was found on Lujor; the lowest, on the most southwesterly islet, Kidrenen (Fig. 3). Their incidence between the 1977 and 1978 Enewetak Atoll samples differed, but not significantly so (Table I).

Furthermore, the fallout contamination levels on the seven islets, the concentration levels of radionuclides in rat tissues from five islets,

TABLE I
Distribution of Aberrant Diastemal Ridges in Polynesian Rat Populations at Enewatak Atoll for 1977 and 1978

Islet	1977		1978		Combined		
						Rats with aberrant ridges	
	Total rats	Number with aberrant ridges	Total rats	Number with aberrant ridges	Total rats	Number	%
Lujor	—	—	34	15	34	15	44.1
Aomon	40	5	105	24	145	29	20.0
Bijire	55	7	106	18	161	25	15.5
Lojwa	55	16	56	16	111	32	28.8
Japtan	86	2	92	6	178	8	4.5
Ikuren	—	—	26	3	26	3	11.5
Kidrenen	—	—	64	3	64	3	4.7
Total	236	30	483	85	719	115	16.0

and the incidence of deformation in the antemolar palatal ridge pattern all are in a parallel descending order (Fig. 3, Table II). Correlations of the incidences of abnormal antemolar palatal ridge patterns from the seven islets with four available estimates of radiation exposure and fallout contamination were positive and statistically significant, except for the total R/h (at $H + 1$ h) fallout (Table III).

4. DISCUSSION

Lujor islet, where the prevalence of antemolar palatal ridge deformation was found to be highest, was denuded in the weapons tests; recent vegetation was in a relatively immature (scrub) successional stage. Japtan and the two islets (Ikuren and Kidrenen) in the southwest quadrant had not been disturbed directly by the testing program, except for the two tests in the adjacent waters just south of the atoll (Koranda et al., 1973), and were partially covered by overgrown coconut plantations.

Polynesian rat palatal ridge data available from these seven islets showed a descending order of abnormalities and radionuclide levels found on these islets (Fig. 3). That statistically significant correlations were found and that palatal ridge aberrations so far have not been seen in samples from other localities support the hypothesis of radiation-induced mutations.

Koranda et al. (1973) reported radiochemical analyses for several radionuclides in plant and animal tissue collected in 1972 for most of the islets at Enewetak Atoll. The total sample from Enewetak Atoll included 249 rats, but islet sample sizes were not given in all cases; and this number included numerous roof rats (*R. rattus*) from other islets. For five of the islets concerned here, Polynesian rat tissues were analyzed and reported in that study.

Correlations between some of these γ energy values and the incidence of abnormal ridge patterns were attempted. The rank of these five islets in accordance with concentration levels of different nuclides in rat tissues is similar to that of the fallout contamination on the islets (Fig. 3). However, correlations between tissue radionuclide concentration levels given by Koranda et al. (1973) and the incidence of abnormal ridges (based on this relatively small number of rats from each islet) were not statistically significant. The best relationship was observed with the radionuclide ^{137}Cs. After pooling the different tissue samples (viscera, liver, muscle, bone, skin, kidney, lung, appendix), a weak correlation was found ($Y = 13.5827 + 0.2762X$, $r = 0.4642$).

TABLE II

Type of Fallout Contamination or Exposure Rates to Some Radionuclides on Islets at Enewetak Atoll where Polynesian Rat Populations Were Sampled in 1978

Islet	Total events[a]	Total R/h (at H + 1h)[a]	Percentage of abnormal ridge patterns	^{137}Cs (μR/r at 1 m)[b]	Total γ (0–3 MeV) (μR/h at 1 m)[b]	Range (μR/r at 1 m)[b]
Lujor	13	4329	44.1	12.0	70	1–400
Aomon	16	1981	22.9	3.5	7	3–110
Bijire	17	774	17.0	4.0	6	2–11
Lojwa	12	651	28.6	3.0	5	1–7
Japtan	3	1	6.5	0.21	(<0.9)[c]	0–5
Ikuren	3	11	11.5	0.33	(<0.9)	0–1
Kidrenen	3	31	4.7	0.15	(<0.9)	0–2

[a]This includes events that produced contamination by fallout or surface ground-zero location (1948–1958) (from Table II in Lynch and Gudiksen, 1973).
[b]Average exposure as measured with Baird-Atomic instrument (November 1972) (from Table 114 in Stuart and Meibaum, 1973).
[c]On islets where activity levels are at the lower limit of the aerial survey equipment dose rates derived from the soil sample data are given in parentheses (November 1972) (Stuart and Meibaum, 1973).

TABLE III
Correlation between Fallout Contamination and Exposure Rates of Some Radionuclides and Prevalence of Abnormal Antemolar Palatal Ridges on Seven Islets at Enewetak Atoll Where Polynesian Rat Populations Were Sampled[a]

Exposure rate	Correlation equation	Correlation coefficient r	Significance level
Total R/h (at $H + 1$ h)[b]	$Y = 10.6150 + 0.7842X$	0.8938	$0.10 > P > 0.05$
^{137}Cs (μR/h at 1 m)[c]	$Y = 9.1688 + 3.0668X$	0.9219	$P < 0.01$
Total (0–3 MeV) (μR/h at 1 m)[c]	$Y = 13.3869 + 0.4587X$	0.8349	$P < 0.02$
Maximum exposure rate (μR/h at 1 m)[c]	$Y = 13.3552 + 0.7801X$	0.8311	$P < 0.001$

[a]Data from Table II.
[b]Fallout contamination at $H + 1$ h, including the events that produced contamination by surface ground-zero location (1948–1958).
[c]Data from the aerial radiological and photographic survey (November 1972).

Templeton et al. (1971) summarized the radiation effects on Enewetak and Bikini Atolls. Lasting effects or noticeable anomalies in land biota or ecosystems had not been found. The aberrations of the antemolar ridge pattern in *R. exulans* are the first somatic tissue mutations found in rats.

All existing evidence and observations suggest that this mutation is heritable but not detrimental, because it has been observed over several years on these islets, in young as well as in old and large specimens. Presumably it has persisted since the weapons testing period. However, breeding experiments to confirm the mode of inheritance have yet to be conducted.

Rattus rattus (Roof rat) populations that lived sympatrically on some of the islets, even those heavily exposed to radiation, did not show any palatal ridge deformations. It is hypothesized that this species may have been more protected at the time of the tests in deep burrows and under concrete structures than the Polynesian rat, which constructs nests in vegetation or surface debris (Jackson, 1967). George Sacher (personal communication) has determined that *R. rattus* is the most radiation-resistant species of rodent studied and thus perhaps able to survive unaffected in the post-test environments.

Only one *R. exulans* specimen from Enyu and two from Airukiiji islets (both Bikini Atoll) showed the same type of deformation (4.6 and

3.8%, respectively) as found at Enewetak Atoll. Although these are low incidences, they are in conformity with the hypothesis presented above.

On Daisu islet (Ujelang Atoll) one rat (1.0%) had a strange diastemal ridge aberration; however, it was without resemblance to the Enewetak Atoll pattern. On Wotho islet (Wotho Atoll) in four specimens the second diastemal ridge was centrally divided, but both humps remained straight as a rectangular traverse. It is considered that these differences are unrelated to the Enewetak aberrations discussed here.

5. SUMMARY

Deformations of antemolar palatal ridges found in Polynesian rat (*Rattus exulans*) populations of seven Enewetak Atoll islets can be considered abnormalities. The highest incidence of such abnormal antemolar ridges has been found on the northernmost islets of the atoll, which were primary sites for nuclear weapons testing. The lowest incidence was noted in rat populations of the southern half, which received the least radiation exposure.

Incidence of abnormal antemolar ridges from seven rat populations at Enewetak Atoll correlated with four different indices to radiation exposure of the respective islets showed positive and significant relationships. The correlations between rat tissue radioisotope concentration levels and the incidence of abnormal ridges were parallel but not statistically significant.

These antemolar ridge deformations in Polynesian rats were absent from populations of other atolls, except for three specimens from Bikini Atoll, also a former nuclear weapons testing site in the Marshall Islands.

It also is hypothesized that these mutations were induced by radiation from the nuclear weapons testing programs between 1948 and 1958. However, definitive breeding tests to determine the mode of inheritance have not been undertaken.

In the roof rat (*R. rattus*), no palatal ridge deformations were found, even from populations on islets with high radiation levels. It is hypothesized that this species was protected in deep burrows under concrete bunkers at the time of tests and high radiation levels.

ACKNOWLEDGMENTS. I gratefully acknowledge the opportunity and support given by the Department of Energy to participate in the survey of the Northern Marshall Islands and the former Mid-Pacific Marine Laboratory for logistic support at Enewetak Atoll. I am further thankful to W. B. Jackson for letting me use his unpublished reports and rat necropsy data from Enewetak Atoll and for his valuable guidance throughout the study.

REFERENCES

Anonymous, 1981, The Radiological Clean Up of Enewetak Atoll, Defense Nuclear Agency, Washington, D. C.

Eisentraut, M., 1976, Das Gaumenfaltenmuster der Säugetiere und seine Bedeutung für stammesgeschichtliche und taxonomische Untersuchungen, *Bonn. Zool. Monogr.* **8**:1–214.

Jackson, W. B., 1967, Rats, bombs and paradise: The story of Eniwetok, in: *Proceedings 3rd Vertebrate Pest Conference*, University of California at Davis, Davis, California, pp. 45–46.

Koranda, J. J., Martin, J. R., Thompson, S. E., Jr., Stuart, M. L., McIntyre, D. R., and Potter, G., 1973, Terrestrial biota survey, in: *Enewetak Radiological Survey*, Volume 1, United States Atomic Energy Commission, Nevada Operations Office, Las Vegas, pp. 225–348.

Lynch, D. D. T., Jr., and Gudiksen, P. H., 1973, Terrestrial soil survey, in: *Enewetak Radiological Survey*, Volume 1, United States Atomic Energy Commission, Nevada Operations Office, Las Vegas, pp. 81–117.

Sokal, R. R., and Rohlf, F. J., 1969, *Biometry*, Freeman, San Francisco.

Stuart, T. P., and Meibaum, R., 1973, Aerial radiological and photographic survey, in: *Enewetak Radiological Survey*, Volume 1, United States Atomic Energy Commission, Nevada Operations Office, Las Vegas, pp. 49–81.

Temme, M., 1979, Polynesian Rat (*Rattus exulans*) Populations in the Northern Marshall Islands, Ph. D. thesis, Bowling Green State University, Bowling Green, Ohio.

Temme, M., 1981, Reproductive parameters of the Polynesian rat (*Rattus exulans*) in the Northern Marshall Islands, *Z. Angew. Zool.* **68**:315–338.

Temme, M., and Jackson, W. B., 1978, Palatal ridges as an epigenetic marker in *Rattus rattus* and *Rattus exulans* populations, *Z. Säugetierkd.* **43**:193–203.

Templeton, W. L., Nakatani, R. E., and Held, E. E., 1971, Radiation Effects, in: *Radioactivity in the Marine Environment*, National Academy of Sciences, Washington, D. C., pp. 223–239.

CHAPTER 13

CURRENT MANAGEMENT STRATEGIES FOR COMMENSAL RODENTS

WILLIAM B. JACKSON

1. INTRODUCTION

The commensal rodents (literally, those with whom we share our table) conventionally are the Norway rat (*Rattus norvegicus*), roof rat (*R. rattus*), and house mouse (*Mus musculus*). All are of Asian origin, introduced to North America by early European immigrants and traders; and these populations now have been augmented from around the world. Native rodents, such as white-footed and deer mice (*Peromyscus* spp.) and various species of squirrels, frequently invade structures, but their management is much the same as for the traditional commensal species.

Some would argue that these commensal rodents would better be termed "parasitic," because they feed at our table to our detriment. Our food is contaminated as well as eaten, our structures are gnawed upon, and we are subject to many of the diseases and parasites they carry.

This paper will review currently used techniques and new products for control of these pests and touch on related areas of zoonotic diseases (those transmittable from animals to humans) and food destruction and contamination. Biological aspects of rodent control recently were re-

WILLIAM B. JACKSON • Department of Biological Sciences, Bowling Green State University, Bowling Green, Ohio 43403.

viewed by Davis and Jackson (1981), and these aspects of commensal rodent management will be touched upon only lightly here. Proceedings from several recent international symposia on commensal rodent biology and control offer additional insights (Al-Sanei, 1983; Dubock, 1984; Myllymaki, 1985).

2. ZOONOSES

We readily associate rats and plague ("Black Death"), but not since 1924 has there been a major outbreak of urban plague in the United States. In contrast, the disease is enzootic (occasionally epizootic) in native rodents of our western states; a few human cases, usually resulting from handling infected rodents, are reported each year. Recently, however, plague-infected fleas (presumably crossing from native rodents) were found on commensal rats in the eastward urban extension of the Los Angeles megalopolis into the chaparral environment (Nelson et al., 1986). Probably such rodent encounters will increase, and we will see renewed concern about urban plague. (Also see Olkowski and Daar, 1985, for a recent review of the problem.)

Leptospirosis is a common and frequently undiagnosed zoonotic disease. In Detroit, for example, over 90% of the adult sewer rats examined were microscopically positive for leptospira infections (Thiermann, 1977). Whether this level of infection in rats was related to the higher-than-average antibody titers in blood samples from children in an adjacent neighborhood is not known, but the association is suspicious. Leptospirosis is a disease that should be looked for more frequently in both rats and people in both urban and rural environments (Thiermann and Frank, 1980).

While rat bite technically is not a disease, such bites can leave both physical and psychological scars. Bites are never adequately reported, but often the statistic of two bites per 100,000 population is used in the United States (Moore et al., 1977). One disease that rats do *not* transmit is rabies. Yet, many physicians seemed unaware of Public Health Service recommendations (PHS, 1980) and continued to prescribe, following rat or mouse bite, the painful Pasteur treatment of 14 inoculations in the abdominal muscle (Jackson and Hann, 1987).

3. RODENTICIDE USAGE

The Environmental Protection Agency (EPA) has responsibility under FIFRA (Federal Insecticide, Fungicide and Rodenticide Act) for rodenticide registration and use. A decade ago the agency waived its efficacy requirement for registration of agricultural pesticides, concluding that the marketplace would eliminate nonefficacious formulations. However, all public-health-related compounds continued to have the efficacy requirement as a hurdle to EPA registration. Rodenticides, even though used for control of rodents in crops, still were included, because commensal rodents were among those doing damage. In a turnabout action in 1982, the requirement was dropped, then effectively reinstated in 1984. This has caused considerable confusion and fostered different criteria for registration of new rodenticides than were used for the currently registered compounds.

Some aspects of rodenticide usage were recently summarized in a description of urban pest control (National Research Council, 1980), but the statistics available at that time from EPA were highly unsatisfactory. More recently, EPA attempted to release a report (Immerman and Drummond, 1984) that contained grossly inaccurate statistics, exemplified by lumping toxic and nontoxic substances together under the category "rodenticides." Furthermore, *any* compound affecting *any* vertebrate was classified as a "rodenticide." (For a summary of these data, see Brandt and Zarrow, 1985.) From these data, it is estimated that the production of technical (pure) rodenticides in 1981 was about 10,000 pounds, with anticoagulants making up about one-third of the total.

3.1. Conversion to Anticoagulants

The story of fermented clover, hemorrhaging cattle, and the Wisconsin biochemist Karl P. Link is now familiar to most (Link, 1959). From this experience came the first of many anticoagulant rodenticides, warfarin (*W*isconsin *A*lumni *R*esearch *F*oundation + *arin*), shortly after the close of WWII. It became the "miracle" rodenticide in the early 1950s and pushed the older, single-dose (acute) compounds into the background.

Research in the United States and Europe produced a number of hydroxycoumarin and then indandione compounds, and most pest control operators (PCOs) today know little else of chemical rodent control but these anticoagulants. Their use resulted in significant reductions of pest rodents. Inevitably, however, because of their extensive and intensive use, selection for resistance occurred. The first evidence came from Scotland (Boyle, 1960). According to World Health Organization criteria, a resistant Norway rat is one able to survive a 6-day, no-choice feeding test with 0.005% warfarin bait (WHO, 1970).

Within another decade the United States had the parallel experience (Jackson et al., 1971) when a population of Norway rats in rural North Carolina was found to be growing fat on warfarin baits that had earlier been highly effective. These farms had been serviced regularly by a pest control operator, and his persistent baiting with warfarin had selected for the resistant allele.

Although the initial identification of resistance in the United States was in a rural area of North Carolina, all other evidence of resistance has been from urban areas or agribusiness sites (for example, stockyards, grain elevators, poultry operations). But it was in the urban areas that we looked, because the Federal *Urban* Rat Control Program included surveillance for resistance among its priorities. It was also in urban areas that intensive, long-term use of anticoagulants often had been promoted by city control programs, so it was to be expected that resistant animals would be selected in these environments (Jackson, 1987). Laboratories at Bowling Green State University and the New York Health Department provided technical assistance to these cities relative to collecting and testing rats.

Forty-five of the 98 populations sampled nationally by these two laboratories contained resistant animals. Chicago became the most famous example, with over 75% of the rats tested being resistant. However, many cities had an incidence below 10%, and most cities showed less than 20%. At these levels the public health hazard was viewed more as potential than real, because most rats could still be controlled, and alternative rodenticides were available for resistant colonies (Jackson et al., 1982; Jackson and Ashton, 1986).

By 1980, resistance had been established in Norway rats and roof rats, and some evidence indicated cross-resistance to all the anticoagulants (Fukui, 1985). This meant that switching rodenticides, a favored PCO tactic, would not be successful. Then federal funds were phased out, and resistance surveillance was halted.

The house mouse was known to be resistant as well, but this phenomenon could not be investigated with federal funds designated for *rat* control. PCOs all over the country reported that they could no longer control some mouse populations. This appeared to be a more important pest management crisis than rat resistance, because mice were more likely to be kitchen and pantry inhabitants; however, no support was available for continued research or surveillance (Ashton and Jackson, 1984).

Selection for anticoagulant resistance appears to have followed a classic pattern. Especially in areas where sanitation was poor and/or food supplies were abundant, the repeated placing of anticoagulant baits selected for animals with the resistant allele. Because only a portion of the area rats had contacted the baits, breeding quickly produced additional individuals carrying the dominant autosomal allele for resistance. Several years of intensive baiting programs produced resistant rat populations in many areas of the country (Jackson, 1986).

3.2. Second-Generation Anticoagulants

The resistance phenomenon was a stimulus to research efforts, first in Europe and then in the United States. Out of that effort came new, significantly more effective hydroxycoumarin compounds that were termed "second-generation" anticoagulants. They had the advantage of killing resistant rats and mice, and a lethal dose could be injested in a single feeding (instead of the three to five feedings required of the first-generation compounds).

Two such compounds, Brodifacoum (Talon) and bromadiolone (Maki), were introduced in the United States early in this decade. Within a couple of years, Talon had captured about 40% of the rodenticide market. Today, many pest control operators use these second-generation compounds whether they know of resistance or not. Furthermore, the single-feeding characteristic increases efficiency because the PCO does not have to return to rebait as frequently.

The concept of "pulsed baiting," in which baits are placed at intervals (for example, weekly) instead of being kept in continuous supply, as was required with the first-generation anticoagulants, was introduced as part of the Talon sales program (Dubock, 1982). Animals that had not fed initially would then have an opportunity to feed on the second baiting, since dominant individuals that may have restricted access to the bait now would be dead. Additional baitings could be applied as needed.

Comparisons of toxicity between groups of anticoagulants can be confusing because of the multiple-feeding requirement (Ashton et al., 1985). For first-generation compounds (like warfarin) the 5-day LD_{50} for rats is about 2 mg/kg (or 0.4 mg/kg per day). For a second-generation material (Talon), where only a single feeding is required, the LD_{50} is about 0.3 mg/kg. (The acute, single-dose, LD_{50} for warfarin is 180 mg/kg, but this is a meaningless statistic in the context of this comparison.) One obvious reason for the relative safety of all anticoagulant compounds is their delayed action and the opportunity for treatment.

The antidote for accidental (or secondary) poisoning with the second-generation compounds is the same as that for the original anticoagulants, vitamin K. However, treatment of poisoning victims must be more intensive, with blood levels of prothrombin being monitored, and repeated doses of vitamin K or blood transfusions given as needed (Batten, 1985; Braithwaite, 1982).

It might be noted that rodenticides were a factor in less than 1% of calls to human poison control centers (National Research Council, 1980). At the University of Illinois Veterinary Poison Control Center, 56% of calls involved dogs, but only 8.1% were related to anticoagulant ingestion (Buck, 1983). Other chemicals found around the house constitute a far greater hazard to children and pets than rodenticides.

3.3. Acute Toxicants

In pre-EPA and pre-warfarin days, many single-dose (acute) toxicants that have since disappeared were available—barium carbonate, thallium sulfate, phosphorus paste (at least largely gone), arsenic, DDT. A few of these older materials hung on because of their low cost and moderate efficacy. Zinc phosphide was one of these. Despite the fact that bait shyness was readily produced, zinc phosphide was an effective rodenticide for outdoor and agricultural use and was registered for use in sugarcane. Even so, the EPA registration requirements prevented its widespread use in the United States. Internationally, it is the most widely used rodenticide.

The announcement by the Occidental Chemical Company (the only producer of zinc phosphide in the United States) of its suspension of production for economic reasons (and EPA registration red tape) in 1984 forced formulators to look elsewhere for supplies of the technical material. Interestingly, the development of a nylon-coated zinc phosphide, which promises to eliminate many of the undesirable features of the chemical, had just occurred (Miller et al., 1985). If developed commercially, this may reverse the chain of events, especially because the new product is odorless and could be used indoors, something that was not possible with the old formulations, which leaked vile-smelling phosphine gas.

Currently, zinc phosphide is undergoing EPA reregistration, and most manufacturers of finished baits have had their labels suspended. Data demands have been arduous, if not impossible, to meet. Only a few labels are likely to remain on the market.

Red Squill, a plant-derived toxicant, was used by many older PCOs, because the emetic action provided safety in the event of accidental ingestion. (Because rodents do not readily vomit, they do not benefit.) However, problems with quality control and limited availability from the Mediterranean region have removed it from most PCO shelves. A comprehensive plant breeding and cultivation program in California for domestic production of the squill bulb has not yet attained commercial operation (Verbiscar et al., 1986). The marketing of Silmurin, a stabilized form of squill, by a European company (Sandoz) has not been undertaken in the United States.

Sodium fluoroacetate (Compound 1080), used in predator and range rodent control in western states, is not now frequently utilized with commensal rodents. However, recent studies have shown it to be effective with native rodents at much lower dose rates than previously considered suitable (Matschke and Hegdal, 1985). Fluoroacetamide (Compound 1081) baits are used for sewer rat control.

3.4. New Rodenticides

Three new compounds have appeared on the U. S. market since 1983. Each had been under development for an extended period and/or been used outside the United States, where registration procedures are more direct.

Bromethalin, a diphenylamine, and different from other rodenticide compounds, is the first true single-feeding, delayed-death rodenticide (Jackson et al., 1982). (With second-generation anticoagulants, feeding continues until near death.) The compound is an oxidative decoupler and so prevents release of cellular energy. The locomotor paralysis resulting from disruption of ionic balances and the consequent myelin sheath edema is readily reversed with diuretic drugs. Animals die 1–3 days after ingestion.

The question of antidote availability is raised with this compound. We are accustomed to the ready availability of vitamin K for accidental anticoagulant ingestions, where the vitamin, required for production of blood-clotting proteins, is administered in large doses to outnumber the blocking anticoagulant molecules. However, an antidote, by definition, can also relieve or reverse symptoms. Thus, an osmotic diuretic (like mannitol) administered to reduce fluid buildup following bromethalin ingestion is antidotal as well.

Extensive field trials indicate a readily accepted bait and a high degree of efficacy for all commensal rodents species (Jackson, 1985; Ashton, 1986). Market release under the trade name "Vengeance" for PCO use by Velsicol Chemical and under "Assault" for use in agricultural (non-crop) areas by Ralston-Purina occurred in late 1985/early 1986.

Calciferol (concentrated vitamin D_2) has long been marketed in Europe, particularly for control of anticoagulant-resistant house mice.

A variant of this (D_3 or cholecalciferol) was marketed in 1985 by Bell Labs as "Quintox" (and by CEVA Laboratories as "Rampage"). Vitamin D_3 is indicated to be more stable than the earlier compound, and the bait is marketed for control of both rats and mice. The compound kills by producing a massive disruption of calcium concentration in body fluids and consequent cardiac arrest, but the rodent will feed several times before death. However, field trial results with this compound have not yet been reported in the literature.

α-Chlorohydrin has been marketed as "Epibloc" in many other countries and finally was registered in the United States in 1982. Originally discovered by the Upjohn Company in their search for human reproductive regulators, it was discarded as being unsatisfactory; it produced sterility only in rats. At lower doses, temporary and then permanent male sterility occurs because of reduced nutrient availability to developing sperm followed by formation of epididymal lesions that block sperm transport. At twice this dose, the compound is lethal to both sexes (Kassa and Jackson, 1984).

In some nontarget species temporary reduction of sperm count can occur after ingestion of the chemical, but the course is readily reversed and cannot be made permanent. Mortality in nontarget species is not likely, because of the dose required (Ericsson, 1982).

Because prebaiting has been recommended, PCOs have not hastened to use the bait, which comes in small, 5-g sachets. A rat needs to eat several of these to get a sterilant dose. In laboratory tests, the bait is not readily accepted by rats, but it is claimed in field trials that this aversion does not occur. Data from few of the successful U. S. field experiments have been published, however (Griffith, 1983; Andrews and Belknap, 1984).

3.5. "Humane" Rodenticides

Especially in British Commonwealth countries, questions of "humaneness" relative to animal-control chemicals are being raised. Rowsell (Rowsell et al., 1979) has attempted an analytical approach to the issue. By analysis of neurologic responses (electroencephalographic patterns), he considers anticoagulants "humane." Hemorrhages largely were into the body cavity rather than into joints (which would be "painful"), and hypoxia results in depression of CNS activity.

It is proposed also to base "judgement on clinical signs of pain, shivering, escape reactions, defensive aggression, painful kicking or muscular rigidity or spasms, and vocalizing." New rodenticides are likely to be judged against such concerns, and this probably will be of increasing importance in the United States in the next decade.

3.6. Bait Stations

A regulatory issue erupted in 1983. Rodenticide labels long have had a statement about placement in "tamper-proof" bait stations, but the definition was vague. It was not clear whether FIFRA gave EPA authority to regulate bait stations per se, but EPA took the view that the rodenticide label (which clearly was within their jurisdiction) could be used to force action on the bait station issue. The EPA bait box criteria (originally developed in 1974) included:

- resistant to weather
- strong enough to prohibit entry of large nontarget animals
- possessing entrances for target species that prevent access by larger animals
- capable of being anchored
- having internal structure for containing bait and preventing accidental spillage
- not considered an "attractive nuisance"

While EPA could not formally endorse a specific manufacturer's products, the Agency did stretch its protocol to cite four specific bait stations that met its criteria. Soon after the publication of this list, EPA quickly had to add additional product names (EPA, 1983). Then controversy over "tamper-proof" and "tamper-resistant" terminology occurred, with the latter seeming to win out (Mix, 1985; Moreland, 1985).

The argument relates to the absolute connotation of "tamper-proof" and whether any device can (or should) be considered as 100% effective. What is desired are devices with a high degree of reliability, whose function can be enhanced by proper placement and use. Obviously rodenticide users are looking over their shoulders at potential liability suits and defending themselves under an EPA "tamper-proof" designation.

After several conferences and many informal discussions with PCOs, university researchers, and industry representatives, criteria to define "tamper-resistant" were generally agreed upon. The National Pest Control Association (NPCA, 1985) developed use guides to help the technician decide under what conditions such bait stations should be used.

How to evaluate bait stations experimentally became a matter of concern. The original EPA tests were simplistic, with only a crude measure of bait consumption being used; and results were not formally published. In more recent observations stations were subjected to replicated observations under standard conditions to obtain data on rodent activity relative to station entrance as well as bait consumption (Kaukeinen, 1986).

In this context, the rodenticide label does not *require* the use of tamper-proof (or tamper-resistant) bait stations. It does require that the bait not be accessible to children, pets, or nontarget animals. Use of tamper-resistant bait stations obviously can be one way to such objectives. However, bait can be pushed down a burrow and the burrow closed to achieve the same restrictions.

4. NONTOXIC TOOLS

Increasingly, nontoxic techniques for rodent elimination are desired as federal or company regulations restrict uses of chemicals, or individuals desire less exposure to chemical use.

4.1. Traps

Snap traps are labor-intensive (and therefore disliked by the PCO), but are ideal for dealing with small infestations by the homeowner. They should be used more commonly, especially the new models that incorporate the expanded, pressure-sensitive trigger, because baiting is not required. However, trap effectiveness depends on proper placement and frequent monitoring.

The multiple-catch traps ("Tin Cat," "Ketch-Al") have the advantage of resetting themselves after each capture. However, some inspectors insist that all traps used in food-processing facilities must be steam-cleaned weekly. Such treatment removes the urine odors that are so essential to attracting mice and essentially requires that the traps be seasoned anew. Routine cleaning with a metal brush or putty knife should be more than satisfactory. New traps cannot be expected to perform well until after an initial capture or unless placed in a mouse colony initially. There is no satisfactory synthetic urine with which to accomplish such seasoning.

4.2. Glue Boards

Glue boards have become important tools for both the PCO and homeowner. They function much like the old-fashioned fly paper, catching mice on a sticky surface. Because they contain no toxic chemicals, they can be recommended for use in many situation. But they are *not* for everybody. They are not for hotel rooms, most business offices, or some homes. Mice do struggle and squeal when they are caught, so their use in areas away from public traffic or occupancy is essential (Frantz and Padula, 1983). Already we are involved in the animal cruelty issue with these devices; their extended promotion may only bring total prohibition of use (Baker, 1983).

Glue boards are especially effective as surveillance devices to warn of new infestations in institutional settings, but many can be placed for the quick removal of large numbers of mice. If large enough in area, they can capture rats as well. These boards can be purchased in finished packs for easy placement and discard, or the bulk glue itself can be obtained for local manufacture.

4.3. Ultrasonics

The use of high-frequency sound (ultrasonics) is controversial, especially because of so many extravagant claims being made (Howard and Marsh, 1985). While testimonials relative to the value of ultrasonics in rodent control are frequently seen in advertisements, actual test data under actual use conditions with before and after installation observations are limited. Some experience (for example, Jackson 1980; McCartney and Jackson, 1986) indicates that devices properly installed and maintained can change rodent behavior patterns and protect sensitive environments, such as food preparation and storage areas. Especially when used in connection with sanitation and other rodent control measures (as part of an integrated management program), some ultrasonic devices are effective. Recent enforcement actions by federal agencies have forced some devices off the market.

5. BIOLOGICAL CONTROL

Diverse biological agents (pathogens, parasites, predators) have been proposed for rodent control, and their employment attempted over the centuries, but with little effect (Davis and Jackson, 1981). Various microorganisms, if effective in killing rats, are also hazardous to man or domestic animals. A disease organism specific to commensal rodents has not been demonstrated. Even if this were so, resistance and/or attenuation could be expected to evolve quickly (as with *Myxoma* virus and rabbits).

While individual cats or dogs can be good "ratters" or "mousers," the use of such predators on a commercial scale has not been successful. Such animals are more successful at preventing entry of emigrating (wandering) rodents than eliminating an established population. The introduction of predators, like the mongoose, has been less than successful for rat control and environmentally disastrous for ground birds (Wodzicki, 1973).

6. INTEGRATED PEST MANAGEMENT

Integrated pest management (IPM) is a current phrase, but the concept has long been a part of rodent control philosophy. The Public Health Service in its program materials of the 1950s placed strong emphasis on garbage storage and removal and environmental improvement. Federally supported programs often had mobile workshops that went along with rat-poisoning crews to rat-proof structures by repairing doors and screening windows and other openings. The concepts of carrying capacity and population growth dramatically illustrated by D. E. Davis (Davis and Jackson, 1981) in Baltimore are imbedded in current rat management literature.

The extensive studies in Baltimore in the 1940s and 1950s documented that outdoor Norway rat populations could be drastically reduced by the removal of harborage and enforcement of garbage storage and collection. Alternatively, use of rodenticides simply reduced the rat population temporarily. Within a few months rat numbers had returned to prepoisoning levels, and the poisoning routine had to be repeated. This was costly and produced no long-term improvement.

In the more recent federal rat control programs, unapproved garbage storage (for example, a can without a lid) was used along with rat infestations for management evaluations. A block could not qualify for environmentally improved block (EIB) status until the premise rat infestation was below 6%, less than 15% of premises had exposed garbage, and unapproved refuse storage occurred on less than 30% of premises.

Unfortunately, concern for personal safety has brought back security fences (which also provide protection to rats) and large dogs (whose excess food feeds the rats). Furthermore, the combination restricts access of crews to census and treat rat infestations (Jackson, 1984).

The PCO often finds it expedient to ignore sanitation problems, and uses a heavy dose of rodenticides in the hope of overriding the environmental shortcomings. One consequence is that the customer does not participate in the management program by undertaking repairs, improving sanitation practices, and monitoring environmental conditions. A second consequence is that selection for resistance in the rodent populations may occur more rapidly.

Environmentally oriented groups increasingly are noting that rodenticide usage can be reduced if the environmental carrying capacity is lowered through improved sanitation. The *IPM Newsletter* recently carried articles on mouse and rat control, but also noted that rodenticides remain valid tools within an IPM program (Anonymous, 1980; Olkowski et al., 1984). The National Park Service, in having to deal with squirrel and rat problems in Lafayette Park across from the White House, was forced by NEPA (National Environmental Policy Act) to develop an IPM document that has many applications to conventional commercial rodent management (National Park Service, 1982). Recommendations included:

1. Food source reduction by retrofitting the bottoms of refuse containers with metal disks to prevent access by rats, emptying refuse at the end of the day, and eliminating overnight refuse storage.
2. Harborage reduction, involving removal of dense yew and boxwood plantings, which permitted better monitoring of burrowing activity; pruning of trees; and careful monitoring of ivy and azalea plantings to prevent a dense, protective mat from forming.
3. Continued use of well-placed anticoagulant bait stations.
4. Encouraging people to stop feeding squirrels (which in themselves had become a nuisance), discouraging people from disposing of lunch remains into the vegetation, and providing displays on urban wildlife and IPM.

7. PEST CONTROL OPERATORS AND PUBLIC ATTITUDES

Small pest control operations are built around monthly service contracts (for example, inspection and renewal of rodenticide bait in bait stations), with little attention paid toward environmental sanitation and elimination of the sources of infestation. After all, the customer may not want to be told to repair and clean up the premises.

Larger PCOs have discarded these small contracts in favor of more profitable industrial accounts with large warehouse operations and groups of retail outlets. Here the PCO frequently functions as an environmental consultant to the client. Large food product manufacturers may employ quality control personnel responsible for pest control. Best results are achieved with a team approach to environmental management.

Some of the institutional concerns for rodent control come from regulatory agencies (both federal and state). The U. S. Food and Drug Administration (FDA) inspects food handling facilities and can confiscate products or close a facility that is rodent-infested or has rodent-infested products. The FDA regulations define food adulteration (in part) as "produced under conditions whereby contamination *may* occur." FDA does not have to find actual contamination of product, simply signs of a rodent infestation that might result in contamination. This is a powerful legal and psychological tool.

FDA does recognize that many food products cannot be produced free from chemical residues or foreign matter. Grain, for example, is difficult to store totally free from rodents and may require fumigation for insect and rodent control. Consequently, FDA establishes tolerances for chemical residues and DALs (detection action levels) for extraneous substances. Years ago a DAL of one rodent dropping per pint of wheat was acceptable. Now the limit is 9 mg of rodent excreta per kg, which is much more quantitative and enforceable.

A curious phenomenon has developed in our litigious society. If a consumer finds a rodent (or rodent contamination) in a food product, the first thought often is to sue the manufacturer (and the retail outlet), because they "have lots of money" (the "deep pocket" concept). That the product had been opened in the home and the mouse might have become entrapped after purchase seems not to occur to such individuals. Then there are individuals who deliberately make such placements with the hope of being paid off handsomely.

Being able to distinguish between cases of actual contamination (which do occur) and those of postpurchase entry often is difficult, especially when the evidence is not preserved well. These kinds of situations place PCOs in a critical role, because they must be able to defend their programs as professionally sound and to support (when appropriate) their clients in demonstrating that contamination would not have occurred during preparation or manufacture of the product. In preparing such a defense, detailed documentation of inspections and control procedures is extremely important. Unfortunately, this aspect of customer/manufacturer relations seems to be on the increase, although, understandably, few statistics are available.

8. CROP LOSSES

Though rodents depredating crops may not regularly be included under the commensal rubric, nevertheless they flourish because of our intensive agriculture, and often the same rodent species are involved.

In the United States the need for field crop (as distinct from tree crop) rodent control is limited, and few rodenticides are labeled for such uses. For example, only zinc phosphide is registered for use in sugarcane, but present formulations are marginally effective at best. Currently EPA registration requirements are a disincentive to additional crop-use registrations.

In contrast, in developing countries (which are usually tropical) within-crop losses to rodents are of major concern. Some multinational companies have elected (and Agency for International Development regulations require) not to use pesticides in developing countries in ways not permitted in the country of origin. Thus, effective control measures may be limited or unavailable for some situations.

Postharvest food losses and contamination by rodents are in addition to field depredations and are of universal concern (Jackson, 1977). Especially in developing countries, where the needs for food grains often are critical, implementation of an integrated, perhaps novel, approach to pest rodent management, especially relative to appropriate storage structures, by the farmer and local elevator (godown) operator is important, but may be difficult.

9. CONCLUSION

Pest rodent management in the future will emphasize an integrated approach, but of necessity will continue to have heavy dependence on rodenticides both for immediate control needs and surveillance/maintenance efforts. New rodenticides—both variants of existing products and totally new compounds—will compete for a share of existing and new, yet undefined markets.

REFERENCES

Al-Sanei, K. S., 1983, *Proceedings of the First Symposium on Recent Advances in Rodent Control* (A. H. Helmy Mohammad, T. M. Zaghloul, and M. Zakaria, eds.), Kuwait Ministry of Public Health.

Andrews, R. V., and Belknap, R. W., 1984, Efficacy of alpha-chlorohydrin in sewer rat control, *J. Hyg.* **91**:359–366.

Anonymous, 1980, IPM for rats in urban areas, *IPM Practitioner* **2**(3):2 p.

Ashton, A. D., 1986, Performance of Vengeance (bromethalin) in urban environments, presented at the *Twelfth Vertebrate Pest Conference*, San Diego, California.

Ashton, A. D., and Jackson, W. B., 1984, Anticoagulant resistance in the house mouse in North America, in: *Proceedings Conference on the Organisation and Practice of Vertebrate Pest Control* (A. Dubock, ed.), ICI, Fernhurst, England, pp. 181–188.

Ashton, A. D., Jackson, W. B., and Peters, H., 1985, Comparative evaluation of LD_{50} values for various anticoagulant rodenticides, in: *Fourth International Theriological Congress, Edmonton, Alberta, Canada, Abstracts of Papers and Posters*, No. 0032.

Baker, B., 1983, About glue traps: Some are better than others, *Pest Control* **51**(2):5.

Batten, P. L., 1985, Antidotes to anticoagulants, *Vet. Record* **117**(3):343.

Boyle, C. M., 1960, Case of apparent resistance of *Rattus norvegicus* Berkenhout to anticoagulant poisons, *Nature* **188:**517.

Braithwaite, G. B., 1982, Vitamin K and brodifacoum, *J. Am. Vet. Med. Assoc.* **181:**531, 534.

Brandt, E., and Zarow, L., 1985, Results of National Urban Pesticide Applicator Survey, by EPA, *Pest Manage.* **4**(1):30–32.

Buck, W. B., 1983, Trends of Toxicoses and Chemical Contamination in Dogs and Cats Encountered by the Animal Poison Control Center, University of Illinois, Urbana, Illinois.

Davis, D. E., and Jackson, W. B., 1981, Rat control, *Adv. Appl. Biol.* **6:**221–277.

Dubock, A., 1982, Pulsed baiting—A new technique for high potency, slow acting rodenticides, in: *Proceedings Tenth Vertebrate Pest Conference* (R. E. Marsh, ed.), University of California, pp. 123–136.

Dubock, A. (ed.), 1984, *Proceedings of a Conference on: The Organisation and Practice of Vertebrate Pest Control*, ICI Plant Protection Division, Fernhurst, England.

EPA, 1983, Tamper-Proof Bait Boxes, Office of Pesticides and Toxic Substances, PR Notice 83-5, pp. 1–6.

Ericsson, R. J., 1982, Alpha-chlorohydrin (Epibloc): A toxicant-sterilant as an alternative in rodent control, in: *Proceedings Tenth Vertebrate Pest Conference* (R. E. Marsh, ed.), University of California, pp. 6–9.

Frantz, S. C., and Padula, C. M., 1983, A laboratory test method for evaluating the efficacy of glueboards for trapping mice, in: *Vertebrate Pest Control and Management Methods, ASTM STP 817* (D. E. Kaukeinen, ed.), American Society for Testing and Materials, Philadelphia, pp. 209–225.

Fukui, H. H., 1985, Pivalyl-resistance in Sprague-Dawley rats (*Rattus norvegicus*), M. S. thesis, Bowling Green State University, Bowling Green, Ohio.

Griffith, W., 1983, Epiblock rodenticide introduces a new concept in long-term rodent control, *Pest Control Technol.* **11**(4):42–44, 67.

Howard, W. E., and Marsh, R. E., 1985, Ultrasonics and electromagnetic control of rodents, *Acta Zool. Fenn.* **173:**187–189.

Immerman, F. W., and Drummond, D. J., 1984, National Urban Pesticide Applicator Survey: Overview and Results, U. S. EPA, Economic Analysis Branch, Office of Pesticide Programs (RTI/2764/08-01F).

Jackson, W. B., 1977, Evaluation of rodent depredations to crops and stored products, *EPPO Bull.* **7**(2):439–458.

Jackson, W. B., 1980, Ultrasonics protect egg farm, *Pest Control* **48**(8):28, 30, 32, 34, 36, 38.

Jackson, W. B., 1984, Urban rodent control in the United States, in: *Proceedings Conference on the Organisation and Practice of Vertebrate Pest Control* (A. Dubock, ed.), ICI, Fernhurst, England, pp. 61–71.

Jackson, W. B., 1985, Single-feeding rodenticides: New chemistry, new formulations, and chemosterilants, *Acta Zool. Fenn.* **173:**167–169.

Jackson, W. B., 1987, Rodenticide resistance, in: *Rodent Pest Management* (I. Prakash, ed.), CRC Press, Boca Raton, Florida, in press.

Jackson, W. B., and Ashton, A. D., 1986, Case histories of anticoagulant resistance, in: *Pesticide Resistance: Strategies and Tactics Management*, National Research Council, National Academy Press, Washington, D. C., pp. 355–369.

Jackson, W. B., and Hann, W. D., 1987, Rats and rabies, in preparation.

Jackson, W. B., Spear, P. T., and Wright, C., 1971, Resistance of Norway rats to anticoagulant rodenticides confirmed in the United States, *Pest Control* **39**(9):13–14.

Jackson, W. B., Spaulding, S. R., van Lier, R. B. L., and Dreikorn, B. A., 1982, Bromethalin—A promising new rodenticide, in: *Proceedings Tenth Vertebrate Pest Conference* (R. E. Marsh, ed.), University of California, pp. 10–16.

Kassa, H., and Jackson, W. B., 1984, Bait acceptance and chemosterilant efficacy of alphachlorohydrin in the Norway rat, *Rattus norvegicus, Int. Pest Control* **26**(1):7–11.

Kaukeinen, D. E., 1986, Systematic evaluations of bait stations under controlled conditions, Presented at Symposium on Test Methods for Vertebrate Pest Control and Management Materials, American Society for Testing and Materials, San Diego, March 7.

Link, K. P., 1959, The discovery of dicumarol and its sequels, *Circulation* **19**:97–107.

Matschke, G. H., and Hegdal, P. L., 1985, Efficacy of Two Lower Concentrations of 1080 bait, 0.022% and 0.035%, Compared to the Standard 1080 Bait, 0.112%, for Controlling Black-tailed Prairie Dog Populations, U. S. Fish and Wildlife Service, Denver Wildlife Research Center.

McCartney, W. C., and Jackson, W. B., 1986, The effects of ultrasound on the movement patterns of Norway rats in a grain mill, *Pest Contr.*, in press.

Miller, G. T., Jackson, W. B., and Ashton, A. D., 1985, Development of coated zinc phosphide rodenticide, in: *Proceedings 12th International Symposium on Controlled Release of Bioactive Materials, 8–12 July, Geneva*, Controlled Release Society, p. 339.

Mix, J., 1985, Tamper-proof stations: Will they stay or go?, *Pest Control* **53**(8):34.

Moore, R. M., Jr., Zelmer, R. B., Moulthrop, J. I., and Parker, R. L., 1977, Surveillance of animal bite cases in the United States, 1971–1972, *Arch. Environ. Health* **32**(6): 267–270.

Moreland, D., 1985, What's all the fuss about "tamper-proof" bait boxes?, *Pest Control Technol.* **11**(9):46–48.

Myllymaki, A. (ed.), 1985, Rodent ecology. Applied theriology. Insectivores. Bats. Part V, Proceedings of the Third International Therological Congress, Helsinki, *Acta Zool. Fenn.* **173.**

National Park Service, 1982, Integrated Pest Management, Year Three, Final Report.

National Research Council, 1980, *Urban Pest Management*, National Academy Press, Washington, D. C.

Nelson, B. C., Madon, M. B., and Tilzer, A., 1986, The complexities at the interface amongst domestic/wild rodents, fleas, pets and man in urban plague ecology in Los Angeles County, in: *Proceeding of Twelfth Vertebrate Pest Conference* (Terrell P. Salmon, ed.), San Diego, California, pp. 88–96.

NPCA, 1985, Tamper-resistant Bait Stations, Technical Release, National Pest Control Association, ESPC 041338-041345.

Olkowski, W., and Daar, S., 1985, Update: Plague on the rise, *IPM Practitioner* **7**(10):1–8.

Olkowski, W., Olkowski, H., and Daar, S., 1984, IPM for the house mouse, *IPM Practitioner* **6**(7):6–10.

PHS, 1980, Rabies prevention, *Morbidity Mortality Weekly Rep.* **29**(23):265–272, 277–280.

Rowsell, H. C., Ritcey, J., and Cox, F., 1979, Assessment of humaneness of vertebrate pesticides, Presented at CALAS Convention, University of Guelph, June 25–28.

Thiermann, A. B., 1977, Incidence of leptospirosis in the Detroit rat population, *Am. J. Trop. Med. Hyg.* **26**(5):970–974.

Thiermann, A. B., and Frank, R. R., 1980, Human leptospirosis in Detroit and the role of rats as chronic carriers, *Int. J. Zoonoses* **7**:62–72.

Verbiscar, A. J., Banigan, T. F., and Gentry, H. S., 1986, Recent research on red squill as a rodenticide potential, in: *Proceeding of the Twelfth Vertebrate Pest Conference* (Terrell P. Salmon, ed.), San Diego, California, pp. 51–56.

WHO, 1970, Instructions for determining the susceptibility or resistance of rodents to anticoagulant rodenticides, *World Health Org. Tech. Rep.* **443**:140–147.

Wodzicki, K., 1973, Prospects for biological control of rodent populations, *Bull. World Health. Org.* **48**:461–467.

INDEX